A First Course in Mathematical Analysis

Mathematical Analysis (often called Advanced Calculus) is generally found by students to be one of their hardest courses in Mathematics. This text uses the so-called sequential approach to continuity, differentiability and integration to make it easier to understand the subject.

Topics that are generally glossed over in the standard Calculus courses are given careful study here. For example, what exactly is a 'continuous' function? And how exactly can one give a careful definition of 'integral'? This latter is often one of the mysterious points in a Calculus course – and it is quite tricky to give a rigorous treatment of integration!

The text has a large number of diagrams and helpful margin notes, and uses many graded examples and exercises, often with complete solutions, to guide students through the tricky points. It is suitable for self study or use in parallel with a standard university course on the subject.

A First Course in Mathematical Analysis

DAVID ALEXANDER BRANNAN

Published in association with The Open University

CAMBRIDGE
UNIVERSITY PRESS

University Printing House, Cambridge CB2 8BS, United Kingdom

One Liberty Plaza, 20th Floor, New York, NY 10006, USA

477 Williamstown Road, Port Melbourne, VIC 3207, Australia

314-321, 3rd Floor, Plot 3, Splendor Forum, Jasola District Centre, New Delhi - 110025, India

79 Anson Road, #06-04/06, Singapore 079906

Cambridge University Press is part of the University of Cambridge.

It furthers the University's mission by disseminating knowledge in the pursuit of education, learning and research at the highest international levels of excellence.

www.cambridge.org
Information on this title: www.cambridge.org/9780521684248

© The Open University 2006

First published 2006
8th printing 2017

A catalogue record for this publication is available from the British Library

ISBN 978-0-521-86439-8 Hardback
ISBN 978-0-521-68424-8 Paperback

To my wife *Margaret*
and my sons *David*, *Joseph* and *Michael*

Contents

Preface

Analysis is a central topic in Mathematics, many of whose branches use key analytic tools. Analysis also has important applications in Applied Mathematics, Physics and Engineering, where a good appreciation of the underlying ideas of Analysis is necessary for a modern graduate.

Changes in the school curriculum over the last few decades have resulted in many students finding Analysis very difficult. The author believes that Analysis nowadays has an unjustified reputation for being hard, caused by the traditional university approach of providing students with a highly polished exposition in lectures and associated textbooks that make it impossible for the average learner to grasp the core ideas. Many students end up agreeing with the German poet and philosopher Goethe who wrote that 'Mathematicians are like Frenchmen: whatever you say to them, they translate into their own language, and forthwith it is something entirely different!'

Since 1971, the Open University in United Kingdom has taught Mathematics to students in their own homes via specially written correspondence texts, and has traditionally given Analysis a central position in its curriculum. Its philosophy is to provide clear and complete explanations of topics, and to teach these in a way that students can understand without much external help. As a result, students should be able to learn, and to enjoy learning, the key concepts of the subject in an uncluttered way. This book arises from correspondence texts for its course *Introduction to Pure Mathematics*, that has now been studied successfully by over ten thousand students.

This book is therefore different from most Mathematics textbooks! It adopts a student-friendly approach, being designed for study by a student on their own OR in parallel with a course that uses as set text either this text or another text. But this is the text that the student is likely to use to learn the subject from. The author hopes that readers will gain enormous pleasure from the subject's beauty and that this will encourage them to undertake further study of Mathematics!

Once a student has grasped the principal notions of *limit* and *continuous function* in terms of inequalities involving the three symbols ε, X and δ, they will quickly understand the unity of areas of Analysis such as limits, continuity, differentiability and integrability. Then they will thoroughly enjoy the beauty of some of the arguments used to prove key theorems – whether their proofs are short or long.

Calculus is the initial study of limits, continuity, differentiation and integration, where functions are assumed to be well-behaved. Thus all functions continuous on an interval are assumed to be differentiable at most points in the interval, and so on. However, Mathematics is not that simple! For example, there exist functions that are continuous everywhere on $\mathbb{R}$, but differentiable

Johann Wolfgang von Goethe (1749–1832) is said to have studied all areas of science of his day except mathematics – for which he had no aptitude.

nowhere on $\mathbb{R}$; this discovery by Karl Weierstrass in 1872 caused a sensation in the mathematical community. In *Analysis* (sometimes called *Advanced Calculus*) we make no assumptions about the behaviour of functions – and the result is that we sometimes come across real surprises!

The book has two principal features in its approach that make it stand out from among other Analysis texts.

Firstly, this book uses the 'sequential approach' to Analysis. All too often students starting on the subject find that they cannot grasp the significance of both ε and δ simultaneously. This means that the whole underlying idea about what is happening is lost, and the student takes a very long time to master the topic – or, in many cases in fact, never masters the topic and acquires a strong dislike of it. In the sequential approach they proceed at a more leisurely pace to understand the notion of limit using ε and X – to handle convergent sequences – before coming across the other symbol δ, used in conjunction with ε to handle continuous functions. This approach avoids the conventional student horror at the perceived 'difficulty of Analysis'. Also, it avoids the necessity to re-prove broadly similar results in a range of settings – for example, results on the sum of two sequences, of two series, of two continuous functions and of two differentiable functions.

Secondly, this book makes great efforts to teach the ε–δ approach too. After students have had a first pass at convergence of sequences and series and at continuity using 'the sequential approach', they then meet 'the ε–δ approach', explained carefully and motivated by a clear 'ε–δ game' discussion. This makes the new approach seem very natural, and this is motivated by using each approach in later work in the appropriate situation. By the end of the book, students should have a good facility at using both the sequential approach and the ε–δ approach to proofs in Analysis, and should be better prepared for later study of Analysis than students who have acquired only a weak understanding of the conventional approach.

Outline of the content of the book

In Chapter 1, we define *real numbers* to be decimals. Rather than give a heavy discussion of *least upper bound* and *greatest lower bound*, we give an introduction to these matters sufficient for our purposes, and the full discussion is postponed to Chapter 7, where it is more timely. We also study *inequalities*, and their properties and proofs.

In Chapter 2, we define *convergent sequences* and examine their properties, basing the discussion on the notion of *null sequences*, which simplifies matters considerably. We also look at *divergent sequences*, sequences defined by *recurrence formulas* and particular sequences which converge to π and e.

In Chapter 3, we define *convergent infinite series*, and establish a number of tests for determining whether a given series is convergent or divergent. We demonstrate the equivalence of the two definitions of the *exponential function* $x \mapsto e^x$, and prove that the number e is irrational.

In Chapter 4, we define carefully what we mean by a *continuous function*, in terms of sequences, and establish the key properties of continuous functions. We also give a rigorous definition of the *exponential function* $x \mapsto a^x$.

We define e^x as $\lim_{n \to \infty} \left(1 + \frac{x}{n}\right)^n$ and as $\sum_{n=0}^{\infty} \frac{x^n}{n!}$.

In Chapter 5, we define the *limit of a function* as x tends to c or as x tends to ∞ in terms of the convergence of sequences. Then we introduce the ε–δ definitions of limit and continuity, and check that these are equivalent to the earlier definitions in terms of sequences. We also look briefly at *uniform continuity*.

In Chapter 6, we define what we mean by a *differentiable function*, using *difference quotients* $Q(h)$; this enables us to use our earlier results on limits to prove corresponding results for differentiable functions. We establish some interesting properties of differentiable functions. Finally, we construct the *Blancmange function* that is continuous everywhere on $\mathbb{R}$, but differentiable nowhere on $\mathbb{R}$.

In Chapter 7, we give a careful definition of what we mean by an *integrable function*, and establish a number of related criteria for establishing whether a given function is integrable or not. Our integral is the so-called *Riemann integral*, defined in terms of upper and lower Riemann sums. We check the standard properties of integrals and verify a number of standard approaches for calculating definite integrals. Then we give a number of applications of integrals to limits of certain sequences and series and prove *Stirling's Formula*.

> Stirling's Formula says that, for large n, $n!$ is 'roughly' $\sqrt{2\pi n}\left(\frac{n}{e}\right)^n$, in a sense that we explain.

Finally, in Chapter 8, we study the convergence and properties of *power series*. The chapter ends with a marvellous proof of the irrationality of the number π that uses a whole range of the techniques that have been met in the previous chapters.

For completeness and for students' convenience, we give a brief guide to our notation for sets and functions, together with a brief indication of the logic involved in proofs in Mathematics (in particular, the Principle of Mathematical Induction) in Appendix 1. Appendix 2 contains a list of standard derivatives and primitives, and Appendix 3 the first 1000 decimal places in the values of the numbers $\sqrt{2}$, e and π. Appendix 4 contains full solutions to all the problems set during each chapter.

Solutions are not given to the exercises at the end of each chapter, however. Lecturers/instructors may wish to use these exercises in homework assignments.

Study guide

This book assumes that students have a fair understanding of Calculus. The assumptions on technical background are deliberately kept slight, however, so that students can concentrate on the newer aspects of the subject 'Analysis'.

Most students will have met some of the material in the early chapters previously. Although this means that they can therefore proceed fairly quickly through some sections, it does NOT mean that those sections can be ignored – each section contains important ideas that are used later on and most include something new or have a different emphasis.

Most chapters are divided into five or six sections (each often further divided into sub-sections); sections are numbered using two digits (such as 'Section 3.2') and sub-sections using three digits (such as 'Sub-section 3.2.4'). Generally a section is considered to be about one evening's work for an average student.

Chapter 7 on Integration is arguably the highlight of the book. However, it contains some rather complicated mathematical arguments and proofs.

Therefore, when reading Chapter 7, it is important not to get bogged down in details, but to keep progressing through the key ideas, and to return later on to reading the things that were left out at the earlier reading. Most students will require three or four passes at this chapter before having a good idea of most of it.

We use wide pages with a large number of margin notes in which we place teaching comments and some diagrams to aid in the understanding of particular points in arguments. We also provide advice on which proofs to omit on a first study of the topic; it is important for the student NOT to get bogged down in a technical discussion or a proof until they have a good idea of the message contained in the result and the situations in which it can be used. Therefore clear encouragement is given on which portions of the text to leave till later, or to simply skim on a first reading.

It is important to read the margin notes!

This signposting benefits students greatly in the author's experience.

The end of the proof of a Theorem is indicated by a solid symbol '■' and the end of the solution of a worked Example by a hollow symbol '□'. There are many worked examples within the text to explain the concepts being taught, together with a good stock of problems to reinforce the teaching. The solutions are a key part of the teaching, and tackling them on your own and then reading our version of the solution is a key part of the learning process.

Tackling the problems is a good use of your time, not something to skimp.

No one can learn Mathematics by simply reading – it is a 'hands on' activity. The reader should not be afraid to draw pictures to illustrate what seems to be happening to a sequence or a function, to get a feeling for its behaviour. A wise old man once said that 'A picture is worth a thousand words!'. A good picture may even suggest a method of proof. However, at the same time it is important not to regard a picture on its own as a proof of anything; it may illustrate just one situation that can arise and miss many other possibilities!

It is important NOT to become discouraged if a topic seems difficult. It took mathematicians hundreds of years to develop Analysis to its current polished state, so it may take the reader a few hours at several sittings to really grasp the more complex or subtle ideas.

Acknowledgements

The material in the Open University course on which this book is based was contributed to in some way by many colleagues, including Phil Rippon, Robin Wilson, Andrew Brown, Hossein Zand, Joan Aldous, Ian Harrison, Alan Best, Alison Cadle and Roberta Cheriyan. Its eventual appearance in book form owes much to Lynne Barber.

Without the forebearance of my family, the writing of the book would have been impossible.

1 Numbers

In this book we study the properties of real functions defined on intervals of the real line (possibly the whole real line) and whose image also lies on the real line. In other words, they map $\mathbb{R}$ into $\mathbb{R}$. Our work will be from a very precise point of view in order to establish many of the properties of such functions which seem intuitively obvious; in the process we will discover that some apparently true properties are in fact not necessarily true!

The types of functions that we shall examine include:

exponential functions, such as $x \mapsto a^x$, where $a, x \in \mathbb{R}$,

trigonometric functions, such as $x \mapsto \sin x$, where $x \in \mathbb{R}$,

root functions, such as $x \mapsto \sqrt{x}$, where $x \geq 0$.

The types of behaviour that we shall examine include *continuity*, *differentiability* and *integrability* – and we shall discover that functions with these properties can be used in a number of surprising applications.

However, to put our study of such functions on a secure foundation, we need first to clarify our ideas of the *real numbers* themselves and their properties. In particular, we need to devote some time to the manipulation of *inequalities*, which play a key role throughout the book.

> For example, what exactly *is* the number $\sqrt{2}$?

In Section 1.1, we start by revising the properties of rational numbers and their *decimal representation*. Then we introduce the real numbers as infinite decimals, and describe the difficulties involved in doing arithmetic with such decimals.

In Section 1.2, we revise the rules for manipulating inequalities and show how to find the *solution set* of an inequality involving a real number, x, by applying the rules. We also explain how to deal with inequalities which involve *modulus* signs.

In Section 1.3, we describe various techniques for proving inequalities, including the very important technique of *Mathematical Induction*.

The concept of a *least upper bound*, which is of great importance in Analysis, is introduced in Section 1.4, and we discuss the *Least Upper Bound Property* of $\mathbb{R}$.

Finally, in Section 1.5, we describe how least upper bounds can be used to define arithmetical operations in $\mathbb{R}$.

> You may omit this section at a first reading.

Even though you may be familiar with much of this material we recommend that you read through it, as we give the system of real numbers a more careful treatment than you may have met before. The material on inequalities and least upper bounds is particularly important for later on.

In later chapters we shall define exactly what the numbers π and e are, and find various ways of calculating them. But, first, we examine numbers in general.

1.1   Real numbers

We start our study of the real numbers with the rational numbers, and investigate their decimal representations, then we proceed to the irrational numbers.

1.1.1   Rational numbers

We assume that you are familiar with the set of **natural numbers**

$$\mathbb{N} = \{1, 2, 3, \ldots\},$$

and with the set of **integers**

$$\mathbb{Z} = \{\ldots, -2, -1, 0, 1, 2, 3, \ldots\}.$$

The set of **rational numbers** consists of all fractions (or ratios of integers)

$$\mathbb{Q} = \left\{ \frac{p}{q} : p \in \mathbb{Z}, q \in \mathbb{N} \right\}.$$

Remember that each rational number has many different representations as a ratio of integers; for example

$$\frac{1}{3} = \frac{2}{6} = \frac{10}{30} = \ldots.$$

We also assume that you are familiar with the usual arithmetical operations of addition, subtraction, multiplication and division of rational numbers.

It is often convenient to represent rational numbers geometrically as points on a **number line**. We begin by drawing a line and marking on it points corresponding to the integers 0 and 1. If the distance between 0 and 1 is taken as a unit of length, then the rationals can be arranged on the line with positive rationals to the right of 0 and negative rationals to the left.

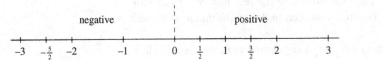

For example, the rational $\frac{3}{2}$ is placed at the point which is one-half of the way from 0 to 3.

This means that rationals have a natural order on the number-line. For example, $\frac{19}{22}$ lies to the left of $\frac{7}{8}$ because

$$\frac{19}{22} = \frac{76}{88} \quad \text{and} \quad \frac{7}{8} = \frac{77}{88}.$$

If a lies to the left of b on the number-line, then

$$a \text{ is } \textit{less than } b \quad \text{or} \quad b \text{ is } \textit{greater than } a,$$

and we write

$$a < b \quad \text{or} \quad b > a.$$

For example

$$\frac{19}{22} < \frac{7}{8} \quad \text{or} \quad \frac{7}{8} > \frac{19}{22}.$$

We write $a \leq b$, or $b \geq a$, if either $a < b$ or $a = b$.

Note that 0 is *not* a natural number.

Problem 1   Arrange the following rational numbers in order:
$$0, 1, -1, \tfrac{17}{20}, -\tfrac{17}{20}, \tfrac{45}{53}, -\tfrac{45}{53}.$$

Problem 2   Show that between any two distinct rational numbers there is another rational number.

1.1.2 Decimal representation of rational numbers

The decimal system enables us to represent all the natural numbers using only the ten integers

$$0, 1, 2, 3, 4, 5, 6, 7, 8 \text{ and } 9,$$

which are called *digits*. We now remind you of the basic facts about the representation of *rational* numbers by decimals.

Definition   A **decimal** is an expression of the form $\pm a_0 \cdot a_1 a_2 a_3 \ldots,$ where a_0 is a non-negative integer and $a_1, a_2, a_3, \ldots$ are digits.

For example
$$0.8500\ldots,$$
$$13.1212\ldots,$$
$$-1.111\ldots.$$

If only a finite number of the digits $a_1, a_2, a_3, \ldots$ are non-zero, then the decimal is called **terminating** or **finite**, and we usually omit the tail of 0s.

For example,
$$0.8500\ldots = 0.85.$$

Terminating decimals are used to represent rational numbers in the following way

$$\pm a_0 \cdot a_1 a_2 a_3 \ldots a_n = \pm \left(a_0 + \frac{a_1}{10^1} + \frac{a_2}{10^2} + \cdots + \frac{a_n}{10^n} \right).$$

For example
$$0.85 = 0 + \frac{8}{10^1} + \frac{5}{10^2}$$
$$= \frac{85}{100} = \frac{17}{20}.$$

It can be shown that any fraction whose denominator contains only powers of 2 and/or 5 (such as $20 = 2^2 \times 5$) can be represented by such a terminating decimal, which can be found by long division.

However, if we apply long division to many other rationals, then the process of long division never terminates and we obtain a **non-terminating** or **infinite** decimal. For example, applying long division to $\frac{1}{3}$ gives $0.333\ldots$, and for $\frac{19}{22}$ we obtain $0.86363\ldots$.

Problem 3   Apply long division to $\frac{1}{7}$ and $\frac{2}{13}$ to find the corresponding decimals.

These non-terminating decimals, which are obtained by applying the long division process, have a certain common property. All of them are **recurring**; that is, they have a recurring block of digits, and so can be written in shorthand form, as follows:

$$0.333\ldots \qquad\qquad = 0.\overline{3},$$
$$0.142857142857\ldots = 0.\overline{142857}\ldots,$$
$$0.86363\ldots \qquad\quad = 0.8\overline{63}.$$

Another commonly used notation is
$$0.\dot{3} \text{ or } 0.\dot{1}4285\dot{7}.$$

It is not hard to show, whenever we apply the long division process to a fraction $\frac{p}{q}$, that the resulting decimal is recurring. To see why we notice that there are only q possible remainders at each stage of the division, and so one of these remainders must eventually recur. If the remainder 0 occurs, then the resulting decimal is, of course, terminating; that is, it ends in recurring 0s.

Non-terminating recurring decimals which arise from the long division of fractions are used to represent the corresponding rational numbers. This representation is not quite so straight-forward as for terminating decimals, however. For example, the statement

$$\frac{1}{3} = 0.\overline{3} = \frac{3}{10^1} + \frac{3}{10^2} + \frac{3}{10^3} + \cdots$$

can be made precise only when we have introduced the idea of the *sum of a convergent infinite series*. For the moment, when we write the statement $\frac{1}{3} = 0.\overline{3}$ we mean simply that the decimal $0.\overline{3}$ arises from $\frac{1}{3}$ by the long division process.

We return to this topic in Chapter 3.

The following example illustrates one way of finding the rational number with a given decimal representation.

Example 1 Find the rational number (expressed as a fraction) whose decimal representation is $0.8\overline{63}$.

Solution First we find the fraction x such that $x = 0.\overline{63}$.

If we multiply both sides of this equation by 10^2 (because the recurring block has length 2), then we obtain

$$100x = 63.\overline{63} = 63 + x.$$

Hence

$$99x = 63 \Rightarrow x = \frac{63}{99} = \frac{7}{11}.$$

Thus

$$0.8\overline{63} = \frac{8}{10} + \frac{x}{10} = \frac{8}{10} + \frac{7}{110} = \frac{95}{110} = \frac{19}{22}. \qquad \square$$

The key idea in the above solution is that multiplication of a decimal by 10^k moves the decimal point k places to the right.

> **Problem 4** Using the above method, find the fractions whose decimal representations are:
>
> (a) $0.\overline{231}$;   (b) $2.2\overline{81}$.

The decimal representation of rational numbers has the advantage that it enables us to decide immediately which of two distinct positive rational numbers is the greater. We need only examine their decimal representations and notice the first place at which the digits differ. For example, to order $\frac{7}{8}$ and $\frac{19}{22}$ we write

$$\frac{7}{8} = 0.875\ldots \quad \text{and} \quad \frac{19}{22} = 0.86363\ldots,$$

and so

$$0.8\overset{\downarrow}{6}363\ldots < 0.8\overset{\downarrow}{7}5 \Rightarrow \frac{19}{22} < \frac{7}{8}.$$

> **Problem 5** Find the first two digits after the decimal point in the decimal representations of $\frac{17}{20}$ and $\frac{45}{53}$, and hence determine which of these two rational numbers is the greater.

Warning Decimals which end in recurring 9s sometimes arise as alternative representations for terminating decimals. For example

$$1 = 0.\overline{9} = 0.999\ldots \quad \text{and} \quad 1.35 = 1.34\overline{9} = 1.34999\ldots.$$

Whenever possible, we avoid using the form of a decimal which ends in recurring 9s.

You may find this rather alarming, but it is important to realise that this is a matter of *convention*. We wish to allow the decimal 0.999... to represent a number x, so x must be less than or equal to 1 and greater than each of the numbers

$$0.9, \ 0.99, \ 0.999, \ldots.$$

The *only* rational with these properties is 1.

1.1.3 Irrational numbers

One of the surprising mathematical discoveries made by the Ancient Greeks was that the system of rational numbers is not adequate to describe all the magnitudes that occur in geometry. For example, consider the diagonal of a square of side 1. What is its length? If the length is x, then, by Pythagoras' Theorem, x must satisfy the equation $x^2 = 2$. However, there is no rational number which satisfies this equation.

$$x^2 = 1^2 + 1^2 = 2$$

Theorem 1 There is no rational number x such that $x^2 = 2$.

Proof Suppose that such a rational number x exists. Then we can write $x = \frac{p}{q}$. By cancelling, if necessary, we may assume that p and q have no common factor. The equation $x^2 = 2$ now becomes

This is a *proof by contradiction*.

$$\frac{p^2}{q^2} = 2, \quad \text{so} \quad p^2 = 2q^2.$$

Now, the square of an odd number is odd, and so p cannot be odd. Hence p is even, and so we can write $p = 2r$, say. Our equation now becomes

For

$$(2k+1)^2 = 4k^2 + 4k + 1$$
$$= 4(k^2 + k) + 1.$$

$$(2r)^2 = 2q^2, \quad \text{so} \quad q^2 = 2r^2.$$

Reasoning as before, we find that q is also even.

Since p and q are both even, they have a common factor 2, which contradicts our earlier statement that p and q have no common factors.

Arguing from our original assumption that x exists, we have obtained two contradictory statements. Thus, our original assumption must be false. In other words, no such x exists. ■

> **Problem 6** By imitating the above proof, show that there is no rational number x such that $x^3 = 2$.

Since we want equations such as $x^2 = 2$ and $x^3 = 2$ to have solutions, we must introduce new numbers which are not rational numbers. We denote these new numbers by $\sqrt{2}$ and $\sqrt[3]{2}$, respectively; thus $\left(\sqrt{2}\right)^2 = 2$ and $\left(\sqrt[3]{2}\right)^3 = 2$. Of course, we must introduce many other new numbers, such as $\sqrt{3}, \sqrt[5]{11}$, and so on. Indeed, it can be shown that, if m, n are natural numbers and $x^m = n$ has no integer solution, then $\sqrt[m]{n}$ cannot be rational. A number which is not rational is called **irrational**.

The case $m = 2$ is treated in Exercise 5 for this section in Section 1.6.

There are many other mathematical quantities which cannot be described exactly by rational numbers. For example, the number π which denotes the area of a disc of radius 1 (or half the length of the perimeter of such a disc) is irrational, as is the number e.

Lambert proved that π is irrational in 1768.

It is natural to ask whether irrational numbers, such as $\sqrt{2}$ and π, can be represented as decimals. Using your calculator, you can check that $(1.41421356)^2$ is very close to 2, and so 1.41421356 is a very good approximate value for $\sqrt{2}$. But is there a decimal which represents $\sqrt{2}$ *exactly*? If such a decimal exists, then it cannot be recurring, because all the recurring decimals correspond to rational numbers.

In fact, it *is* possible to represent all the irrational numbers mentioned so far by non-recurring decimals. For example, there are non-recurring decimals such that

$$\sqrt{2} = 1.41421356\ldots \quad \text{and} \quad \pi = 3.14159265\ldots.$$

It is also natural to ask whether non-recurring decimals, such as

$$0.101001000100001\ldots \quad \text{and} \quad 0.123456789101112\ldots,$$

represent irrational numbers. In fact, a decimal corresponds to a rational number if and only if it is recurring; so a non-recurring decimal must correspond to an irrational number.

We may summarise this as:

recurring decimal ⇔ *rational number*

non-recurring decimal ⇔ *irrational number*

In fact
$$(1.41421356)^2$$
$$= 1.9999999932878736.$$

We explain why $\sqrt{2}$ has a decimal representation in Section 1.5.

1.1.4   The real number system

Taken together, the rational numbers (recurring decimals) and irrational numbers (non-recurring decimals) form the set of **real numbers**, denoted by $\mathbb{R}$.

As with rational numbers, we can determine which of two real numbers is greater by comparing their decimals and noticing the first pair of corresponding digits which differ. For example

$$0.10\overset{\downarrow}{1}00100010000\ldots < 0.12\overset{\downarrow}{3}456789101112\ldots.$$

We now associate with each irrational number a point on the number-line. For example, the irrational number $x = 0.123456789101112\ldots$ satisfies each of the inequalities

$$0.1 < x < 0.2$$
$$0.12 < x < 0.13$$
$$0.123 < x < 0.124$$
$$\vdots$$

We assume that there is a point on the number-line corresponding to x, which lies to the right of each of the (rational) numbers $0.1, 0.12, 0.123\ldots$, and to the left of each of the (rational) numbers $0.2, 0.13, 0.124, \ldots$.

When comparing decimals in this way, we do not allow either decimal to end in recurring 9s.

$$x = 0.12345\ldots$$

As usual, negative real numbers correspond to points lying to the left of 0; and the 'number-line', complete with both rational and irrational points, is called the **real line**.

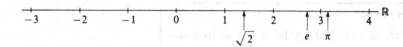

There is thus a one–one correspondence between the points on the real line and the set $\mathbb{R}$ of real numbers (or decimals).

We now state several properties of $\mathbb{R}$, with which you will be already familiar, although you may not have met their names before. These properties are used frequently in Analysis, and we do not always refer to them explicitly by name.

Order Properties of $\mathbb{R}$

1. **Trichotomy Property**   If $a, b \in \mathbb{R}$, then *exactly one* of the following inequalities holds

$$a < b \quad \text{or} \quad a = b \quad \text{or} \quad a > b.$$

2. **Transitive Property**   If $a, b, c \in \mathbb{R}$, then

$$a < b \quad \text{and} \quad b < c \Rightarrow a < c.$$

3. **Archimedean Property**   If $a \in \mathbb{R}$, then there is a positive integer n such that

$$n > a.$$

4. **Density Property**   If $a, b \in \mathbb{R}$ and $a < b$, then there is a rational number x and an irrational number y such that

$$a < x < b \quad \text{and} \quad a < y < b.$$

The first three of these properties are almost self-evident, but the Density Property is not so obvious.

Remark

The Archimedean Property is sometimes expressed in the following equivalent way: for any positive real number a, there is a positive integer n such that $\frac{1}{n} < a$.

The following example illustrates how we can prove the Density Property.

Example 2   Find a rational number x and an irrational number y satisfying

$$a < x < b \quad \text{and} \quad a < y < b,$$

where $a = 0.12\overline{3}$ and $b = 0.12345\ldots$.

Solution   The two decimals

$$a = 0.123\overset{\downarrow}{3}\ldots \quad \text{and} \quad b = 0.1234\overset{\downarrow}{5}\ldots$$

differ first at the fourth digit. If we truncate b after this digit, we obtain the rational number $x = 0.1234$, which satisfies the requirement that $a < x < b$.

To find an irrational number y between a and b, we attach to x a (sufficiently small) non-recurring tail such as $010010001\ldots$ to give $y = 0.1234|010010001\ldots$. It is then clear that y is irrational (because its decimal is non-recurring) and that $a < y < b$.   □

> **Problem 7**   Find a rational number x and an irrational number y such that $a < x < b$ and $a < y < b$, where $a = 0.\overline{3}$ and $b = 0.3401$.

Theorem 2   Density Property of $\mathbb{R}$

If $a, b \in \mathbb{R}$ and $a < b$, then there is a rational number x and an irrational number y such that

$$a < x < b \quad \text{and} \quad a < y < b.$$

You may omit the following proof on a first reading.

Proof   For simplicity, we assume that $a, b \geq 0$. So, let a and b have decimal representations

$$a = a_0 \cdot a_1 a_2 a_3 \ldots \quad \text{and} \quad b = b_0 \cdot b_1 b_2 b_3 \ldots,$$

where we arrange that a does not end in recurring 9s, whereas b does not terminate (this latter can be arranged by replacing a terminating representation by an equivalent representation that ends in recurring 9s).

Here a_0, b_0 are non-negative integers, and $a_1, b_1, a_2, b_2, \ldots$ are digits.

Since $a < b$, there must be some integer n such that

$$a_0 = b_0, \; a_1 = b_1, \; \ldots, \; a_{n-1} = b_{n-1}, \; \text{but } a_n < b_n.$$

Then $x = a_0 \cdot a_1 a_2 a_3 \ldots a_{n-1} b_n$ is rational, and $a < x < b$ as required.

Finally, since $x < b$, it follows that we can attach a sufficiently small non-recurring tail to x to obtain an irrational number y for which $a < y < b$. ∎

Remark

One consequence of the Density Property is that between any two real numbers there are infinitely many rational numbers and infinitely many irrational numbers.

Problem 8   Prove that between any two non-negative real numbers a and b there are at least two distinct rational numbers.

A proof of the previous remark would involve ideas similar to those involved in tackling Problem 8.

1.1.5   Arithmetic in $\mathbb{R}$

We can do arithmetic with *recurring* decimals by first converting the decimals to fractions. However, it is not obvious how to perform arithmetical operations with *non-recurring* decimals.

Assuming that we can represent $\sqrt{2}$ and π by the non-recurring decimals

$$\sqrt{2} = 1.41421356\ldots \quad \text{and} \quad \pi = 3.14159265\ldots,$$

can we also represent the sum $\sqrt{2} + \pi$ and the product $\sqrt{2} \times \pi$ as decimals? Indeed, what do we mean by the operations of addition and multiplication when non-recurring decimals (irrationals) are involved, and do these operations satisfy the same properties as addition and multiplication of rationals?

It would take many pages to answer these questions fully. Therefore, we shall *assume* that it is possible to define all the usual arithmetical operations with decimals, and that they do satisfy the usual properties. For definiteness, we now list these properties.

<table>
<tr><td colspan="3">Arithmetic in $\mathbb{R}$</td></tr>
</table>

Addition	*Multiplication*	

A1 If $a, b \in \mathbb{R}$, then
   $a + b \in \mathbb{R}$.
	M1 If $a, b \in \mathbb{R}$, then
	   $a \times b \in \mathbb{R}$.
		CLOSURE

A2 If $a \in \mathbb{R}$, then
   $a + 0 = 0 + a = a$.
	M2 If $a \in \mathbb{R}$, then
	   $a \times 1 = 1 \times a = a$.
		IDENTITY

A3 If $a \in \mathbb{R}$, then there is a
   number $-a \in \mathbb{R}$ such that
   $a + (-a) = (-a) + a = 0$.
	M3 If $a \in \mathbb{R} - \{0\}$, then there is a
	   number $a^{-1} \in \mathbb{R}$ such that
	   $a \times a^{-1} = a^{-1} \times a = 1$.
		INVERSES

A4 If $a, b, c \in \mathbb{R}$, then
   $(a + b) + c = a + (b + c)$.
	M4 If $a, b, c \in \mathbb{R}$, then
	   $(a \times b) \times c = a \times (b \times c)$.
		ASSOCIATIVITY

A5 If $a, b \in \mathbb{R}$, then
   $a + b = b + a$.
	M5 If $a, b \in \mathbb{R}$, then
	   $a \times b = b \times a$.
		COMMUTATIVITY

D If $a, b, c \in \mathbb{R}$, then $a \times (b + c) = a \times b + a \times c$.		DISTRIBUTATIVITY

To summarise the contents of this table:

- $\mathbb{R}$ is an Abelian group under the operation of addition $+$;   Properties A1–A5
- $\mathbb{R} - \{0\}$ is an Abelian group under the operation of multiplication $\times$;   Properties M1–M5
- These two group structures are linked by the Distributive Property.   Property D

It follows from the above properties that we can perform addition, subtraction (where $a - b = a + (-b)$), multiplication and division (where $\frac{a}{b} = a \times b^{-1}$) in $\mathbb{R}$, and that these operations satisfy all the usual properties.

Any system satisfying the properties listed in the table is called a **field**. Both $\mathbb{Q}$ and $\mathbb{R}$ are fields.

Furthermore, we shall assume that the set $\mathbb{R}$ contains the nth roots and rational powers of positive real numbers, with their usual properties. In Section 1.5 we describe one way of justifying the existence of nth roots.

1.2   Inequalities

Much of Analysis is concerned with inequalities of various kinds; the aim of this section and the next section is to provide practice in the manipulation of inequalities.

1.2.1   Rearranging inequalities

The fundamental rule, upon which much manipulation of inequalities is based, is that the statement $a < b$ means exactly the same as the statement $b - a > 0$. This fact can be stated concisely in the following way:

Recall that the symbol '$\Leftrightarrow$' means 'if and only if' or 'implies and is implied by'.

> **Rule 1**   For any $a, b \in \mathbb{R}$, $a < b \Leftrightarrow b - a > 0$.

Put another way, the inequalities $a < b$ and $b - a > 0$ are *equivalent*.

There are several other standard rules for rearranging a given inequality into an equivalent form. Each of these can be deduced from our first rule above. For example, we obtain an equivalent inequality by adding the same number to both sides.

> **Rule 2** For any $a, b, c \in \mathbb{R}$, $\quad a < b \Leftrightarrow a + c < b + c$.

Another way to rearrange an inequality is to multiply both sides by a non-zero number, making sure to *reverse* the inequality if the number is negative.

> **Rule 3**
> - For any $a, b \in \mathbb{R}$ and any $c > 0$, $\quad a < b \Leftrightarrow ac < bc$;
> - For any $a, b \in \mathbb{R}$ and any $c < 0$, $\quad a < b \Leftrightarrow ac > bc$.

For example
$$2 < 3 \Leftrightarrow 20 < 30 \, (c = 10),$$
$$2 < 3 \Leftrightarrow -20 > -30$$
$$(c = -10).$$

Sometimes the most effective way to rearrange an inequality is to take reciprocals. However, in this case, both sides of the inequality must be positive, and the direction of the inequality must be *reversed*.

> **Rule 4   (Reciprocal Rule)**
> For any positive $a, b \in \mathbb{R}$, $\quad a < b \Leftrightarrow \frac{1}{a} > \frac{1}{b}$.

For example
$$2 < 4 \Leftrightarrow \frac{1}{2} \, (= 0.5)$$
$$> \frac{1}{4} \, (= 0.25).$$

Some inequalities can be simplified only by taking powers. However, in order to do this, both sides must be non-negative and must be raised to a *positive* power.

> **Rule 5   (Power Rule)**
> For any non-negative $a, b \in \mathbb{R}$, and any $p > 0$, $\quad a < b \Leftrightarrow a^p < b^p$.

For example
$$4 < 9 \Leftrightarrow 4^{\frac{1}{2}} (= 2)$$
$$< 9^{\frac{1}{2}} (= 3).$$
We shall discuss the meaning of non-integer powers in Section 1.5.

For positive integers p, Rule 5 follows from the identity
$$b^p - a^p = (b - a)\left(b^{p-1} + b^{p-2}a + \cdots + ba^{p-2} + a^{p-1}\right);$$
thus, since the right-hand bracket is positive, we have
$$b - a > 0 \Leftrightarrow b^p - a^p > 0,$$
which is equivalent to our desired result.

For example,
$$b^3 - a^3 = (b - a) \times$$
$$(b^2 + ba + a^2).$$

Remark

There are corresponding versions of Rules 1–5 in which the *strict* inequality $a < b$ is replaced by the *weak* inequality $a \leq b$.

> **Problem 1** State (without proof) the versions of Rules 1–5 for weak inequalities.

We shall give one more rule for rearranging inequalities in Sub-section 1.2.3.

1.2.2 Solving inequalities

Solving an inequality involving an unknown real number x means determining those values of x for which the given inequality holds; that is, finding the *solution set* of the inequality. We can often do this by rewriting the inequality in an equivalent, but simpler form, using the rules given in the last sub-section.

The *solution set* is the set of those values of x for which the inequality holds.

In this activity we frequently use the usual rules for the sign of a product, and the fact that the square of any real number is non-negative. Also, we need to remember the difference between the logical statements: 'implies', 'is implied by' and 'implies and is implied by'.

$\times$	$+$	$-$
$+$	$+$	$-$
$-$	$-$	$+$

Example 1 Solve the inequality $\frac{x+2}{x+4} > \frac{x-3}{2x-1}$.

Solution We rearrange this inequality to give a somewhat simpler inequality, using Rule 1

$$\frac{x+2}{x+4} > \frac{x-3}{2x-1} \Leftrightarrow \frac{x+2}{x+4} - \frac{x-3}{2x-1} > 0$$

It is a common strategy to bring all terms to one side.

$$\Leftrightarrow \frac{x^2 + 2x + 10}{(x+4)(2x-1)} > 0$$

We bring everything to a common denominator.

$$\Leftrightarrow \frac{(x+1)^2 + 9}{(x+4)(2x-1)} > 0.$$

Here we complete the square in the numerator, since we cannot factorise it.

Now, the numerator is always positive. The denominator vanishes when $x = -4$ or $x = \frac{1}{2}$. By examining separately the sign of the denominator when $x < -4$, $-4 < x < \frac{1}{2}$ and $x > \frac{1}{2}$, we can deduce that the last fraction is positive precisely when $x < -4$ or $x > \frac{1}{2}$. Hence the solution set of the original inequality is

This is because the final displayed inequality is *equivalent* to the inequality we are solving. The logical implication symbols between the displayed inequalities were all 'implies and is implied by'.

$$\left\{ x : \frac{x+2}{x+4} > \frac{x-3}{2x-1} \right\} = (-\infty, -4) \cup \left(\tfrac{1}{2}, \infty \right). \qquad \square$$

Example 2 Solve the inequality $\frac{1}{2x^2+2} < \frac{1}{4}$.

Solution Since $2x^2 + 2 > 0$, we have

$$\frac{1}{2x^2+2} < \frac{1}{4} \Leftrightarrow 2x^2 + 2 > 4 \qquad \text{(by Rule 4)}$$

$$\Leftrightarrow x^2 + 1 > 2 \qquad \text{(by Rule 3)}$$

$$\Leftrightarrow x^2 - 1 > 0 \qquad \text{(by Rule 1)}$$

$$\Leftrightarrow (x-1)(x+1) > 0.$$

Here we factorise the left-hand side of the inequality to examine the signs of its factors.

This last inequality holds precisely when $x < -1$ or $x > 1$. It follows that the solution set of the original inequality is

$$\left\{ x : \frac{1}{2x^2+2} < \frac{1}{4} \right\} = (-\infty, -1) \cup (1, \infty). \qquad \square$$

Problem 2 Use each of the following expressions to write down an inequality with the given expression on its left-hand side which is equivalent to the inequality $x > 2$:

(a) $x + 3$;   (b) $2 - x$;   (c) $5x + 2$;   (d) $\frac{-1}{5x+2}$.

Problem 3 Solve the following inequalities:

(a) $\frac{4x - x^2 - 7}{x^2 - 1} \geq 3$;   (b) $2x^2 \geq (x+1)^2$.

Warning   Great care is needed when solving inequalities which involve rational powers. In particular, when applying Rule 5 both sides of the inequality *must* be non-negative.

Example 3   Solve the inequality $\sqrt{2x+3} > x$.

Solution   The expression $\sqrt{2x+3}$ is defined only when $2x+3 \geq 0$; that is, when $x \geq -\frac{3}{2}$. Hence we need only consider those x in $\left[-\frac{3}{2}, \infty\right)$.

We can obtain an equivalent inequality by squaring, provided that both $\sqrt{2x+3}$ and x are non-negative. Thus, for $x \geq 0$, we obtain

$$\sqrt{2x+3} > x \Leftrightarrow 2x + 3 > x^2 \qquad \text{(by Rule 5, with } p = 2\text{)}$$
$$\Leftrightarrow x^2 - 2x - 3 < 0$$
$$\Leftrightarrow (x-3)(x+1) < 0.$$

So the part of the solution set in $[0, \infty)$ is $[0, 3)$.

We now examine those x for which $-\frac{3}{2} \leq x < 0$. For such x, $\sqrt{2x+3} \geq 0$ and $x < 0$, so that $\sqrt{2x+3}\,(\geq 0) > x$ for all such x. It follows that all these x, namely the set $\left[-\frac{3}{2}, 0\right)$, belong to the solution set too.

Combining these results, the solution set of the original inequality is

$$\left\{x : \sqrt{2x+3} > x\right\} = \left[-\frac{3}{2}, 0\right) \cup [0, 3)$$
$$= \left[-\frac{3}{2}, 3\right). \qquad \square$$

> Note that, for the moment, we are examining only those x for which $x \geq 0$.

> Notice the use of the Transitive Property here.

Problem 4   Solve the inequality $\sqrt{2x^2 - 2} > x$.

1.2.3   Inequalities involving modulus signs

We now turn our attention to inequalities involving the *modulus*, or *absolute value*, of a real number. Recall that, if $a \in \mathbb{R}$, then its modulus $|a|$ is defined by

$$|a| = \begin{cases} a, & \text{if } a \geq 0, \\ -a, & \text{if } a < 0. \end{cases}$$

It is often useful to think of $|a|$ as the distance along the real line from 0 to a.

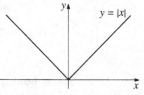

For example

$$|3| = |-3| = 3.$$

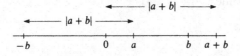

In the same way, $|a - b|$ is the distance along the real line from 0 to $a - b$, which is the same as the distance from a to b.

We sometimes write

$$|a - b| = d(a, b).$$

Notice also that $|a + b| = |a - (-b)|$ is the distance from a to $-b$.

For example, the distance from -2 to 3 is

$$|(-2) - 3| = |-5| = 5.$$

We now list some basic properties of the modulus, which follow immediately from the definition:

> **Properties of the modulus**   For any real numbers a and b:
> 1. $|a| \geq 0$, with equality if and only if $a = 0$;
> 2. $-|a| \leq a \leq |a|$;
> 3. $|a|^2 = a^2$;
> 4. $|a - b| = |b - a|$;
> 5. $|ab| = |a| \times |b|$.

For example, if $a = -2$, $b = 1$, then:
1. $|-2| > 0$;
2. $-|-2| \leq -2 \leq |-2|$;
3. $|-2|^2 = (-2)^2$;
4. $|(-2) - 1| = |1 - (-2)|$;
5. $|(-2) \times 1| = |-2| \times |1|$.

There is a basic rule for rearranging inequalities involving modulus signs:

> **Rule 6**   For any real numbers a and b, where $b > 0$: $|a| < b \Leftrightarrow -b < a < b$.

Note that, in a similar way
$$|a| \leq b \Leftrightarrow -b \leq a \leq b.$$

Also, it is often possible, and sometimes easier, to use Rule 5 with $p = 2$ than to use Rule 6. The following example illustrates the use of both rules.

Example 4   Solve the inequality $|x - 2| < 1$.

Solution   Using Rule 6, we obtain
$$|x - 2| < 1 \Leftrightarrow -1 < x - 2 < 1$$
$$\Leftrightarrow 1 < x < 3.$$

We take $a = x - 2$, $b = 1$ in Rule 6.

So the solution set of the original inequality is
$$\{x : |x - 2| < 1\} = (1, 3).$$

Alternatively, using Rule 5 (with $p = 2$), we obtain
$$|x - 2| < 1 \Leftrightarrow (x - 2)^2 < 1$$
$$\Leftrightarrow x^2 - 4x + 3 < 0$$
$$\Leftrightarrow (x - 1)(x - 3) < 0.$$

For $|x - 2|^2 = (x - 2)^2$.

Again, this shows that the required solution set is $(1, 3)$. $\square$

Example 5   Solve the inequality $|x - 2| \leq |x + 1|$.

Solution   Using Rule 5 (with $p = 2$), we obtain
$$|x - 2| \leq |x + 1| \Leftrightarrow (x - 2)^2 \leq (x + 1)^2$$
$$\Leftrightarrow x^2 - 4x + 4 \leq x^2 + 2x + 1$$
$$\Leftrightarrow 3 \leq 6x$$
$$\Leftrightarrow \frac{1}{2} \leq x.$$

So the solution set of the original inequality is
$$\{x : |x - 2| \leq |x + 1|\} = \left[\frac{1}{2}, \infty\right).$$ $\square$

The inequalities in Examples 4 and 5 can easily be interpreted geometrically.

In Example 4, the inequality $|x - 2| < 1$ holds when the distance from x to 2 is strictly less than 1. So it holds for all points on either side of 2 at a distance less than 1 from 2 – namely, in the open interval $(1, 3)$.

In Example 5, the inequality $|x-2| \leq |x+1|$ holds when the distance from x to 2 is less than or equal to the distance from x to -1, since $|x+1| = |x-(-1)|$. The mid-point of 2 and -1 (that is, the point x where the distance from x to 2 equals the distance from x to -1) is $\frac{1}{2}$. So the inequality holds when x lies in $[\frac{1}{2}, \infty)$.

Some good ideas when tackling problems involving inequalities of these types are:

- use your geometrical intuition, where possible, to give yourself an idea of the sets involved;
- test one or two values of x in your final solution set to see if they are valid – this often detects errors in manipulating inequality signs!

Problem 5   Solve the following inequalities:

(a) $|2x^2 - 13| < 5$;   (b) $|x-1| \leq 2|x+1|$.

1.3   Proving inequalities

In this section we show you how to *prove* inequalities of various types. We shall use the rules for rearranging inequalities given in Section 1.2, and also use other rules which enable us to *deduce new inequalities from old*. We have already met the first rule in Section 1.1, where it was called the Transitive Property of $\mathbb{R}$.

Transitive Rule $a < b$ and $b < c \Rightarrow a < c$.

We use the Transitive Rule when we want to prove that $a < c$, and we know that $a < b$ and $b < c$.

The following rules are also useful:

Combination Rules
If $a < b$ and $c < d$, then:
Sum Rule $a+c < b+d$;
Product Rule $ac < bd$ (provided that $a, c \geq 0$).

For example, since $2 < 3$ and $4 < 5$, then

$$2 + 4 < 3 + 5;$$
$$2 \times 4 < 3 \times 5.$$

There are also weak and weak/strict versions of the Transitive Rule and Combination Rules, which we will ask you to work out and use as they arise.

For example, if $a < b$ and $c \leq d$, then

$$a + c < b + d$$

and

$$ac < bd, \text{provided } a, c > 0.$$

Remark

It is important to appreciate that the Transitive Rule and the Combination Rules have a different nature from Rules 1–6 in Section 1.2. Rules 1–6 tell you how to rearrange inequalities into equivalent forms, whereas the Transitive Rule and the Combination Rules enable you to deduce new inequalities which are *not* equivalent to the old ones.

1.3.1   The Triangle Inequality

If **a** and **b** are both vectors in $\mathbb{R}^2$, then the vector $\mathbf{a} + \mathbf{b}$ is obtained from the 'parallelogram construction':

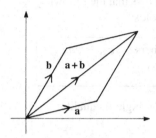

By elementary geometry, the length of any side of a triangle is less than or equal to the sum of the lengths of the other two sides. In the special case of $\mathbb{R}$, when all the vectors lie on a line, this can be interpreted as the *Triangle Inequality*, which involves the absolute value of the real numbers a, b and $a + b$.

Triangle Inequality   If $a, b \in \mathbb{R}$, then:

1. $|a + b| \leq |a| + |b|$;
2. $|a - b| \geq \big||a| - |b|\big|$ (the 'reverse form' of the Triangle Inequality).

Proof   In order to prove part 1, we use Rule 5, with $p = 2$

$$|a + b| \leq |a| + |b| \Leftrightarrow (a + b)^2 \leq (|a| + |b|)^2$$
$$\Leftrightarrow a^2 + 2ab + b^2 \leq a^2 + 2|a||b| + b^2$$
$$\Leftrightarrow 2ab \leq 2|a||b|.$$

The final inequality is certainly true for all $a, b \in \mathbb{R}$, and so the first inequality must also be true for all $a, b \in \mathbb{R}$. Hence we have proved part 1.

We prove part 2 by using the same method

$$|a - b| \geq \big||a| - |b|\big| \Leftrightarrow (a - b)^2 \geq (|a| - |b|)^2$$
$$\Leftrightarrow a^2 - 2ab + b^2 \geq a^2 - 2|a||b| + b^2$$
$$\Leftrightarrow -2ab \geq -2|a||b|$$
$$\Leftrightarrow 2ab \leq 2|a||b|,$$

which is again true for all $a, b \in \mathbb{R}$.   ∎

Remarks

1. Although we have used double-headed arrows here, the actual proof requires only the arrows going *from right to left*. For example, in the proof of part 1 the important implication is

$$|a + b| \leq |a| + |b| \Leftarrow 2ab \leq 2|a||b|.$$

2. Part 1 of the Triangle Inequality can also be proved by using Rule 6, as follows. By Rule 6

$$|a + b| \leq |a| + |b| \Leftrightarrow -(|a| + |b|) \leq a + b \leq (|a| + |b|). \quad (1)$$

Here we discuss $\mathbb{R}^2$, rather than $\mathbb{R}$, simply because the argument is then geometrically clearer.

In the 'parallelogram construction', you draw the vector **a** from the origin to some point, then the vector **b** from that point to a final point. The vector $\mathbf{a} + \mathbf{b}$ is then the vector from the origin to that final point.

For example, with $a = -1$ and $b = 3$:

1. $|-1 + 3| \leq |-1| + |3|$;
2. $|(-1) - 3| \geq \big||-1| - |3|\big|$.

Remember that $|a|^2 = a^2$.

Part 2 is sometimes called the 'cunning form' or 'backwards form' of the Triangle Inequality.

Now, we know that $-|a| \leq a \leq |a|$ and $-|b| \leq b \leq |b|$; so, by the Sum Rule

$$-(|a| + |b|) \leq a + b \leq (|a| + |b|).$$

It follows that the left-hand inequality in (1) must also hold, as required.

3. An obvious modification of the proof in Remark 2 shows that the following more general form of the Triangle Inequality also holds:

Triangle Inequality for n terms   For any real numbers $a_1, a_2, \ldots, a_n$
$$|a_1 + a_2 + \cdots + a_n| \leq |a_1| + |a_2| + \cdots + |a_n|.$$

The following example is a typical application of the Triangle Inequality.

Example 1   Use the Triangle Inequality to prove that:

(a) $|a| \leq 1 \Rightarrow |3 + a^3| \leq 4$;   (b) $|b| < 1 \Rightarrow |3 - b| > 2$.

Solution

(a) Suppose that $|a| \leq 1$. The Triangle Inequality then gives

$$\begin{aligned} |3 + a^3| &\leq |3| + |a^3| \\ &= 3 + |a|^3 \\ &\leq 3 + 1 \quad \text{(since } |a| \leq 1) \\ &= 4. \end{aligned}$$

Note the use of the Transitive Rule here.

(b) Suppose that $|b| < 1$. The 'reverse form' of the Triangle Inequality then gives

$$\begin{aligned} |3 - b| &\geq \big||3| - |b|\big| \\ &= \big|3 - |b|\big| \\ &\geq 3 - |b|. \end{aligned}$$

Now $|b| < 1$, so that $-|b| > -1$. Thus

$$\begin{aligned} 3 - |b| &> 3 - 1 \\ &= 2, \end{aligned}$$

Again, we use the Transitive Rule.

and we can then deduce from the previous chain of inequalities that $|3 - b| > 2$, as desired.   □

Remarks

1. The results of Example 1 can also be stated in the form:

(a) $|3 + a^3| \leq 4$, for $|a| \leq 1$;

(b) $|3 - b| > 2$, for $|b| < 1$.

2. The reverse implications

$$|3 + a^3| \leq 4 \Rightarrow |a| \leq 1 \quad \text{and} \quad |3 - b| > 2 \Rightarrow |b| < 1$$

are FALSE. For example, try putting $a = -\frac{3}{2}$ and $b = -2$!

Problem 1   Use the Triangle Inequality to prove that:

(a) $|a| \leq \frac{1}{2} \Rightarrow |a + 1| \leq \frac{3}{2}$;   (b) $|b| < \frac{1}{2} \Rightarrow |b^3 - 1| > \frac{7}{8}$.

1.3.2   Inequalities involving *n*

In Analysis we often need to prove inequalities involving an integer *n*. It is a common convention in mathematics that the symbol *n* is used to denote an integer (frequently a natural number).

It is often possible to deal with inequalities involving *n* by using the rearrangement rules given in Section 1.2. Here is such an example.

Example 2   Prove that $2n^2 \geq (n+1)^2$,   for $n \geq 3$.

n	1	2	3	4
$2n^2$	2	8	18	32
$(n+1)^2$	4	9	16	25

Solution   Rearranging this inequality into an equivalent form, we obtain

$$2n^2 \geq (n+1)^2 \Leftrightarrow 2n^2 - (n+1)^2 \geq 0$$

$$\Leftrightarrow n^2 - 2n - 1 \geq 0$$

$$\Leftrightarrow (n-1)^2 - 2 \geq 0 \quad \text{(by 'completing the square')}$$

$$\Leftrightarrow (n-1)^2 \geq 2.$$

This final inequality is clearly true for $n \geq 3$, and so the original inequality $2n^2 \geq (n+1)^2$ is true for $n \geq 3$.   ☐

Remarks

1. In Problem 3 of Section 1.2, we asked you to solve the inequality $2x^2 \geq (x+1)^2$; its solution set was $\left(-\infty, 1 - \sqrt{2}\right] \cup \left[1 + \sqrt{2}, \infty\right)$. In Example 2, above, we found those natural numbers *n* lying in this solution set.

2. An alternative solution to Example 2 is as follows

$$2n^2 \geq (n+1)^2 \Leftrightarrow 2 \geq \left(\frac{n+1}{n}\right)^2 \quad \text{(by Rule 3, with } c = 1/n^2 \text{)}$$

Actually, an extended version of Rule 3.

$$\Leftrightarrow \sqrt{2} \geq 1 + \frac{1}{n} \quad \text{(by Rule 5, with } p = \tfrac{1}{2} \text{)};$$

and this final inequality certainly holds for $n \geq 3$.

Problem 2   Prove that $\frac{3n}{n^2+2} < 1$, for $n > 2$.

1.3.3   More on inequalities

We now look at a number of inequalities and methods for proving inequalities that will be useful later on.

Example 3   Prove that $ab \leq \left(\frac{a+b}{2}\right)^2$, for all $a, b \in \mathbb{R}$.

Solution   We tackle this inequality using the various rearrangement rules and a chain of equivalent inequalities until we obtain an inequality that we know must be true

$$ab \leq \left(\frac{a+b}{2}\right)^2 \Leftrightarrow ab \leq \frac{a^2 + 2ab + b^2}{4}$$

$$\Leftrightarrow 4ab \leq a^2 + 2ab + b^2$$

$$\Leftrightarrow 0 \leq a^2 - 2ab + b^2$$

$$\Leftrightarrow 0 \leq (a-b)^2.$$

This has the following geometric interpretation: The area of a rectangle with sides of length *a* and *b* is less than or equal to the area of a square with sides of length $\frac{a+b}{2}$.

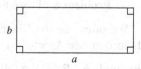

This final inequality is certainly true, since all squares are non-negative. It follows that the original inequality $ab \leq \left(\frac{a+b}{2}\right)^2$ is also true, for all $a, b \in \mathbb{R}$. □

Remark

A close examination of the above chain of equivalent statements shows that in fact $ab = \left(\frac{a+b}{2}\right)^2$ if and only if $a = b$.

> **Problem 3**   Prove that $\frac{a+b}{\sqrt{2}} \leq \sqrt{a^2 + b^2}$, for all $a, b \in \mathbb{R}$.
>
> **Problem 4**   Suppose that $a > \sqrt{2}$. Prove the following inequalities:
>
> (a) $\frac{1}{2}\left(a + \frac{2}{a}\right) < a$;   (b) $\left(\frac{1}{2}\left(a + \frac{2}{a}\right)\right)^2 > 2$.
>
> *Hint:* In part (b), use the result of Example 3 and the subsequent remark.

In the form $\sqrt{ab} \leq \frac{a+b}{2}$ this inequality is sometimes called the Arithmetic–Geometric Mean Inequality for non-negative $a, b \in \mathbb{R}$.

Example 4   Prove that $\sqrt{a^2 + b^2} \leq a + b$, for $a, b \geq 0$.

Solution   We tackle this inequality using the various rearrangement rules and a chain of equivalent inequalities until we obtain an inequality that we know must be true

$$\sqrt{a^2 + b^2} \leq a + b \Leftrightarrow a^2 + b^2 \leq (a+b)^2$$
$$\Leftrightarrow a^2 + b^2 \leq a^2 + 2ab + b^2$$
$$\Leftrightarrow 0 \leq 2ab.$$

This final inequality is certainly true, since $a, b \geq 0$. It follows that the original inequality $\sqrt{a^2 + b^2} \leq a + b$ is also true, for $a, b \geq 0$.   □

This has the following geometric interpretation: The length of the hypotenuse of a right-angled triangle whose other sides are of lengths a and b is less than or equal to the sum of the lengths of those two sides.

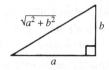

> **Problem 5**   Use the result of Example 4 to prove that $\sqrt{c + d} \leq \sqrt{c} + \sqrt{d}$, for $c, d \geq 0$.

Example 5   Prove that $\left|\sqrt{a} - \sqrt{b}\right| \leq \sqrt{|a - b|}$, for $a, b \geq 0$.

Solution   Notice first that interchanging the roles of a and b leaves the inequality unaltered. It follows that it is sufficient to prove the inequality under the assumption that $a \geq b$.

So, assume that $a \geq b$. Then we know that $\sqrt{a} \geq \sqrt{b}$ and $|a - b| = a - b$. Hence

$$\left|\sqrt{a} - \sqrt{b}\right| \leq \sqrt{|a - b|} \Leftrightarrow \sqrt{a} - \sqrt{b} \leq \sqrt{a - b}$$
$$\Leftrightarrow \sqrt{a} \leq \sqrt{a - b} + \sqrt{b}.$$

This final inequality is certainly true, and is obtained from the result of Problem 5 by simply substituting $a - b$ in place of c and b in place of d. It follows that the original inequality $\left|\sqrt{a} - \sqrt{b}\right| \leq \sqrt{|a - b|}$ is also true, for $a, b \geq 0$.   □

This will simplify the details of our chain of inequalities.

We avoid one modulus as a result of our simplifying assumption!

Never be ashamed to utilise every tool at your disposal! (Why do the same work twice?)

> **Problem 6**   Prove that $\sqrt{a + b + c} \leq \sqrt{a} + \sqrt{b} + \sqrt{c}$, for $a, b, c \geq 0$.

We often use the Binomial Theorem and the Principle of Mathematical Induction (see Appendix 1) to prove inequalities.

Example 6   Prove the following inequalities, for $n \geq 1$:

(a) $2^n \geq 1 + n$;   (b) $2^{\frac{1}{n}} \leq 1 + \frac{1}{n}$.

Solution

(a) By the Binomial Theorem for $n \geq 1$

$$(1+x)^n = 1 + nx + \frac{n(n-1)}{2!}x^2 + \cdots + x^n$$

$$\geq 1 + nx, \quad \text{for } x \geq 0.$$

n	1	2	3	4
2^n	2	4	8	16
$1+n$	2	3	4	5

We decrease the sum by omitting subsequent non-negative terms.

Then, if we substitute $x = 1$ in this last inequality, we get

$$2^n \geq 1 + n, \quad \text{for } n \geq 1.$$

(b) We start by rewriting the required result in an equivalent form

$$2^{\frac{1}{n}} \leq 1 + \frac{1}{n} \Leftrightarrow 2 \leq \left(1 + \frac{1}{n}\right)^n \quad \text{(by the Power Rule)}.$$

n	1	2	3	4
$2^{\frac{1}{n}}$	2	1.41	1.26	1.19
$1+\frac{1}{n}$	2	1.5	1.33	1.25

Now, if we substitute $x = \frac{1}{n}$ in the Binomial Theorem for $(1+x)^n$, we get

$$\left(1 + \frac{1}{n}\right)^n = 1 + n\left(\frac{1}{n}\right) + \frac{n(n-1)}{2!}\left(\frac{1}{n}\right)^2 + \cdots + \left(\frac{1}{n}\right)^n$$

$$\geq 1 + 1 = 2.$$

We decrease the sum by omitting all but the first two terms.

It then follows, from the fact that $2 \leq \left(1 + \frac{1}{n}\right)^n$, that the original inequality $2^{\frac{1}{n}} \leq 1 + \frac{1}{n}$, for $n \geq 1$, is also true, as required. $\qquad\square$

Problem 7 Prove the inequality $\left(1 + \frac{1}{n}\right)^n \geq \frac{5}{2} - \frac{1}{2n}$, for $n \geq 1$.

Hint: consider the first three terms in the binomial expansion.

Example 7 Prove that $2^n \geq n^2$, for $n \geq 4$.

Solution Let $P(n)$ be the statement

$$P(n): \ 2^n \geq n^2.$$

First we show that $P(4)$ is true: $2^4 \geq 4^2$.

n	1	2	3	4	5
2^n	2	4	8	16	32
n^2	1	4	9	16	25

STEP 1 Since $2^4 = 16$ and $4^2 = 16$, $P(4)$ is certainly true.

STEP 2 We now assume that $P(k)$ holds for some $k \geq 4$, and deduce that $P(k+1)$ is then true.

So, we are assuming that $2^k \geq k^2$. Multiplying this inequality by 2 we get

$$2^{k+1} \geq 2k^2,$$

This assumption is just $P(k)$.

so it is therefore sufficient for our purposes to prove that $2k^2 \geq (k+1)^2$. Now

Since $P(k+1)$ is: $2^{k+1} \geq (k+1)^2$.

$$2k^2 \geq (k+1)^2 \Leftrightarrow 2k^2 \geq k^2 + 2k + 1$$

$$\Leftrightarrow k^2 - 2k - 1 \geq 0 \quad \text{(by 'completing the square')}$$

$$\Leftrightarrow (k-1)^2 - 2 \geq 0.$$

This last inequality certainly holds for $k \geq 4$, and so $2^{k+1} \geq (k+1)^2$ also holds for $k \geq 4$.

In other words: $P(k)$ true for some $k \geq 4 \Rightarrow P(k+1)$ true.

It follows, by the Principle of Mathematical Induction, that $2^n \geq n^2$, for $n \geq 4$. $\qquad\square$

Problem 8 Prove that $4^n > n^4$, for $n \geq 5$.

Three important inequalities in Analysis

Our first inequality, called *Bernoulli's Inequality*, will be of regular use in later chapters.

Theorem 1   Bernoulli's Inequality
For any real number $x \geq -1$ and any natural number n, $(1+x)^n \geq 1+nx$.

The value of this result will come from making suitable choices of x and n for particular purposes.

Remark

In the solution to part (a) of Example 6, you saw that $(1+x)^n \geq 1+nx$, for $x \geq 0$ and n a natural number. Theorem 1 asserts that the same result holds under the *weaker* assumption that $x \geq -1$.

Proof   Let $P(n)$ be the statement
$$P(n): (1+x)^n \geq 1+nx, \quad \text{for } x \geq -1.$$

We prove the result using Mathematical Induction.

STEP 1   First we show that $P(1)$ is true: $(1+x)^1 \geq 1+x$. This is obviously true.

STEP 2   We now assume that $P(k)$ holds for some $k \geq 1$, and prove that $P(k+1)$ is then true.

So, we are assuming that $(1+x)^k \geq 1+kx$, for $x \geq -1$. Multiplying this inequality by $(1+x)$, we get

This assumption is $P(k)$. This multiplication is valid since $(1+x) \geq 0$.

$$\begin{aligned}(1+x)^{k+1} &\geq (1+x)(1+kx)\\ &= 1+(k+1)x+kx^2\\ &\geq 1+(k+1)x.\end{aligned}$$

We decrease the expression if we omit the final non-negative term.

Thus, we have $(1+x)^{k+1} \geq 1+(k+1)x$; in other words the statement $P(k+1)$ holds.

So, $P(k)$ true for some $k \geq 1 \Rightarrow P(k+1)$ true.

It follows, by the Principle of Mathematical Induction, that $(1+x)^n \geq 1+nx$, for $x \geq -1$, $n \in \mathbb{N}$. $\square$

Problem 9   By applying Bernoulli's Inequality with $x = -\frac{1}{(2n)}$, prove that $2^{\frac{1}{n}} \geq 1+\frac{1}{2n-1}$, for any natural number n.

You saw in part (b) of Example 6 that $2^{\frac{1}{n}} \leq 1+\frac{1}{n}$.

Our second inequality is of considerable use in various branches of Analysis. In Problem 3 you proved that $\frac{a+b}{\sqrt{2}} \leq \sqrt{a^2+b^2}$, for all $a, b \in \mathbb{R}$. We can rewrite this inequality in the equivalent form $(a+b)^2 \leq 2(a^2+b^2)$ or $(a+b)^2 \leq (a^2+b^2)(1^2+1^2)$. The Cauchy–Schwarz Inequality is a generalisation of this result to $2n$ real numbers.

Theorem 2   Cauchy–Schwarz Inequality
For any real numbers $a_1, a_2, \ldots, a_n$ and $b_1, b_2, \ldots, b_n$, we have
$$(a_1b_1+a_2b_2+\cdots+a_nb_n)^2 \leq (a_1^2+a_2^2+\cdots+a_n^2)(b_1^2+b_2^2+\cdots+b_n^2).$$

We give the proof of Theorem 2 at the end of the sub-section.

Problem 10   Use Theorem 2 to prove that for any positive real numbers $a_1, a_2, \ldots, a_n$, then $(a_1 + a_2 + \cdots + a_n)\left(\frac{1}{a_1} + \frac{1}{a_2} + \cdots + \frac{1}{a_n}\right) \geq n^2$.

For example, with $n = 3$

$$(1 + 2 + 3)\left(\frac{1}{1} + \frac{1}{2} + \frac{1}{3}\right) = 6 \times \frac{11}{6}$$

$$= 11 \geq 3^2.$$

Our final result also has many useful applications. In Example 3 you proved that $ab \leq \left(\frac{a+b}{2}\right)^2$, for all $a, b \in \mathbb{R}$; it follows that, if a and b are positive, then $(ab)^{\frac{1}{2}} \leq \frac{a+b}{2}$. The Arithmetic Mean–Geometric Mean Inequality is a generalisation of this result for two real numbers to n real numbers.

Theorem 3   Arithmetic Mean–Geometric Mean Inequality

For any positive real numbers $a_1, a_2, \ldots, a_n$, we have

$$(a_1 a_2 \ldots a_n)^{\frac{1}{n}} \leq \frac{a_1 + a_2 + \cdots + a_n}{n}.$$

We give the proof of Theorem 3 at the end of the sub-section.

Problem 11   Use Theorem 3 with the $n+1$ positive numbers $1, 1+\frac{1}{n}$, $1+\frac{1}{n}, \ldots, 1+\frac{1}{n}$ to prove that, for any positive integer n

$$\left(1 + \frac{1}{n}\right)^n \leq \left(1 + \frac{1}{n+1}\right)^{n+1}.$$

For example, with $n = 3$

$$\left(1 + \frac{1}{3}\right)^3 = \frac{64}{27} = 2.37\ldots$$

$$< \left(1 + \frac{1}{4}\right)^4 = \frac{625}{256} = 2.44\ldots$$

Proofs of Theorems 2 and 3

You may omit these proofs at a first reading.

Theorem 2   Cauchy–Schwarz Inequality

For any real numbers $a_1, a_2, \ldots, a_n$ and $b_1, b_2, \ldots, b_n$, we have

$$(a_1 b_1 + a_2 b_2 + \cdots + a_n b_n)^2 \leq \left(a_1^2 + a_2^2 + \cdots + a_n^2\right)\left(b_1^2 + b_2^2 + \cdots + b_n^2\right).$$

Proof   If all the as are zero, the result is obvious; so we need only examine the case when not all the as are zero. It follows that, if we denote the sum $\sum_{k=1}^n a_k^2 = a_1^2 + a_2^2 + \cdots + a_n^2$ by A, then $A > 0$. Also, denote $\sum_{k=1}^n b_k^2 = b_1^2 + b_2^2 + \cdots + b_n^2$ by B and $\sum_{k=1}^n a_k b_k = a_1 b_1 + a_2 b_2 + \cdots + a_n b_n$ by C.

Note that we will use the Σ notation to keep the argument brief.

Now, for any real number λ, we have that $(\lambda a_k + b_k)^2 \geq 0$, so that

$$\sum_{k=1}^n \left(\lambda^2 a_k^2 + 2\lambda a_k b_k + b_k^2\right) = \sum_{k=1}^n (\lambda a_k + b_k)^2 \geq 0,$$

which we may rewrite in the form

$$\lambda^2 A + 2\lambda C + B \geq 0.$$

But this inequality is equivalent to the inequality

$$(\lambda A + C)^2 + AB \geq C^2, \text{ for any real number } \lambda.$$

Since A is non-zero, we may now choose $\lambda = -\frac{C}{A}$. It follows from the last inequality that $AB \geq C^2$, which is exactly what we had to prove. ∎

A is non-zero, by assumption, so that $1/A$ makes sense.

Remark

If not all the as are zero, equality can only occur if $\sum_{k=1}^n (\lambda a_k + b_k)^2 = 0$; that is, if all the numbers a_k are proportional to all the numbers b_k, $1 \leq k \leq n$.

Theorem 3   Arithmetic Mean–Geometric Mean Inequality

For any positive real numbers a_1, a_2, ..., a_n, we have

$$(a_1 a_2 \ldots a_n)^{\frac{1}{n}} \leq \frac{a_1 + a_2 + \cdots + a_n}{n}. \qquad (2)$$

Proof   Since the a_i are positive, we can rewrite (2) in the equivalent form

$$\frac{(a_1 a_2 \ldots a_n)^{\frac{1}{n}}}{(a_1 + a_2 + \cdots + a_n)/n} \leq 1. \qquad (3)$$

We denote the typical term by a_i rather than a_k to avoid confusion with a different use of the letter k in the Mathematical Induction argument below.

Now, replacing each term a_i by λa_i for any non-zero number λ does not alter the left-hand side of the inequality (3). It follows that it is sufficient to prove the inequality (2) in the special case when the product of the terms a_i is 1. Hence it is sufficient to prove the following statement $P(n)$ for each natural number n:

$P(n)$: For any positive real numbers a_i with $a_1 a_2 \ldots a_n = 1$, then $a_1 + a_2 + \cdots + a_n \geq n$.

We will prove this by Mathematical Induction.

First, the statement $P(1)$ is obviously true.

Next, we assume that $P(k)$ holds for some $k \geq 1$, and prove that $P(k+1)$ is then true.

Now, if all the terms a_1, a_2, ..., a_{k+1} are equal to 1, the result $P(k+1)$ certainly holds. Otherwise, at least two of the terms differ from 1, say a_1 and a_2, such that $a_1 > 1$ and $a_2 < 1$ (say). Hence

$$(a_1 - 1) \times (a_2 - 1) \leq 0,$$

The argument is exactly the same *whichever* two terms actually differ from 1.

which after some manipulation we may rewrite as

$$a_1 + a_2 \geq 1 + a_1 a_2. \qquad (4)$$

You should check this yourself.

We are now ready to tackle $P(k+1)$. Then

$$a_1 + a_2 + \cdots + a_{k+1} \geq 1 + a_1 a_2 + a_3 + a_4 + \cdots + a_{k+1}$$
$$\geq k + 1,$$

By (4).

since we may apply the assumption that $P(k)$ holds to the k quantities $a_1 a_2$, a_3, a_4, ..., a_{k+1}. This last inequality is simply the statement that $P(k+1)$ is indeed true.

That is
$(a_1 a_2) a_3 a_4 \ldots a_{k+1} = 1$
$\Rightarrow (a_1 a_2) + a_3 + a_4 + \cdots + a_{k+1} \geq k.$

It follows by the Principle of Mathematical Induction that $P(n)$ holds for all natural numbers n, and so the inequality (2) must also hold. ∎

Remark

A careful examination of the proof of Theorem 3 shows that equality can only occur if all the terms a_i are equal.

1.4   Least upper bounds and greatest lower bounds

1.4.1   Upper and lower bounds

Any finite set $\{x_1, x_2, \ldots, x_n\}$ of real numbers obviously has a greatest element and a least element, but this property does not necessarily hold for infinite sets.

For example, the interval (0, 2] has greatest element 2, but neither of the sets $\mathbb{N} = \{1, 2, 3, \ldots\}$ nor $[0, 2)$ has a greatest element. However the set $[0, 2)$ is bounded above by 2, since all points of $[0, 2)$ are less than or equal to 2.

Definitions   A set $E \subseteq \mathbb{R}$ is **bounded above** if there is a real number, M say, called an **upper bound** of E, such that

$$x \leq M, \quad \text{for all } x \in E.$$

If the upper bound M belongs to E, then M is called the **maximum element** of E, denoted by max E.

Geometrically, the set E is bounded above by M if no point of E lies to the right of M on the real line.

For example, if $E = [0, 2)$, then the numbers 2, 3, 3.5 and 157.1 are all upper bounds of E, whereas the numbers 1.995, 1.5, 0 and -157.1 are not upper bounds of E. Although it seems obvious that $[0, 2)$ has no maximum element, you may find it difficult to write down a formal proof. The following example shows you how to do this:

Example 1   Determine which of the following sets are bounded above, and which have a maximum element:

(a) $E_1 = [0, 2)$;   (b) $E_2 = \{\frac{1}{n} : n = 1, 2, \ldots\}$;   (c) $E_3 = \mathbb{N}$.

Solution

(a) The set E_1 is bounded above. For example, $M = 2$ is an upper bound of E_1, since

$$x \leq 2, \quad \text{for all } x \in E_1.$$

However, E_1 has no maximum element. For each x in E_1, we have $x < 2$, and so there is some real number y such that

$$x < y < 2,$$

by the Density Property of $\mathbb{R}$.
Hence $y \in E_1$, and so x cannot be a maximum element.

For example, y can be of the form $1.99\ldots9$ or $y = \frac{1}{2}(x + 2)$.

2 is not a maximum element of E_1, since $2 \notin E_1$.

(b) The set E_2 is bounded above. For example, $M = 1$ is an upper bound of E_2, since

$$\frac{1}{n} \leq 1, \quad \text{for all } n = 1, 2, \ldots.$$

Also, since $1 \in E_2$

$$\max E_2 = 1.$$

(c) The set E_3 is not bounded above. For each real number M, there is a positive integer n such that $n > M$, by the Archimedean Property of $\mathbb{R}$.
Hence M cannot be an upper bound of E_3.
This also means that E_3 cannot have a maximum element. □

Problem 1   Sketch the following sets, and determine which are bounded above, and which have a maximum element:

(a) $E_1 = (-\infty, 1]$;   (b) $E_2 = \{1 - \frac{1}{n} : n = 1, 2, \ldots\}$;
(c) $E_3 = \{n^2 : n = 1, 2, \ldots\}$.

Similarly, we define *lower bounds*. For example, the interval (0, 2) is bounded below by 0, since

$$0 \leq x, \quad \text{for all } x \in (0, 2).$$

However, 0 does not belong to (0, 2), and so 0 is not a minimum element of (0, 2). In fact, (0, 2) has no minimum element.

Definitions   A set $E \subseteq \mathbb{R}$ is **bounded below** if there is a real number, m say, called a **lower bound** of E, such that

$$m \leq x, \quad \text{for all } x \in E.$$

If the lower bound m belongs to E, then m is called the **minimum element** of E, denoted by min E.

Geometrically, the set E is bounded below by m if no point of E lies to the left of m on the real line.

>   **Problem 2**   Determine which of the following sets are bounded below, and which have a minimum element:
>
>   (a) $E_1 = (-\infty, 1]$;   (b) $E_2 = \{1 - \frac{1}{n}: n = 1, 2, \ldots\}$;
>   (c) $E_3 = \{n^2: n = 1, 2, \ldots\}$.

The following terminology is also useful:

Definition   A set $E \subseteq \mathbb{R}$ is **bounded** if it is bounded above *and* bounded below.

For example, the set $E_2 = \{1 - \frac{1}{n}: n = 1, 2, \ldots\}$ is bounded, but the sets $E_1 = (-\infty, 1]$ and $E_3 = \{n^2: n = 1, 2, \ldots\}$ are not bounded.

Similar terminology applies to functions.

Definitions   A function f defined on an interval $I \subseteq \mathbb{R}$ is said to:

- be **bounded above** by M if $f(x) \leq M$, for all $x \in I$; M is an **upper bound** of f;
- be **bounded below** by m if $f(x) \geq m$, for all $x \in I$; m is a **lower bound** of f;
- have a **maximum** (or **maximum value**) M if M is an upper bound of f and $f(x) = M$, for at least one $x \in I$;
- have a **minimum** (or **minimum value**) m if m is a lower bound of f and $f(x) = m$, for at least one $x \in I$.

> Strictly speaking, M and m are the upper bound and lower bound *of the image set* $\{f(x): x \in I\}$.

Example 2   Let f be the function defined by $f(x) = x^2$, $x \in [\frac{1}{2}, 3)$. Determine whether f is bounded above or below, and any maximum or minimum value of f.

Solution   First, f is increasing on the interval $[\frac{1}{2}, 3)$, so that since $\frac{1}{2} \leq x < 3$ it follows that $\frac{1}{4} \leq f(x) < 9$. Hence f is bounded above and bounded below.

Next, since $f(\frac{1}{2}) = \frac{1}{4}$ and $\frac{1}{4}$ is a lower bound for f on the interval $[\frac{1}{2}, 3)$, it follows that f has a minimum value of $\frac{1}{4}$ on this interval.

Finally, 9 is an upper bound for f on the interval $[\frac{1}{2}, 3)$ but there is no point x in $[\frac{1}{2}, 3)$ for which $f(x) = 9$. So 9 cannot be a maximum of f on the interval. However, if y is any number in (8, 9) there is a number $x > \sqrt{y}$ in $(\sqrt{y}, 3) \subset (2\sqrt{2}, 3) \subset [\frac{1}{2}, 3)$ such that $f(x) = x^2 > y$, so that no number in (8, 9) will serve

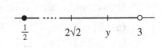

as a maximum of f on the interval. It follows that f has no maximum value on $[\frac{1}{2}, 3)$. □

Problem 3 Let f be the function defined by $f(x) = \frac{1}{x^2}$, $x \in [-3, -2)$. Determine whether f is bounded above or below, and any maximum or minimum value of f.

1.4.2 Least upper bounds and greatest lower bounds

We have seen that the interval $[0, 2]$ has a maximum element 2, but $[0, 2)$ has no maximum element. However, the number 2 is 'rather like' a maximum element of $[0, 2)$, because 2 is an upper bound of $[0, 2)$ and any number less than 2 is not an upper bound of $[0, 2)$. In other words, 2 is the *least* upper bound of $[0, 2)$.

Definition A real number M is the **least upper bound**, or **supremum**, of a set $E \subseteq \mathbb{R}$ if:

1. M is an upper bound of E;
2. if $M' < M$, then M' is not an upper bound of E.

In this case, we write $M = \sup E$.

Part 1 says that M is *an* upper bound.

Part 2 says that *no smaller number* can be an upper bound.

If E has a maximum element, max E, then $\sup E = \max E$. For example, the closed interval $[0, 2]$ has least upper bound 2. We can think of the least upper bound of a set, when it exists, as a kind of 'generalised maximum element'.

If a set does not have a maximum element, but is bounded above, then we may be able to guess the value of its least upper bound. As in the case $E = [0, 2)$, there may be an obvious 'missing point' at the upper end of the set. However it is important to *prove* that your guess is correct. We now show you how to do this.

Example 3 Prove that the least upper bound of $[0, 2)$ is 2.

Solution We know that $M = 2$ is an upper bound of $[0, 2)$, because
$$x \le 2, \quad \text{for all } x \in [0, 2).$$

To show that 2 is the *least* upper bound, we must prove that each number $M' < 2$ is *not* an upper bound of $[0, 2)$. To do this, we must find an element x in $[0, 2)$ which is greater than M'. But, if $M' < 2$, then there is a real number x such that
$$M' < x < 2$$
and also
$$0 < x < 2.$$

For example, x can be of the form 1.99...9 for a suitably large number of digits, or it can be $\frac{1}{2}(M' + 2)$ since $M' < \frac{1}{2}(M' + 2) < 2$.

Since $x \in [0, 2)$, the number M' cannot be an upper bound of $[0, 2)$. Hence $M = 2$ is the least upper bound, or supremum, of $[0, 2)$. □

Although the conclusion of Example 3 may seem painfully obvious, we have written out the solution in detail because it illustrates the strategy for determining the least upper bound of a set, if it has one.

> **Strategy**   Given a subset E of $\mathbb{R}$, to show that M is the least upper bound, or supremum, of E, check that:
>
> 1. $x \leq M$, for *all* $x \in E$;
> 2. if $M' < M$, then there is *some* $x \in E$ such that $x > M'$.

GUESS the value of M, then CHECK parts 1 and 2.

Notice that, if M is an upper bound of E and $M \in E$, then part 2 is automatically satisfied, and so $M = \sup E = \max E$.

Example 4   Determine the least upper bound of $E = \{1 - \frac{1}{n^2} : n = 1, 2, \ldots\}$.

Solution   We guess that the least upper bound of E is $M = 1$. Certainly, 1 is an upper bound of E, since

$$1 - \frac{1}{n^2} \leq 1, \quad \text{for } n = 1, 2, \ldots.$$

To check part 2 of the strategy, we need to show that, if $M' < 1$, then there is some natural number n such that

$$1 - \frac{1}{n^2} > M'. \tag{1}$$

However

$$1 - \frac{1}{n^2} > M' \iff 1 - M' > \frac{1}{n^2}$$

$$\iff \frac{1}{1 - M'} < n^2 \qquad (\text{since } 1 - M' > 0)$$

$$\iff \sqrt{\frac{1}{1 - M'}} < n \qquad (\text{since } \frac{1}{1 - M'} > 0$$
$$\text{and } n > 0).$$

We can certainly choose n so that this final inequality holds, by the Archimedean Property of $\mathbb{R}$, and so we can choose n so that inequality (1) holds.

Hence 1 is the least upper bound of E.   ☐

That is, $1 = \sup E$.

Remark

Although we used double-headed arrows in this solution, the actual proof required only the implications going from *right* to *left*. In other words, the proof uses only the fact that

$$1 - \frac{1}{n^2} > M' \impliedby \sqrt{\frac{1}{1 - M'}} < n.$$

Problem 4   Determine $\sup E$, if it exists, for each of the following sets:

(a) $E_1 = (-\infty, 1]$;   (b) $E_2 = \{1 - \frac{1}{n} : n = 1, 2, \ldots\}$;
(c) $E_3 = \{n^2 : n = 1, 2, \ldots\}$.

Similarly, we define the notion of a *greatest lower bound*.

> **Definition**   A real number m is the **greatest lower bound**, or **infimum**, of a set $E \subseteq \mathbb{R}$ if:
>
> 1. m is a lower bound of E;
> 2. if $m' > m$, then m' is not a lower bound of E.
>
> In this case, we write $m = \inf E$.

Part 1 says that m is *a* lower bound.
Part 2 says that *no larger number* can be a lower bound.

If E has a minimum element, min E, then inf $E = $ min E. For example, the closed interval [0, 2] has greatest lower bound 0. We can think of the greatest lower bound of a set, when it exists, as a kind of 'generalised minimum element'.

The strategy for establishing that a number is the greatest lower bound of a set is very similar to that for proving that a number is the least upper bound of a set.

Strategy   Given a subset E of $\mathbb{R}$, to show that m is the greatest lower bound, or infimum, of E, check that:

1. $x \geq m$, for *all* $x \in E$;

2. if $m' > m$, then there is *some* $x \in E$ such that $x < m'$.

GUESS the value of m, then CHECK parts 1 and 2.

Notice that, if m is a lower bound of E and $m \in E$, then part 2 is automatically satisfied, and so $m = $ inf $E = $ min E.

Problem 5   Determine inf E, if it exists, for each of the following sets:

(a) $E_1 = (1, 5]$;   (b) $E_2 = \{\frac{1}{n^2} : n = 1, 2, \ldots\}$.

Remarks

1. For any subset E of $\mathbb{R}$, inf $E \leq$ sup E. This follows from the fact that, for any $x \in E$, we have inf $E \leq x \leq$ sup E.

2. For any bounded interval I of $\mathbb{R}$, let a be its left end-point and b its right end-point. Then inf $I = a$ and sup $I = b$.

Least upper bounds and greatest lower bounds of functions

Similar terminology applies to bounds for functions.

Definitions   Let f be a function defined on an interval $I \subseteq \mathbb{R}$. Then:

- A real number M is the **least upper bound**, or **supremum**, of f on I if:
  1. M is an upper bound of $f(I)$;
  2. if $M' < M$, then M' is not an upper bound of $f(I)$.

  In this case, we write $M = \sup_I f$ or $\sup f$ or $\sup\{f(x) : x \in I\}$ or $\sup_{x \in I} f(x)$.

- A real number m is the **greatest lower bound**, or **infimum**, of f on I if:
  1. m is a lower bound of $f(I)$;
  2. if $m' > m$, then m' is not a lower bound of $f(I)$.

  In this case, we write $m = \inf_I f$ or $\inf f$ or $\inf\{f(x) : x \in I\}$ or $\inf_{x \in I} f(x)$.

There are similar definitions for the least upper bound and the greatest lower bound of f on a general set S in $\mathbb{R}$.

These are really the definitions for the least upper bound or the greatest lower bound of the set $\{f(x): x \in I\}$.

Notice, for instance, that:

- if M is an upper bound for f on I, then $\sup_I f \leq M$,

- if m is a lower bound for f on I, then $\inf_I f \geq m$.

For $\sup_I f$ is the *least* upper bound, and $\inf_I f$ is the *greatest* lower bound, for f on I.

The strategies for *proving* that M is the least upper bound or m the greatest lower bound of f on I are similar to the corresponding strategies for the least upper bound or the greatest lower bound of a set E.

Strategies   Let f be a function defined on an interval $I \subseteq \mathbb{R}$. Then:

- To show that m is the **greatest lower bound**, or **infimum**, of f on I, check that:
    1. $f(x) \geq m$, for *all* $x \in I$;
    2. if $m' > m$, then there is *some* $x \in I$ such that $f(x) < m'$.

- To show that M is the **least upper bound**, or *supremum*, of f on I, check that:
    1. $f(x) \leq M$, for *all* $x \in I$;
    2. if $M' < M$, then there is *some* $x \in I$ such that $f(x) > M'$.

Example 5   Let f be the function defined by $f(x) = x^2, x \in \left[\frac{1}{2}, 3\right)$. Determine the least upper bound and the greatest lower bound of f on $\left[\frac{1}{2}, 3\right)$.

Solution   We have already seen that 9 is an upper bound for f on $\left[\frac{1}{2}, 3\right)$, and that no smaller number will serve as an upper bound. It follows that 9 must be the least upper bound of f on $\left[\frac{1}{2}, 3\right)$.

You saw this in Example 2.

Similarly, we have already seen that $\frac{1}{4}$ is the minimum value of f on $\left[\frac{1}{2}, 3\right)$; it follows that $\frac{1}{4}$ is the greatest lower bound of f on $\left[\frac{1}{2}, 3\right)$, and this is actually attained (at the point $\frac{1}{2}$). $\qquad\qquad\qquad\qquad\qquad\qquad\square$

Example 2.

> **Problem 6**   Let f be the function defined by $f(x) = \frac{1}{x^2}$, $x \in [1, 4)$. Determine the least upper bound and the greatest lower bound of f on $[1, 4)$.

Remark

For any interval I of $\mathbb{R}$, $\inf_I f \leq \sup_I f$. This follows from the fact that, for any $x \in I$, we have $\inf_I f \leq f(x) \leq \sup_I f$.

The least upper bound and the greatest lower bound of a function on an interval will be particularly significant in our later work on continuity and integrability of functions.

Chapters 4 and 7.

1.4.3   The Least Upper Bound Property

In the examples in the previous sub-section, it was easy to guess the values of sup E and inf E. At times, however, we shall meet sets for which these values are not so easy to determine. For example, if

$$E = \left\{ \left(1 + \frac{1}{n}\right)^n : n = 1, \ 2, \dots \right\},$$

We will study this set closely in Section 2.5.

then it can be shown that E is bounded above by 3, but it is not easy to guess the least upper bound of E.

In such circumstances, it is reassuring to know that sup E does exist, even though it may be difficult to find. This existence is guaranteed by the following fundamental result.

> **The Least Upper Bound Property of** $\mathbb{R}$ Let E be a non-empty subset of $\mathbb{R}$. If E is bounded above, then E has a least upper bound.

We leave the proof of the Least Upper Bound Property of $\mathbb{R}$ to the next sub-section. However, the Property itself is intuitively obvious. If the set E lies entirely to the left of some number M, then you can imagine moving M steadily to the left until you meet E. At this point, sup E has been reached.

The Least Upper Bound Property of $\mathbb{R}$ can be used to show that $\mathbb{R}$ does include decimals which represent irrational numbers such as $\sqrt{2}$, as we claimed in Section 1.1.

In Sections 1.2 and 1.3 we have taken for granted the existence of rational powers and their properties, without giving formal definitions. How can we supply these definitions? For example, how can we define $\sqrt{2}$ as a decimal?

Consider the set

$$E = \{x \in \mathbb{Q} : x > 0, x^2 < 2\}.$$

This is the set of positive rational numbers whose squares are less than 2. Intuitively, $\sqrt{2}$ lies on the number line to the right of the numbers in E, but 'only just'! In fact, we should expect $\sqrt{2}$ to be the least upper bound of E. Certainly E has a least upper bound, by the Least Upper Bound Property, because E is bounded above, by 1.5 for example. Thus it seems likely that sup E is the decimal representation of $\sqrt{2}$. But how can we *prove* that $(\sup E)^2 = 2$?

We shall prove this in Section 1.5, once we have described how to do arithmetic with real numbers (decimals).

Finally, note that there is a corresponding result about lower bounds.

> **The Greatest Lower Bound Property of** $\mathbb{R}$ Let E be a non-empty subset of $\mathbb{R}$. If E is bounded below, then E has a greatest lower bound.

The Least Upper Bound Property of $\mathbb{R}$ is an example of an *existence theorem*, one which asserts that a real number exists having a certain property. Analysis contains many such results which depend on the Least Upper Bound Property of $\mathbb{R}$. While these results are often very general, and their proofs elegant, they do not always provide the most efficient methods of calculating good approximate values for the numbers in question.

1.4.4 Proof of the Least Upper Bound Property

You may omit this proof at a first reading.

We know that E is a non-empty set, and we shall assume for simplicity that E contains at least one positive number. We also know that E is bounded above. The following procedure gives us the successive digits in a particular decimal, which we then prove to be the least upper bound of E.

> **Procedure to find** $a = a_0 \cdot a_1 a_2 \ldots = \sup E$
> Choose in succession:
> - the greatest integer a_0 such that a_0 is not an upper bound of E;
> - the greatest digit a_1 such that $a_0 \cdot a_1$ is not an upper bound of E;
> - the greatest digit a_2 such that $a_0 \cdot a_1 a_2$ is not an upper bound of E;
> $\vdots$
> - the greatest digit a_n such that $a_0 \cdot a_1 a_2 \ldots a_n$ is not an upper bound of E;
> $\vdots$

For example, if
$$E = \{x \in \mathbb{Q} : x > 0, \\ x^2 < 2\},$$
then
$$a_0 = 1, \text{ since } 1^2 < 2 < 2^2;$$
$$a_0 \cdot a_1 = 1.4, \text{ since}$$
$$1.4^2 < 2 < 1.5^2;$$
$$a_0 \cdot a_1 a_2 = 1.41, \text{ since}$$
$$1.41^2 < 2 < 1.42^2;$$
$\vdots$

Thus, at the nth stage we choose the digit a_n so that:

- $a_0 \cdot a_1 a_2 \ldots a_n$ *is not* an upper bound of E;
- $a_0 \cdot a_1 a_2 \ldots a_n + \frac{1}{10^n}$ *is* an upper bound of E.

We now prove that the least upper bound of E is $a = a_0 \cdot a_1 a_2 \ldots$.

First, we have to prove that a is an upper bound of E. To do this, we prove that, if $x > a$, then $x \notin E$ (this is equivalent to proving, that, if $x \in E$, then $x \leq a$). We begin by representing x as a non-terminating decimal $x = x_0 \cdot x_1 x_2 \ldots$. Since $x > a$, there is an integer n such that

$$a < x_0 \cdot x_1 x_2 \ldots x_n.$$

Hence

$$x_0 \cdot x_1 x_2 \ldots x_n \geq a_0 \cdot a_1 a_2 \ldots a_n + \frac{1}{10^n},$$

and so, by our choice of a_n, $x = x_0 \cdot x_1 x_2 \ldots x_n$ is an upper bound of E. Since $x > x_0 \cdot x_1 x_2 \ldots x_n$, we have that $x \notin E$, as required.

Next, we have to show that, if $x < a$, then x is *not* an upper bound of E. Since $x < a$, there is an integer n such that

$$x < a_0 \cdot a_1 a_2 \ldots a_n,$$

and so x is *not* an upper bound of E, by our choice of a_n.

Thus we have proved that a is the least upper bound of E. □

Remark

Notice that this proof does not use any arithmetical properties of the real numbers but only their order properties, together with the arithmetical properties of *rational numbers*. In the next section, we use the Least Upper Bound Property to define some of the arithmetical operations on $\mathbb{R}$.

1.5   Manipulating real numbers

1.5.1   Arithmetic in $\mathbb{R}$

At the end of Section 1.1 we discussed the decimals

$$\sqrt{2} = 1.41421356\ldots \quad \text{and} \quad \pi = 3.14159265\ldots,$$

Here we are assuming that $\sqrt{2}$ and π *can* be represented as decimals.

and asked whether it is possible to add and multiply these numbers to obtain another real number. We now explain how this can be done, using the Least Upper Bound Property of $\mathbb{R}$.

A natural way to obtain a sequence of approximations to the sum $\sqrt{2} + \pi$ is to truncate each of the above decimals, and form the sums of the truncations.

If each of the decimals is truncated at the same decimal place, this gives a sequence of approximations which is increasing:

$\sqrt{2}$	π	$\sqrt{2} + \pi$
1	3	4
1.4	3.1	4.5
1.41	3.14	4.55
1.414	3.141	4.555
1.4142	3.1415	4.5557
$\vdots$	$\vdots$	$\vdots$

Intuitively, we should expect that the sum $\sqrt{2} + \pi$ is greater than each of the numbers in the right-hand column, but 'only just'! To accord with our intuition, therefore, we *define* the sum $\sqrt{2} + \pi$ to be the least upper bound of the set of numbers in the right-hand column; that is

$$\sqrt{2} + \pi = \sup\{4, \ 4.5, \ 4.55, \ 4.555, \ 4.5557, \ldots\}.$$

To be sure that this definition makes sense, we need to show that this set is bounded above. But all the truncations of $\sqrt{2}$ are less than 1.5, and all the truncations of π are less than, say, 4. Hence, all the sums in the right-hand column are less than $1.5 + 4 = 5.5$. So, by the Least Upper Bound Property, the set of numbers in the right-hand column *does* have a least upper bound, and we *can* define $\sqrt{2} + \pi$ in this way.

This method can be used to define the sum of any pair of positive real numbers.

Let us check that this method of adding decimals gives the correct answer when we use it in a familiar case. Consider the simple calculation

We do not expect *you* to use this method to add decimals!

$$\frac{1}{3} + \frac{2}{3} = 0.333\ldots + 0.666\ldots.$$

Truncating each of these decimals and forming the sums, we obtain the set

$$\{0, \ 0.9, \ 0.99, \ 0.999, \ldots\}.$$

The supremum of this set is, of course, the number $0.999\ldots = 1$, which is the correct answer.

Similarly, we can define the *product* of any two positive real numbers. For example, to define $\sqrt{2} \times \pi$, we can form the sequence of products of their truncations:

$\sqrt{2}$	π	$\sqrt{2} \times \pi$
1	3	3
1.4	3.1	4.34
1.41	3.14	4.4274
1.414	3.141	4.441374
1.4142	3.1415	4.4427093
$\vdots$	$\vdots$	$\vdots$

As before, we define $\sqrt{2} \times \pi$ to be the least upper bound of the set of numbers in the right-hand column.

We omit the details.

Similar ideas can be used to define the operations of subtraction and division.

Thus we can define arithmetic with real numbers in terms of the familiar arithmetic with rationals, using the Least Upper Bound Property of $\mathbb{R}$. Moreover, it can be proved that these operations in $\mathbb{R}$ satisfy all the usual properties of a field.

These properties were listed in Sub-section 1.1.5.

1.5.2 The existence of roots

Just as we usually take for granted the basic arithmetical operations with real numbers, so we usually assume that, given any positive real number a, there is a unique positive real number $b = \sqrt{a}$ such that $b^2 = a$. We now discuss the justification for this assumption.

First, here is a geometrical justification. Given line segments of lengths 1 and a, we can construct a semi-circle with diameter $a + 1$ as shown.

For each positive integer n, we can also construct $\sqrt{n}$ as follows:

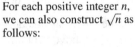

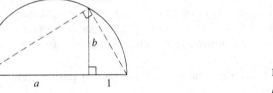

Using similar triangles, we see that

$$\frac{a}{b} = \frac{b}{1},$$

and so

$$b^2 = a.$$

This shows that there should be a positive real number b such that $b^2 = a$, so that the length of the vertical line segment in the figure can be described exactly by the expression $\sqrt{a}$. But does $b = \sqrt{a}$ exist *exactly* as a real number? In fact it does, and a more general result is true.

Theorem 1 For each positive real number a and each integer $n > 1$, there is a unique positive real number b such that

$$b^n = a.$$

We shall prove Theorem 1 in Sub-section 4.3.3.

We call this number b the nth root of a, and we write $b = \sqrt[n]{a}$. We also define $\sqrt[n]{0} = 0$, since $0^n = 0$, and if n is odd we define $\sqrt[n]{(-a)} = -\sqrt[n]{a}$, since $(-\sqrt[n]{a})^n = -a$ if n is odd.

For example, $\sqrt[3]{(-8)} = -2$.

Let us illustrate Theorem 1 with the special case $a = 2$ and $n = 2$. In this case, Theorem 1 asserts the existence of a real number b such that $b^2 = 2$. In other words, it asserts the existence of a decimal b which can be used to *define* $\sqrt{2}$ precisely.

Here is a direct proof of Theorem 1 in this special case. We choose the numbers $1, 1.4, 1.41, 1.414, \ldots$ to satisfy the inequalities

$$1^2 < 2 < 2^2$$
$$(1.4)^2 < 2 < (1.5)^2$$
$$(1.41)^2 < 2 < (1.42)^2$$
$$(1.414)^2 < 2 < (1.415)^2$$
$$\vdots$$

(1)

This process gives an infinite decimal

$$b = 1.414\ldots,$$

and we claim that

$$b^2 = (1.414\ldots)^2 = 2.$$

This can be proved using our method of multiplying decimals:

Notice that
$$b = 1.414\ldots$$
is the decimal that we obtained as the least upper bound of the set
$$\{x \in \mathbb{Q} : x > 0,\ x^2 < 2\}$$
in Sub-section 1.5.1.

b	b	b^2
1	1	1
1.4	1.4	1.96
1.41	1.41	1.9881
1.414	1.414	1.999396
$\vdots$	$\vdots$	$\vdots$

We have to prove that the least upper bound of the set E of numbers in the right-hand column is 2, in other words that

$$\sup E = \sup \{1,\ (1.4)^2,\ (1.41)^2,\ (1.414)^2,\ldots\} = 2.$$

To do this, we employ the strategy given in Sub-section 1.4.2.

First, we check that $M = 2$ is an upper bound of E. This follows from the left-hand inequalities in (1).

Next, we check that, if $M' < 2$, then there is a number in E which is greater than M'. To prove this, put

$$x_0 = 1,\ x_1 = 1.4,\ x_2 = 1.41,\ x_3 = 1.414,\ldots.$$

Then, by the right-hand inequalities in (1), we have that

$$\left(x_n + \frac{1}{10^n}\right)^2 > 2.$$

For example, if $n = 1$, then

$$\left(1.4 + \frac{1}{10}\right)^2 = 1.5^2 > 2.$$

Also

$$\left(x_n + \frac{1}{10^n}\right)^2 - x_n^2 = \frac{1}{10^n}\left(2x_n + \frac{1}{10^n}\right)$$
$$< \frac{1}{10^n}(2 \times 2 + 1) = \frac{5}{10^n},$$

and so

$$x_n^2 > \left(x_n + \frac{1}{10^n}\right)^2 - \frac{5}{10^n}$$
$$> 2 - \frac{5}{10^n}$$
$$= \underbrace{1.99\ldots95}_{n \text{ digits}}.$$

For example, if $n = 2$, then

$$(1.41)^2 > 1.95.$$

34   1: Numbers

So, if $M' < 2$, then we *can* choose n so large that $x_n^2 > M'$ (while still having $x_n \in E$). This proves that the least upper bound of E is 2, and so $(1.414\ldots)^2 = 2$. Thus we can define

$$\sqrt{2} = 1.414\ldots,$$

which justifies our earlier claim that $\sqrt{2}$ can be represented exactly by a decimal.

1.5.3   Rational powers

Having discussed nth roots, we are now in a position to define the expression a^x, where a is positive and x is rational.

> **Definition**   If $a > 0$, $m \in \mathbb{Z}$ and $n \in \mathbb{N}$, then
> $$a^{\frac{m}{n}} = \left(\sqrt[n]{a}\right)^m.$$

For example, for $a > 0$, with $m = 1$ we have $a^{\frac{1}{n}} = \sqrt[n]{a}$, and with $m = 2$ and $n = 3$ we have $a^{\frac{2}{3}} = \left(\sqrt[3]{a}\right)^2$.

This notation is particularly useful, because *rational powers* (or *rational exponents*) satisfy the following *exponent laws* (whose proofs depend on Theorem 1):

> **Exponent Laws**
> * If $a, b > 0$ and $x \in \mathbb{Q}$,   then $a^x b^x = (ab)^x$.
> * If $a > 0$ and $x, y \in \mathbb{Q}$, then $a^x a^y = a^{x+y}$.
> * If $a > 0$ and $x, y \in \mathbb{Q}$, then $(a^x)^y = a^{xy}$.

For example
$$2^{\frac{1}{2}} \times 3^{\frac{1}{2}} = 6^{\frac{1}{2}},$$
$$2^{\frac{1}{2}} \times 2^{\frac{1}{3}} = 2^{\frac{5}{6}},$$
$$\left(2^{\frac{1}{2}}\right)^{\frac{1}{3}} = 2^{\frac{1}{6}}.$$

If x and y are *integers*, these laws actually hold for all non-zero real numbers a and b. However, if x and y are not integers, then we must have $a, b > 0$. For example, $(-1)^{\frac{1}{2}}$ is not defined as a real number.

However, if a is a negative real number, then $a^{\frac{m}{n}}$ can be defined whenever $m \in \mathbb{Z}$, $n \in \mathbb{N}$ and $\frac{m}{n}$ is reduced to its lowest terms with n odd, as follows

$$a^{\frac{m}{n}} = \left(\sqrt[n]{a}\right)^m.$$

This extends our above definition of $a^{\frac{m}{n}}$; for instance, it defines $a^{\frac{1}{n}}$ whenever $n \in \mathbb{N}$ and n is odd. For example, $(-8)^{\frac{1}{3}} = 4$.

Finally, you may have wondered why we did not mention that each positive number has two nth roots when n is even. For example, $2^2 = (-2)^2 = 4$. We shall adopt the convention that, for $a > 0$, $\sqrt[n]{a}$ and $a^{\frac{1}{n}}$ always refer to the *positive* nth root of a. If we wish to refer to both roots (for example, when solving equations), we write $\pm \sqrt[n]{a}$.

1.5.4   Real powers

We conclude this section by briefly discussing the meaning of a^x when $a > 0$ and x is an arbitrary real number. We have defined this expression when x is rational, but the same definition does not work if x is irrational. However, it is common practice to write down expressions such as $\sqrt{2}^{\sqrt{2}}$, and even to apply the Exponent Laws to give equalities such as

$$\left(\sqrt{2}^{\sqrt{2}}\right)^{\sqrt{2}} = \left(\sqrt{2}\right)^{\sqrt{2}\times\sqrt{2}} = \left(\sqrt{2}\right)^{2} = 2.$$

Can such manipulations be justified?

In fact, it *is* possible to define a^x, for $a > 0$ and $x \in \mathbb{R}$, using the Least Upper Bound Property of $\mathbb{R}$, but it is then rather tricky to check that the Exponent Laws work. In Chapters 2 and 3 we shall explain how to define the expression e^x, and in Chapter 4 we use e^x to define the real powers in general and show that the Exponent Laws hold. For the time being, whenever the expression a^x appears, you should assume that x is rational.

1.6 Exercises

Section 1.1

1. Arrange the following numbers in increasing order:

   (a) $\frac{7}{36}, \frac{3}{20}, \frac{1}{6}, \frac{7}{45}, \frac{11}{60}$;

   (b) $0.\overline{465}, 0.4\overline{65}, 0.46\overline{5}, 0.4655, 0.4656$.

2. Find the fractions whose decimal expansions are:

   (a) $0.\overline{481}$; (b) $0.48\overline{1}$.

3. Let $x = 0.\overline{21}$ and $y = 0.\overline{2}$. Find $x + y$ and xy (in decimal form).

4. Find a rational number x and an irrational number y in the interval $(0.119, 0.12)$.

5. Prove that, if n is a positive integer which is not a perfect square, then $\sqrt{n}$ is irrational.

   Hint: If p, q are positive integers with no factor > 1 in common such that $\left(\frac{p}{q}\right)^2 = n$, and k is the positive integer such that $k < \frac{p}{q} < k + 1$ (why does such a positive integer k exist?), show that $0 < p - kq < q$ and $\frac{nq - kp}{p - kq} = \frac{p}{q}$, and hence obtain a contradiction.

Section 1.2

1. Solve the following inequalities:

   (a) $\frac{x-1}{x^2+4} < \frac{x+1}{x^2-4}$; (b) $\sqrt{4x - 3} > x$;

   (c) $\left|17 - 2x^4\right| \le 15$; (d) $|x + 1| + |x - 1| < 4$.

Section 1.3

1. Use the Triangle Inequality to prove that

   $$|a| \le 1 \Rightarrow |a - 3| \ge 2.$$

2. Prove that $(a^2 + b^2)(c^2 + d^2) \ge (ac + bd)^2$, for any $a, b, c, d \in \mathbb{R}$.

3. Prove the inequality $3^n \geq 2n^2 + 1$, for $n = 1, 2, \ldots$:

    (a) by using the Binomial Theorem, applied to $(1+x)^n$ with $x = 2$;

    (b) by using the Principle of Mathematical Induction.

4. Use the Principle of Mathematical Induction to prove that, for $n = 1, 2, \ldots$:

    (a) $1^2 + 2^2 + 3^2 + \cdots + n^2 = \frac{n(n+1)(2n+1)}{6}$;

    (b) $\sqrt{\frac{5/4}{4n+1}} \leq \frac{1 \cdot 3 \cdot 5 \cdot \ldots \cdot (2n-1)}{2 \cdot 4 \cdot 6 \cdot \ldots \cdot (2n)} \leq \sqrt{\frac{3/4}{2n+1}}$.

5. Apply Bernoulli's Inequality, first with $x = \frac{2}{n}$ and then with $x = \frac{-2}{(3n)}$ to prove that

$$1 + \frac{2}{3n-2} \leq 3^{\frac{1}{n}} \leq 1 + \frac{2}{n}, \quad \text{for } n = 1, 2, \ldots.$$

6. By applying the Arithmetic Mean–Geometric Mean Inequality to the $n+1$ positive numbers $1, 1 - \frac{1}{n}, 1 - \frac{1}{n}, 1 - \frac{1}{n}, \ldots, 1 - \frac{1}{n}$, prove that

$$\left(1 - \tfrac{1}{n}\right)^n \leq \left(1 - \tfrac{1}{n+1}\right)^{n+1}, \quad \text{for } n = 1, 2, \ldots.$$

7. Use the Cauchy–Schwarz Inequality to prove that, if $a_1, a_2, \ldots, a_n$ are positive numbers with $a_1 + a_2 + \cdots + a_n = 1$, then

$$\sqrt{a_1} + \sqrt{a_2} + \cdots + \sqrt{a_n} \leq \sqrt{n}.$$

8. Use the Cauchy–Schwarz Inequality to prove that

$$3\sqrt[4]{\cos x} + 4\sqrt{1 - \sqrt{\cos x}} \leq 5, \quad \text{for } x \in \left[0, \tfrac{\pi}{2}\right].$$

Section 1.4

In Exercises 1–4, take $E_1 = \{x: x \in \mathbb{Q}, 0 \leq x < 1\}$ and $E_2 = \left\{\left(1 + \frac{1}{n}\right)^2: n = 1, 2, \ldots\right\}$.

1. Prove that each of the sets E_1 and E_2 is bounded above. Which of them has a maximum element?

2. Prove that each of the sets E_1 and E_2 is bounded below. Which of them has a minimum element?

3. Determine the least upper bound of each of the sets E_1 and E_2.

4. Determine the greatest lower bound of each of the sets E_1 and E_2.

5. For each of the following functions, determine whether it has a maximum or a minimum, and determine its supremum and infimum:

    (a) $f(x) = \frac{1}{1+x^2}$, $x \in [0, 1)$;   (b) $f(x) = 1 - x + x^2$, $x \in [0, 2)$.

6. Prove that, for any two numbers $a, b \in \mathbb{R}$

$$\min\{a, b\} = \tfrac{1}{2}(a + b - |a - b|) \quad \text{and}$$

$$\max\{a, b\} = \tfrac{1}{2}(a + b + |a + b|).$$

Here $\min\{a, b\}$ means the lesser of a and b, and $\max\{a, b\}$ the greater.

2 Sequences

This chapter deals with sequences of real numbers, such as

$$1, \frac{1}{2}, \frac{1}{3}, \frac{1}{4}, \frac{1}{5}, \frac{1}{6}, \ldots,$$

$$0, 1, 0, 1, 0, 1, \ldots,$$

$$1, 2, 4, 8, 16, 32, \ldots.$$

Three dots are used to indicate that the sequence continues indefinitely.

It describes in detail various properties that a sequence may possess, the most important of which is *convergence*. Roughly speaking, a sequence is *convergent*, or *tends to a limit*, if the numbers, or *terms*, in the sequence approach arbitrarily close to a unique real number, which is called the *limit* of the sequence. For example, we shall see that the sequence

$$1, \frac{1}{2}, \frac{1}{3}, \frac{1}{4}, \frac{1}{5}, \frac{1}{6}, \ldots$$

is convergent with limit 0. On the other hand, the terms of the sequence

$$0, 1, 0, 1, 0, 1, \ldots$$

do not approach arbitrarily close to any unique real number, and so this sequence is not convergent. Likewise, the sequence

$$1, 2, 4, 8, 16, 32, \ldots$$

is not convergent.

A sequence which is not convergent is called *divergent*. The sequence

$$0, 1, 0, 1, 0, 1, \ldots,$$

is a *bounded* divergent sequence. The sequence

$$1, 2, 4, 8, 16, 32, \ldots$$

is *unbounded*; its terms become arbitrarily large and positive, and we say that it *tends to infinity*.

Intuitively, it seems plausible that some sequences are convergent, whereas others are not. However, the above description of convergence, involving the phrase 'approach arbitrarily close to', lacks the precision required in Pure Mathematics. If we wish to work in a serious way with convergent sequences, prove results about them and decide whether a given sequence is convergent, then we need a rigorous definition of the concept of convergence.

Historically, such a definition emerged only in the late nineteenth century, when mathematicians such as Cantor, Cauchy, Dedekind and Weierstrass sought to place Analysis on a rigorous non-intuitive footing. It is not surprising, therefore, that the definition of convergence seems at first sight rather obscure, and it may take you a little time to master the logic that it involves.

In Section 2.1 we show how to picture the behaviour of a sequence by drawing a *sequence diagram*. We also introduce *monotonic* sequences; that is, sequences which are either increasing or decreasing.

In Section 2.2 we explain the definition of a *null* sequence; that is, a sequence which is convergent with limit 0. We then establish various properties of null sequences, and list some *basic null sequences*.

In Section 2.3 we discuss general convergent sequences (that is, sequences which converge but whose limit is not necessarily 0), together with techniques for calculating their limits.

In Section 2.4 we study divergent sequences, giving particular emphasis to sequences which tend to infinity or tend to minus infinity. We also show that convergent sequences are bounded; it follows that unbounded sequences are necessarily divergent.

In Section 2.5 we prove the Monotone Convergence Theorem, which states that any increasing sequence which is bounded above must be convergent, and, similarly, that any decreasing sequence which is bounded below must be convergent. We use this theorem to study simple examples of sequences defined by *recurrence formulas*, and particular sequences which converge to e and π.

2.1 Introducing sequences

2.1.1 What is a sequence?

Ever since learning to count you have been familiar with the sequence of natural numbers

$$1, 2, 3, 4, 5, 6, \ldots.$$

You have also encountered many other sequences of numbers, such as

$$2, 4, 6, 8, 10, 12, \ldots,$$

$$\frac{1}{2}, \frac{1}{4}, \frac{1}{8}, \frac{1}{16}, \frac{1}{32}, \frac{1}{64}, \ldots.$$

We begin our study of sequences with a definition and some notation.

Definition   A **sequence** is an unending list of real numbers

$$a_1, a_2, a_3, \ldots.$$

The real number a_n is called the **nth term** of the sequence, and the sequence is denoted by

$$\{a_n\}.$$

Alternative notations are $\{a_n\}_1^\infty$ and $\{a_n\}_{n=1}^\infty$.

In each of the sequences above, we wrote down the first few terms and left you to assume that subsequent terms were obtained by continuing the pattern in an obvious way. It is sometimes better, however, to give a precise description of a typical term of a sequence, and we do this by stating an explicit formula for the nth term. Thus the expression $\{2n - 1\}$ denotes the sequence

$$1, 3, 5, 7, 9, 11, \ldots,$$

and the sequence $\{a_n\}$ defined by the statement

$$a_n = (-1)^n, \quad n = 1, 2, \ldots,$$

has terms

$$a_1 = -1, \quad a_2 = 1, \quad a_3 = -1, \quad a_4 = 1, \quad a_5 = -1, \ldots.$$

Problem 1

(a) Calculate the first five terms of each of the following sequences:

    (i) $\{3n + 1\}$;   (ii) $\{3^{-n}\}$;   (iii) $\{(-1)^n n\}$.

(b) Calculate the first five terms of each of the following sequences $\{a_n\}$:

    (i) $a_n = n!, \quad n = 1, 2, \ldots$;

    (ii) $a_n = \left(1 + \frac{1}{n}\right)^n, \quad n = 1, 2, \ldots$ (to 2 decimal places).

Sequences often begin with a term corresponding to $n = 1$. Sometimes, however, it is necessary to begin a sequence with some other value of n. We indicate this by writing, for example, $\{a_n\}_3^\infty$ to represent the sequence

$$a_3, a_4, a_5, \ldots.$$

For example, the sequence $\{\frac{1}{n! - n}\}$ cannot begin with $n = 1$ or $n = 2$.

Sequence diagrams

It is often helpful to picture how a given sequence $\{a_n\}$ behaves by drawing a **sequence diagram**; that is, a graph of the sequence in $\mathbb{R}^2$. To do this, we mark the values $n = 1, 2, 3, \ldots$ on the x-axis and, for each value of n, we plot the point (n, a_n). For example, the sequence diagrams for the sequences $\{2n - 1\}$, $\{\frac{1}{n}\}$ and $\{(-1^n)\}$ are as follows:

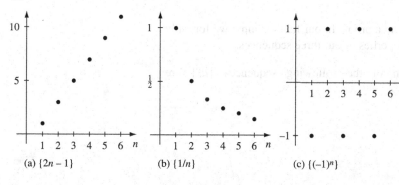

(a) $\{2n - 1\}$ (b) $\{1/n\}$ (c) $\{(-1)^n\}$

In Figure (a), the points plotted all lie on the straight line $y = 2x - 1$.

In Figure (b), they all lie on the hyperbola $y = \frac{1}{x}$.

Problem 2 Draw a sequence diagram, showing the first five points, for each of the following sequences:

(a) $\{n^2\}$;   (b) $\{3\}$;   (c) $\left\{\left(1 + \frac{1}{n}\right)^n\right\}$;   (d) $\left\{\frac{(-1)^n}{n}\right\}$.

(In part (c), use the result of Problem 1, part (b).)

2.1.2 Monotonic sequences

Many of the sequences considered so far have the property that, as n increases, their terms are either *increasing* or *decreasing*. For example, the sequence $\{2n - 1\}$ has terms $1, 3, 5, 7, \ldots$, which are increasing, whereas the sequence

$\{\frac{1}{n}\}$ has terms $1, \frac{1}{2}, \frac{1}{3}, \frac{1}{4}, \ldots$, which are decreasing. The sequence $\{(-1)^n\}$ is neither increasing nor decreasing. All this can be seen clearly on the above sequence diagrams.

We now give a precise meaning to these words *increasing* and *decreasing*, and introduce the term *monotonic*.

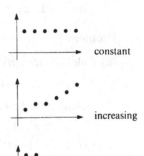

constant

increasing

decreasing

> **Definition**   A sequence $\{a_n\}$ is:
> - **constant**, if $a_{n+1} = a_n$,   for $n = 1, 2, \ldots$;
> - **increasing**, if $a_{n+1} \geq a_n$,   for $n = 1, 2, \ldots$;
> - **decreasing**, if $a_{n+1} \leq a_n$,   for $n = 1, 2, \ldots$;
> - **monotonic**, if $\{a_n\}$ is either increasing or decreasing.

Remarks

1. Note that, for a sequence $\{a_n\}$ to be increasing, it is essential that $a_{n+1} \geq a_n$, for *all* $n \geq 1$. However, we do not require strict inequalities, because we wish to describe sequences such as

$$1, 1, 2, 2, 3, 3, 4, 4, \ldots \quad \text{and} \quad 1, 2, 2, 3, 4, 4, 5, 6, 6, \ldots$$

   as increasing. One slightly bizarre consequence of the definition is that constant sequences are both increasing and decreasing!

2. A sequence $\{a_n\}$ is said to be:

   strictly increasing, if $a_{n+1} > a_n$,   for $n = 1, 2, \ldots$;

   strictly decreasing, if $a_{n+1} < a_n$,   for $n = 1, 2, \ldots$;

   strictly monotonic, if $\{a_n\}$ is either strictly increasing or strictly decreasing.

3. A diagram does NOT constitute a proof! In our first example we formally establish the monotonicity properties of our three sequences.

Example 1   Determine which of the following sequences $\{a_n\}$ are monotonic:

(a) $a_n = 2n - 1, n = 1, 2, \ldots$;

(b) $a_n = \frac{1}{n}, n = 1, 2, \ldots$;

(c) $a_n = (-1)^n, n = 1, 2, \ldots$

Solution

(a) The sequence $\{2n - 1\}$ is monotonic because

$$a_n = 2n - 1 \quad \text{and} \quad a_{n+1} = 2(n + 1) - 1 = 2n + 1,$$

so that

$$a_{n+1} - a_n = (2n + 1) - (2n - 1) = 2 > 0, \quad \text{for } n = 1, 2, \ldots.$$

Thus $\{2n - 1\}$ is increasing.

In fact, strictly increasing.

(b) The sequence $\{\frac{1}{n}\}$ is monotonic because

$$a_n = \frac{1}{n} \quad \text{and} \quad a_{n+1} = \frac{1}{n + 1},$$

so that

$$a_{n+1} - a_n = \frac{1}{n+1} - \frac{1}{n} = \frac{n-(n+1)}{(n+1)n} = \frac{-1}{(n+1)n} < 0, \quad \text{for } n = 1, 2, \ldots.$$

Thus $\{\frac{1}{n}\}$ is decreasing.

Alternatively, since $a_n > 0$, for all n, and

$$\frac{a_{n+1}}{a_n} = \frac{n}{n+1} < 1, \quad \text{for } n = 1, 2, \ldots,$$

it follows that

$$a_{n+1} < a_n, \quad \text{for } n = 1, 2, \ldots.$$

Thus $\{a_n\}$ is decreasing.

In fact, strictly decreasing.

(c) The sequence $\{(-1)^n\}$ is not monotonic. In fact, $a_1 = -1$, $a_2 = 1$ and $a_3 = -1$.

Hence $a_3 < a_2$, which means that $\{a_n\}$ is not increasing.

Also, $a_2 > a_1$, which means that $\{a_n\}$ is not decreasing.

Thus $\{(-1)^n\}$ is neither increasing nor decreasing, and so is not monotonic. $\square$

A *single counter-example* is sufficient to show that $\{(-1)^n\}$ is not increasing; similarly a (different!) *single counter-example* is sufficient to show that $\{(-1)^n\}$ is not decreasing.

Example 1 illustrates the use of the following strategies:

Strategy To show that a given sequence $\{a_n\}$ is monotonic, consider the expression $a_{n+1} - a_n$.

- If $a_{n+1} - a_n \geq 0$, for $n = 1, 2, \ldots$, then $\{a_n\}$ is increasing.
- If $a_{n+1} - a_n \leq 0$, for $n = 1, 2, \ldots$, then $\{a_n\}$ is decreasing.

If $a_n > 0$ for all n, it may be more convenient to use the following version of the strategy:

Strategy To show that a given sequence of *positive* terms, $\{a_n\}$, is monotonic, consider the expression $\frac{a_{n+1}}{a_n}$.

- If $\frac{a_{n+1}}{a_n} \geq 1$, for $n = 1, 2, \ldots$, then $\{a_n\}$ is increasing.
- If $\frac{a_{n+1}}{a_n} \leq 1$, for $n = 1, 2, \ldots$, then $\{a_n\}$ is decreasing.

Problem 3 Show that the following sequences $\{a_n\}$ are monotonic:

(a) $a_n = n!$, $n = 1, 2, \ldots$;

(b) $a_n = 2^{-n}$, $n = 1, 2, \ldots$;

(c) $a_n = n + \frac{1}{n}$, $n = 1, 2, \ldots$.

It is often possible to *guess* whether a sequence given by a specific formula is monotonic by calculating the first few terms. For example, consider the sequence $\{a_n\}$ given by

$$a_n = \left(1 + \frac{1}{n}\right)^n, \quad n = 1, 2, \ldots.$$

We will study this important sequence in detail in Section 2.5.

In Problem 1 you found that the first five terms of this sequence are approximately

$$2, 2.25, 2.37, 2.44 \text{ and } 2.49.$$

These terms suggest that the sequence $\{a_n\}$ is increasing, and in fact it is.

However, the first few terms of a sequence are *not always* a reliable guide to the sequence's behaviour. Consider, for example, the sequence

$$a_n = \frac{10^n}{n!}, \quad n = 1, 2, \ldots.$$

The first two terms certainly prove that the sequence is neither decreasing nor constant.

The first five terms of this sequence are approximately

$$10, 50, 167, 417 \text{ and } 833;$$

this *suggests* that the sequence $\{a_n\}$ is increasing. However, calculation of more terms shows that this is not so, and the sequence diagram for $\{a_n\}$ looks like this:

In fact

$$a_6 \simeq 1389, a_7 \simeq 1984,$$
$$a_8 \simeq 2480, a_9 \simeq 2756,$$
$$a_{10} \simeq 2756, a_{11} \simeq 2505,$$
$$a_{12} \simeq 2088.$$

In particular, the fact that $a_{12} < a_{11}$ shows that the sequence $\{a_n\}$ cannot be increasing.

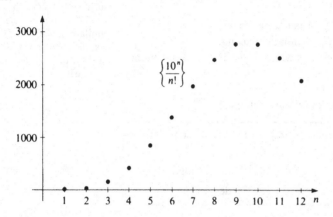

Simplifying $\frac{a_{n+1}}{a_n}$, we find that

$$\frac{a_{n+1}}{a_n} = \frac{10^{n+1}}{(n+1)!} \bigg/ \frac{10^n}{n!} = \frac{10}{n+1}.$$

But $\frac{10}{n+1} \leq 1$, for $n = 9, 10, \ldots$, so that $\frac{a_{n+1}}{a_n} \leq 1$, for $n = 9, 10, \ldots$; it follows that the sequence $\{a_n\}$ is decreasing, if we ignore the first eight terms.

In fact, $\frac{a_{10}}{a_9} = 1$ and $\frac{a_{n+1}}{a_n} < 1$, for $n = 10, 11, \ldots$.

In a situation like this, when a given sequence has a certain property *provided that we ignore a finite number of terms*, we say that the sequence *eventually* has the property. Thus we have just seen that the sequence $\left\{\frac{10^n}{n!}\right\}$ is *eventually decreasing*.

Another example of this usage is the following statement:

the terms of the sequence $\{n^2\}$ are eventually greater than 100.

This statement is true because $n^2 > 100$, for all $n > 10$.

Problem 4 Classify each of the following statements as TRUE or FALSE, and justify your answers (that is, if a statement is TRUE, prove it; if a statement is FALSE, give a specific counter-example):

Note this general approach to 'TRUE' and 'FALSE'.

(a) The terms of the sequence $\{2^n\}$ are eventually greater than 1000.

(b) The terms of the sequence $\{(-1)^n\}$ are eventually positive.

(c) The terms of the sequence $\left\{\frac{1}{n}\right\}$ are eventually less than 0.025.

(d) The sequence $\left\{\frac{n^4}{4^n}\right\}$ is eventually decreasing.

2.2 Null sequences

2.2.1 What is a null sequence?

In this sub-section we give a precise definition of a *null sequence* (that is, a sequence which converges to 0) and introduce some properties of null sequences.

We shall frequently use the rules for rearranging inequalities which you met in Chapter 1, so you may find it helpful to reread that section quickly before starting here. We shall also use the following inequalities which were proved in Sub-section 1.3.3

$$2^n \geq 1 + n, \quad \text{for } n = 1, 2, \ldots,$$

and

$$2^n \geq n^2, \quad \text{for } n \geq 4.$$

Section 1.2

Problem 1 For each of the following statements, find a number X such that the statement is true:

(a) $\frac{1}{n} < \frac{1}{100}$, for all $n > X$;   (b) $\frac{1}{n} < \frac{3}{1000}$, for all $n > X$.

Problem 2 For each of the following statements, find a number X such that the statement is true:

(a) $\left|\frac{(-1)^n}{n^2}\right| < \frac{1}{100}$, for all $n > X$;   (b) $\left|\frac{(-1)^n}{n^2}\right| < \frac{3}{1000}$, for all $n > X$.

The solutions of Problems 1 and 2 both suggest that the larger and larger we choose n, the closer and closer to 0 the terms of the sequences $\left\{\frac{1}{n}\right\}$ and $\left\{\frac{(-1)^n}{n^2}\right\}$ become.

We can express this in terms of the Greek letter ε (pronounced 'epsilon'), which we introduce to denote a positive number that may be as small as we please in any given particular instance. In terms of the sequence diagrams for $\left\{\frac{1}{n}\right\}$ and $\left\{\frac{(-1)^n}{n^2}\right\}$, this means that the terms of these sequences eventually lie inside a horizontal strip in the sequence diagram from $-\varepsilon$ up to ε. However, the smaller we choose ε, the further to the right we have to go before we can be sure that all the terms of the sequence from that point onwards lie inside the strip. That is, the smaller we choose ε the larger we have to choose X if we wish to have

$$\frac{1}{n} < \varepsilon, \quad \text{for all } n > X, \quad \text{or} \quad \left|\frac{(-1)^n}{n^2}\right| < \varepsilon, \quad \text{for all } n > X.$$

Thus in Problems 1 and 2 we chose the particular examples of $\frac{1}{100}$ and $\frac{3}{1000}$ in place of the 'general' positive number ε.

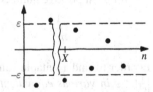

Definition A sequence $\{a_n\}$ is a **null sequence** if:

for each positive number ε, there is a number X such that

$$|a_n| < \varepsilon, \quad \text{for all } n > X. \tag{1}$$

Note that here X *need not* be an integer; *any* appropriate real number will serve. X does NOT depend on n, but does in general depend on ε.

Using this terminology, it follows from our previous discussion that the sequences $\left\{\frac{1}{n}\right\}$ and $\left\{\frac{(-1)^n}{n^2}\right\}$ are both null.

We can interpret our finding of a suitable number X for (1) to hold as an '$\varepsilon - X$ game' in which player A chooses a positive number ε and challenges player B to find some number X such that the property (1) holds.

For example, consider the sequence $\{a_n\}$ where $a_n = \frac{(-1)^n}{n^2}$, $n = 1, 2, \ldots$. This sequence is null. For

$$|a_n| = \left|\frac{(-1)^n}{n^2}\right| = \frac{1}{n^2};$$

so that, if we make the choice $X = \frac{1}{\sqrt{\varepsilon}}$, it is certainly true that

$$|a_n| < \varepsilon, \quad \text{for all } n > X.$$

In terms of the $\varepsilon - X$ game, this simply means that whatever choice of ε is made by player A, player B can always win by making the choice $X = \frac{1}{\sqrt{\varepsilon}}$!

Remarks

1. In the sequence $\left\{\frac{(-1)^n}{n^2}\right\}$, the signs of the terms made no difference to whether the sequence was null, for they disappeared immediately we took the modulus of the terms in order to examine whether the definition (1) of a null sequence was satisfied. Indeed, in general, the signs of terms in a sequence make no difference to whether a sequence is null. We can express this formally as follows:

> A sequence $\{a_n\}$ is null if and only if the corresponding sequence $\{|a_n|\}$ is null.

2. A null sequence $\{a_n\}$ remains null if we *add, delete or alter a finite number of terms* in the sequence.

   Similarly, a non-null sequence remains non-null if we add, delete or alter a finite number of terms.

3. If one number serves as a suitable value of X for the inequality in (1) to hold, then any *larger* number will also serve as a suitable X. Hence, for simplicity in some proofs, we may assume if we wish that our initial choice of X in (1) is a positive integer.

Example 1   Prove that the sequence $\left\{\frac{1}{n^3}\right\}$ is a null sequence.

Solution   We have to prove that

for each positive number ε, there is a number X such that

$$\left|\frac{1}{n^3}\right| < \varepsilon, \quad \text{for all } n > X. \tag{2}$$

In order to find a suitable value of X for (2) to hold, we *rewrite the inequality* $\left|\frac{1}{n^3}\right| < \varepsilon$ *in various equivalent ways* until we spy a value for X that will suit our purpose. Now

$$\left|\frac{1}{n^3}\right| < \varepsilon \Leftrightarrow \frac{1}{n^3} < \varepsilon$$

$$\Leftrightarrow n^3 > \frac{1}{\varepsilon}$$

$$\Leftrightarrow n > \frac{1}{\sqrt[3]{\varepsilon}}.$$

So, let us choose X to be $\frac{1}{\sqrt[3]{\varepsilon}}$. With this choice of X, the above chain of equivalent inequalities shows us that, if $n > X$ (the last line in the chain), then $\left|\frac{1}{n^3}\right| < \varepsilon$ (in the first line). Thus, with this choice of X, (2) holds; so $\left\{\frac{1}{n^3}\right\}$ is indeed null.   □

This is the case because, if
$$n > \frac{1}{\sqrt{\varepsilon}}[= X],$$
then we have $n^2 > \frac{1}{\varepsilon}$ or $\varepsilon > \frac{1}{n^2}$; hence, for $n > X$, we have
$$|a_n| = \frac{1}{n^2} < \frac{1}{X^2} = \varepsilon.$$

All that happens is that the definition (1) remains valid with a possibly different value of X having to be chosen. This is often expressed in the following memorable way: '*a finite number of terms do not matter*'.

For example, if $7 + \pi$ will serve as a suitable value for X, then so will 12 or 37; but 10 might not.

The term inside the modulus is positive.

Here we use the Reciprocal Rule for inequalities that you met in Sub-section 1.2.1.

Here we use the Power Rule for inequalities that you met in Sub-section 1.2.1.

Example 2 Prove that the following sequence is not null

$$a_n = \begin{cases} 1, & \text{if } n \text{ is odd,} \\ 0, & \text{if } n \text{ is even.} \end{cases}$$

Solution To prove that the sequence is not null, we have to show that it does *not* satisfy the definition. In other words, we must show that the following statement is not true:

> for each positive number ε, there is a number X such that
> $|a_n| < \varepsilon$, for all $n > X$.

So, what we have to show is that the following is true:

> for *some* positive number ε, *whatever* X one chooses it is not true that
> $|a_n| < \varepsilon$, for all $n > X$.

which we may rephrase as:

> there is *some* positive number ε, such that there is NO number X such that
> $|a_n| < \varepsilon$, for all $n > X$.

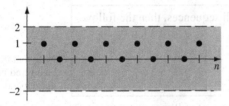

Strip of half-width 2

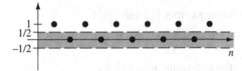

Strip of half-width $\frac{1}{2}$

So, we need to find some positive number ε with such a property. The sequence diagram provides the clue! If we choose $\varepsilon = \frac{1}{2}$, then the point (n, a_n) lies outside the strip $\left(-\frac{1}{2}, \frac{1}{2}\right)$ for every odd n. In other words, whatever X one then chooses, the statement

$$|a_n| < \varepsilon, \quad \text{for all } n > X,$$

is false. It follows that the sequence is not a null sequence. □

Note

Notice that any positive value of ε less than 1 will serve for our purpose here; there is nothing special about the number $\frac{1}{2}$.

However, any value of ε greater than 1 provides no information!

These two examples illustrate the following strategy:

Strategy for using the definition of null sequence

1. To show that $\{a_n\}$ is null, solve the inequality $|a_n| < \varepsilon$ to find a number X (generally depending on ε) such that $|a_n| < \varepsilon$, for all $n > X$.

2. To show that $\{a_n\}$ is *not* null, find ONE value of ε for which there is NO number X such that $|a_n| < \varepsilon$, for all $n > X$.

Problem 3   Use the above strategy to determine which of the following sequences are null:

(a) $\left\{\frac{1}{2n-1}\right\}$;   (b) $\left\{\frac{(-1)^n}{10}\right\}$;   (c) $\left\{\frac{(-1)^n}{n^4+1}\right\}$.

We now look at a number of Rules for 'getting new null sequences from old'.

> We shall prove these Rules later, in Sub-section 2.2.2.

Power Rule   If $\{a_n\}$ is a null sequence, where $a_n \geq 0$, for $n = 1, 2, \ldots$, and $p > 0$, then $\{a_n^p\}$ is a null sequence.

> Recall that $a_n^p = (a_n)^p$.

Thus, for example, the sequence $\left\{\frac{1}{\sqrt{n}}\right\}$ is null – we simply apply the Power Rule to the sequence $\left\{\frac{1}{n}\right\}$ that we saw earlier to be null, using the positive power $p = \frac{1}{2}$.

> For $\frac{1}{\sqrt{n}} = \left(\frac{1}{n}\right)^{\frac{1}{2}}$.

Remark

Notice that the Power Rule also holds without the requirement that $a_n \geq 0$, so long as a_n^p is defined – for example, if $p = \frac{1}{m}$ where m is a positive integer.

Combination Rules   If $\{a_n\}$ and $\{b_n\}$ are null sequences, then the following are also null sequences:

Sum Rule $\{a_n + b_n\}$;
Multiple Rule $\{\lambda a_n\}$, for any real number λ;
Product Rule $\{a_n b_n\}$.

> Note that the number λ has to be a fixed number that does *not* depend on n.

Thus, for example, we may use known examples of null sequences to verify that the following sequences are also null:

$\left\{\frac{1}{n} + \frac{1}{n^3}\right\}$ – by applying the Sum Rule to the null sequences $\left\{\frac{1}{n}\right\}$ and $\left\{\frac{1}{n^3}\right\}$;

$\left\{\frac{47\pi}{n^3}\right\}$ – by applying the Multiple Rule to the null sequence $\left\{\frac{1}{n^3}\right\}$ with $\lambda = 47\pi$;

$\left\{\frac{1}{n(2n-1)}\right\}$ – by applying the Product Rule to the null sequences $\left\{\frac{1}{n}\right\}$ and $\left\{\frac{1}{2n-1}\right\}$.

Problem 4   Use the above rules to show that the following sequences are null:

(a) $\left\{\frac{1}{(2n-1)^3}\right\}$;   (b) $\left\{\frac{6}{\sqrt[5]{n}} + \frac{5}{(2n-1)^7}\right\}$;   (c) $\left\{\frac{1}{3n^4(2n-1)^{\frac{1}{3}}}\right\}$.

> In your solution, you may use any of the sequences that you have proved to be null so far in this sub-section.

Our next rule, the Squeeze Rule, also enables us to 'get new null sequences from old' – but in a slightly different way. To illustrate this rule, we look first at the sequence diagrams of the two sequences $\left\{\frac{1}{\sqrt{n}}\right\}$ and $\left\{\frac{1}{1+\sqrt{n}}\right\}$.

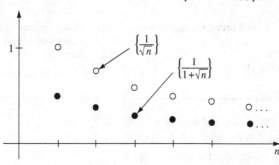

The points corresponding to the sequence $\left\{\frac{1}{1+\sqrt{n}}\right\}$ are squeezed in between the horizontal axis and the points corresponding to the null sequence $\left\{\frac{1}{\sqrt{n}}\right\}$, since $\frac{1}{1+\sqrt{n}} < \frac{1}{\sqrt{n}}$, for $n = 1, 2, \ldots$. Hence, if from some point onwards all the points corresponding to $\left\{\frac{1}{\sqrt{n}}\right\}$ lie in a narrow strip in the sequence diagram of half-width ε about the axis, then (from the same point onwards) all the points corresponding to $\left\{\frac{1}{1+\sqrt{n}}\right\}$ will also lie in the same strip. So, since ε may be *any* positive number, it certainly looks from this sequence diagram argument that the sequence $\left\{\frac{1}{1+\sqrt{n}}\right\}$ must be a null sequence too.

Squeeze Rule   If $\{b_n\}$ is a null sequence and
$$|a_n| \leq b_n, \quad \text{for } n = 1, 2, \ldots,$$
then $\{a_n\}$ is a null sequence.

That is, $\{a_n\}$ is *dominated* by $\{b_n\}$.

The trick in using the Squeeze Rule to prove that a given sequence $\{a_n\}$ is null is to think of a *suitable* sequence $\{b_n\}$ that dominates $\{a_n\}$ and is *itself* null. Thus, for example, since the sequence $\left\{\frac{1}{\sqrt{n}}\right\}$ is null and $\frac{1}{1+\sqrt{n}} < \frac{1}{\sqrt{n}}$, it follows from the Squeeze Rule that $\left\{\frac{1}{1+\sqrt{n}}\right\}$ is also a null sequence.

Proof of the Squeeze Rule   We want to prove that $\{a_n\}$ is null; that is: for each positive number ε, there is a number X such that
$$|a_n| < \varepsilon, \quad \text{for all } n > X. \tag{3}$$
We know that $\{b_n\}$ is null, so there is some number X such that
$$|b_n| < \varepsilon, \quad \text{for all } n > X. \tag{4}$$

And this is the X that we shall use to verify (3).

We also know that $|a_n| \leq b_n$, for $n = 1, 2, \ldots$, and hence it follows from (4) that
$$|a_n| \, (< |b_n|) < \varepsilon, \quad \text{for all } n > X.$$
Thus inequality (3) holds, as required.   ∎

Remark

In applying the Squeeze Rule, it is often useful to remember the following fact: *The behaviour of a finite number of terms does not matter: it is sufficient to check that $|a_n| \leq b_n$ holds eventually.*

That is, that $\{a_n\}$ is *eventually dominated* by $\{b_n\}$.

Example 3   Use the Squeeze Rule to prove that the sequence $\left\{\left(\frac{1}{2}\right)^n\right\}$ is null.

Proof   We want to prove that $\left\{\left(\frac{1}{2}\right)^n\right\}$ is null.

What sequence can we find that dominates $\left\{\left(\frac{1}{2}\right)^n\right\}$? Well, we saw at the start of this sub-section that $2^n \geq n+1$ and so $2^n \geq n$, for $n = 1, 2, \ldots$; it follows from this, by the Reciprocal Rule for inequalities, that $\frac{1}{2^n} \leq \frac{1}{n}$, for $n = 1, 2, \ldots$.

Thus the sequence $\left\{\frac{1}{n}\right\}$ dominates the sequence $\left\{\left(\frac{1}{2}\right)^n\right\}$, and is itself null. It follows from the Squeeze Rule that $\left\{\left(\frac{1}{2}\right)^n\right\}$ is null.   □

Problem 5   Use the inequality $2^n \geq n^2$, for $n \geq 4$, and the Squeeze Rule to prove that the sequence $\left\{n\left(\frac{1}{2}\right)^n\right\}$ is null.

Example 4   Use the Squeeze Rule to prove that the sequence $\left\{\frac{10^n}{n!}\right\}$ is null.

Solution   We want to prove that $\left\{\frac{10^n}{n!}\right\}$ is null.

With a bit of inspiration, we guess that the sequence $\left\{\frac{10^n}{n!}\right\}$ is eventually dominated by $\left\{\frac{\lambda}{n}\right\}$, for some constant λ.

Writing out the expression $\frac{10^n}{n!}$ in full, we see that

$$\frac{10^n}{n!} = \left(\frac{10}{1}\right)\left(\frac{10}{2}\right)\cdots\left(\frac{10}{10}\right) \times \left(\frac{10}{11}\right)\cdots\left(\frac{10}{n-1}\right)\left(\frac{10}{n}\right)$$

$$< 3000 \times \frac{10}{n}, \quad \text{for all } n > 10,$$

$$= \frac{30000}{n}.$$

For $\frac{10}{1}\cdot\frac{10}{2}\cdot\ldots\cdot\frac{10}{10} \simeq 2,755$.

In other words, $\left\{\frac{10^n}{n!}\right\}$ is eventually dominated by $\left\{\frac{\lambda}{n}\right\}$, for $\lambda = 30000$. This latter sequence is null, by the Multiple Rule applied to the null sequence $\left\{\frac{1}{n}\right\}$.

It follows, from the Squeeze Rule and the fact that $\left\{\frac{30000}{n}\right\}$ is a null sequence, that the sequence $\left\{\frac{10^n}{n!}\right\}$ is also null.   □

Problem 6   Prove that the following sequences are null:

(a) $\left\{\frac{1}{n^2+n}\right\}$;   (b) $\left\{\frac{(-1)^n}{n!}\right\}$;   (c) $\left\{\frac{\sin n^2}{n^2+2^n}\right\}$.

We can now list a good number of generic types of null sequences, of which we have seen specific instances already in this sub-section. We call these *basic null sequences*, since we shall use them commonly together with the Combination Rules and other rules for null sequences in order to prove that particular sequences that we meet are themselves null.

Basic null sequences

(a) $\left\{\frac{1}{n^p}\right\}$,   for $p > 0$;

(b) $\{c^n\}$,   for $|c| < 1$;

(c) $\{n^p c^n\}$,   for $p > 0$ and $|c| < 1$;

(d) $\left\{\frac{c^n}{n!}\right\}$,   for any real c;

(e) $\left\{\frac{n^p}{n!}\right\}$,   for $p > 0$.

Instances of these that we have seen are

$$\left\{\frac{1}{\sqrt{n}}\right\};$$

$$\left\{\left(\frac{1}{2}\right)^n\right\};$$

$$\left\{n\left(\frac{1}{2}\right)^n\right\};$$

$$\left\{\frac{10^n}{n!}\right\}.$$

2.2.2 Proofs

We now give a number of proofs which were omitted from the previous sub-section so as not to slow down your gaining an understanding of the key ideas there. First, recall the definition of a *null* sequence.

We suggest that you omit these proofs on a first reading, and return to them when you are confident that you understand the basic ideas.

Definition   A sequence $\{a_n\}$ is **null** if

for each positive number ε, there is a number X such that

$$|a_n| < \varepsilon, \quad \text{for all } n > X.$$

Recall that X *need not* be an integer; *any* appropriate real number will serve.

Proofs of the Power Rule and the Combination Rules

In the previous sub-section we proved the Squeeze Rule, but did not prove the Power Rule or the Combination Rules. We now supply these proofs.

Power Rule   If $\{a_n\}$ is a null sequence, where $a_n \geq 0$, for $n = 1, 2, \ldots$, and, if $p > 0$, then $\{a_n^p\}$ is a null sequence.

Recall that $a_n^p = (a_n)^p$.

Proof We want to prove that $\{a_n^p\}$ is null; that is:

for each positive number ε, there is a number X such that

$$a_n^p < \varepsilon, \quad \text{for all } n > X. \tag{5}$$

We know that $\{a_n\}$ is null, so there is some number X such that

$$a_n < \varepsilon^{\frac{1}{p}}, \quad \text{for all } n > X. \tag{6}$$

Taking the pth power of both sides of (6), we obtain the desired result (5), with the same value of X. ∎

Remark

Notice how we used the (positive) number $\varepsilon^{\frac{1}{p}}$ in place of ε in (6) in order to obtain ε in the final result (5). In the proofs which follow, we again apply the definition of null sequence, using positive numbers which depend in some way on ε, in order to obtain ε in the inequality that we are aiming to prove.

> **Sum Rule** If $\{a_n\}$ and $\{b_n\}$ are null sequences, then $\{a_n + b_n\}$ is a null sequence.

Proof We want to prove that the sum $\{a_n + b_n\}$ is null; that is:

for each positive number ε, there is a number X such that

$$|a_n + b_n| < \varepsilon, \quad \text{for all } n > X. \tag{7}$$

We know that $\{a_n\}$ and $\{b_n\}$ are null, so there are numbers X_1 and X_2 such that

$$|a_n| < \frac{1}{2}\varepsilon, \quad \text{for all } n > X_1,$$

and

$$|b_n| < \frac{1}{2}\varepsilon, \quad \text{for all } n > X_2.$$

Hence, if $X = \max\{X_1, X_2\}$, then both the two previous inequalities hold; so if we add them we obtain, by the Triangle Inequality, that

$$|a_n + b_n| \leq |a_n| + |b_n|$$
$$< \frac{1}{2}\varepsilon + \frac{1}{2}\varepsilon = \varepsilon, \quad \text{for all } n > X.$$

Thus inequality (7) holds, with this value of X. ∎

Before going on, first a comment about the number $\frac{1}{2}$ that appears several times in the above proof. It is used twice in expressions $\frac{1}{2}\varepsilon$ at the start of the proof in order to end up with a final inequality that says that some expression is '$< \varepsilon$'. While this means that we end up with an inequality that shows at once that the desired result holds, in fact it is not strictly necessary to end up with precisely '$< \varepsilon$', as the following result shows:

> **Lemma   The '$K\varepsilon$ Lemma'** Let $\{a_n\}$ be a sequence. Suppose that, for each positive number ε, there is a number X such that
>
> $$|a_n| < K\varepsilon, \quad \text{for all } n > X,$$
>
> where K is a positive real number that does not depend on ε or n. Then $\{a_n\}$ is a null sequence.

Margin notes:

For $|a_n^p| = a_n^p$, since $a_n \geq 0$.

Here we use the number $\varepsilon^{\frac{1}{p}}$ in place of ε in the definition of null sequence.

We use $\frac{1}{2}\varepsilon$ here rather than ε, in order to end up with the symbol ε on its own eventually in the desired inequality (7).

You met the Triangle Inequality in Sub-section 1.3.1.

In this case, the result that a particular sequence is null.

Loosely speaking, we may express this result as '$K\varepsilon$ is just as good as ε' in the definition of *null sequence*.

For example, K might be 2 or $\frac{\pi}{7}$ or 259, but it could not be $2n$ or $\frac{X}{259}$.

Proof We want to prove that the sequence $\{a_n\}$ is null; that is:

for each positive number ε, there is a number X such that

$$|a_n| < \varepsilon, \quad \text{for all } n > X.$$

You may omit this proof on a first reading.

Now, whatever this positive number ε may be, the number $\frac{\varepsilon}{K}$ is also a positive number. It follows from the hypothesis stated in the Lemma that therefore there is some number X such that

$$|a_n| < K \times \frac{\varepsilon}{K}$$
$$= \varepsilon, \quad \text{for all } n > X.$$

This is precisely the condition for $\{a_n\}$ to be null.   ∎

From time to time we shall use this Lemma in order to avoid arithmetic complexity in our proofs.

Multiple Rule If $\{a_n\}$ is a null sequence, then $\{\lambda a_n\}$ is a null sequence for any real number λ.

Proof We want to prove that the multiple $\{\lambda a_n\}$ is null; that is:

for each positive number ε, there is a number X such that

$$|\lambda a_n| < \varepsilon, \quad \text{for all } n > X. \tag{8}$$

If $\lambda = 0$, this is obvious, and so we may assume that $\lambda \neq 0$.

We know that $\{a_n\}$ is null, so there is some number X such that

$$|a_n| < \frac{1}{|\lambda|}\varepsilon, \quad \text{for all } n > X.$$

Multiplying both sides of this inequality by the positive number $|\lambda|$, this gives us that

$$|\lambda a_n| < \varepsilon, \quad \text{for all } n > X.$$

Thus the desired result (8) holds.   ∎

We use $\frac{1}{|\lambda|}\varepsilon$ here rather than ε in order to end up eventually with the symbol ε on its own in the desired inequality (8).

Product Rule If $\{a_n\}$ and $\{b_n\}$ are null sequences, then $\{a_n b_n\}$ is a null sequence.

Proof We want to prove that the product $\{a_n b_n\}$ is null; that is:

for each positive number ε, there is a number X such that

$$|a_n b_n| < \varepsilon, \quad \text{for all } n > X. \tag{9}$$

We know that $\{a_n\}$ and $\{b_n\}$ are null, so there are numbers X_1 and X_2 such that

$$|a_n| < \sqrt{\varepsilon}, \quad \text{for all } n > X_1,$$

and

$$|b_n| < \sqrt{\varepsilon}, \quad \text{for all } n > X_2.$$

We use $\sqrt{\varepsilon}$ here rather than ε in order to end up eventually with the symbol ε on its own in the desired inequality (9).

Hence, if $X = \max\{X_1, X_2\}$, then both the two previous inequalities hold; so if we multiply them we obtain that

$$|a_n b_n| = |a_n| \times |b_n|$$
$$< \sqrt{\varepsilon} \times \sqrt{\varepsilon} = \varepsilon, \quad \text{for all } n > X.$$

Thus inequality (9) holds with this value of X.   ∎

Basic null sequences

At the end of the previous sub-section we gave a list of basic null sequences. We end this sub-section by proving that these sequences are indeed null sequences.

For example

> **Basic null sequences**   The following sequences are null sequences:
>
> (a) $\left\{\frac{1}{n^p}\right\}$,   for $p > 0$;
>
> (b) $\{c^n\}$,   for $|c| < 1$;
>
> (c) $\{n^p c^n\}$,   for $p > 0$, $|c| < 1$;
>
> (d) $\left\{\frac{c^n}{n!}\right\}$,   for any real c;
>
> (e) $\left\{\frac{n^p}{n!}\right\}$,   for $p > 0$.

$\left\{\dfrac{1}{n^{10}}\right\}$;

$\{(0.9)^n\}$;

$\{n^3 (0.9)^n\}$;

$\left\{\dfrac{10^n}{n!}\right\}$;

$\left\{\dfrac{n^{10}}{n!}\right\}$

Proof

(a) To prove that $\left\{\frac{1}{n^p}\right\}$ is null, for $p > 0$, we simply apply the Power Rule to the sequence $\left\{\frac{1}{n}\right\}$, which we know is null.

(b) To prove that $\{c^n\}$ is null, for $|c| < 1$, it is sufficient to consider only the case $0 \le c < 1$. If $c = 0$, the sequence is obviously null; so we may assume that $0 < c < 1$.

For a sequence $\{a_n\}$ is null if and only if $\{|a_n|\}$ is null.

With this assumption, we can write c in the form

$$c = \frac{1}{1+a}, \quad \text{for some number } a > 0.$$

Now, by the Binomial Theorem

$$(1+a)^n \ge 1 + na$$
$$\ge na, \quad \text{for } n = 1, 2, \ldots,$$

You met the Binomial Theorem in Sub-section 1.3.3.

and hence

$$c^n = \frac{1}{(1+a)^n}$$
$$\le \frac{1}{na}, \quad \text{for } n = 1, 2, \ldots.$$

Since $\left\{\frac{1}{n}\right\}$ is null, we deduce that $\left\{\frac{1}{na}\right\}$ is also null by the Multiple Rule; hence $\{c^n\}$ is null, by the Squeeze Rule.

Here we take $\lambda = \frac{1}{a}$ in the Multiple Rule.

(c) To prove that $\{n^p c^n\}$ is null, for $p > 0$ and $|c| < 1$, we may again assume that $0 < c < 1$. Hence

$$c = \frac{1}{1+a}, \quad \text{for some number } a > 0.$$

First, we deal with the case $p = 1$. By the Binomial Theorem

$$(1 + a)^n \geq 1 + na + \frac{1}{2}n(n - 1)a^2$$

$$\geq \frac{1}{2}n(n - 1)a^2, \quad \text{for } n = 2, 3, \ldots,$$

and hence

$$nc^n = \frac{n}{(1 + a)^n}$$

$$\leq \frac{n}{\frac{1}{2}n(n - 1)a^2} = \frac{2}{(n - 1)a^2}, \quad \text{for } n = 2, 3, \ldots.$$

Since $\left\{\frac{1}{n-1}\right\}_2^\infty$ is null, we deduce that $\left\{\frac{2}{(n-1)a^2}\right\}_2^\infty$ is also null, by the Multiple Rule. Hence $\{nc^n\}$ is null, by the Squeeze Rule. This proves part (c) in the case $p = 1$.

To deduce that $\{n^p c^n\}$ is null for any $p > 0$ and $0 < c < 1$, we note that

$$n^p c^n = (nd^n)^p, \quad \text{for } n = 1, 2, \ldots,$$

where $d = c^{\frac{1}{p}}$. Since $0 < d < 1$, we know that $\{nd^n\}$ is null, and so $\{n^p c^n\}$ is null, by the Power Rule.

Here we use the special case of part (c) that we have already proved,

(d) To prove that $\left\{\frac{c^n}{n!}\right\}$ is null for any real c, we may again assume that $c > 0$. If we choose any integer m such that $m + 1 > c$, then we have, for $n > m + 1$, that

$$\frac{c^n}{n!} = \left(\frac{c}{1}\right)\left(\frac{c}{2}\right)\cdots\left(\frac{c}{m}\right) \times \left(\frac{c}{m+1}\right)\cdots\left(\frac{c}{n-1}\right)\left(\frac{c}{n}\right)$$

$$\leq \left(\frac{c}{1}\right)\left(\frac{c}{2}\right)\cdots\left(\frac{c}{m}\right) \times \frac{c}{n}$$

$$= K \times \frac{c}{n},$$

where $K = \frac{c^m}{m!}$ is a constant.

Since $\left\{\frac{1}{n}\right\}$ is null, we deduce that $\left\{\frac{Kc}{n}\right\}$ is also null, by the Multiple Rule; it follows that $\left\{\frac{c^n}{n!}\right\}$ is also null, by the Squeeze Rule.

For c is a constant and m is some constant, and hence $\frac{c^m}{m!}$ is also some constant. Note that what is varying is n, nothing else.

(e) To prove that $\left\{\frac{n^p}{n!}\right\}$ is null, for $p > 0$, we write

$$\frac{n^p}{n!} = \left(\frac{n^p}{2^n}\right) \times \left(\frac{2^n}{n!}\right), \quad \text{for } n = 1, 2, \ldots.$$

Since $\left\{\frac{n^p}{2^n}\right\}$ and $\left\{\frac{2^n}{n!}\right\}$ are both null sequences, by parts (c) and (d) respectively, we deduce that $\left\{\frac{n^p}{n!}\right\}$ is null, by the Product Rule. ∎

2.3 Convergent sequences

2.3.1 What is a convergent sequence?

In the previous section we looked at *null sequences*; that is, sequences which converge to 0. We now turn our attention to sequences which converge to limits other than 0.

Problem 1 Consider the sequence $a_n = \frac{n+1}{n}$, $n = 1, 2, \ldots$.

(a) Draw the sequence diagram for $\{a_n\}$, and describe (informally) how this sequence behaves.

(b) What can you say (formally) about the behaviour of the sequence

$$b_n = a_n - 1,\ n = 1, 2, \ldots?$$

The terms of the sequence $\{a_n\}$ in Problem 1 appear to get arbitrarily close to 1; that is, the sequence $\{a_n\}$ appears to *converge to* 1. If we subtract 1 from each term a_n to form the sequence $\{b_n\}$, then we obtain a null sequence. This example suggests the following definition of a *convergent sequence*:

Definition The sequence $\{a_n\}$ is **convergent** with **limit** ℓ, or **converges** to the **limit** ℓ, if $\{a_n - \ell\}$ is a null sequence. In this case, we say that $\{a_n\}$ **converges to** ℓ, and we write:

EITHER $\quad \lim\limits_{n \to \infty} a_n = \ell$,

OR $\quad a_n \to \ell$ as $n \to \infty$.

These statements are read as: 'the limit of a_n, as n tends to infinity, is ℓ' and 'a_n tends to ℓ, as n tends to infinity'. Often, we omit 'as $n \to \infty$'. Do not let this use of the *symbol* ∞ tempt you to think that ∞ is a real number.

The following are examples of convergent sequences:

every null sequence converges to 0;

every constant sequence $\{c\}$ converges to c;

the sequence $\left\{\frac{n+1}{n}\right\}$ is convergent with limit 1.

See Problem 1 above.

Problem 2 For each of the following sequences $\{a_n\}$, draw its sequence diagram and show that $\{a_n\}$ converges to ℓ by considering $a_n - \ell$:

(a) $a_n = \frac{n^2-1}{n^2+1}$, $\ell = 1$; (b) $a_n = \frac{n^3+(-1)^n}{2n^3}$, $\ell = \frac{1}{2}$.

The definition of convergence of a sequence is often given in the following equivalent form:

Equivalent definition The sequence $\{a_n\}$ is **convergent** with **limit** ℓ if:

for each positive number ε, there is a number X such that

$$|a_n - \ell| < \varepsilon, \quad \text{for all } n > X. \tag{1}$$

Note that here X *need not* be an integer; any appropriate real number will serve.

Remarks

1. In terms of the sequence diagram for $\{a_n\}$, this definition states that, for each positive number ε, the terms a_n *eventually* lie *inside* the horizontal strip from $\ell - \varepsilon$ up to $\ell + \varepsilon$.

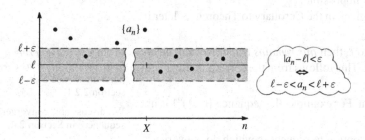

2. Just as for null sequences, we can interpret '$a_n \to \ell$ as $n \to \infty$' as an $\varepsilon - X$ game in which player A chooses a positive number ε and challenges player B to find some number X such that (1) holds.

   For example, consider the sequence $\{a_n\}$ where $a_n = \frac{8n+6}{2n}$, $n = 1, 2, \ldots$. This sequence is convergent, with limit 4. For

   $$|a_n - 4| = \left| \frac{8n+6}{2n} - 4 \right| = \frac{6}{2n} = \frac{3}{n};$$

   so that, if we make the choice $X = \frac{3}{\varepsilon}$, it is certainly true that

   $$|a_n - 4| < \varepsilon, \quad \text{for all } n > X.$$

   In terms of the $\varepsilon - X$ game, this simply means that whatever choice of ε is made by player A, player B can always win by making the choice $X = \frac{3}{\varepsilon}$!

   This is the case because, if

   $$n > \frac{3}{\varepsilon} \, [= X],$$

   then we have $\varepsilon > \frac{3}{n}$ or $\frac{3}{n} < \varepsilon$; hence, for $n > X$, we have

   $$|a_n - 4| = \frac{3}{n} < \frac{3}{X} = \varepsilon.$$

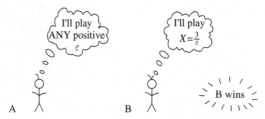

A   B   B wins

3. If a sequence is convergent, then it has a unique limit. We can see this by using a sequence diagram. Suppose that the sequence $\{a_n\}$ has two limits, ℓ and m, where $\ell \neq m$. Then it is possible to choose a positive number ε such that horizontal strips from $\ell - \varepsilon$ up to $\ell + \varepsilon$ and from $m - \varepsilon$ up to $m + \varepsilon$ do not overlap. For example, we can take $\varepsilon = \frac{1}{3}|\ell - m|$.

   The vertical distance between ℓ and m is $|\ell - m|$, so that any choice of ε that is *less than half* of this quantity will serve.

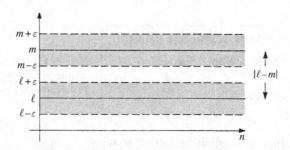

However, since ℓ and m are both limits of $\{a_n\}$, the terms a_n must eventually lie inside *both* strips, and this is impossible.

   A formal proof of this fact is given in the Corollary to Theorem 3, later in this section.

4. If a given sequence converges to ℓ, then *this remains true if we add, delete or alter a finite number of terms*. This follows from the corresponding result for null sequences.

   In other words 'altering a finite number of terms does not matter'. See Sub-section 2.2.1.

5. Not all sequences are convergent. For example, the sequence $\{(-1)^n\}$ is not convergent.

   We discuss non-convergent sequences in Section 2.4.

   In this section we restrict our attention to sequences which do converge.

2.3.2   Combination Rules for convergent sequences

So far you have tested the convergence of a given sequence $\{a_n\}$ by calculating $a_n - \ell$ and showing that $\{a_n - \ell\}$ is null. This presupposes that you know in advance the value of ℓ. Usually, however, you are not given the value of ℓ. You are given only a sequence $\{a_n\}$ and asked to decide whether or not it converges and, if it does, to find its limit. Fortunately many sequences can be dealt with by using the following Combination Rules, which extend the Combination Rules for null sequences:

Sub-section 2.2.1.

Theorem 1   Combination Rules

If $\lim\limits_{n\to\infty} a_n = \ell$ and $\lim\limits_{n\to\infty} b_n = m$, then:

Sum Rule $\lim\limits_{n\to\infty} (a_n + b_n) = \ell + m$;

Multiple Rule $\lim\limits_{n\to\infty} (\lambda a_n) = \lambda \ell$,   for any real number λ;

Product Rule $\lim\limits_{n\to\infty} (a_n b_n) = \ell m$;

Quotient Rule

$$\lim\limits_{n\to\infty} \left(\frac{a_n}{b_n} \right) = \frac{\ell}{m}, \quad \text{provided that } m \neq 0.$$

Remarks

1. In applications of the Quotient Rule, it may happen that some of the terms b_n take the value 0, in which case $\frac{a_n}{b_n}$ is not defined. However, we shall see (in Lemma 1) that this can happen only for *finitely many* b_n (because $m \neq 0$), and so $\{b_n\}$ is eventually non-zero. Thus the statement of the rule does make sense.

A finite number of terms do not matter.

2. The following rule is a special case of the Quotient Rule:

Corollary 1   If $\lim\limits_{n\to\infty} a_n = \ell$ and $\ell \neq 0$, then:

Reciprocal Rule $\lim\limits_{n\to\infty} \dfrac{1}{a_n} = \dfrac{1}{\ell}$.

We shall prove the Combination Rules at the end of this sub-section, but first we illustrate how to apply them.

Applying the Combination Rules

Example 1   Show that each of the following sequences $\{a_n\}$ is convergent, and find its limit:

(a) $a_n = \frac{(2n+1)(n+2)}{3n^2 + 3n}$;   (b) $a_n = \frac{2n^2 + 10^n}{n! + 3n^3}$.

Solution   Although the expressions for a_n are quotients, we cannot apply the Quotient Rule immediately, because the sequences defined by the numerators and denominators are not convergent. In each case, however, we can rearrange the expressions for a_n and then apply the Combination Rules.

(a) In this case we divide both the numerator and denominator by n^2 to give

$$a_n = \frac{(2n+1)(n+2)}{3n^2 + 3n} = \frac{\left(2 + \frac{1}{n}\right)\left(1 + \frac{2}{n}\right)}{3 + \frac{3}{n}}.$$

Since $\{\frac{1}{n}\}$ is a basic null sequence, we find by the Combination Rules that

$$\lim_{n \to \infty} a_n = \frac{(2+0)(1+0)}{3+0} = \frac{2}{3}.$$

(b) This time we divide both the numerator and denominator by $n!$ to give

$$a_n = \frac{2n^2 + 10^n}{n! + 3n^3} = \frac{\frac{2n^2}{n!} + \frac{10^n}{n!}}{1 + \frac{3n^3}{n!}}$$

Since $\{\frac{n^2}{n!}\}$, $\{\frac{10^n}{n!}\}$ and $\{\frac{n^3}{n!}\}$ are all basic null sequences, we find by the Combination Rules that

$$\lim_{n \to \infty} a_n = \frac{0+0}{1+0} = 0. \qquad \square$$

See the list of basic null sequences (introduced in Sub-section 2.2.1) which is repeated in the margin below.

Remark

We simplified each of the above sequences by dividing both numerator and denominator by the *dominant term*. In part (a), we divided by n^2, which is the highest power of n in the expression. In part (b), the choice of dominant term was a little harder, but the choice of $n!$ ensured that the resulting quotients in the numerator and denominator were all typical forms of convergent sequences.

In choosing the dominant term the following ordering is often useful:

Domination Hierarchy

A factorial term $n!$ dominates a power term c^n for any $c \in \mathbb{R}$.
A power term c^n dominates a term n^p for $p > 0$, $c > 1$.

The above examples illustrate the following general strategy:

Strategy   To evaluate the limit of a complicated quotient:

1. Identify the dominant term, bearing in mind the basic null sequences.
2. Divide both the numerator and the denominator by the dominant term.
3. Apply the Combination Rules.

Basic null sequences:

$\left\{\frac{1}{n^p}\right\}$,   for $p > 0$;

$\{c^n\}$,   for $|c| < 1$;

$\{n^p c^n\}$,   for $p > 0, |c| < 1$;

$\left\{\frac{c^n}{n!}\right\}$,   for $c \in \mathbb{R}$;

$\left\{\frac{n^p}{n!}\right\}$,   for $p > 0$.

Problem 3   Show that each of the following sequences $\{a_n\}$ is convergent, and find its limit:

(a) $a_n = \frac{n^3 + 2n^2 + 3}{2n^3 + 1}$;   (b) $a_n = \frac{n^2 + 2^n}{3^n + n^3}$;   (c) $a_n = \frac{n! + (-1)^n}{2^n + 3n!}$.

Warning Notice that '$3n!$' means '$3 \times (n!)$' and not '$(3n)!$'.

Proofs of the Combination Rules   We prove the Sum Rule, the Multiple Rule and the Product Rule by using the corresponding Combination Rules for null sequences. Remember that

$$\lim_{n \to \infty} a_n = \ell$$

You may omit these proofs at a first reading.

means that

$$\{a_n \to \ell\} \text{ is a null sequence.}$$

> **Sum Rule**   If $\lim_{n\to\infty} a_n = \ell$ and $\lim_{n\to\infty} b_n = m$, then
> $$\lim_{n\to\infty} (a_n + b_n) = \ell + m.$$

Proof   By assumption, $\{a_n - \ell\}$ and $\{b_n - m\}$ are null sequences; since
$$(a_n + b_n) - (\ell + m) = (a_n - \ell) + (b_n - m)$$
we deduce that $\{(a_n + b_n) - (\ell + m)\}$ is null, by the Sum Rule for null sequences. ∎

> **Product Rule**   If $\lim_{n\to\infty} a_n = \ell$ and $\lim_{n\to\infty} b_n = m$, then
> $$\lim_{n\to\infty} (a_n b_n) = \ell m.$$

Proof   The idea here is to express $a_n b_n - \ell m$ in terms of $a_n - \ell$ and $b_n - m$
$$a_n b_n - \ell m = (a_n - \ell)(b_n - m) + m(a_n - \ell) + \ell(b_n - m).$$
Since $\{a_n - \ell\}$ and $\{b_n - m\}$ are null, we deduce that $\{a_n b_n - \ell m\}$ is null, by the Combination Rules for null sequences. ∎

Remark

Note that the Multiple Rule is just a special case of the Product Rule in which the sequence $\{b_n\}$ is a constant sequence.

To prove the Quotient Rule, we need to use the following lemma, which will also be needed in the next sub-section:

> **Lemma 1**   If $\lim_{n\to\infty} a_n = \ell$ and $\ell > 0$, then there is a number X such that
> $$a_n > \frac{1}{2}\ell, \quad \text{for all } n > X.$$

Proof   Since $\frac{1}{2}\ell > 0$, the terms a_n must eventually lie within a distance $\frac{1}{2}\ell$ of the limit ℓ.

Here we are taking $\varepsilon = \frac{1}{2}\ell$ in the definition of convergence.

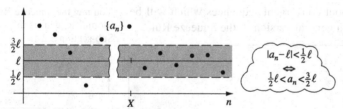

In other words, there is a number X such that
$$|a_n - \ell| < \frac{1}{2}\ell, \quad \text{for all } n > X.$$
Hence
$$-\frac{1}{2}\ell < a_n - \ell < \frac{1}{2}\ell, \quad \text{for all } n > X,$$
and so the left-hand inequality gives
$$\frac{1}{2}\ell < a_n, \quad \text{for all } n > X,$$
as required. ∎

Quotient Rule If $\lim\limits_{n\to\infty} a_n = \ell$ and $\lim\limits_{n\to\infty} b_n = m$, then

$$\lim_{n\to\infty}\left(\frac{a_n}{b_n}\right) = \frac{\ell}{m}, \text{ provided that } m \neq 0.$$

Proof We assume that $m > 0$; the proof for the case $m < 0$ is similar. Once again the idea is to write the required expression in terms of $a_n - \ell$ and $b_n - m$

$$\frac{a_n}{b_n} - \frac{\ell}{m} = \frac{m(a_n - \ell) - \ell(b_n - m)}{b_n m}.$$

Now, however, there is a slight problem: $\{m(a_n - \ell) - \ell(b_n - m)\}$ is certainly a null sequence, but the denominator is rather awkward. Some of the terms b_n may take the value 0, in which case the expression is undefined.

However, by Lemma 1, we know that that for some X we have

$$b_n > \frac{1}{2}m, \quad \text{for all } n > X.$$

In particular, this implies that the terms of $\{b_n\}$ are eventually positive.

Thus, for all $n > X$

$$\left|\frac{a_n}{b_n} - \frac{\ell}{m}\right| = \frac{|m(a_n - \ell) - \ell(b_n - m)|}{b_n m}$$

$$\leq \frac{|m(a_n - \ell) - \ell(b_n - m)|}{\frac{1}{2}m^2}$$

$$\leq \frac{|m| \times |a_n - \ell| + |\ell| \times |b_n - m|}{\frac{1}{2}m^2}.$$

Here we use Lemma 1.

Here we apply the Triangle Inequality to the numerator.

Since this last expression defines a null sequence, it follows by the Squeeze Rule that $\left\{\frac{a_n}{b_n} - \frac{\ell}{m}\right\}$ is null. ∎

2.3.3 Further properties of convergent sequences

There are several other theorems about convergent sequences which will be needed in later chapters. The first is a general version of the Squeeze Rule.

You met the Squeeze Rule for null sequences in Sub-section 2.2.1.

Theorem 2   Squeeze Rule

If:

1. $b_n \leq a_n \leq c_n$, for $n = 1, 2, \ldots$,
2. $\lim\limits_{n\to\infty} b_n = \lim\limits_{n\to\infty} c_n = \ell$,

then $\lim\limits_{n\to\infty} a_n = \ell$.

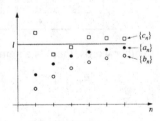

Proof By the Combination Rules

$$\lim_{n\to\infty}(c_n - b_n) = \ell - \ell = 0,$$

so that $\{c_n - b_n\}$ is a null sequence. Also, by condition 1

$$0 \leq a_n - b_n \leq c_n - b_n, \quad \text{for } n = 1, 2, \ldots,$$

and so $\{a_n - b_n\}$ is null, by the Squeeze Rule for null sequences.

That is, $\{a_n - b_n\}$ is *squeezed* by $\{c_n - b_n\}$.

Now we write a_n in the form

$$a_n = (a_n - b_n) + b_n.$$

Hence by the Combination Rules

$$\lim_{n \to \infty} a_n = \lim_{n \to \infty} (a_n - b_n) + \lim_{n \to \infty} b_n$$
$$= 0 + \ell = \ell.$$ ∎

Remark

In applications of the Squeeze Rule, it is sufficient to check that condition 1 applies *eventually*. This is because the values of a *finite* number of terms do not affect convergence.

A finite number of terms do not matter.

The following example and problem illustrate the use of the Squeeze Rule and the Binomial Theorem in the derivation of two important limits.

Example 2

(a) Prove that, if $c > 0$, then

$$(1 + c)^{\frac{1}{n}} \leq 1 + \frac{c}{n}, \quad \text{for } n = 1, 2, \dots.$$

(b) Deduce that

$$\lim_{n \to \infty} a^{\frac{1}{n}} = 1, \text{ for any positive number } a.$$

We proved this inequality for the case $c = 1$ in Example 6(b) of Sub-section 1.3.3.

Solution

(a) Using the rules for inequalities, we obtain

$$(1 + c)^{\frac{1}{n}} \leq 1 + \frac{c}{n} \Leftrightarrow 1 + c \leq \left(1 + \frac{c}{n}\right)^n, \quad \text{since } c > 0.$$

The right-hand inequality holds, because

$$\left(1 + \frac{c}{n}\right)^n \geq 1 + n\left(\frac{c}{n}\right) = 1 + c,$$

by the Binomial Theorem. It follows that the required inequality also holds.

(b) We consider the cases $a > 1$, $a = 1$, $a < 1$ separately.

If $a > 1$, then we can write $a = 1 + c$, where $c > 0$. Then, by part (a)

$$1 \leq a^{\frac{1}{n}} = (1 + c)^{\frac{1}{n}} \leq 1 + \frac{c}{n}, \quad \text{for } n = 1, 2, \dots.$$

All the terms in the Binomial Theorem expansion of $(1 + \frac{c}{n})^n$ are positive, so we get a smaller number if we ignore all the terms after the first two.

Since $\lim_{n \to \infty} \left(1 + \frac{c}{n}\right) = 1$, we deduce that

$$\lim_{n \to \infty} a^{\frac{1}{n}} = 1,$$

by the Squeeze Rule.

If $a = 1$, then $a^{\frac{1}{n}} = 1$, for $n = 1, 2, \dots$, so

$$\lim_{n \to \infty} a^{\frac{1}{n}} = 1.$$

In this application of the Squeeze Rule, the 'lower' sequence is $\{1\}$.

If $0 < a < 1$, then $\frac{1}{a} > 1$, so that $\lim_{n \to \infty} \left(\frac{1}{a}\right)^{\frac{1}{n}} = 1$ by the first case in part (b). Hence, by the Reciprocal Rule

$$\lim_{n \to \infty} a^{\frac{1}{n}} = \frac{1}{\lim_{n \to \infty} \left(\frac{1}{a}\right)^{\frac{1}{n}}} = \frac{1}{1} = 1.$$ □

Problem 4

(a) Use the Binomial Theorem to prove that

$$n^{\frac{1}{n}} \leq 1 + \sqrt{\frac{2}{n-1}}, \quad \text{for } n = 2, 3, \ldots.$$

Hint: $(1+x)^n \geq \frac{n(n-1)}{2!}x^2, \quad \text{for } n \geq 2, x \geq 0.$

(b) Use the Squeeze Rule to deduce that

$$\lim_{n \to \infty} n^{\frac{1}{n}} = 1.$$

Next we show that taking limits preserves weak inequalities.

Theorem 3   Limit Inequality Rule

If $\lim_{n \to \infty} a_n = \ell$ and $\lim_{n \to \infty} b_n = m$, and also

$$a_n \leq b_n, \quad \text{for } n = 1, 2, \ldots,$$

then $\ell \leq m$.

Proof   Suppose that $a_n \to \ell$ and $b_n \to m$, where $a_n \leq b_n$ for $n = 1, 2, \ldots$, but that it is not true that $\ell \leq m$. Then $\ell > m$ and so, by the Combination Rules

$$\lim_{n \to \infty}(a_n - b_n) = \ell - m > 0.$$

This is a 'proof by contradiction'.

Hence, by Lemma 1, there is an X such that

$$a_n - b_n > \frac{1}{2}(\ell - m), \quad \text{for all } n > X. \qquad (2)$$

However, we assumed that $a_n - b_n \leq 0$, for $n = 1, 2, \ldots$, so statement (2) is a contradiction.

Hence it *is* true that $\ell \leq m$.   ∎

Warning   Taking limits does NOT preserve strict inequalities. That is, if $\lim_{n \to \infty} a_n = \ell$ and $\lim_{n \to \infty} b_n = m$, and also $a_n < b_n$, for $n = 1, 2, \ldots$, then it follows from Theorem 3 that $\ell \leq m$ – but it does NOT necessarily follow that $\ell < m$. For example, $\frac{1}{2n} < \frac{1}{n}$, for $n = 1, 2, \ldots$, but $\lim_{n \to \infty} \frac{1}{2n} = \lim_{n \to \infty} \frac{1}{n} = 0$.

We can now give the formal proof, promised earlier, that a convergent sequence has only one limit.

Corollary 2   If $\lim_{n \to \infty} a_n = \ell$ and $\lim_{n \to \infty} a_n = m$, then $\ell = m$.

Proof   Applying Theorem 3 with $b_n = a_n$, we deduce that $\ell \leq m$ and $m \leq \ell$. Hence $\ell = m$.   ∎

In Sub-section 2.2.1, we saw that a sequence $\{a_n\}$ is null if and only if the sequence $\{|a_n|\}$ is null. Our next result is a partial generalisation of this fact.

Theorem 4   If $\lim_{n \to \infty} a_n = \ell$, then $\lim_{n \to \infty} |a_n| = |\ell|$.

Proof   Using the 'reverse form' of the Triangle Inequality, we obtain

$$\big||a_n| - |\ell|\big| \leq |a_n - \ell|, \quad \text{for } n = 1, 2, \ldots.$$

Since $\{a_n - \ell\}$ is null, we deduce from the Squeeze Rule for null sequences that $\{|a_n| - |\ell|\}$ is null, as required. ∎

Remark

Theorem 4 is only a partial generalisation of the earlier result about null sequences, because the converse statement does not hold. That is, if $|a_n| \to |\ell|$, it does NOT necessarily follow that $a_n \to \ell$. For example, consider the sequence $a_n = (-1)^n$, for $n = 1, 2, \ldots$; in this case, $|a_n| \to 1$, but $\{a_n\}$ does not even converge.

You met this in Sub-section 1.3.1, expressed in the form $|a - b| \geq \big||a| - |b|\big|$. Here we are writing a_n in place of a and ℓ in place of b.

2.4   Divergent sequences

2.4.1   What is a divergent sequence?

We have commented several times that not all sequences are convergent. We now investigate the behaviour of sequences which do not converge.

> **Definition**   A sequence is **divergent** if it is not convergent.

Here are the sequence diagrams for $\{(-1)^n\}$, $\{2n\}$ and $\{(-1)^n n\}$. Each of these sequences is divergent but, as you can see, they behave differently.

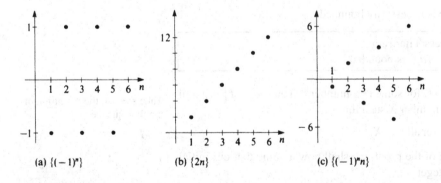

(a) $\{(-1)^n\}$ (b) $\{2n\}$ (c) $\{(-1)^n n\}$

It is rather tricky to prove, directly from the definition, that these sequences are divergent.

The aim of this section is to obtain criteria for divergence, which avoid having to argue directly from the definition. At the end of the section, we give a strategy involving two criteria which deal with all types of divergence. We obtain these criteria by establishing certain properties, which are necessarily possessed by a *convergent* sequence; if a sequence does *not* have these properties, then it must be *divergent*.

For example, to show that the sequence $\{(-1)^n\}$ is divergent, we have to show that $\{(-1)^n\}$ is not convergent; that is, for every real number ℓ, the sequence $\{(-1)^n - \ell\}$ is not null.

2.4.2 Bounded and unbounded sequences

One property possessed by a convergent sequence is that it must be *bounded*.

Definition   A sequence $\{a_n\}$ is **bounded** if there is a number K such that

$$|a_n| \leq K, \quad \text{for } n = 1, 2, \ldots.$$

A sequence is **unbounded** if it is not bounded.

Thus a sequence $\{a_n\}$ is bounded if *all* the terms a_n lie on the sequence diagram in a horizontal strip from $-K$ up to K, for some positive number K.

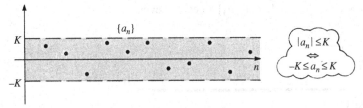

For example, the sequence $\{(-1)^n\}$ is bounded, because

$$|(-1)^n| \leq 1, \quad \text{for } n = 1, 2, \ldots.$$

However the sequences $\{2n\}$ and $\{n^2\}$ are unbounded, since, for each number K, we can find terms of these sequences whose absolute values are greater than K.

> **Problem 1**   Classify the following sequences as *bounded* or *unbounded*:
> (a) $\{1+(-1)^n\}$;   (b) $\{(-1)^n n\}$;   (c) $\left\{\frac{2n+1}{n}\right\}$;   (d) $\left\{\left(1-\frac{1}{n}\right)^n\right\}$.

The sequence $\{(-1)^n\}$ shows that:

> a bounded sequence is not necessarily convergent.

However we can prove that:

> a convergent sequence is necessarily bounded.

Theorem 1   Boundedness Theorem
If $\{a_n\}$ is convergent, then $\{a_n\}$ is bounded.

Proof   We know that $a_n \to \ell$, for some real number ℓ. Thus $\{a_n - \ell\}$ is a null sequence, and so there is a number X such that

$$|a_n - \ell| < 1, \quad \text{for all } n > X.$$

Take $\varepsilon = 1$ in the definition of a null sequence.

For simplicity in the rest of the proof, we shall now assume that our initial choice of X is a positive integer.

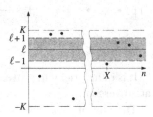

Now

$$|a_n| = |(a_n - \ell) + \ell|$$
$$\leq |a_n - \ell| + |\ell|, \quad \text{by the Triangle Inquality.}$$

It follows that

$$|a_n| \leq 1 + |\ell|, \quad \text{for all } n > X.$$

This is the type of inequality needed to prove that $\{a_n\}$ is bounded, but it does not include the terms $a_1, a_2, \ldots, a_X$. To complete the proof, we let K be the maximum of the numbers $|a_1|, |a_2|, \ldots, |a_X|, 1 + |\ell|$.

That is, $K = \max\{|a_1|, |a_2|, \ldots, |a_X|, 1 + |\ell|\}$.

It follows that
$$|a_n| \leq K, \quad \text{for } n = 1, 2, \ldots,$$
as required.   ■

From Theorem 1, we obtain the following test for the *divergence* of a sequence:

Corollary 1   If $\{a_n\}$ is unbounded, then $\{a_n\}$ is divergent.

For example, the sequences $\{2n\}$ and $\{(-1)^n n\}$ are both unbounded, so they are both divergent.

> **Problem 2**   Classify the following sequences as *convergent* or *divergent*, and as *bounded* or *unbounded*:
>
> (a) $\{\sqrt{n}\}$;   (b) $\left\{\frac{n^2 + n}{n^2 + 1}\right\}$;   (c) $\{(-1)^n n^2\}$;   (d) $\{n^{(-1)^n}\}$.

2.4.3   Sequences which tend to infinity

Although the sequences $\{2n\}$ and $\{(-1)^n n\}$ are both unbounded (and hence divergent), there is a marked difference in their behaviour. The terms of both sequences become arbitrarily large, but those of the sequence $\{2n\}$ become arbitrarily large and positive. To make this precise, we must explain what we mean by 'arbitrarily large and positive'.

Definition   The sequence $\{a_n\}$ **tends to infinity** if:

for each positive number K, there is a number X such that
$$a_n > K, \quad \text{for all } n > X.$$

In this case, we write
$$a_n \to \infty \text{ as } n \to \infty.$$ Often, we omit 'as $n \to \infty$'.

Remarks

1. In terms of the sequence diagram for $\{a_n\}$, this definition states that, for each positive number K, the terms a_n *eventually* lie *above* the line at height K.

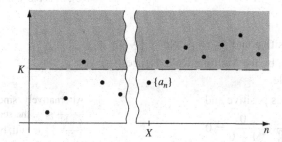

2. If a sequence tends to infinity, then it is unbounded – and hence divergent, by Corollary 1.

3. If a sequence tends to infinity, then this remains true *if we add, delete or alter a finite number of terms*.

A finite number of terms do not matter.

4. In the definition we can replace the phrase 'for each positive number K' by 'for each number K'; for, if the inequality '$a_n > K$, for all $n > X$' holds for each number K, then in particular it certainly holds for each *positive* number K.

There is a version of the Reciprocal Rule for sequences which tend to infinity. This enables us to use our knowledge of null sequences to identify sequences which tend to infinity.

Theorem 2   Reciprocal Rule

(a) If the sequence $\{a_n\}$ satisfies both of the following conditions:

   1. $\{a_n\}$ is eventually positive,

   2. $\{\frac{1}{a_n}\}$ is a null sequence,

   then $a_n \to \infty$.

(b) If $a_n \to \infty$, then $\frac{1}{a_n} \to 0$.

Proof of part (a)   To prove that $a_n \to \infty$, we have to show that:

for each positive number K, there is a number X such that
$$a_n > K, \quad \text{for all } n > X. \tag{1}$$

We prove only part (a): the proof of part (b) is similar.

Since $\{a_n\}$ is eventually positive, we can choose a number X_1 such that
$$a_n > 0, \quad \text{for all } n > X_1.$$

Since $\{\frac{1}{a_n}\}$ is null, we can choose a number X_2 such that
$$\left| \frac{1}{a_n} \right| < \frac{1}{K}, \quad \text{for all } n > X_2.$$

Here we are taking $\varepsilon = \frac{1}{K}$ in the definition of a null sequence.

Now, let $X = \max \{X_1 X_2\}$; then
$$0 < \frac{1}{a_n} < \frac{1}{K}, \quad \text{for all } n > X.$$

We make this choice of X so that BOTH of the preceding inequalities hold for all $n > X$.

This statement is equivalent to the statement (1), so $a_n \to \infty$, as required. ∎

Example 1   Use the Reciprocal Rule to prove that the following sequences tend to infinity:

(a) $\left\{ \frac{n^3}{2} \right\}$;   (b) $\{n! + 10^n\}$;   (c) $\{n! - 10^n\}$

Notice that, in parts (b) and (c), $n!$ is the dominant term.

Solution

(a) Each term of the sequence $\left\{ \frac{n^3}{2} \right\}$ is positive and $\frac{1}{n^3/2} = \frac{2}{n^3}$. Now, $\left\{\frac{1}{n^3}\right\}$ is a basic null sequence and so $\left\{\frac{2}{n^3}\right\}$ is null, by the Multiple Rule. Hence $\left\{\frac{n^3}{2}\right\}$ tends to infinity, by the Reciprocal Rule.

(b) Each term of the sequence $\{n! + 10^n\}$ is positive and
$$\lim_{n \to \infty} \frac{1}{n! + 10^n} = \lim_{n \to \infty} \frac{\frac{1}{n!}}{1 + \frac{10^n}{n!}} = \frac{0}{1 + 0} = 0,$$

Alternatively, since $\frac{1}{n! + 10^n} \leq \frac{1}{n!}$, the sequence $\left\{\frac{1}{n! + 10^n}\right\}$ is null, by the Squeeze Rule for null sequences.

by the Combination Rules.

Hence $\{n! + 10^n\}$ tends to infinity, by the Reciprocal Rule.

(c) First note that

$$n! - 10^n = n!\left(1 - \frac{10^n}{n!}\right), \quad \text{for } n = 1, 2, \ldots.$$

Since $\left\{\frac{10^n}{n!}\right\}$ is a basic null sequence, we know that $\frac{10^n}{n!}$ is eventually less than 1, and so $n! - 10^n$ is eventually positive. Also

$$\lim_{n \to \infty} \frac{1}{n! - 10^n} = \lim_{n \to \infty} \frac{\frac{1}{n!}}{1 - \frac{10^n}{n!}} = \frac{0}{1 - 0} = 0,$$

by the Combination Rules.

Hence $\{n! - 10^n\}$ tends to infinity, by the Reciprocal Rule. $\square$

This follows by taking $\varepsilon = 1$ in the definition of a null sequence.

There are also versions of the Combination Rules and Squeeze Rule for sequences which tend to infinity. We state these without proof.

Theorem 3   Combination Rules

If $\{a_n\}$ tends to infinity and $\{b_n\}$ tends to infinity, then:

Sum Rule $\{a_n + b_n\}$ tends to infinity;

Multiple Rule $\{\lambda a_n\}$ tends to infinity, for $\lambda > 0$;

Product Rule $\{a_n b_n\}$ tends to infinity.

Theorem 4   Squeeze Rule

If $\{b_n\}$ tends to infinity, and

$$a_n \geq b_n, \quad \text{for } n = 1, 2, \ldots,$$

then $\{a_n\}$ tends to infinity.

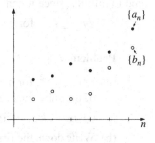

Problem 3   For each of the following sequences $\{a_n\}$, prove that $a_n \to \infty$:

(a) $\left\{\frac{2^n}{n}\right\}$;   (b) $\left\{2^n - n^{100}\right\}$;   (c) $\left\{\frac{2^n}{n} + 5n^{100}\right\}$;   (d) $\left\{\frac{2^n + n^2}{n^{10} + n}\right\}$.

We can also define $\{a_n\}$ *tends to minus infinity*.

Definition   The sequence $\{a_n\}$ **tends to minus infinity** if

$$-a_n \to \infty \quad \text{as } n \to \infty.$$

In this case, we write

$$a_n \to -\infty \quad \text{as } n \to \infty.$$

For example, the sequences $\{-n^2\}$ and $\{10^n - n!\}$ both tend to minus infinity, because $\{n^2\}$ and $\{n! - 10^n\}$ tend to infinity. Sequences which tend to minus infinity are unbounded, and hence divergent. However, the sequence $\{(-1)^n n\}$ shows that an unbounded sequence need not tend to infinity or to minus infinity.

This follows from the fact that sequences which tend to infinity are unbounded.

2.4.4 Subsequences

We now give two useful criteria for establishing that a sequence diverges; both of them involve the idea of a *subsequence*. For example, consider the bounded divergent sequence $\{(-1)^n\}$. This sequence splits naturally into two:

the *even* terms $a_2, a_4, \ldots, a_{2k}, \ldots$, each of which equals 1;

the *odd* terms $a_1, a_3, \ldots, a_{2k-1}, \ldots$, each of which equals -1.

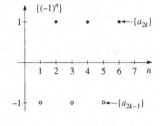

Both of these are sequences in their own right, and we call them the **even subsequence** $\{a_{2k}\}$ and the **odd subsequence** $\{a_{2k-1}\}$.

In general, given a sequence $\{a_n\}$ we may consider many different subsequences, such as:

$\{a_{3k}\}$, comprising the terms $a_3, a_6, a_9, \ldots$;

$\{a_{4k+1}\}$, comprising the terms $a_5, a_9, a_{13}, \ldots$;

$\{a_{2k!}\}$, comprising the terms $a_2, a_4, a_{12}, \ldots$.

Definition The sequence $\{a_{n_k}\}$ is a **subsequence** of the sequence $\{a_n\}$ if $\{n_k\}$ is a strictly increasing sequence of positive integers; that is, if

$$n_1 < n_2 < n_3 < \ldots.$$

The sequence $\{a_{n_k}\}$ is the sequence

$$a_{n_1}, a_{n_2}, a_{n_3}, \ldots.$$

Note that the sequence $\{a_n\}$ is a subsequence of itself.

For example, the subsequence $\{a_{5k+2}\}$ corresponds to the sequence of positive integers

$$n_k = 5k + 2, \quad k = 1, 2, \ldots.$$

The first term of $\{a_{5k+2}\}$ is a_7, the second is a_{12}, the third is a_{17}, and so on.

Notice that any strictly increasing sequence $\{n_k\}$ of positive integers must tend to infinity, since it can be proved by Mathematical Induction that

$$n_k \geq k, \quad \text{for } k = 1, 2, \ldots.$$

Problem 4

(a) Let $a_n = n^2$, for $n = 1, 2, \ldots$. Write down the first five terms of each of the subsequences $\{a_{n_k}\}$, where:

(i) $n_k = 2k$;   (ii) $n_k = 4k - 1$;   (iii) $n_k = k^2$.

(b) Write down the first three terms of the odd and even subsequences of the following sequence: $a_n = n^{(-1)^n}$, $n = 1, 2, \ldots$.

Our next theorem shows that certain properties of sequences are inherited by their subsequences.

Theorem 5   Inheritance Property of Subsequences

For any subsequence $\{a_{n_k}\}$ of $\{a_n\}$:

(a) if $a_n \to \ell$ as $n \to \infty$, then $a_{n_k} \to \ell$ as $k \to \infty$;

(b) if $a_n \to \infty$ as $n \to \infty$, then $a_{n_k} \to \infty$ as $k \to \infty$.

A result similar to part (b) holds for sequences where $a_n \to -\infty$.

Proof of part (a)   We want to show that for each positive number ε, there is a number K such that

$$|a_{n_k} - \ell| < \varepsilon, \quad \text{for all } k > K. \tag{2}$$

However, since $\{a_n - \ell\}$ is null, we know that there is a number X such that

$$|a_n - \ell| < \varepsilon, \quad \text{for all } n > X.$$

You may omit this proof at a first reading. We prove only part (a): the proof of part (b) is similar.

For simplicity in the rest of the proof, we shall now assume that our initial choice of X is a positive integer.

So, if we take K so large that

$$n_K \geq X,$$

then

$$n_k > n_K \geq X, \quad \text{for all } k > K,$$

and so

$$|a_{n_k} - \ell| < \varepsilon, \quad \text{for all } k > K,$$

which proves (2). ∎

We will occasionally make this assumption if we want to refer to terms such as a_X rather than mess about with nasty expressions such as $a_{[X]}$, where $[X]$ is the integral part of X.

The following criteria for establishing that a sequence is *divergent* are immediate consequences of Theorem 5, part (a):

Corollary 2

1. **First Subsequence Rule** The sequence $\{a_n\}$ is divergent if it has two convergent subsequences with different limits.

2. **Second Subsequence Rule** The sequence $\{a_n\}$ is divergent if it has a subsequence which tends to infinity or a subsequence which tends to minus infinity.

Though we do not prove the fact, ANY divergent sequence is of one (or both) of these two types.

We can now formulate the strategy promised at the beginning of this section.

Strategy To prove that a sequence $\{a_n\}$ is divergent:

EITHER
1. show that $\{a_n\}$ has two convergent subsequences with different limits

OR
2. show that $\{a_n\}$ has a subsequence which tends to infinity or a subsequence which tends to minus infinity.

The fact that this strategy will always apply is simply a reformulation of the result mentioned in the above margin note.

For example, the sequence $\{(-1)^n\}$ has two convergent subsequences which have different limits: namely, the even subsequence with limit 1 and the odd subsequence with limit -1. So the sequence $\{(-1)^n\}$ is divergent, by the First Subsequence Rule.

On the other hand, the sequence $\{n^{(-1)^n}\}$ has a subsequence (the even subsequence) which tends to infinity. So $\{n^{(-1)^n}\}$ is divergent by the Second Subsequence Rule.

Remark

In order to apply the above strategy successfully to prove that a sequence is divergent, you need to be able to spot convergent subsequences with different limits or subsequences which tend to infinity or to minus infinity. It is not always easy to do this, and some experimentation may be required! If the formula for a_n involves the expression $(-1)^n$, it is a good idea to consider the odd and even subsequences, although this may not always work. It may be helpful to calculate the values of the first few terms in order to try to spot some subsequences whose behaviour can be identified.

Problem 5 Use the above strategy to prove that each of the following sequences $\{a_n\}$ is divergent:

(a) $\left\{(-1)^n + \frac{1}{n}\right\}$;   (b) $\left\{\frac{1}{3}n - \left[\frac{1}{3}n\right]\right\}$;   (c) $\left\{n \sin\left(\frac{1}{2}n\pi\right)\right\}$.

In part (b) the square brackets denote 'the integer part' function.

We end this section by giving a result about subsequences which will be needed in later chapters.

Theorem 6   If the odd and even subsequences of $\{a_n\}$ both tend to the *same* limit ℓ, then
$$\lim_{n \to \infty} a_n = \ell.$$

Proof   We want to show that:

for each positive number ε, there is a number X such that
$$|a_n - \ell| < \varepsilon, \quad \text{for all } n > X. \tag{3}$$

We know that there are integers K_1 and K_2 such that
$$|a_{2k-1} - \ell| < \varepsilon, \quad \text{for all } k > K_1,$$

$\{a_{2k-1}\}$ and $\{a_{2k}\}$ are the odd and even subsequences of $\{a_n\}$.

and
$$|a_{2k} - \ell| < \varepsilon, \quad \text{for all } k > K_2.$$

So we now let
$$X = \max\{2K_1 - 1, 2K_2\}.$$

With this choice of X, each $n > X$ is either of the form $2k - 1$, with $k > K_1$, or of the form $2k$, with $k > K_2$; it follows then that (3) holds with this value of X. ∎

2.5   The Monotone Convergence Theorem

2.5.1   Monotonic sequences

In Section 2.3, we gave various techniques for finding the limit of a convergent sequence. As a result, you may be under the impression that, if we know that a sequence converges, then there is some way of finding its limit explicitly. However, it is sometimes possible to prove that a sequence is convergent, even though we do not know its limit. For example, this situation occurs with a given sequence $\{a_n\}$, which has the following two properties:

1. $\{a_n\}$ is an *increasing* sequence;

2. $\{a_n\}$ is *bounded above*; that is, there is a real number M such that
$$a_n \leq M, \quad \text{for } n = 1, 2, \ldots.$$

Likewise, if $\{a_n\}$ is a sequence which is *decreasing* and *bounded below*, then $\{a_n\}$ must be convergent.

We combine these two results into one statement.

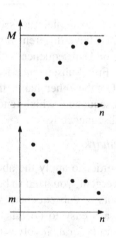

Theorem 1   Monotone Convergence Theorem

If the sequence $\{a_n\}$ is:

EITHER

1. increasing *and* 2. bounded above

OR

1. decreasing *and* 2. bounded below,

then $\{a_n\}$ is convergent.

Proof of the first case   Let $\{a_n\}$ be a sequence that is increasing and bounded above. Since $\{a_n\}$ is bounded above, the set $\{a_n\colon n=1,2,\ldots\}$ has a least upper bound, ℓ say; this is true by the Least Upper Bound Property of $\mathbb{R}$. We now prove that $\lim\limits_{n\to\infty} a_n = \ell$.

We want to show that:

for each positive number ε, there is a number X such that:
$$|a_n - \ell| < \varepsilon, \quad \text{for all } n > X. \tag{1}$$

We know that, if $\varepsilon > 0$, then, since ℓ is the least upper bound of the set $\{a_n\colon n=1,2,\ldots\}$, there is an integer X such that

$$a_X > \ell - \varepsilon.$$

Since $\{a_n\}$ is increasing, $a_n \ge a_X$, for $n > X$, and so

$$a_n > \ell - \varepsilon, \quad \text{for all } n > X.$$

Thus

$$|a_n - \ell| = \ell - a_n < \varepsilon, \quad \text{for all } n > X,$$

which proves (1).

Hence $\{a_n\}$ converges to ℓ.   ∎

The Monotone Convergence Theorem tells us that a sequence such as $\left\{1 - \frac{1}{n}\right\}$, which is increasing and bounded above (by 1, for example), must be convergent. In this case, of course, we know already that $\left\{1 - \frac{1}{n}\right\}$ is convergent (with limit 1) without using the Monotone Convergence Theorem.

The Monotone Convergence Theorem is often used when we suspect that a sequence is convergent, but cannot find the actual limit. It can also be used to give precise definitions of numbers, such as π, about which we have only an intuitive idea, and we do this later in this section.

For completeness, we point out that:

> If $\{a_n\}$ is increasing but is not bounded above, then $a_n \to \infty$ as $n \to \infty$.

Indeed, for any real number K, we can find an integer X such that

$$a_X > K,$$

because $\{a_n\}$ is not bounded above. Since $\{a_n\}$ is increasing, $a_n \ge a_X$ for $n > X$, and so

$$a_n > K, \quad \text{for all } n > X.$$

Hence $a_n \to \infty$ as $n \to \infty$.

Similarly, we have the following result:

> If $\{a_n\}$ is decreasing but is not bounded below, then $a_n \to -\infty$ as $n \to \infty$.

We summarise these results about monotonic sequences in the following useful theorem:

Theorem 2   Monotonic Sequence Theorem

If the sequence $\{a_n\}$ is monotonic, then:

EITHER $\{a_n\}$ is convergent

OR $a_n \to \pm\infty$.

Side notes:

We prove only the increasing version; the proof of the decreasing version is similar. You met this Property in Sub-section 1.4.3.

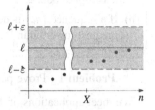

This follows from the definition of least upper bound.

$|a_n - \ell| = \ell - a_n$, because $a_n \le \ell$.

We will use the Monotone Convergence Theorem in precisely this way in Chapter 3. Sub-section 2.5.4.

The expression '$a_n \to \pm\infty$' is shorthand for 'either $a_n \to \infty$ or $a_n \to -\infty$'.

Next we note a consequence of the Monotone Convergence Theorem that is sometimes useful.

Corollary 1

(a) If a sequence $\{a_n\}$ is increasing and bounded above, then it tends to its least upper bound, sup $\{a_n : n = 1, 2, \ldots\}$.

(b) If a sequence $\{a_n\}$ is decreasing and bounded below, then it tends to its greatest lower bound, inf $\{a_n : n = 1, 2, \ldots\}$.

The proof of part (a) is contained in the proof of Theorem 1 above.

Problem 1   Prove part (b) of the Corollary.

This proof is similar to the earlier discussion.

We meet applications of these results in the rest of this section, to functions defined recursively and to the study of the numbers e and π. But our first application is to a result of considerable value in Analysis, Topology and other parts of Mathematics.

We shall use it in Section 4.2.

The Bolzano–Weierstrass Theorem

We have seen that if a sequence is convergent, then certainly it must be bounded. However if a sequence is bounded, it does not follow that it is necessarily convergent. For example, the sequence $\{a_n\}$, where $a_n = \frac{1}{n}$, is bounded (by 1, since $|a_n| = |\frac{1}{n}| \leq 1, 2, \ldots$) and is also convergent (to the limit 0). Yet the sequence $\{a_n\}$, where $a_n = (-1)^n$, is also bounded (since $|a_n| = |(-1)^n| = 1$, for $n = 1, 2, \ldots$) but is divergent (since its odd and even subsequences tend to different limits).

This was the Boundedness Theorem – Theorem 1, Sub-section 2.4.2.

Our next result shows that, if a sequence is bounded, then, even though it may diverge, it cannot behave 'too badly'!

Theorem 3   Bolzano–Weierstrass Theorem

A bounded sequence must contain a convergent subsequence.

   That is, if a sequence $\{a_n\}$ is such that $|a_n| \leq M$, for all n, then there exist some number ℓ in $[-M, M]$ and some subsequence $\{a_{n_k}\}$ such that $\{a_{n_k}\}$ converges to ℓ as $k \to \infty$.

For example, the sequence $\{\sin n\}$ is bounded. With the tools at our disposal, it is not at all clear what the behaviour of this sequence might be as $n \to \infty$! However, the Bolzano–Weierstrass Theorem asserts that there is at least one number ℓ in $[-1, 1]$ such that some subsequence $\{\sin n_k\}$ converges to ℓ. This is itself a surprising result! In fact, however, using quite sophisticated mathematics we can prove that every number ℓ in $[-1, 1]$ has this property!

For $|\sin n| \leq 1$.

Sadly, this is beyond the range of this book.

The proof of the Bolzano–Weierstrass Theorem that we give is of interest in its own right; it depends on the notion of *repeated bisection*, which is a standard technique in many areas of Analysis.

Proof   We shall assume that the bounded sequence $\{a_n\}$ has the property that $|a_n| \leq M$, for all n. Then all the terms a_n lie in the closed interval $[-M, M]$, which we will denote as $[A_1, B_1]$.

   Next, denote by p the midpoint of the interval $[A_1, B_1]$. Then at least one of the two intervals $[A_1, p]$ and $[p, B_1]$ must contain infinitely many terms in the

$A_1 = -M$ and $B_1 = M$.

$p = \frac{1}{2}(A_1 + B_1)$.

sequence $\{a_n\}$ – for otherwise the whole sequence would only contain finitely many terms. If only one of the two intervals has this property, denote that interval by the notation $[A_2, B_2]$; if both intervals have this property, choose the left one (that is, $[A_1, p]$) to be $[A_2, B_2]$.

If both intervals contain infinitely many terms in the sequence, it *does not matter* which choice we take; but we specify the left interval just for definiteness.

In either case, we obtain:

1. $[A_2, B_2] \subset [A_1, B_1]$;
2. $B_2 - A_2 = \frac{1}{2}(B_1 - A_1)$;
3. $[A_1, B_1]$ and $[A_2, B_2]$ both contain infinitely many terms in the sequence $\{a_n\}$.

We now repeat this process indefinitely often, bisecting $[A_2, B_2]$ to obtain $[A_3, B_3]$, and so on. This gives a sequence of closed intervals $\{[A_n, B_n]\}$ for which:

1. $[A_{n+1}, B_{n+1}] \subset [A_n, B_n]$, for each $n \in \mathbb{N}$;
2. $B_n - A_n = \left(\frac{1}{2}\right)^{n-1}(B_1 - A_1)$, for each $n \in \mathbb{N}$;
3. each interval $[A_n, B_n]$ contains infinitely many terms in the sequence $\{a_n\}$.

Property 1 implies that the sequence $\{A_n\}$ is increasing and bounded above by $B_1 = M$. Hence, by the Monotone Convergence Theorem, $\{A_n\}$ is convergent; denote by A its limit. By the Limit Inequality Rule, we must have that $A \leq M$.

Theorem 3, Sub-section 2.3.3.

Similarly, Property 1 implies that the sequence $\{B_n\}$ is decreasing and bounded below by $A_1 = -M$. Hence, by the Monotone Convergence Theorem, $\{B_n\}$ is convergent; denote by B its limit. By the Limit Inequality Rule, we must have that $B \geq -M$.

We may then deduce, by letting $n \to \infty$ in Property 2 and using the Combination Rules for sequences, that

$$B - A = \lim_{n \to \infty} B_n - \lim_{n \to \infty} A_n = \lim_{n \to \infty} (B_n - A_n)$$

$$= \lim_{n \to \infty} \left(\frac{1}{2}\right)^{n-1}(B_1 - A_1)$$

$$= (B_1 - A_1) \lim_{n \to \infty} \left(\frac{1}{2}\right)^{n-1} = 0.$$

In other words, the sequences $\{A_n\}$ and $\{B_n\}$ both converge to a common limit. Denote this limit by ℓ. Since $\ell = A = B$, we must have $-M \leq \ell \leq M$.

That is, $\ell \in [-M, M]$.

We find a suitable subsequence of $\{a_n\}$ as follows. Choose any a_{n_1} that lies in $[A_1, B_1]$; this is possible since $[A_1, B_1]$ contains infinitely terms in $\{a_n\}$. Next, choose a term a_{n_2} in the sequence, with $n_2 > n_1$, such that $a_{n_2} \in [A_2, B_2]$; this is possible since $[A_2, B_2]$ contains infinitely terms in $\{a_n\}$, by Property 3. And so on.

In this way, we construct a subsequence $\{a_{n_k}\}$ of $\{a_n\}$, with $n_{k+1} > n_k$, for each k. Since $A_k \leq a_{n_k} \leq B_k$, it follows from the Squeeze Rule for sequences (that is, by letting $k \to \infty$) that $\{a_{n_k}\}$ converges to ℓ, as required. ∎

Since $n_{k+1} > n_k$, we must have $n_k \geq k$, for each k.

For $A_k \to A = \ell$ and $B_k \to B = \ell$.

2.5.2 Sequences defined by recursion formulas

As we have seen, often sequences are defined by formulas; that is, for a given n, we substitute that value of n into a formula and obtain the term a_n in the sequence $\{a_n\}$. Another way of specifying a sequence is to define its terms 'inductively', or 'recursively'; here we specify the first term (or several terms) in the sequence, and then have a formula that enables us to calculate all successive terms.

For example, the following sequences are defined recursively:

$\{a_n\}$, where $a_1 = 2$ and $a_{n+1} = \frac{1}{4}\left(a_n^2 + 3\right)$, for $n \geq 1$;

$\{a_n\}$, where $a_1 = 2$, $a_2 = 4$ and $a_n = \frac{1}{2}(a_{n-1} + a_{n-2})$, for $n \geq 3$.

Do these sequences converge? If they do converge, what are their limits? If we choose a different value for a_1 (or for a_1 and a_2, in the second sequence), do their behaviours change? We can use the Monotone Convergence Theorem to answer many such questions.

First, consider the sequence with recursion formula

$$a_{n+1} = \frac{1}{4}\left(a_n^2 + 3\right), \quad \text{for } n \geq 1. \tag{2}$$

If indeed $\{a_n\}$ is convergent, what value could its limit take? If we let ℓ denote $\lim_{n \to \infty} a_n$ and let $n \to \infty$ in equation (2), we obtain $\ell = \frac{1}{4}(\ell^2 + 3)$. This can be rearranged as $4\ell = \ell^2 + 3$, and so as $\ell^2 - 4\ell + 3 = 0$. Hence $(\ell - 1)(\ell - 3) = 0$, so that $\ell = 1$ or $\ell = 3$.

Of course we have not yet proved that $\{a_n\}$ is convergent. This has been a 'what if?' discussion so far, but a useful one.

This suggests that we should look at the differences $a_{n+1} - 1$ and $a_{n+1} - 3$. Using equation (2), we deduce that

Because 1 and 3 are the possible limits.

$$a_{n+1} - 1 = \frac{1}{4}\left(a_n^2 - 1\right) = \frac{1}{4}(a_n + 1)(a_n - 1), \quad \text{for } n \geq 1, \tag{3}$$

$$a_{n+1} - 3 = \frac{1}{4}\left(a_n^2 - 9\right) = \frac{1}{4}(a_n + 3)(a_n - 3), \quad \text{for } n \geq 1. \tag{4}$$

Next, it is useful to examine whether the sequence is monotonic by looking at the difference $a_{n+1} - a_n$

$$a_{n+1} - a_n = \frac{1}{4}\left(a_n^2 - 4a_n + 3\right) = \frac{1}{4}(a_n - 1)(a_n - 3). \tag{5}$$

So the sign of $a_{n+1} - a_n$ depends on where a_n lies on the number-line in relation to the numbers 1 and 3.

For example, it follows from (5) that

if $1 < a_n < 3$, for some $n \geq 1$, then $a_{n+1} - a_n < 0$,

so that $a_{n+1} < a_n$.

(6) *We shall use this fact below.*

If $a_n = 1$, for some number $n \geq 1$, it follows from equation (3) that $a_{n+1} = 1$. Consequently, if our initial term is $a_1 = 1$, then it follows that $a_n = 1$, for all $n \geq 1$; in other words, the sequence is simply the constant sequence $\{1\}$.

Similarly, if $a_n = 3$, for some number $n \geq 1$, it follows from equation (4) that $a_{n+1} = 3$. Consequently, if our initial term is $a_1 = 3$, then it follows that $a_n = 3$, for all $n \geq 1$; in other words, the sequence is simply the constant sequence $\{3\}$.

Next, suppose that $1 < a_n < 3$, for some number $n \geq 1$. It follows from equation (3) that $a_{n+1} - 1$ is positive, so that $1 < a_{n+1}$. On the other hand, it follows from equation (4) that $a_{n+1} - 3$ is negative, so that $a_{n+1} < 3$.

Thus, if $1 < a_n < 3$, for some number $n \geq 1$, then $1 < a_{n+1} < 3$.

We can then prove by Mathematical Induction that, if $1 < a_1 < 3$, then $1 < a_n < 3$, for all $n > 1$. (We omit the details.)

In particular this discussion covers the case $a_1 = 2$ that we mentioned at the start of this sub-section.

Next, it follows, from this last fact and (6) above that, if $1 < a_1 < 3$, then the sequence $\{a_n\}$ is strictly decreasing.

Hence, if $1 < a_1 < 3$ the sequence $\{a_n\}$ is strictly decreasing and is bounded below. Hence, by the Monotone Convergence Theorem, $\{a_n\}$ is convergent.

Now comes the coup de grace!

Since $\{a_n\}$ is decreasing, whatever its limit might be (and we know that the only two possibilities for the limit are 1 and 3), its limit must be less than or equal to its first term a_1. It follows that the limit of the sequence must be 1.

> We saw that 1 and 3 were the only possible limits at the start of the discussion.

Problem 2   Let the sequence $\{a_n\}$ be defined by the recursion formula $a_{n+1} = \frac{1}{4}\left(a_n^2 + 3\right)$, for $n \geq 1$.

(a) Prove that, if $a_1 > 3$, then $a_n \to \infty$ as $n \to \infty$.

(b) In the case that $0 \leq a_1 < 1$, determine whether $\{a_n\}$ is convergent and, if so, to what limit.

(c) Describe the behaviour of the sequence $\{a_n\}$ in the case that $a_1 < 0$.

2.5.3   The number *e*

We will define e to be the limit of the sequence $\left\{\left(1 + \frac{1}{n}\right)^n\right\}$. If we plot the first few terms of this sequence on a sequence diagram, then it seems that the sequence is increasing and converges to a limit, which is less than 3.

> For example, to three decimal places the first two terms are 2 and 2.25, the sixth term is 2.521, and the hundredth term is 2.704.

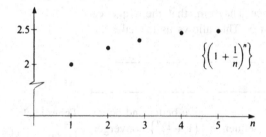

To show that $\left\{\left(1 + \frac{1}{n}\right)^n\right\}$ is convergent, using the Monotone Convergence Theorem, we prove that the sequence is increasing and is bounded above.

1. $\left\{\left(1 + \frac{1}{n}\right)^n\right\}$ *is increasing*

   By the Binomial Theorem

   $$\left(1 + \frac{1}{n}\right)^n = 1 + n\left(\frac{1}{n}\right) + \frac{n(n-1)}{2!}\left(\frac{1}{n}\right)^2 + \cdots + \left(\frac{1}{n}\right)^n.$$

   As n increases, the number of terms in this sum increases, and the new terms are all positive. Also, for each fixed $k \geq 1$ and any $n \geq k$, the $(k+1)$th term of this sum is

   $$\frac{n(n-1)\ldots(n-k+1)}{k!}\left(\frac{1}{n}\right)^k$$

   $$= \frac{1}{k!}\left(1 - \frac{1}{n}\right)\left(1 - \frac{2}{n}\right)\ldots\left(1 - \frac{k-1}{n}\right). \qquad (7)$$

   Here the product in (7) increases as n increases (since each of its factors is itself increasing). It follows that the sequence $\left(1 + \frac{1}{n}\right)^n$ is increasing.

2. $\left\{\left(1 + \frac{1}{n}\right)^n\right\}$ *is bounded above*

   Here, note that the general term (7) satisfies the inequality

   $$\frac{1}{k!}\left(1 - \frac{1}{n}\right)\left(1 - \frac{2}{n}\right)\ldots\left(1 - \frac{k-1}{n}\right) \leq \frac{1}{k!},$$

since each of the brackets is at most 1. Hence

$$\left(1+\frac{1}{n}\right)^n \le 1 + 1 + \frac{1}{2!} + \frac{1}{3!} + \cdots + \frac{1}{n!}$$

$$\le 1 + 1 + \frac{1}{2^1} + \frac{1}{2^2} + \cdots + \frac{1}{2^{n-1}},$$

since $k! = k(k-1)\ldots 2.1 \ge 2^{k-1}$.

Now

$$1 + \frac{1}{2^1} + \frac{1}{2^2} + \cdots + \frac{1}{2^{n-1}} = 1 \times \frac{1 - \left(\frac{1}{2}\right)^n}{1 - \frac{1}{2}} = 2 \times \left(1 - \left(\frac{1}{2}\right)^n\right)$$

$$= 2 - \frac{1}{2^{n-1}},$$

For the sum of the geometric progression $a + ar + ar^2 + \cdots + ar^{n-1}$ is $a\frac{1-r^n}{1-r}$, if $r \ne 1$; here we have $a = 1, r = \frac{1}{2}$.

and so

$$\left(1+\frac{1}{n}\right)^n \le 3 - \frac{1}{2^{n-1}}, \quad \text{for } n = 1, 2, \ldots.$$

Thus the sequence $\left\{\left(1+\frac{1}{n}\right)^n\right\}$ is bounded above, by 3.

It follows, by the Monotone Convergence Theorem, that the sequence $\left\{\left(1+\frac{1}{n}\right)^n\right\}$ is convergent with limit at most 3. This allows us to make the following definition:

Definition $e = \lim_{n\to\infty}\left(1+\frac{1}{n}\right)^n.$

For larger and larger values of n, the terms $\left(1+\frac{1}{n}\right)^n$ give better and better approximate values for e. Unfortunately, the sequence $\left\{\left(1+\frac{1}{n}\right)^n\right\}$ converges to e rather slowly, and we need to take very large integers n to get a reasonable approximation to $e = 2.71828\ldots$.

For example

$$\left(1+\tfrac{1}{1000}\right)^{1000} = 2.716\ldots.$$

Now we would like in general to make the definition of e^x as $e^x = \lim_{n\to\infty}\left(1+\frac{x}{n}\right)^n$, but first we need to know that this limit actually exists.

Then $e = e^1$.

Problem 3   Let $x > 0$.

(a) Prove that $\left\{\left(1+\frac{x}{n}\right)^n\right\}$ is an increasing sequence, by adapting the method above where $x = 1$.

(b) Verify that $1 + \frac{k}{n} \le \left(1+\frac{1}{n}\right)^k$, for $k = 1, 2, \ldots$. Using this fact, prove that $\left\{\left(1+\frac{x}{n}\right)^n\right\}$ is bounded above.

(c) Deduce that $\left\{\left(1+\frac{x}{n}\right)^n\right\}$ is convergent.

So $\lim_{n\to\infty}\left(1+\frac{x}{n}\right)^n$ exists for $x > 0$. Note that the limit is at least 1, and so cannot be zero.

Problem 4   By considering the product of the first n terms of the sequence $\left\{\left(1+\frac{1}{n}\right)^n\right\}$, prove that $n! > \left(\frac{n+1}{e}\right)^n$, for $n = 1, 2, \ldots$.

We now examine the convergence of the sequence $\left\{\left(1+\frac{x}{n}\right)^n\right\}$, for $x < 0$.

Recall that, when $x < 0$, then $-x > 0$; in particular, from the above discussion, the sequence $\left\{\left(1-\frac{x}{n}\right)^n\right\}$ converges. We shall use Bernoulli's inequality $(1+c)^n \ge 1 + nc$, for $c \ge -1$.

You met Bernoulli's Inequality in Subsection 1.3.3.

In investigating the convergence of the sequence $\left\{\left(1+\frac{x}{n}\right)^n\right\}$, we are only interested in what happens when n is large. So, we need consider only the situation when $n > -x$. Then $n^2 > x^2$, so that $\frac{1}{n^2} < \frac{1}{x^2}$; thus $\frac{x^2}{n^2} < 1$, or $-\frac{x^2}{n^2} > -1$.

Recall that $-x > 0$.

It follows that we may substitute $-\frac{x^2}{n^2}$ for c in Bernoulli's Inequality. This gives that

$$\left(1 - \frac{x^2}{n^2}\right)^n \geq 1 + n\left(-\frac{x^2}{n^2}\right)$$

$$= 1 - \frac{x^2}{n}, \quad \text{for } n > -x.$$

It follows that

$$1 \geq \left(1 - \frac{x^2}{n^2}\right)^n \geq 1 - \frac{x^2}{n}, \quad \text{for } n > -x. \tag{8}$$

Now the sequence $\left\{1 - \frac{x^2}{n}\right\}$ converges to the limit 1, by the Combination Rules. Hence, by applying the Squeeze Rule to inequality (8), the sequence $\left\{\left(1 - \frac{x^2}{n^2}\right)^n\right\}$ converges to the limit 1 as $n \to \infty$.

Here we use the fact that $\{\frac{1}{n}\}$ is a basic null sequence.

But

$$\left(1 + \frac{x}{n}\right)^n = \frac{\left(1 + \frac{x}{n}\right)^n \left(1 - \frac{x}{n}\right)^n}{\left(1 - \frac{x}{n}\right)^n}$$

$$= \frac{\left(1 - \frac{x^2}{n^2}\right)^n}{\left(1 - \frac{x}{n}\right)^n}. \tag{9}$$

We have just seen that the numerator is convergent, and we saw above that the denominator is convergent (to a non-zero limit) for $-x > 0$; it follows from (9), by the Quotient Rule, that the sequence $\left\{\left(1 + \frac{x}{n}\right)^n\right\}$ is convergent.

Recall that here we are considering the case $x < 0$.

So whatever value we choose for the real number x, the sequence $\left\{\left(1 + \frac{x}{n}\right)^n\right\}$ is convergent. This means that we can now make the following legitimate definition:

Note that when $x = 0$, the sequence is simply the constant sequence $\{1\}$.

Definition For any real value of x,
$$e^x = \lim_{n \to \infty} \left(1 + \frac{x}{n}\right)^n.$$

The function $x \mapsto e^x$, for $x \in \mathbb{R}$, is called *the exponential function*.

We can deduce more than this from equation (9), if we rewrite it in the convenient form

$$\left(1 + \frac{x}{n}\right)^n \left(1 - \frac{x}{n}\right)^n = \left(1 - \frac{x^2}{n^2}\right)^n.$$

Letting $n \to \infty$, we deduce from this last equation that
$$e^x e^{-x} = 1.$$

Here we also use the fact that
$$\lim_{n \to \infty} \left(1 - \frac{x^2}{n^2}\right)^n = 1,$$
that we proved above.

We have shown that this holds for $x < 0$; however, by simply interchanging x and $-x$, it is clear that this holds for $x > 0$ also.

Theorem 4   Inverse Property of e^x
For any real value of x, $e^x e^{-x} = 1$

The next property of the exponential function that we need to verify is that $e^x e^y = e^{x+y}$ for any real values of x and y. We shall prove this fact later.

In Sub-section 3.4.3.

2.5.4 The number π

One of the oldest mathematical problems is to determine the area of a disc of radius r, and the length of its perimeter. It is well known that these magnitudes are πr^2 and $2\pi r$, respectively. But what exactly is π? Is there a real number π which makes these formulas correct? – and, if so, how is it formally defined?

Problem 5   Verify that the following triangles have the stated areas:

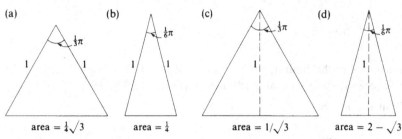

(a)   (b)   (c)   (d)

Here π is used only as a symbol to represent angles; its value is not required.

area $= \frac{1}{4}\sqrt{3}$ area $= \frac{1}{4}$ area $= 1/\sqrt{3}$ area $= 2 - \sqrt{3}$

Hint: In part (d), use the half-angle formula $\tan \vartheta = \frac{2 \tan \frac{1}{2}\vartheta}{1 - \tan^2 \frac{1}{2}\vartheta}$.

We now give a precise definition of π as the area of a disc of radius 1, using a method originally devised by Archimedes to calculate approximate values for the area of this disc. The idea is to calculate the areas of regular polygons inscribed in the disc. Archimedes found an easy way to calculate the areas of such regular polygons with 6 sides, 12 sides, 24 sides and so on. The results of parts (a) and (b) of the above problem help to give the first two of these areas.

Here 'inscribed' means that all the vertices of the polygon lie on the circle, so that the inside of the polygon is contained in the inside of the circle.

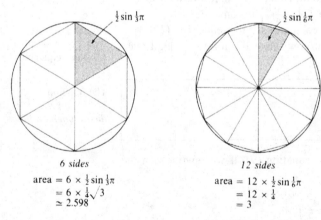

$\frac{1}{2} \sin \frac{1}{3}\pi$ $\frac{1}{2} \sin \frac{1}{6}\pi$

6 sides   *12 sides*

area $= 6 \times \frac{1}{2} \sin \frac{1}{3}\pi$ area $= 12 \times \frac{1}{2} \sin \frac{1}{6}\pi$
$\quad = 6 \times \frac{1}{4}\sqrt{3}$ $\quad = 12 \times \frac{1}{4}$
$\quad \simeq 2.598$ $\quad = 3$

Let s_n denote the number of sides of the nth such inner polygon (so $s_1 = 6$, $s_2 = 12$, $s_3 = 24$, and, in general, $s_n = 3 \times 2^n$) and let a_n denote the area of the nth inner polygon. Then

$$a_n = \frac{1}{2} s_n \sin\left(\frac{2\pi}{s_n}\right), \quad \text{for } n = 1, 2, \ldots. \tag{10}$$

Geometrically, it is obvious that each time we double the number of sides of the inner polygon the area increases, and so

$$a_1 < a_2 < a_3 < \ldots < a_n < a_{n+1} < \ldots.$$

Hence $\{a_n\}$ is (strictly) increasing. But is $\{a_n\}$ convergent?

For example, to three decimal places

$a_1 = 2.598,$

$a_2 = 3,$

$\ldots$

$a_6 = 3.141.$

Notice that each of the polygons lies inside a square of side 2, which has area 4. This means that

$$a_n \le 4, \quad \text{for } n = 1, 2, \ldots,$$

and so $\{a_n\}$ is increasing and bounded above (by 4). Hence, by the Monotone Convergence Theorem, $\{a_n\}$ is convergent.

Our intuitive idea of the area of the disc suggests that its area is greater than each of the areas a_n, but 'only just'! Put another way, the area of the disc should be the limit of the increasing sequence $\{a_n\}$. This leads us to make the following definition:

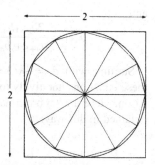

Definition $\quad \pi = \lim\limits_{n \to \infty} a_n.$

We shall explain in a moment how to calculate the terms a_n without assuming a value for π.

First, however, we describe how to estimate the area of the disc using *outer* polygons. Once again we start with a regular hexagon and repeatedly double the number of sides. The results of part (c) and (d) of the last problem help to give the first two such areas.

This will enable us to *squeeze* π between the area of the inner polygons and the area of the outer polygons.

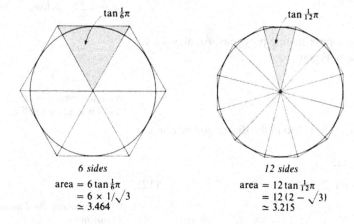

tan $\frac{1}{6}\pi$ tan $\frac{1}{12}\pi$

6 sides
area $= 6 \tan \frac{1}{6}\pi$
$= 6 \times 1/\sqrt{3}$
$\simeq 3.464$

12 sides
area $= 12 \tan \frac{1}{12}\pi$
$= 12(2 - \sqrt{3})$
$\simeq 3.215$

As before, let $s_n = 3 \times 2^n$, for $n = 1, 2, \ldots$, and let b_n denote the area of the nth outer polygon. This nth outer polygon consists of s_n isosceles triangles, each of height 1 and base $2\tan(\frac{\pi}{s_n})$. Thus

$$b_n = s_n \tan\left(\frac{\pi}{s_n}\right), \quad \text{for } n = 1, 2, \ldots. \tag{11}$$

Geometrically, it is obvious that each time we double the number of sides of the outer polygon the area decreases, and so

$$b_1 > b_2 > b_3 > \cdots > b_n > b_{n+1} > \cdots.$$

For example, to three decimal places

$b_1 = 3.464,$

$b_2 = 3.215,$

$\ldots$

$b_6 = 3.142.$

So the sequence $\{b_n\}$ is (strictly) decreasing and bounded below (by 0, for example). Thus, by the Monotone Convergence Theorem, $\{b_n\}$ is also convergent.

Intuitively, we expect that $\{b_n\}$ has the same limit as $\{a_n\}$, which we have defined to be π. But how can we prove this?

The terms a_n and b_n can be calculated by using the following equations, which are known jointly as *the Euclidean algorithm*

$$a_{n+1} = \sqrt{a_n b_n}, \quad \text{for } n = 1, 2, \ldots, \tag{12}$$

and

$$b_{n+1} = \frac{2a_{n+1} b_n}{a_{n+1} + b_n}, \quad \text{for } n = 1, 2, \ldots. \tag{13}$$

Equations (12) and (13) can be verified from equations (10) and (11); we give the details in Subsection 2.5.5 below.

Starting with $a_1 = \frac{3}{2}\sqrt{3} = 2.598\ldots$ and $b_1 = 2\sqrt{3} = 3.464\ldots$, we use these equations iteratively to calculate first $a_2 = \sqrt{a_1 b_1}$, then $b_2 = \frac{2a_2 b_1}{a_2 + b_1}$, and so on. Here are (approximations to) the first few values of each sequence obtained in this way:

s_n	6	12	24	48	96	192
a_n	2.598	3	3.106	3.133	3.139	3.141
b_n	3.464	3.215	3.160	3.146	3.143	3.142

It does appear that both sequences do converge to a common limit, namely π.

We prove this in Subsection 2.5.5 below.

Remark

We have defined π using the areas of approximating polygons. An alternative approach uses the perimeters of these polygons.

2.5.5 Proofs

You may omit this subsection at a first reading.

We now prove several of the results referred to in the discussion of π in the previous sub-section.

First, we prove equations (12) and (13)

$$a_{n+1} = \sqrt{a_n b_n}, \quad \text{for } n = 1, 2, \ldots, \tag{12}$$

and

$$b_{n+1} = \frac{2a_{n+1} b_n}{a_{n+1} + b_n}, \quad \text{for } n = 1, 2, \ldots. \tag{13}$$

These comprise *the Euclidean algorithm*.

Using the half-angle formulas for $\sin \vartheta$ and $\cos \vartheta$, it is easy to check that

$$\sin \frac{1}{2}\vartheta = \sqrt{\frac{1}{2}\sin \vartheta \tan \frac{1}{2}\vartheta}$$

and

$$\tan \frac{1}{2}\vartheta = \frac{\sin \vartheta \tan \vartheta}{\sin \vartheta + \tan \vartheta}.$$

We omit the details in both cases.

Hence, since $s_{n+1} = 2s_n$

$$\begin{aligned} a_{n+1} &= \frac{1}{2} s_{n+1} \sin\left(\frac{2\pi}{s_{n+1}}\right) \\ &= \sqrt{\frac{1}{2} s_n \sin\left(\frac{2\pi}{s_n}\right) \times s_n \tan\left(\frac{\pi}{s_n}\right)} \\ &= \sqrt{a_n b_n}. \end{aligned}$$

This proves equation (12).

Similarly

$$b_{n+1} = s_{n+1} \tan\left(\frac{\pi}{s_{n+1}}\right)$$

$$= \frac{2s_n \sin\left(\frac{\pi}{s_n}\right) \times s_n \tan\left(\frac{\pi}{s_n}\right)}{s_n \sin\left(\frac{\pi}{s_n}\right) + s_n \tan\left(\frac{\pi}{s_n}\right)}$$

$$= \frac{2a_{n+1}b_n}{a_{n+1} + b_n}.$$

This proves equation (13).

Finally we prove that $\lim\limits_{n\to\infty} b_n = \lim\limits_{n\to\infty} a_n$.

To do this, we rearrange equation (12) as follows

$$b_n = \frac{a_{n+1}^2}{a_n}.$$

Then, by the Combination Rules, we can deduce that

$$\lim_{n\to\infty} b_n = \frac{\left(\lim\limits_{n\to\infty} a_{n+1}\right)^2}{\lim\limits_{n\to\infty} a_n} = \frac{\pi^2}{\pi} = \pi.$$

Then, since we defined π to be $\lim\limits_{n\to\infty} a_n$, it follows that $\lim\limits_{n\to\infty} b_n = \lim\limits_{n\to\infty} a_n$, as required.

Definition Let a_n denote the area of the regular polygon with 3×2^n sides inscribed in a disc of radius 1, and b_n the area of the regular polygon with 3×2^n sides circumscribing the disc. Then

π is the common value of the limits $\lim\limits_{n\to\infty} a_n$ and $\lim\limits_{n\to\infty} b_n$.

Remark

It can be shown that every two applications of equations (12) and (13) give an extra decimal place in the decimal expansion for π. This decimal expansion is now known to many millions of decimal places, using sequences which converge to π much more rapidly than $\{a_n\}$ or $\{b_n\}$. Since π is irrational, there is no possibility of the decimal expansion of π recurring.

We shall return in Section 8.5 to the question of approximating π, by the use of power series.

2.6 Exercises

Section 2.1

1. Calculate the first five terms of each of the following sequences, and draw a sequence diagram in each case:

   (a) $\{n^2 - 4n + 4\}$;   (b) $\left\{\frac{(-1)^{n+1}}{n!}\right\}$;   (c) $\left\{\sin\left(\frac{1}{4}n\pi\right)\right\}$.

2. Determine which of the following sequences are monotonic:

   (a) $\left\{\frac{n+1}{n+2}\right\}$;   (b) $\left\{\frac{(-1)^n}{n}\right\}$;   (c) $\left\{2^{\frac{1}{n}}\right\}$.

3. Prove that the following sequences are eventually monotonic:

   (a) $\left\{\frac{5^n}{n!}\right\}$;   (b) $\left\{n + \frac{8}{n}\right\}$.

Section 2.2

1. For each of the following sequences $\{a_n\}$ and numbers ε, find a number X such that $|a_n| < \varepsilon$, for all $n > X$:

   (a) $a_n = \frac{(-1)^n}{n^5}$, $\varepsilon = 0.001$;   (b) $a_n = \frac{1}{(2n+1)^2}$, $\varepsilon = 0.002$.

2. Use the definition of *null sequence* to prove that the two sequences in the previous exercise are null.

3. Prove that the following sequences are not null:

   (a) $\{\sqrt{n}\}$;   (b) $\left\{1 + \frac{(-1)^n}{n}\right\}$.

4. Assuming only that $\left\{\frac{1}{n}\right\}$ is null, deduce that the following sequences are null. State which rules you use.

   (a) $\left\{\frac{2}{\sqrt{n}} + \frac{3}{n^7}\right\}$;   (b) $\left\{\frac{\sin n}{n^2+1}\right\}$;   (c) $\left\{\frac{\tan^{-1}(2n)}{n+1+\sin n}\right\}$.

5. Prove that $\left\{\sqrt{n+1} - \sqrt{n}\right\}$ is null.
   Hint: Use the fact that $a - b = \frac{a^2-b^2}{a+b}$, for $a+b \neq 0$.

6. Use the list of basic null sequences to prove that the following sequences are null. State which rules you use.   You met this list in Sub-section 2.2.1.

   (a) $\left\{\frac{3}{4^n} + \frac{2n}{3^n}\right\}$;   (b) $\left\{\frac{6n^{10}}{n!}\right\}$;   (c) $\left\{\frac{n^{10}10^n}{n!}\right\}$.

7. Prove that, if the sequence $\{a_n\}$ of positive numbers is null, then $\left\{\sqrt{a_n}\right\}$ is null.

Section 2.3

1. Show that the following sequences converge to 1, by calculating $a_n - 1$ in each case:

   (a) $\left\{\frac{n-1}{n+3}\right\}$;   (b) $\left\{\frac{n^2}{n^2+n+1}\right\}$.

2. Use the Combination Rules to find the limits of the following sequences:

   (a) $\left\{\frac{n^3}{2n^3+3n+\frac{4}{n}}\right\}$;   (b) $\left\{\frac{n^{10}-2^n}{2^n+n^{100}}\right\}$;   (c) $\left\{\frac{5n!+5^n}{n^{100}+n!}\right\}$.

3. (a) Prove that, if the sequence $\{a_n\}$ of positive numbers is convergent with limit ℓ, where $\ell > 0$, then $\left\{\sqrt{a_n}\right\}$ is convergent with limit $\left\{\sqrt{\ell}\right\}$.

Hint: Use the fact that $a - b = \frac{a^2 - b^2}{a+b}$, for $a + b \neq 0$.

(b) Prove that, if the sequence $\{a_k\}$ is convergent with limit ℓ, $\ell \neq 0$, and $a_k^{\frac{1}{n}}$ is defined, where $n \in \mathbb{N}$, then $\left\{a_k^{\frac{1}{n}}\right\}$ is convergent with limit $\ell^{\frac{1}{n}}$.

Hint: Use the identity $a^q - b^q = (a - b)(a^{q-1} + a^{q-2}b + a^{q-3}b^2 + \cdots + b^{q-1})$ with suitable choices for a, b and q.

The Power Rule in Subsection 2.2.1 and the subsequent remark there show that this result holds in the case that $\ell = 0$.

Section 2.4

1. Classify the following sequences as *convergent* or *divergent*, and as *bounded* or *unbounded*:

   (a) $\left\{n^{\frac{1}{4}}\right\}$; (b) $\left\{\frac{(-1)^n 100^n}{n!}\right\}$.

2. Use the Reciprocal Rule to prove that the following sequences tend to infinity:

   (a) $\left\{\frac{n!}{n^3}\right\}$; (b) $\{n^2 + 2n\}$; (c) $\{n^2 - 2n\}$; (d) $\{n! - n^3 - 3^n\}$.

3. Use the Subsequence Rules to prove that the following sequences are divergent:

   (a) $\{(-1)^n 2^n\}$; (b) $\left\{\frac{(-1)^n n^2}{2n^2 + 1}\right\}$.

4. Each of the following general statements concerning sequences $\{a_n\}$ and $\{b_n\}$ is true or false. For each statement that is true, prove it. For each statement that is false, write down examples of sequences to illustrate your assertion:

   (a) If $\{a_n\}$ and $\{b_n\}$ are divergent, then $\{a_n + b_n\}$ is divergent.

   (b) If $\{a_n\}$ and $\{b_n\}$ are divergent, then $\{a_n b_n\}$ is divergent.

   (c) If $\{a_n\}$ is convergent and $\{b_n\}$ is divergent, then $\{a_n b_n\}$ is divergent.

   (d) If $\{a_n^2\}$ is convergent, then $\{a_n\}$ is convergent.

   (e) If $\{a_n^2\}$ is convergent, and $a_n > 0$, then $\{a_n\}$ is convergent.

   (f) If $\{b_n\}$ is convergent, where $b_n = a_{n+1} a_n$, then $\{a_n\}$ is convergent.

   (g) If $a_n \to 0$ as $n \to \infty$, then $\left\{\frac{a_{n+1} + a_{n+2} + \cdots + a_{2n}}{\sqrt{n}}\right\}$ is convergent.

5. Prove that every sequence has a monotonic subsequence.

Section 2.5

1. Prove that, if $\{a_n\}$ is increasing and has a subsequence $\{a_{n_k}\}$ which is convergent, then $\{a_n\}$ is convergent.

2. Let $\{a_n\}_{n=1}^{\infty}$ be a sequence for which $a_{n+1} = \frac{3a_n + 1}{a_n + 3}$, for $n \geq 1$.

   (a) Prove that, if $a_1 > -1$, the sequence $\{a_n\}$ converges.
   Hint: consider the cases $-1 < a_1 < 1$, $a_1 = 1$ and $a_1 > 1$ separately.

   (b) If $a_1 < -3$, does the sequence converge or diverge?

3. Let $\{a_n\}_{n=1}^{\infty}$ be a sequence for which $a_{n+1} = \frac{1}{2}\left(a_n + \frac{\ell^2}{a_n}\right)$, for $n \geq 1$, where $a_1 > 0$ and $\ell > 0$.

   (a) Prove that $\{a_n\}$ is decreasing for $n \geq 2$, and hence that $a_n \to \ell$ as $n \to \infty$.

   (b) With $\ell^2 = 2$ and $a_1 = 1.5$, calculate the terms a_2, a_3, a_4 and a_5.

4. Obtain formulas for the perimeters of the inner and outer polygons introduced in Sub-section 2.5.4. Prove that these sequences tend to 2π as $n \to \infty$.

5. Bernoulli's Inequality states that $(1+x)^n \geq 1 + nx$, for $x \geq -1$, $n = 1, 2, \ldots$. You met this in Sub-section 1.3.3.

   (a) Apply Bernoulli's Inequality, with $x = \frac{-1}{n^2}$, to give an alternative proof that $\left\{\left(1 + \frac{1}{n}\right)^n\right\}$ is increasing.

   (b) Apply Bernoulli's Inequality, with $x = \frac{1}{n^2-1}$, to prove that $\left\{\left(1 + \frac{1}{n}\right)^{n+1}\right\}$ is decreasing.

   (c) Deduce from parts (a) and (b) that
   $$\left(1 + \frac{1}{n}\right)^n \leq e \leq \left(1 + \frac{1}{n}\right)^{n+1}, \quad \text{for } n = 1, 2, \ldots.$$

6. Let $\{[a_n, b_n]\}_{n=1}^{\infty}$ be a sequence of closed intervals with the property that $[a_{n+1}, b_{n+1}] \subseteq [a_n, b_n]$, for $n = 1, 2, \ldots$.

   (a) Prove that there exists some point c that lies in *all* the intervals $[a_n, b_n]$, for $n = 1, 2, \ldots$. This result is sometimes called *The Nested Intervals Theorem*.

   (b) Prove that, if $b_n - a_n \to 0$ as $n \to \infty$, then there exists *only one* such point c.

3 Series

The Ancient Greek philosopher Zeno of Elea proposed a number of paradoxes of the infinite, which are said to have had a profound influence on Greek mathematics. For example, Zeno claimed that it is impossible for an object to travel any given distance, since it must first travel half the distance, then half of the remaining distance, then half of what remains, and so on. There must always remain some distance left to travel, and so the journey cannot be completed.

This paradox relies partly on the intuitive feeling that it is impossible to add up an infinite number of positive quantities and obtain a finite answer. However, the following illustration of the paradox suggests that such an infinite sum is plausible.

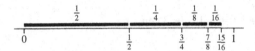

The distance from 0 to 1 can be split up into the infinite sequence of distances $\frac{1}{2}, \frac{1}{4}, \frac{1}{8}, \ldots$, and so it seems reasonable to write

$$\frac{1}{2} + \frac{1}{4} + \frac{1}{8} + \frac{1}{16} + \cdots = 1.$$

This chapter is devoted to the study of such expressions, which are called *infinite series*.

The following example shows that infinite series need to be treated with care. Suppose that it is possible to add up 2, 4, 8, ..., and that the answer is s

$$2 + 4 + 8 + 16 + \cdots = s.$$

If we multiply through by $\frac{1}{2}$, we obtain

$$1 + 2 + 4 + 8 + \cdots = \frac{1}{2}s,$$

which we can rewrite in the form

$$1 + s = \frac{1}{2}s.$$

Because the left-hand side of the previous equation can be expressed as

$$1 + (2 + 4 + 8 + \cdots) = 1 + s.$$

It follows from this equation that $s = -2$, which is obviously non-sensical.

We can avoid reaching such absurd conclusions by performing arithmetical operations only with *convergent* infinite series.

In Section 3.1 we define the concept of convergent infinite series in terms of convergent sequences, and give some examples. We also describe various properties which are common to all convergent series.

Section 3.2 is devoted to series with non-negative terms, and we give several tests for the convergence of such series.

In Section 3.3 we deal with the much harder problem of determining whether a series is convergent when it contains both positive and negative terms. We also look at what can happen if we *rearrange* the terms of a convergent series – does the new series still converge? This section is quite long, so you might wish to tackle it in several sessions.

In Section 3.4 we explain how e^x can be represented as an infinite series of powers of x, and we use this representation to prove that the number e is irrational, and that $e^x e^y = e^{x+y}$.

You first met e^x in Section 2.5.

3.1 Introducing series

3.1.1 What is a convergent series?

We begin by defining precisely what is meant by the statement that

$$\frac{1}{2} + \frac{1}{4} + \frac{1}{8} + \frac{1}{16} + \cdots = 1.$$

Let s_n be the sum of the first n terms on the left-hand side. Then

$$s_1 = \frac{1}{2},$$

$$s_2 = \frac{1}{2} + \frac{1}{4} = \frac{3}{4},$$

$$s_3 = \frac{1}{2} + \frac{1}{4} + \frac{1}{8} = \frac{7}{8}, \ldots, \text{ and so on.}$$

In general, using the formula for the sum of a geometric progression, with $a = r = \frac{1}{2}$, we obtain

$$s_n = \frac{1}{2} + \frac{1}{4} + \frac{1}{8} + \cdots + \frac{1}{2^n} = \frac{1}{2} + \frac{1}{2^2} + \frac{1}{2^3} + \cdots + \frac{1}{2^n}$$

$$= \frac{1}{2} \times \frac{1 - \left(\frac{1}{2}\right)^n}{1 - \frac{1}{2}} = 1 - \left(\frac{1}{2}\right)^n.$$

The sum of a geometric progression $a + ar + ar^2 + \cdots + ar^{n-1}$ is $a\frac{1-r^n}{1-r}$, if $r \neq 1$.

The sequence $\left\{\left(\frac{1}{2}\right)^n\right\}$ is a null sequence, and so

$$\lim_{n \to \infty} s_n = \lim_{n \to \infty} \left(1 - \left(\frac{1}{2}\right)^n\right)$$

Here we use the Combination Rules for sequences.

$$= 1 - \lim_{n \to \infty} \left(\frac{1}{2}\right)^n = 1.$$

It is the precise mathematical statement that $s_n \to 1$ as $n \to \infty$ which justifies our earlier (intuitive) statement that

$$\frac{1}{2} + \frac{1}{4} + \frac{1}{8} + \frac{1}{16} + \cdots = 1.$$

We use this approach to define a *convergent infinite series*.

Definition   Given a sequence $\{a_n\}$ of real numbers, the expression

$$a_1 + a_2 + a_3 + \cdots$$

is called an **infinite series**, or simply a **series**. The **nth partial sum** of this series is

$$s_n = a_1 + a_2 + a_3 + \cdots + a_n.$$

The behaviour of the infinite series $a_1 + a_2 + a_3 + \cdots$ is determined by the behaviour of $\{s_n\}$, its sequence of partial sums.

Definition   The series $a_1 + a_2 + a_3 + \cdots$ is **convergent** with **sum** s if the sequence $\{s_n\}$ of partial sums converges to s. In this case, we say that the series **converges to s**, and we write

$$a_1 + a_2 + a_3 + \cdots = s.$$

The series **diverges** if the sequence $\{s_n\}$ diverges.

Thus we prove results about series by applying results for sequences proved in Chapter 2 to the sequence of partial sums $\{s_n\}$.

Example 1   For each of the following infinite series, calculate its nth partial sum, and determine whether the series is convergent or divergent:

(a) $1 + 1 + 1 + \cdots$;   (b) $\frac{1}{3} + \frac{1}{3^2} + \frac{1}{3^3} + \cdots$;   (c) $2 + 4 + 8 + \cdots$.

Solution

(a) In this case

$$s_n = 1 + 1 + \cdots + 1 = n.$$

The sequence $\{s_n\}$ tends to infinity, and so this series is divergent.

(b) Using the formula for summing a finite geometric progression, with $a = r = \frac{1}{3}$, we obtain

$$s_n = \frac{1}{3} + \frac{1}{3^2} + \frac{1}{3^3} + \cdots + \frac{1}{3^n}$$

$$= \frac{1}{3} \times \frac{1 - \left(\frac{1}{3}\right)^n}{1 - \frac{1}{3}} = \frac{1}{2}\left(1 - \left(\frac{1}{3}\right)^n\right).$$

Since $\left\{\left(\frac{1}{3}\right)^n\right\}$ is a basic null sequence

$$\lim_{n \to \infty} s_n = \frac{1}{2},$$

and so this series is convergent, with sum $\frac{1}{2}$.

(c) In this case

$$s_n = 2 + 4 + 8 + \cdots + 2^n$$

$$= 2\left(\frac{2^n - 1}{2 - 1}\right) = 2^{n+1} - 2.$$

The sequence $\{2^{n+1} - 2\}$ tends to infinity, and so this series is divergent. $\square$

Sigma notation

Next, we explain how to use the *sigma notation* to represent infinite series.

Recall that a finite sum such as

$$\frac{1}{2}+\frac{1}{4}+\frac{1}{8}+\cdots+\frac{1}{1024}=\frac{1}{2^1}+\frac{1}{2^2}+\frac{1}{2^3}+\cdots+\frac{1}{2^{10}}$$

can be represented using sigma notation as

$$\sum_{n=1}^{10}\frac{1}{2^n}.$$

This notation can be adapted to represent infinite series, as follows

$$\sum_{n=1}^{\infty}a_n=a_1+a_2+a_3+\cdots.$$

> **Convention**   When using the sigma notation to represent the *n*th partial sum s_n of a series, we write
> $$s_n=a_1+a_2+a_3+\cdots+a_n=\sum_{k=1}^{n}a_k.$$

We have written k as the 'dummy variable' here, to avoid using n for two different purposes in the same expression. We could have used any letter (other than n) for the dummy variable; for example, the expressions $\sum_{i=1}^{n}a_i$ and $\sum_{r=1}^{n}a_r$ also stand for $a_1+a_2+a_3+\cdots+a_n$.

If we need to begin a series with a term other than a_1, then we write, for example

$$\sum_{n=0}^{\infty}a_n=a_0+a_1+a_2+\cdots \quad \text{or} \quad \sum_{n=3}^{\infty}a_n=a_3+a_4+a_5+\cdots.$$

For any series, the *n*th partial sum s_n is obtained by adding the first n terms. For example, the *n*th partial sums of the above two series are

$$s_n=a_0+a_1+a_2+\cdots+a_{n-1}=\sum_{k=0}^{n-1}a_k \quad \text{and}$$

$$s_n=a_3+a_4+a_5+\cdots+a_{n+2}=\sum_{k=3}^{n+2}a_k.$$

Problem 1   For each of the following infinite series, calculate the *n*th partial sum s_n, and determine whether the series is convergent or divergent:

(a) $\sum_{n=1}^{\infty}\left(-\frac{1}{3}\right)^n$;   (b) $\sum_{n=0}^{\infty}(-1)^n$;   (c) $\sum_{n=-1}^{\infty}\left(\frac{1}{2}\right)^n$.

Remark

Notice that inserting into a series, or omitting or altering, a *finite* number of terms does not affect the convergence of the series, but may affect the sum. For example, since the series $\frac{1}{2}+\frac{1}{4}+\frac{1}{8}+\cdots$ converges with limit 1, it follows that

Σ is the Greek capital letter 'sigma'.

For example
$$\sum_{n=1}^{\infty}\frac{1}{2^n}=\frac{1}{2}+\frac{1}{4}+\frac{1}{8}+\cdots,$$
$$\sum_{n=1}^{\infty}\frac{1}{n}=1+\frac{1}{2}+\frac{1}{3}+\frac{1}{4}+\cdots.$$

For the above examples
$$s_n=\frac{1}{2}+\frac{1}{4}+\frac{1}{8}+\cdots+\frac{1}{2^n}$$
$$=\sum_{k=1}^{n}\frac{1}{2^k},$$
$$s_n=1+\frac{1}{2}+\frac{1}{3}+\frac{1}{4}+\cdots+\frac{1}{n}$$
$$=\sum_{k=1}^{n}\frac{1}{k}.$$

For example, we write
$$\sum_{n=3}^{\infty}\frac{1}{n!-n}$$
for the series
$$\frac{1}{3!-3}+\frac{1}{4!-4}+\cdots.$$

Recall that the convergence or divergence of sequences was not affected by omitting or altering a *finite* number of terms in the sequence.

the series $8+4+2+1+\frac{1}{2}+\frac{1}{4}+\frac{1}{8}+\cdots = (8+4+2+1)+\frac{1}{2}+\frac{1}{4}+\frac{1}{8}+\cdots$
converges with limit $(8+4+2+1)+1=16$.   For $\frac{1}{2}+\frac{1}{4}+\frac{1}{8}+\cdots=1$.

3.1.2   Geometric series

All the series considered so far are examples of *geometric* series. The standard
geometric series with first term a and common ratio r is

$$\sum_{n=0}^{\infty} ar^n = a + ar + ar^2 + \cdots.$$

The following theorem enables us to decide whether any given geometric
series is convergent or divergent.

> **Theorem 1   Geometric Series**
>
> (a) If $|r| < 1$, then $\displaystyle\sum_{n=0}^{\infty} ar^n$ is convergent, with sum $\frac{a}{1-r}$.
>
> (b) If $|r| \geq 1$ and $a \neq 0$, then $\displaystyle\sum_{n=0}^{\infty} ar^n$ is divergent.

Proof

(a) If $r \neq 1$, then the nth partial sum s_n is given by

$$s_n = a + ar + ar^2 + \cdots + ar^{n-1} = a\frac{1-r^n}{1-r}.$$

This value for s_n is easily
verified by Mathematical
Induction.

Now, if $|r| < 1$, then $\{r^n\}$ is a basic null sequence, and so

$$\lim_{n\to\infty} s_n = \lim_{n\to\infty} a\frac{1-r^n}{1-r} = \frac{a}{1-r}\lim_{n\to\infty}(1-r^n)$$
$$= \frac{a}{1-r},$$

by the Combination Rules for sequences. Thus, if $|r| < 1$, then $\displaystyle\sum_{n=0}^{\infty} ar^n$ is
convergent, with sum $\frac{a}{1-r}$.

(b) Part (b) can be deduced immediately from the Non-null Test, which
    appears later in this section. We defer the proof until then. ■   Sub-section 3.1.5, Example 3.

Decimal representation of rational numbers

Geometric series provide another interpretation of the decimal representation
of rational numbers. Recall that rational numbers have terminating, or recurring, decimal representations. For example   Sub-section 1.1.3.

$$\frac{3}{10} = 0.3 \quad \text{and} \quad \frac{1}{3} = 0.333\ldots = 0.\overline{3}.$$

Another way to interpret the symbol $0.333\ldots$ is as an infinite series

$$\frac{3}{10^1} + \frac{3}{10^2} + \frac{3}{10^3} + \cdots.$$

This is a geometric series, with first term $a = \frac{3}{10}$ and common ratio $r = \frac{1}{10}$.
   Since $0 < \frac{1}{10} < 1$, this series is convergent, with sum

$$\frac{a}{1-r} = \frac{\frac{3}{10}}{1-\frac{1}{10}} = \frac{3}{9} = \frac{1}{3}.$$

Thus we have another method of finding the fraction which is equal to a given recurring decimal, by calculating the sum of the corresponding geometric series.

Problem 2 Interpret the following decimals as infinite series, and hence represent them as fractions:

(a) $0.111\ldots$;   (b) $0.86363\ldots$;   (c) $0.999\ldots$.

You should at least read the solution to part (c), even if you do not tackle it first.

3.1.3 Telescoping series

Geometric series are easy to deal with because it is possible to find a formula for the nth partial sum s_n. The next problem deals with another series for which we can calculate s_n explicitly.

Problem 3 Calculate the first four partial sums of the following series, giving your answers as fractions

$$\sum_{n=1}^{\infty}\frac{1}{n(n+1)}=\frac{1}{1\times 2}+\frac{1}{2\times 3}+\frac{1}{3\times 4}+\cdots.$$

The results obtained in Problem 3 suggest the general formula

$$s_n=\frac{1}{1\times 2}+\frac{1}{2\times 3}+\frac{1}{3\times 4}+\cdots+\frac{1}{n(n+1)}=\frac{n}{n+1}.$$

This formula can be proved by Mathematical Induction, or we can use the identity

$$\frac{1}{n(n+1)}=\frac{1}{n}-\frac{1}{n+1},\quad\text{for }n=1,2,\ldots,$$

which implies that

$$s_n=\frac{1}{1\times 2}+\frac{1}{2\times 3}+\frac{1}{3\times 4}+\cdots+\frac{1}{(n-1)n}+\frac{1}{n(n+1)}$$
$$=\left(\frac{1}{1}-\frac{1}{2}\right)+\left(\frac{1}{2}-\frac{1}{3}\right)+\left(\frac{1}{3}-\frac{1}{4}\right)+\cdots+\left(\frac{1}{n-1}-\frac{1}{n}\right)+\left(\frac{1}{n}-\frac{1}{n+1}\right)$$
$$=1-\frac{1}{n+1}$$
$$=\frac{n}{n+1}.$$

This cancellation of adjacent terms explains why this series is said to be telescoping.

Since

$$\lim_{n\to\infty}s_n=\lim_{n\to\infty}\frac{n}{n+1}=\lim_{n\to\infty}\frac{1}{1+\frac{1}{n}}=1,$$

we deduce that the given series is convergent, with sum

$$\sum_{n=1}^{\infty}\frac{1}{n(n+1)}=1.$$

Problem 4 Find the nth partial sum of the series $\sum_{n=1}^{\infty}\frac{1}{n(n+2)}$, using the fact that

$$\frac{2}{n(n+2)}=\frac{1}{n}-\frac{1}{n+2},\quad\text{for }n=1,2,\ldots.$$

Deduce that this series is convergent, and find its sum.

3.1.4 Combination Rules for convergent series

At the start of this chapter we saw that performing arithmetical operations on the divergent series $2 + 4 + 8 + \cdots$ can lead to absurd conclusions. However, the following result shows that there are Combination Rules for convergent series, which follow directly from the Combination Rules for convergent sequences:

Theorem 2   Combination Rules

If $\displaystyle\sum_{n=1}^{\infty} a_n = s$ and $\displaystyle\sum_{n=1}^{\infty} b_n = t$, then:

Sum Rule $\displaystyle\sum_{n=1}^{\infty} (a_n + b_n) = s + t;$

Multiple Rule $\displaystyle\sum_{n=1}^{\infty} \lambda a_n = \lambda s$, for $\lambda \in \mathbb{R}$.

Proof   Consider the sequences of partial sums $\{s_n\}$ and $\{t_n\}$, where

$$s_n = \sum_{k=1}^{n} a_k \quad \text{and} \quad t_n = \sum_{k=1}^{n} b_k.$$

By assumption, $s_n \to s$ and $t_n \to t$ as $n \to \infty$.

Sum Rule

The nth partial sum of the series $\displaystyle\sum_{n=1}^{\infty} (a_n + b_n)$ is

$$\sum_{k=1}^{n} (a_k + b_k) = (a_1 + b_1) + (a_2 + b_2) + \cdots + (a_n + b_n)$$

$$= (a_1 + a_2 + \cdots + a_n) + (b_1 + b_2 + \cdots + b_n) = s_n + t_n.$$

We may rearrange the terms here, because this is the sum of a *finite* number of terms.

By the Sum Rule for sequences

$$\lim_{n \to \infty} (s_n + t_n) = \lim_{n \to \infty} s_n + \lim_{n \to \infty} t_n = s + t,$$

and so the sequence $\{s_n + t_n\}$ of partial sums of $\displaystyle\sum_{n=1}^{\infty} (a_n + b_n)$ has limit $s + t$.
Hence this series is convergent and

$$\sum_{n=1}^{\infty} (a_n + b_n) = s + t.$$

Multiple Rule

The nth partial sum of the series $\displaystyle\sum_{n=1}^{\infty} \lambda a_n$ is

$$\sum_{k=1}^{n} \lambda a_k = \lambda a_1 + \lambda a_2 + \cdots + \lambda a_n$$

$$= \lambda (a_1 + a_2 + \cdots + a_n) = \lambda s_n.$$

By the Multiple Rule for sequences

$$\lim_{n \to \infty} (\lambda s_n) = \lambda \lim_{n \to \infty} s_n = \lambda s,$$

and so the sequence $\{\lambda s_n\}$ of partial sums of $\sum_{n=1}^{\infty} \lambda a_n$ has limit λs. Hence this series is convergent and

$$\sum_{n=1}^{\infty} \lambda a_n = \lambda s. \qquad \blacksquare$$

Example 2   Prove that the following series is convergent and calculate its sum

$$\sum_{n=1}^{\infty} \left(\frac{1}{2^n} + \frac{3}{n(n+1)} \right).$$

Solution   We know that $\sum_{n=1}^{\infty} \frac{1}{2^n}$ is convergent, with sum 1, and that $\sum_{n=1}^{\infty} \frac{1}{n(n+1)}$ is convergent, with sum 1.

> You met the second series in Sub-section 3.1.3.

Hence, by the Sum Rule and the Multiple Rule

$$\sum_{n=1}^{\infty} \left(\frac{1}{2^n} + \frac{3}{n(n+1)} \right) \text{ is convergent, with sum } 1 + (3 \times 1) = 4. \quad \square$$

> **Problem 5**   Prove that the following series is convergent and calculate its sum
> $$\sum_{n=1}^{\infty} \left(\left(\frac{3}{4}\right)^n - \frac{2}{n(n+1)} \right).$$

3.1.5   The Non-null Test

For all the infinite series we have so far considered, it is possible to derive a simple formula for the nth partial sum. For most series, however, this is very difficult or even impossible.

Nevertheless, it may still be possible to decide whether such series are convergent or divergent by applying various tests. The first test that we give arises from the following result:

Theorem 3   If $\sum_{n=1}^{\infty} a_n$ is a convergent series, then $\{a_n\}$ is a null sequence.

Proof   Let $s_n = \sum_{k=1}^{n} a_k$ denote the nth partial sum of $\sum_{n=1}^{\infty} a_n$. Because $\sum_{n=1}^{\infty} a_n$ is convergent, we know that $\{s_n\}$ is convergent. Suppose that $\lim_{n \to \infty} s_n = s$.

We want to deduce that $\{a_n\}$ is null. To do this, we write

$$s_n = s_{n-1} + a_n,$$

and so

$$a_n = s_n - s_{n-1}.$$

Thus, by the Combination Rules for sequences

$$\lim_{n \to \infty} a_n = \lim_{n \to \infty} (s_n - s_{n-1}) = \lim_{n \to \infty} s_n - \lim_{n \to \infty} s_{n-1}$$
$$= s - s = 0,$$

and so $\{a_n\}$ is a null sequence, as required. $\blacksquare$

The following useful test for divergence is an immediate corollary of Theorem 3:

Corollary   **Non-null Test**   If $\{a_n\}$ is *not* a null sequence, then $\sum\limits_{n=1}^{\infty} a_n$ is divergent.

The virtue of the Non-null Test is its ease of application. For example, it enables us to decide immediately that

$$\sum_{n=1}^{\infty} 1, \quad \sum_{n=1}^{\infty} (-1)^n \quad \text{and} \quad \sum_{n=1}^{\infty} n \text{ are divergent,}$$

because the corresponding sequences $\{1\}$, $\{(-1)^n\}$ and $\{n\}$ are not null.

There are various ways to show that a given sequence $\{a_n\}$ *does not* tend to zero. For example, we can show that

$$a_n \to \ell \quad \text{as} \quad n \to \infty,$$

where $\ell \neq 0$, or that $\{a_n\}$ tends to infinity or to minus infinity. More generally, we can use the result about subsequences which states that

$$\text{if } a_n \to 0 \text{ as } n \to \infty, \text{ then } a_{n_k} \to 0 \text{ as } k \to \infty,$$

for any subsequence $\{a_{n_k}\}$ of $\{a_n\}$. This leads to the following strategy:

> You met this Inheritance Property of subsequences in Sub-section 2.4.4, Theorem 5, for *any* limit ℓ.

Strategy   To show that $\sum\limits_{n=1}^{\infty} a_n$ is divergent, using the Non-null Test:

EITHER

1. show that $\{a_n\}$ has a convergent subsequence with non-zero limit;

OR

2. show that $\{a_n\}$ has a subsequence which tends to infinity, or a subsequence which tends to minus infinity.

This strategy can be used to prove part (b) of Theorem 1, as follows:

Example 3   Prove that, if $|r| \geq 1$ and $a \neq 0$, then $\sum\limits_{n=0}^{\infty} ar^n$ is divergent.

Solution   We want to show that, if $|r| \geq 1$ and $a \neq 0$, then $\{ar^n\}$ is not a null sequence. Now, the even subsequence

$$\{ar^{2k}\} = \{a, ar^2, ar^4, ar^6, \ldots\}$$

converges to $a \neq 0$ if $|r| = 1$, and tends to infinity or to minus infinity if $|r| \geq 1$. Thus, $\{ar^{2k}\}$ is not null, and so $\{ar^n\}$ is not null.

> It tends to infinity if $a > 0$ and to minus infinity if $a < 0$.

Hence, by the Non-null Test, $\sum\limits_{n=0}^{\infty} ar^n$ is divergent if $|r| \geq 1$ and $a \neq 0$. $\quad\square$

Warning   The converse of the Non-null Test is FALSE. In other words:

If $\{a_n\}$ is a null sequence, it is *not necessarily true* that $\sum\limits_{n=1}^{\infty} a_n$ is convergent.

For example, the sequence $\{\frac{1}{n}\}$ is a null sequence, but $\sum\limits_{n=1}^{\infty} \frac{1}{n} = 1 + \frac{1}{2} + \frac{1}{3} + \cdots$ is divergent.

> We shall prove this rather surprising fact in Sub-section 3.2.1.

The important thing to remember is that *you can never use the Non-null Test to prove that a series is convergent*, although you *may* be able to use it to prove that a series is divergent.

Problem 6 Prove that the series $\sum_{n=1}^{\infty} \frac{n^2}{2n^2+1}$ is divergent.

3.2 Series with non-negative terms

In this section we consider only series $\sum_{n=1}^{\infty} a_n$ with non-negative terms. In other words we assume that $a_n \geq 0$, for $n = 1, 2, \ldots$.

It follows that the partial sums of $\sum_{n=1}^{\infty} a_n$, which are given by

$$s_1 = a_1, \quad s_2 = a_1 + a_2, \quad s_3 = a_1 + a_2 + a_3, \ldots,$$
$$s_n = a_1 + a_2 + a_3 + \cdots + a_n, \ldots,$$

form an *increasing* sequence $\{s_n\}$.

The fact that $\{s_n\}$ is increasing makes it easier to deal with series having non-negative terms. If we can prove that the sequence $\{s_n\}$ of partial sums is bounded above, then $\{s_n\}$ is convergent, by the Monotone Convergence Theorem, and so $\sum_{n=1}^{\infty} a_n$ is convergent. On the other hand, if we can prove that the sequence $\{s_n\}$ of partial sums is *not* bounded above, then $\{s_n\}$ cannot be convergent, and so we must have that $s_n \to \infty$ as $n \to \infty$. We can rephrase these facts in the following convenient way:

Sub-section 2.5.1, Theorem 1.

Boundedness Theorem for series A series $\sum_{n=1}^{\infty} a_n$ of non-negative terms is convergent if and only if its sequence $\{s_n\}$ of partial sums is bounded above.

This is essentially the same result as that of the Boundedness Theorem for sequences, Sub-section 2.4.2, Theorem 1.

We shall use this, for example, to prove the Cauchy Condensation Test in Sub-section 3.2.1.

3.2.1 Tests for convergence

We now explore several tests for the convergence of series with non-negative terms.

Problem 1 Use your calculator to find the first eight partial sums of each of the following series

$$\sum_{n=1}^{\infty} \frac{1}{n^2} = 1 + \frac{1}{2^2} + \frac{1}{3^2} + \cdots \quad \text{and} \quad \sum_{n=1}^{\infty} \frac{1}{n} = 1 + \frac{1}{2} + \frac{1}{3} + \cdots,$$

giving your answers to 2 decimal places. Plot your answers on a sequence diagram.

Example 1 Prove that the series $\sum_{n=1}^{\infty} \frac{1}{n^2} = 1 + \frac{1}{2^2} + \frac{1}{3^2} + \cdots$ is convergent.

Solution All the terms of the series are positive, so we shall use the Monotone Convergence Theorem.

Let s_n be the nth partial sum of the series. Then

$$s_n = 1 + \frac{1}{2^2} + \frac{1}{3^2} + \frac{1}{4^2} + \cdots + \frac{1}{n^2}$$

$$< 1 + \frac{1}{1 \times 2} + \frac{1}{2 \times 3} + \frac{1}{3 \times 4} + \cdots + \frac{1}{(n-1) \times n}$$

For any $k > 1$, $\frac{1}{k^2} < \frac{1}{(k-1)k}$.

$$= 1 + \left(\frac{1}{1} - \frac{1}{2}\right) + \left(\frac{1}{2} - \frac{1}{3}\right) + \left(\frac{1}{3} - \frac{1}{4}\right) + \cdots + \left(\frac{1}{n-1} - \frac{1}{n}\right)$$

Here we have telescoping cancellation.

$$= 2 - \frac{1}{n} < 2.$$

It follows that $\{s_n\}$ is both increasing and bounded above (by 2), so that by the Monotone Convergence Theorem $\{s_n\}$ is convergent. Hence the series itself is also convergent. $\qquad\square$

Remark

In fact the sum of the series is $\sum_{n=1}^{\infty} \frac{1}{n^2} = \frac{\pi^2}{6}$, surprisingly. However a proof of this goes beyond the scope of this book.

The proof depends on use of trigonometric series or of Complex Analysis.

Not all series are convergent, though!

Example 2 Prove that the series $\sum_{n=1}^{\infty} \frac{1}{n} = 1 + \frac{1}{2} + \frac{1}{3} + \cdots$ is divergent.

This series is often called the *harmonic series*, since its terms are proportional to the lengths of strings that produce harmonic tones in music.

Solution All the terms of the series are positive.

Let s_n be the nth partial sum of the series. Then

$$s_n = 1 + \frac{1}{2} + \frac{1}{3} + \frac{1}{4} + \frac{1}{5} + \frac{1}{6} + \frac{1}{7} + \frac{1}{8} + \frac{1}{9} + \frac{1}{10} + \cdots$$

$$= 1 + \frac{1}{2} + \left(\frac{1}{3} + \frac{1}{4}\right) + \left(\frac{1}{5} + \frac{1}{6} + \frac{1}{7} + \frac{1}{8}\right) + \cdots$$

We think of n as being fairly large, in order to see the pattern.

$$\cdots + \left(\frac{1}{2^k + 1} + \frac{1}{2^k + 2} + \cdots + \frac{1}{2^k + 2^k}\right) + \cdots.$$

Here each bracket adds up to more than $\frac{1}{2}$.

It follows that a subsequence of partial sums is

$$s_1 = 1,$$

$$s_2 = 1 + \frac{1}{2},$$

$$s_4 = 1 + \frac{1}{2} + \left(\frac{1}{3} + \frac{1}{4}\right) > 1 + \frac{1}{2} + \frac{1}{2},$$

$$s_8 = 1 + \frac{1}{2} + \left(\frac{1}{3} + \frac{1}{4}\right) + \left(\frac{1}{5} + \frac{1}{6} + \frac{1}{7} + \frac{1}{8}\right) > 1 + \frac{1}{2} + \frac{1}{2} + \frac{1}{2},$$

$$\vdots$$

$$s_{2^k} > 1 + \underbrace{\frac{1}{2} + \frac{1}{2} + \cdots + \frac{1}{2}}_{k \text{ terms}} = 1 + \frac{1}{2}k.$$

Hence the sequence $\{s_n\}$ is increasing and not bounded above. Therefore it cannot be convergent, so that by the Monotone Convergence Theorem the sequence $\{s_n\}$ is divergent. Hence the series itself is also divergent.   □

We can extend this type of domination approach to test further series for convergence, in the following way:

Theorem 1   Comparison Test

(a) If $0 \leq a_n \leq b_n$, for $n = 1, 2, \ldots$, and $\sum\limits_{n=1}^{\infty} b_n$ is convergent, then $\sum\limits_{n=1}^{\infty} a_n$ is convergent.

(b) If $0 \leq b_n \leq a_n$, for $n = 1, 2, \ldots$, and $\sum\limits_{n=1}^{\infty} b_n$ is divergent, then $\sum\limits_{n=1}^{\infty} a_n$ is divergent.

In the proof of part (a), in Subsection 3.2.2, we shall in fact see that

$$\sum_{n=1}^{\infty} a_n \leq \sum_{n=1}^{\infty} b_n.$$

Remark

As with the Squeeze Rule for non-negative null sequences, it is sufficient to prove that the necessary inequalities in parts (a) and (b) hold *eventually*.

In applications, we use this result in the following way:

That is, it is sufficient that $0 \leq a_n \leq b_n$ or $0 \leq b_n \leq a_n$ for all $n > X$, for some number X.

Strategy   To test a series $\sum\limits_{n=1}^{\infty} a_n$ of non-negative terms for convergence using the Comparison Test:

EITHER   find a convergent series $\sum\limits_{n=1}^{\infty} b_n$ of non-negative terms where the b_n dominate the a_n,

OR   find a divergent series $\sum\limits_{n=1}^{\infty} b_n$ of non-negative terms where the a_n dominate the b_n.

That is, where $a_n \leq b_n$.

That is, where $b_n \leq a_n$.

Example 3   Prove that the series $\sum\limits_{n=1}^{\infty} \frac{1}{n^3}$ is convergent.

Solution   We shall use the Comparison Test.

Let $b_n = \frac{1}{n^2}$. Then $0 \leq \frac{1}{n^3} \leq b_n$, for $n = 1, 2, \ldots$, and we know that the series $\sum\limits_{n=1}^{\infty} \frac{1}{n^2}$ is convergent (this was Example 1 above). It follows from part (a) of the Comparison Test that the series $\sum\limits_{n=1}^{\infty} \frac{1}{n^3}$ is convergent.   □

For $\frac{1}{n^3} \leq \frac{1}{n^2}$.

Example 4   Prove that the series $\sum\limits_{n=1}^{\infty} \frac{1}{\sqrt{n}}$ is divergent.

Solution   We shall use the Comparison Test.

Let $b_n = \frac{1}{n}$. Then $0 \leq b_n \leq \frac{1}{\sqrt{n}}$, for $n = 1, 2, \ldots$, and we know that the series $\sum\limits_{n=1}^{\infty} \frac{1}{n}$ is divergent (this was Example 2 above). It follows from part (b) of the Comparison Test that the series $\sum\limits_{n=1}^{\infty} \frac{1}{\sqrt{n}}$ is divergent.   □

For $\frac{1}{n} \leq \frac{1}{\sqrt{n}}$.

Next, consider the series $\sum\limits_{n=1}^{\infty} \frac{1}{\sqrt{n}+3\sqrt[4]{n}+1}$. Its terms are somewhat similar to those of the series $\sum\limits_{n=1}^{\infty} \frac{1}{\sqrt{n}}$, since we can express $\frac{1}{\sqrt{n}+3\sqrt[4]{n}+1}$ in the form

In other words, the term $\sqrt{n}$ dominates the term $3\sqrt[4]{n}+1$.

$\frac{1}{\sqrt{n}} \times \frac{1}{1+3n^{-\frac{1}{4}}+n^{-\frac{1}{2}}}$, where the second fraction tends to 1 as $n \to \infty$. So, since the series $\sum_{n=1}^{\infty} \frac{1}{\sqrt{n}}$ is divergent, it seems likely that the series $\sum_{n=1}^{\infty} \frac{1}{\sqrt{n}+3\sqrt[4]{n}+1}$ is also divergent.

By Example 4.

We can pin down the underlying idea here in the following useful result:

Theorem 2   Limit Comparison Test

Suppose that $\sum_{n=1}^{\infty} a_n$ and $\sum_{n=1}^{\infty} b_n$ have positive terms, and that

$$\frac{a_n}{b_n} \to L \quad \text{as} \quad n \to \infty, \quad \text{where } L \neq 0.$$

(a) If $\sum_{n=1}^{\infty} b_n$ is convergent, then $\sum_{n=1}^{\infty} a_n$ is convergent.

(b) If $\sum_{n=1}^{\infty} b_n$ is divergent, then $\sum_{n=1}^{\infty} a_n$ is divergent.

In other words, b_n behaves 'rather like a_n,' for large n. Note that it is IMPORTANT that L is *non-zero*.

So, take $a_n = \frac{1}{\sqrt{n}+3\sqrt[4]{n}+1}$ and $b_n = \frac{1}{\sqrt{n}}$, for $n = 1, 2, \ldots$. Then both a_n and b_n are positive, and

$$\frac{a_n}{b_n} = \frac{(1/(\sqrt{n}+3\sqrt[4]{n}+1))}{(1/\sqrt{n})}$$

$$= \frac{\sqrt{n}}{\sqrt{n}+3\sqrt[4]{n}+1}$$

$$= \frac{1}{1+3n^{-\frac{1}{4}}+n^{-\frac{1}{2}}} \to 1 \text{ as } n \to \infty.$$

It then follows from part (b) of the Limit Comparison Test that, since the series $\sum_{n=1}^{\infty} \frac{1}{\sqrt{n}}$ is divergent, the series $\sum_{n=1}^{\infty} \frac{1}{\sqrt{n}+3\sqrt[4]{n}+1}$ is also divergent.

Since $1 \neq 0$.

By Example 4.

Example 5   Use the Limit Comparison Test to prove that the series $\sum_{n=1}^{\infty} \frac{n^2-3n+4}{5n^4-n}$ is convergent.

Solution   Set $a_n = \frac{n^2-3n+4}{5n^4-n}$ and $b_n = \frac{1}{n^2}$, for $n = 1, 2, \ldots$. Then both a_n and b_n are positive, and

$$\frac{a_n}{b_n} = \frac{n^2-3n+4}{5n^4-n} \times \frac{n^2}{1}$$

$$= \frac{n^4-3n^3+4n^2}{5n^4-n}$$

$$= \frac{1-3n^{-1}+4n^{-2}}{5-n^{-3}} \to \frac{1}{5} \quad \text{as } n \to \infty.$$

We make this choice for b_n since, for large n, the expression $\frac{n^2-3n+4}{5n^4-n}$ is 'more or less the same as' $\frac{n^2}{5n^4} = \frac{1}{5n^2}$ – where the factor '5' will not affect our argument.

Dividing numerator and denominator by the dominant term n^4.

Since $\frac{1}{5} \neq 0$ and the series $\sum_{n=1}^{\infty} \frac{1}{n^2}$ is known to be convergent, then the series $\sum_{n=1}^{\infty} \frac{n^2-3n+4}{5n^4-n}$ is also convergent. $\square$

By Example 1.

Problem 2 Use the Comparison Test or the Limit Comparison Test to determine the convergence of the following series:

(a) $\sum_{n=1}^{\infty} \frac{1}{n^3+n}$; (b) $\sum_{n=1}^{\infty} \frac{1}{n+\sqrt{n}}$; (c) $\sum_{n=1}^{\infty} \frac{n+4}{2n^3-n+1}$; (d) $\sum_{n=1}^{\infty} \frac{\cos^2(2n)}{n^3}$.

Our next test is motivated in part by the geometric series – recall that the geometric series $a + ar + ar^2 + \cdots = \sum_{n=0}^{\infty} ar^n$, where $a \neq 0$, is convergent if $|r| < 1$ but divergent if $|r| \geq 1$. The reason that this converges, if $|r| < 1$, is that the terms are then forced to tend to 0 so quickly that the partial sums s_n increase so slowly as n increases that they remain bounded above. On the other hand, if $|r| \geq 1$, the terms do not tend to 0, so that the series must diverge.

By the Non-null Test.

Theorem 3 Ratio Test

Suppose that $\sum_{n=1}^{\infty} a_n$ has positive terms, and that $\frac{a_{n+1}}{b_n} \to \ell$ as $n \to \infty$.

(a) If $0 \leq \ell < 1$, then $\sum_{n=1}^{\infty} a_n$ is convergent.

(b) If $\ell > 1$, then $\sum_{n=1}^{\infty} a_n$ is divergent.

Note that with the Ratio Test, we concentrate on the series $\sum_{n=1}^{\infty} a_n$ itself and do not need to 'think of' some other series $\sum_{n=1}^{\infty} b_n$.

Part (b) includes the case $\frac{a_{n+1}}{a_n} \to \infty$.

Remark

If $\ell = 1$, then the test gives us no information on whether the series converges. For example, if $a_n = \frac{1}{n}$, then $\frac{a_{n+1}}{a_n} = \frac{n}{n+1} = \frac{1}{1+\frac{1}{n}} \to 1$ as $n \to \infty$; we have seen that the series $\sum_{n=1}^{\infty} a_n = \sum_{n=1}^{\infty} \frac{1}{n}$ diverges. On the other hand, if $a_n = \frac{1}{n^2}$ then $\frac{a_{n+1}}{a_n} = \frac{n^2}{(n+1)^2} = \frac{1}{1+\frac{2}{n}+\frac{1}{n^2}} \to 1$ as $n \to \infty$; but the series $\sum_{n=1}^{\infty} a_n = \sum_{n=1}^{\infty} \frac{1}{n^2}$ converges.

That is, the Ratio Test is *inconclusive* if $\ell = 1$.

By Example 2.

By Example 1.

Example 6 Use the Ratio Test to determine the convergence of the following series:

(a) $\sum_{n=1}^{\infty} \frac{n}{2^n}$; (b) $\sum_{n=1}^{\infty} \frac{10^n}{n!}$.

Solution

(a) Let $a_n = \frac{n}{2^n}$, $n = 1, 2, \ldots$. Then a_n is positive, and

$$\frac{a_{n+1}}{a_n} = \frac{n+1}{2^{n+1}} \times \frac{2^n}{n}$$

$$= \frac{1+\frac{1}{n}}{2} \to \frac{1}{2} \quad \text{as } n \to \infty.$$

Since $0 < \frac{1}{2} < 1$, it follows from the Ratio Test that the series $\sum_{n=1}^{\infty} a_n = \sum_{n=1}^{\infty} \frac{n}{2^n}$ is convergent.

(b) Let $a_n = \frac{10^n}{n!}$, $n = 1, 2, \ldots$. Then a_n is positive, and

$$\frac{a_{n+1}}{a_n} = \frac{10^{n+1}}{(n+1)!} \times \frac{n!}{10^n}$$

$$= \frac{10}{n+1} \to 0 \quad \text{as } n \to \infty.$$

It follows from the Ratio Test that the series $\sum_{n=1}^{\infty} a_n = \sum_{n=1}^{\infty} \frac{10^n}{n!}$ is convergent. $\square$

Problem 3   Use the Ratio Test to determine whether the following series are convergent

(a) $\sum_{n=1}^{\infty} \frac{n^3}{n!}$;   (b) $\sum_{n=1}^{\infty} \frac{n^2 2^n}{n!}$;   (c) $\sum_{n=1}^{\infty} \frac{(2n)!}{n^n}$.

We have now seen examples of many different convergent and divergent series, including examples of various generic types. We call these *basic series*, since we shall use them commonly together with various other tests in order to prove that particular series are themselves convergent or divergent.

Basic series   The following series are **convergent**:

(a) $\sum_{n=1}^{\infty} \frac{1}{n^p}$,   for $p \geq 2$;

(b) $\sum_{n=1}^{\infty} c^n$,   for $0 \leq c < 1$;

(c) $\sum_{n=1}^{\infty} n^p c^n$,   for $p > 0, 0 \leq c < 1$;

(d) $\sum_{n=1}^{\infty} \frac{c^n}{n!}$,   for $c \geq 0$.

The following series is **divergent**:

(e) $\sum_{n=1}^{\infty} \frac{1}{n^p}$,   for $p \leq 1$.

Instances of these that we have seen are

$\sum_{n=1}^{\infty} \frac{1}{n^2}$;

Geometric Series, $\sum_{n=1}^{\infty} \left(\frac{1}{2}\right)^n$;

$\sum_{n=1}^{\infty} \frac{n}{2^n} = \sum_{n=1}^{\infty} n\left(\frac{1}{2}\right)^n$;

$\sum_{n=1}^{\infty} \frac{10^n}{n!}$;

$\sum_{n=1}^{\infty} \frac{1}{\sqrt{n}} = \sum_{n=1}^{\infty} \frac{1}{n^{1/2}}$.

Another test for convergence

When we examined the convergence or divergence of the series $\sum_{n=1}^{\infty} \frac{1}{n}$, we found it convenient to group the terms into blocks that contained in turn $1, 2, 4, 8, \ldots$ successive individual terms of the series. In fact, a similar technique applies to a wide range of series, and the following result is thus often invaluable in order to save repeating that same type of argument:

You will see this technique used in the proof of Theorem 4, in Subsection 3.2.2.

Theorem 4   Cauchy Condensation Test

Let $\{a_n\}$ be a decreasing sequence of positive terms. Then, if $b_n = 2^n a_{2^n}$, for $n = 1, 2, \ldots$

$$\sum_{n=1}^{\infty} a_n \text{ is convergent if and only if } \sum_{n=1}^{\infty} b_n \text{ is convergent.}$$

For example, consider the series $\sum_{n=1}^{\infty} \frac{1}{n}$. Here, let $a_n = \frac{1}{n}$, so that

$$b_n = 2^n a_{2^n} = 2^n \times \frac{1}{2^n} = 1.$$

For, if $a_n = \frac{1}{n}$, then $a_{2^n} = \frac{1}{2^n}$.

Then the Condensation Test tells us that the series $\sum_{n=1}^{\infty} \frac{1}{n}$ is convergent if and only if the series $\sum_{n=1}^{\infty} b_n = \sum_{n=1}^{\infty} 1$ is convergent. This is divergent, by the Non-null Test, so that the original series $\sum_{n=1}^{\infty} \frac{1}{n}$ must itself have been divergent.

This use of the Condensation Test is much simpler than setting out to perform a grouping and estimation exercise on such occasions! The Condensation Test is also often useful when the individual terms of a series include expressions such as $\log_e n$.

Example 7 Use the Condensation Test to determine the convergence of the series $\sum_{n=2}^{\infty} \frac{1}{n\sqrt{\log_e n}}$.

Solution Let $a_n = \frac{1}{n\sqrt{\log_e n}}$, $n = 2, \ldots$. Then a_n is positive; and, since $\{n\sqrt{\log_e n}\}$ is an increasing sequence, $\{a_n\} = \left\{\frac{1}{n\sqrt{\log_e n}}\right\}$ is a decreasing sequence.

Next, let $b_n = 2^n a_{2^n}$; thus

$$b_n = 2^n \times \frac{1}{2^n \sqrt{\log_e (2^n)}}$$
$$= \frac{1}{\sqrt{n \log_e 2}} = \frac{1}{\sqrt{\log_e 2}} \times \frac{1}{n^{\frac{1}{2}}}.$$

Since $\sum_{n=2}^{\infty} \frac{1}{n^{\frac{1}{2}}}$ is a basic divergent series, it follows by the Multiple Rule that $\sum_{n=2}^{\infty} b_n$ must be divergent. Hence by the Condensation Test, the original series $\sum_{n=2}^{\infty} \frac{1}{n\sqrt{\log_e n}}$ must also be divergent. $\qquad \square$

Note that here we only sum from $n = 2$, since it makes no sense to talk about $\frac{1}{1\sqrt{\log_e 1}}$, as $\log_e 1 = 0$. Again we are using the fact that the convergence of a series is not affected if we add or alter a finite of terms.

Problem 4 Use the Condensation Test to determine the convergence of the following series:

(a) $\sum_{n=2}^{\infty} \frac{1}{n \log_e n}$; (b) $\sum_{n=2}^{\infty} \frac{1}{n(\log_e n)^2}$.

3.2.2 Proofs

You may omit these proofs at a first reading.

In the previous sub-section we gave a number of tests for the convergence of series with non-negative terms. We now supply the proofs of these tests.

Theorem 1   Comparison Test

(a) If $0 \leq a_n \leq b_n$, for $n = 1, 2, \ldots$, and $\sum_{n=1}^{\infty} b_n$ is convergent, then $\sum_{n=1}^{\infty} a_n$ is convergent.

(b) If $0 \leq b_n \leq a_n$, for $n = 1, 2, \ldots$, and $\sum_{n=1}^{\infty} b_n$ is divergent, then $\sum_{n=1}^{\infty} a_n$ is divergent.

Proof

(a) Consider the nth partial sums

$$s_n = a_1 + a_2 + \cdots + a_n, \quad \text{for } n = 1, 2, \ldots,$$

and
$$t_n = b_1 + b_2 + \cdots + b_n, \quad \text{for } n = 1, 2, \ldots.$$
We know that
$$a_1 \leq b_1, a_2 \leq b_2, \ldots, a_n \leq b_n,$$
and so
$$s_n \leq t_n, \quad \text{for } n = 1, 2, \ldots.$$

By adding up the previous inequalities.

We also know that $\sum_{n=1}^{\infty} b_n$ is convergent, and so the increasing sequence $\{t_n\}$ is convergent, with limit t, say. Hence
$$s_n \leq t_n \leq t, \quad \text{for } n = 1, 2, \ldots,$$
and so the increasing sequence $\{s_n\}$ is bounded above by t. By the Monotone Convergence Theorem, $\{s_n\}$ is therefore also convergent, and so $\sum_{n=1}^{\infty} a_n$ is a convergent series.

Note: Moreover, by the Limit Inequality Rule
$$\sum_{n=1}^{\infty} a_n \leq \sum_{n=1}^{\infty} b_n.$$

You met this Rule in Sub-section 2.3.3, Theorem 3. This additional inequality will sometimes be useful.

(b) We can deduce part (b) from part (a), as follows. Suppose that $\sum_{n=1}^{\infty} a_n$ is convergent. Then, by part (a), $\sum_{n=1}^{\infty} b_n$ must also be convergent. However, $\sum_{n=1}^{\infty} b_n$ is assumed to be divergent, and so $\sum_{n=1}^{\infty} a_n$ must also be divergent. ∎

Such 'proofs by contradiction' can sometimes save a great deal of detailed arguments.

Theorem 2   Limit Comparison Test

Suppose that $\sum_{n=1}^{\infty} a_n$ and $\sum_{n=1}^{\infty} b_n$ have positive terms, and that
$$\frac{a_n}{b_n} \to L \text{ as } n \to \infty, \quad \text{where } L \neq 0.$$

(a) If $\sum_{n=1}^{\infty} b_n$ is convergent, then $\sum_{n=1}^{\infty} a_n$ is convergent.

(b) If $\sum_{n=1}^{\infty} b_n$ is divergent, then $\sum_{n=1}^{\infty} a_n$ is divergent.

Proof

(a) Because $\left\{\frac{a_n}{b_n}\right\}$ is convergent, it must be bounded. Thus there is a constant K such that
$$\frac{a_n}{b_n} \leq K, \quad \text{for } n = 1, 2, \ldots,$$
and so
$$a_n \leq K b_n, \quad \text{for } n = 1, 2, \ldots.$$

By the Multiple Rule, $\sum_{n=1}^{\infty} K b_n$ is convergent. Hence, by the Comparison Test, $\sum_{n=1}^{\infty} a_n$ is also convergent.

See Sub-section 2.4.2, Theorem 1 (the *Boundedness Theorem*).

(b) We can deduce part (b) from part (a), as follows.

Suppose that $\sum_{n=1}^{\infty} a_n$ is convergent. Then, by part (a), $\sum_{n=1}^{\infty} b_n$ must also be convergent, because

$$\frac{b_n}{a_n} \to \frac{1}{L} \quad \text{as } n \to \infty,$$

by the Quotient Rule for sequences.

However, $\sum_{n=1}^{\infty} b_n$ is assumed to be divergent, and so $\sum_{n=1}^{\infty} a_n$ must also be divergent. ∎

We are following a 'proof by contradiction' approach here.

Remember that $L \neq 0$.

Theorem 3   Ratio Test

Suppose that $\sum_{n=1}^{\infty} a_n$ has positive terms, and that $\frac{a_{n+1}}{a_n} \to \ell$ as $n \to \infty$.

(a) If $0 \leq \ell < 1$, then $\sum_{n=1}^{\infty} a_n$ is convergent.

(b) If $\ell > 1$, then $\sum_{n=1}^{\infty} a_n$ is divergent.

Remember that part (b) includes the case $\frac{a_{n+1}}{a_n} \to \infty$.

Proof

(a) Because $\ell < 1$, we can choose $\varepsilon > 0$ so that

$$r = \ell + \varepsilon < 1.$$

Hence there is a positive number X, which we may assume to be a positive integer, such that

$$\frac{a_{n+1}}{a_n} \leq r, \quad \text{for all } n \geq X.$$

Thus, for $n \geq X$, we have

$$\frac{a_n}{a_X} = \left(\frac{a_n}{a_{n-1}}\right)\left(\frac{a_{n-1}}{a_{n-2}}\right) \cdots \left(\frac{a_{X+1}}{a_X}\right) \leq r^{n-X},$$

since each of the $n - X$ brackets is less than or equal to r. Hence

$$a_n \leq a_X r^{n-X}, \quad \text{for all } n \geq X. \tag{1}$$

Now

$$\sum_{n=X}^{\infty} a_X r^{n-X} = a_X + a_X r + a_X r^2 + \cdots,$$

which is a geometric series with first term a_X and common ratio r. Since $0 < r < 1$, this series is convergent, and so, by inequality (1) and the Comparison Test, $\sum_{n=X}^{\infty} a_n$ is also convergent. Hence $\sum_{n=1}^{\infty} a_n$ is convergent.

(b) Since

$$\frac{a_{n+1}}{a_n} \to \ell \quad \text{as } n \to \infty,$$

and $\ell > 1$, there is a positive number X, which we may assume to be a positive integer, such that

$$\frac{a_{n+1}}{a_n} \geq 1, \quad \text{for all } n \geq X.$$

For example, take $\varepsilon = \frac{1}{2}(1 - \ell)$.

Recall that we occasionally make the convenient assumption that X is an integer to avoid notational complications.

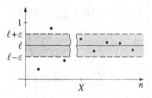

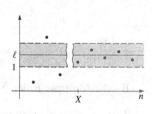

(This also holds if $\frac{a_{n+1}}{a_n} \to \infty$ as $n \to \infty$.)

Thus, for $n \geq X$, we have

$$\frac{a_n}{a_X} = \left(\frac{a_n}{a_{n-1}}\right)\left(\frac{a_{n-1}}{a_{n-2}}\right)\cdots\left(\frac{a_{X+1}}{a_X}\right) \geq 1,$$

since each of the brackets is greater than or equal to 1. Hence

$$a_n \geq a_X > 0, \quad \text{for all } n \geq X,$$

and so $\{a_n\}$ cannot be a null sequence. Thus, by the Non-null Test, $\sum\limits_{n=X}^{\infty} a_n$ is

divergent. Hence $\sum\limits_{n=1}^{\infty} a_n$ is also divergent. ∎

Theorem 4   Cauchy Condensation Test

Let $\{a_n\}$ be a decreasing sequence of positive terms. Then, if $b_n = 2^n a_{2^n}$ for $n = 1, 2, \ldots$

$$\sum_{n=1}^{\infty} a_n \text{ is convergent if and only if } \sum_{n=1}^{\infty} b_n \text{ is convergent.}$$

Proof   Denote by s_n and t_n the nth partial sums of the series $\sum\limits_{n=1}^{\infty} a_n$ and $\sum\limits_{n=1}^{\infty} b_n$, respectively

$$s_n = a_1 + a_2 + \cdots + a_n \quad \text{and} \quad t_n = 2^1 a_2 + 2^2 a_4 + \cdots + 2^n a_{2^n}.$$

Since the terms a_n are decreasing and $a_1 \geq 0$, it follows that

$$s_{2^n} = a_1 + (a_2) + (a_3 + a_4) + \cdots + (a_{2^{n-1}+1} + \cdots + a_{2^n})$$

The kth bracket contains 2^{k-1} terms.

$$\geq 0 + (a_2) + (a_4 + a_4) + \cdots + (a_{2^n} + \cdots + a_{2^n})$$

The kth bracket contains 2^{k-1} occurrences of a_{2^k}.

$$= 0 + (a_2) + (2a_4) + \cdots + (2^{n-1} a_{2^n})$$

$$= \frac{1}{2} t_n. \tag{2}$$

We may also write s_{2^n} in *another* way

$$s_{2^n} = a_1 + (a_2 + a_3) + (a_4 + a_5 + a_6 + a_7) + \cdots$$
$$+ (a_{2^{n-1}} + \cdots + a_{2^n-1}) + a_{2^n}$$

The kth bracket contains 2^k terms.

$$\leq a_1 + (a_2 + a_2)$$
$$+ (a_4 + a_4 + a_4 + a_4) + \cdots + (a_{2^{n-1}} + \cdots + a_{2^{n-1}}) + a_{2^n}$$

The kth bracket contains 2^k occurrences of a_{2^k}.

$$= a_1 + (2a_2) + (4a_4) + \cdots + (2^{n-1} a_{2^{n-1}}) + a_{2^n}$$

In obtaining (3), we have replaced the term a_{2^n} in the previous line by a larger term $2^n a_{2^n}$.

$$\leq a_1 + t_n. \tag{3}$$

Now suppose that $\sum\limits_{n=1}^{\infty} a_n$ is convergent. Then, by the Boundedness Theorem for series, its sequence of partial sums $\{s_n\}$ must be bounded above and so the sequence $\{s_{2^n}\}$ must also be bounded above – since it is a subsequence of $\{s_n\}$. It therefore follows from inequality (2) that the sequence $\{\frac{1}{2} t_n\}$ must be bounded above, so the sequence $\{t_n\}$ must also be bounded above. Another application of the Boundedness Theorem shows that therefore the series $\sum\limits_{n=1}^{\infty} b_n$ must be convergent.

The Boundedness Theorem for series appeared in Section 3.2, above.

Next, suppose that $\sum_{n=1}^{\infty} a_n$ is divergent. Then, by the Boundedness Theorem for series, its sequence of partial sums $\{s_n\}$ must be unbounded above, and so the sequence $\{s_{2^n}\}$ must also be unbounded above – since it is a subsequence of $\{s_n\}$. It therefore follows from inequality (3) that the sequence $\{a_1 + t_n\}$ must be unbounded above, so the sequence $\{t_n\}$ must also be unbounded above. Another application of the Boundedness Theorem shows that therefore the series $\sum_{n=1}^{\infty} b_n$ must be divergent. ∎

We end this section by proving the convergence/divergence of the basic series given earlier.

In Sub-section 3.2.1.

Basic series   The following series are **convergent**:

(a) $\sum_{n=1}^{\infty} \frac{1}{n^p}$,   for $p \geq 2$;

(b) $\sum_{n=1}^{\infty} c^n$,   for $0 \leq c < 1$;

(c) $\sum_{n=1}^{\infty} n^p c^n$,   for $p > 0, 0 \leq c < 1$;

(d) $\sum_{n=1}^{\infty} \frac{c^n}{n!}$,   for $c \geq 0$.

The following series is **divergent**:

(e) $\sum_{n=1}^{\infty} \frac{1}{n^p}$,   for $p \leq 1$.

In Chapter 7 we shall prove that $\sum_{n=1}^{\infty} \frac{1}{n^p}$ is convergent, for all $p > 1$.

Proof

(a) This series is convergent, by the Comparison Test, since, if $p \geq 2$, then

$$\frac{1}{n^p} \leq \frac{1}{n^2}, \quad \text{for } n = 1, 2, \ldots,$$

and the series $\sum_{n=1}^{\infty} \frac{1}{n^2}$ is convergent.

Sub-section 3.2.1, Example 1.

(b) The series $\sum_{n=1}^{\infty} c^n$ is a geometric series with common ratio c, and so it converges if $0 \leq c < 1$.

Sub-section 3.1.2.

(c) Let $a_n = n^p c^n$, for $n = 1, 2, \ldots$. Then, for $0 < c < 1$

If $c = 0$, the series is clearly convergent.

$$\frac{a_{n+1}}{a_n} = \frac{(n+1)^p c^{n+1}}{n^p c^n} = \left(1 + \frac{1}{n}\right)^p c.$$

Thus, if k is any integer greater than or equal to p, then

We introduce the integer k here, so that we can use the Product Rule for sequences.

$$c \leq \frac{a_{n+1}}{a_n} \leq \left(1 + \frac{1}{n}\right)^k c.$$

Now, by the Product Rule for sequences

$$\left(1 + \frac{1}{n}\right)^k \to 1 \quad \text{as } n \to \infty,$$

and so, by the Squeeze Rule for sequences

$$\frac{a_{n+1}}{a_n} \to c \quad \text{as } n \to \infty.$$

Since $0 \leq c < 1$, we deduce, from the Ratio Test, that $\sum_{n=1}^{\infty} n^p c^n$ is convergent.

(d) Let $a_n = \frac{c^n}{n!}$, for $n = 1, 2, \ldots$. Then, for $c > 0$,

$$\frac{a_{n+1}}{a_n} = \frac{c^{n+1}}{(n+1)!} \Big/ \frac{c^n}{n!} = \frac{c}{n+1}.$$

If $c = 0$, the series is clearly convergent.

Thus $\frac{a_{n+1}}{a_n} \to 0$ as $n \to \infty$, and we deduce from the Ratio Test that $\sum_{n=1}^{\infty} \frac{c^n}{n!}$ is convergent.

(e) We saw earlier that the series $\sum_{n=1}^{\infty} \frac{1}{n}$ is divergent. It follows that the series $\sum_{n=1}^{\infty} \frac{1}{n^p}$ is also divergent, by the Comparison Test, since if $p \leq 1$ then $\frac{1}{n^p} \geq \frac{1}{n}$, for $n = 1, 2, \ldots$. ∎

In Sub-section 3.2.1, Example 2.

3.3 Series with positive and negative terms

The study of series $\sum_{n=1}^{\infty} a_n$, with $a_n \geq 0$ for all values of n, is relatively straightforward, because the sequence of partial sums $\{s_n\}$ is increasing. Similarly, if $a_n \leq 0$ for all values of n, then $\{s_n\}$ is decreasing.

However, it is harder to determine the behaviour of a series with both positive and negative terms, because $\{s_n\}$ is neither increasing nor decreasing. However, if the sequence $\{a_n\}$ contains *only finitely many* negative terms, then the sequence $\{s_n\}$ is *eventually* increasing, and we can apply the methods of Section 3.2. Similarly, if $\{a_n\}$ contains *only finitely many* positive terms, then the sequence $\{s_n\}$ is *eventually* decreasing, and we can again apply the methods of Section 3.2, after making a sign change.

In this section we look at series such as

$$1 - \frac{1}{2} + \frac{1}{3} - \frac{1}{4} + \frac{1}{5} - \frac{1}{6} + \cdots \quad \text{and} \quad 1 + \frac{1}{2^2} - \frac{1}{3^2} + \frac{1}{4^2} + \frac{1}{5^2} - \frac{1}{6^2} + \cdots,$$

which contain infinitely many terms of either sign. We give several methods which can often be used to prove that such series are convergent.

For example, the convergence of

$$1 + 2 + 3 - \frac{1}{4^2} - \frac{1}{5^2} - \frac{1}{6^2} - \cdots$$

follows from that of $\sum_{n=1}^{\infty} \frac{1}{n^2}$.

3.3.1 Absolute convergence

Suppose that we want to determine the behaviour of the infinite series

$$\sum_{n=1}^{\infty} \frac{(-1)^{n+1}}{n^2} = 1 - \frac{1}{2^2} + \frac{1}{3^2} - \frac{1}{4^2} + \frac{1}{5^2} - \frac{1}{6^2} + \cdots. \qquad (1)$$

We know that the series

$$\sum_{n=1}^{\infty} \frac{1}{n^2} = 1 + \frac{1}{2^2} + \frac{1}{3^2} + \frac{1}{4^2} + \frac{1}{5^2} + \frac{1}{6^2} + \cdots \qquad (2)$$

The series (2) is a basic convergent series.

is convergent. Does this mean that the series (1) is also convergent? In fact it does, as we now prove.

Consider the two related series

$$1 + 0 + \frac{1}{3^2} + 0 + \frac{1}{5^2} + 0 + \cdots$$

and

$$0 + \frac{1}{2^2} + 0 + \frac{1}{4^2} + 0 + \frac{1}{6^2} + \cdots.$$

Each of these series is dominated by the series (2), and so they are both convergent, by the Comparison Test. Applying the Multiple Rule with $\lambda = -1$, and the Sum Rule, we deduce that the series

$$1 - \frac{1}{2^2} + \frac{1}{3^2} - \frac{1}{4^2} + \frac{1}{5^2} - \frac{1}{6^2} + \cdots \quad \text{is convergent.}$$

The argument just given is the basis for a concept called *absolute convergence*, which we now define.

Definitions

A series $\sum_{n=1}^{\infty} a_n$ is **absolutely convergent** if $\sum_{n=1}^{\infty} |a_n|$ is convergent.

A series $\sum_{n=1}^{\infty} a_n$ that is convergent but not absolutely convergent is **conditionally convergent**.

If the terms a_n are all non-negative, then absolute convergence and convergence have the same meaning.

For example, the series (1) is absolutely convergent, because the series $\sum_{n=1}^{\infty} \frac{1}{n^2}$ is convergent.

However, the series

$$\sum_{n=1}^{\infty} \frac{(-1)^{n+1}}{n} = 1 - \frac{1}{2} + \frac{1}{3} - \frac{1}{4} + \frac{1}{5} - \frac{1}{6} + \cdots$$

is not absolutely convergent, because the series $\sum_{n=1}^{\infty} \frac{1}{n}$ is divergent; the series is, in fact, conditionally convergent.

We shall examine the behaviour of conditionally convergent series later, in Sub-sections 3.3.2 and 3.3.3.

You saw this in Example 2, Sub-section 3.2.1.

Theorem 1   Absolute Convergence Test

If $\sum_{n=1}^{\infty} a_n$ is absolutely convergent, then $\sum_{n=1}^{\infty} a_n$ is convergent.

The proofs of the results in this sub-section appear in Sub-section 3.3.6.

It follows that the series (1) is convergent (as we have already seen), and also that the series

$$1 + \frac{1}{2^2} - \frac{1}{3^2} + \frac{1}{4^2} + \frac{1}{5^2} - \frac{1}{6^2} + \cdots$$

is convergent, because the series $\sum_{n=1}^{\infty} \frac{1}{n^2}$ is convergent. Indeed, no matter how we distribute the plus and minus signs among the terms of $\left\{\frac{1}{n^2}\right\}$, the resulting series is convergent.

However, the Absolute Convergence Test tells us nothing about the behaviour of the series

$$1 - \frac{1}{2} + \frac{1}{3} - \frac{1}{4} + \frac{1}{5} - \frac{1}{6} + \cdots \tag{3}$$

and

$$1 + \frac{1}{2} - \frac{1}{3} + \frac{1}{4} + \frac{1}{5} - \frac{1}{6} + \frac{1}{7} + \frac{1}{8} - \frac{1}{9} + \cdots. \tag{4}$$

In fact, series (3) is convergent (by the Alternating Test, which we introduce in Sub-section 3.3.2), and series (4) is divergent (see Exercise 2(a) for Section 3.3).

The series $\sum\limits_{n=1}^{\infty} \frac{1}{n}$ is divergent, and so the two series (3) and (4) are not absolutely convergent.

Example 1 Prove that the following series are convergent:

(a) $\sum\limits_{n=1}^{\infty} \frac{(-1)^{n+1}}{n^3}$;   (b) $1 + \frac{1}{2} - \frac{1}{4} + \frac{1}{8} + \frac{1}{16} - \frac{1}{32} \cdots$.

Solution

(a) Let $a_n = \frac{(-1)^{n+1}}{n^3}$, for $n = 1, 2, \ldots$; then $|a_n| = \frac{1}{n^3}$. We know that $\sum\limits_{n=1}^{\infty} \frac{1}{n^3}$ is convergent, so it follows that $\sum\limits_{n=1}^{\infty} \frac{(-1)^{n+1}}{n^3}$ is absolutely convergent. Hence, by the Absolute Convergence Test, $\sum\limits_{n=1}^{\infty} \frac{(-1)^{n+1}}{n^3}$ is convergent.

> For $\sum\limits_{n=1}^{\infty} \frac{1}{n^3}$ is an example of a basic convergent series.

(b) The series $\sum\limits_{n=0}^{\infty} \frac{1}{2^n}$ is a convergent geometric series, so that the series
$$1 + \frac{1}{2} - \frac{1}{4} + \frac{1}{8} + \frac{1}{16} - \frac{1}{32} \cdots$$
is absolutely convergent. Then, by the Absolute Convergence Test, this series is also convergent.   □

Problem 1 Prove that the following series are convergent:

(a) $\sum\limits_{n=1}^{\infty} \frac{(-1)^{n+1} n}{n^3+1}$;   (b) $\sum\limits_{n=1}^{\infty} \frac{\cos n}{2^n}$.

The Absolute Convergence Test states that, if $\sum |a_n|$ is convergent, then $\sum a_n$ is also convergent, but it does not indicate any explicit connection between the sums of these two series. Clearly, $\sum a_n$ is less than $\sum |a_n|$ if any of the terms a_n are negative.

For example, $\sum\limits_{n=1}^{\infty} \frac{1}{2^n} = \frac{1}{2} + \frac{1}{4} + \frac{1}{8} + \cdots = 1$, whereas $\sum\limits_{n=1}^{\infty} \frac{(-1)^{n+1}}{2^n} = \frac{1}{2} - \frac{1}{4} + \frac{1}{8} - \cdots = \frac{1}{3}$.

The following result relates the values of $\sum a_n$ and $\sum |a_n|$:

Triangle Inequality (infinite form)

If $\sum\limits_{n=1}^{\infty} a_n$ is absolutely convergent, then $\left| \sum\limits_{n=1}^{\infty} a_n \right| \leq \sum\limits_{n=1}^{\infty} |a_n|$.

> Recall that you met the Triangle Inequality for $a_1, a_2, \ldots, a_n$
> $$\left| \sum_{k=1}^{n} a_k \right| \leq \sum_{k=1}^{n} |a_k|,$$
> in Sub-section 1.3.1.

Problem 2 Show that the series $\frac{1}{2} + \frac{1}{4} - \frac{1}{8} + \frac{1}{16} + \frac{1}{32} - \frac{1}{64} + \cdots$ is convergent, and that its sum lies in $[-1,1]$. (You do NOT need to find the sum of the series.)

3.3.2 The Alternating Test

Suppose that we want to determine the behaviour of the following infinite series, in which the terms have alternating signs
$$\sum_{n=1}^{\infty} \frac{(-1)^{n+1}}{n} = 1 - \frac{1}{2} + \frac{1}{3} - \frac{1}{4} + \frac{1}{5} - \frac{1}{6} + \cdots.$$

The Absolute Convergence Test does not help us with this series, because $\sum_{n=1}^{\infty} \frac{1}{n}$ is divergent.

In fact, the series $\sum_{n=1}^{\infty} \frac{(-1)^{n+1}}{n}$ is convergent. To see why, we first calculate some of its partial sums and plot them on a sequence diagram

$$s_1 = 1;$$

$$s_2 = 1 - \frac{1}{2} = 0.5;$$

$$s_3 = 1 - \frac{1}{2} + \frac{1}{3} = 0.8\overline{3};$$

$$s_4 = 1 - \frac{1}{2} + \frac{1}{3} - \frac{1}{4} = 0.58\overline{3};$$

$$s_5 = 1 - \frac{1}{2} + \frac{1}{3} - \frac{1}{4} + \frac{1}{5} = 0.78\overline{3};$$

$$s_6 = 1 - \frac{1}{2} + \frac{1}{3} - \frac{1}{4} + \frac{1}{5} - \frac{1}{6} = 0.61\overline{6}.$$

The sequence diagram for $\{s_n\}$ suggests that

$$s_1 \geq s_3 \geq s_5 \geq \ldots \geq s_{2k-1} \geq \ldots$$

and

$$s_2 \leq s_4 \leq s_6 \leq \ldots \leq s_{2k} \leq \ldots,$$

for all k. In other words:

the *odd subsequence* $\{s_{2k-1}\}$ is decreasing

and:

the *even subsequence* $\{s_{2k}\}$ is increasing.

Also, the terms of $\{s_{2k-1}\}$ all exceed the terms of $\{s_{2k}\}$, and both subsequences appear to converge to a common limit s, which lies between the odd and even partial sums.

You met subsequences earlier, in Sub-section 2.4.4.

To prove this, we write the even partial sums $\{s_{2k}\}$ as follows

$$s_{2k} = \left(1 - \frac{1}{2}\right) + \left(\frac{1}{3} - \frac{1}{4}\right) + \cdots + \left(\frac{1}{2k-1} - \frac{1}{2k}\right).$$

All the brackets are positive, and so the subsequence $\{s_{2k}\}$ is increasing.

We can also write

$$s_{2k} = 1 - \left(\frac{1}{2} - \frac{1}{3}\right) - \left(\frac{1}{4} - \frac{1}{5}\right) - \cdots - \left(\frac{1}{2k-2} - \frac{1}{2k-1}\right) - \frac{1}{2k}.$$

Again, all the brackets are positive, and so $\{s_{2k}\}$ is bounded above, by 1.

Hence $\{s_{2k}\}$ is convergent, by the Monotone Convergence Theorem.

Let

$$\lim_{k \to \infty} s_{2k} = s.$$

Since

$$s_{2k} = s_{2k-1} - \frac{1}{2k}$$

and $\left\{\frac{1}{2k}\right\}$ is null, we have

$$\lim_{k \to \infty} s_{2k-1} = \lim_{k \to \infty} \left(s_{2k} + \frac{1}{2k} \right) = \lim_{k \to \infty} s_{2k} + \lim_{k \to \infty} \left(\frac{1}{2k} \right)$$
$$= s + 0 = s,$$

by the Combination Rules for sequences. Thus the odd and even subsequences of $\{s_n\}$ both tend to the same limit s, and so $\{s_n\}$ itself tends to s. Hence

By Theorem 6 of Sub-section 2.4.4.

$$\sum_{n=1}^{\infty} \frac{(-1)^{n+1}}{n} = 1 - \frac{1}{2} + \frac{1}{3} - \frac{1}{4} + \frac{1}{5} - \frac{1}{6} + \cdots \text{ is convergent, with sum } s.$$

In fact, $s = \log_e 2 \simeq 0.69$.

The same method can be used to prove the following general result:

Theorem 2   Alternating Test

If

$$a_n = (-1)^{n+1} b_n, \quad n = 1, 2, \ldots,$$

where $\{b_n\}$ is a decreasing null sequence with positive terms, then

$$\sum_{n=1}^{\infty} a_n = b_1 - b_2 + b_3 - b_4 + \cdots \text{ is convergent.}$$

This test is sometimes called the *Leibniz Test*.

When we apply the Alternating Test, there are a number of conditions which must be checked. We now describe these in the form of a strategy.

Strategy   To prove that $\sum\limits_{n=1}^{\infty} a_n$ is convergent, using the Alternating Test, check that

$$a_n = (-1)^{n+1} b_n, \quad n = 1, 2, \ldots,$$

where:

1. $b_n \geq 0, \quad$ for $n = 1, 2, \ldots$;

2. $\{b_n\}$ is a null sequence;

3. $\{b_n\}$ is decreasing.

To show that $\{b_n\}$ is null, use the techniques introduced in Sub-section 2.2.1.

Remark

It is often easiest to check that $\{b_n\}$ is decreasing by verifying that $\{\frac{1}{b_n}\}$ is increasing.

Here are some examples.

Example 2   Determine which of the following series are convergent:

(a) $\sum\limits_{n=1}^{\infty} \frac{(-1)^{n+1}}{\sqrt{n}}$;   (b) $\sum\limits_{n=1}^{\infty} \frac{(-1)^{n+1}}{n^4}$;   (c) $\sum\limits_{n=1}^{\infty} (-1)^{n+1}$.

Solution

(a) The sequence $\left\{ \frac{(-1)^{n+1}}{\sqrt{n}} \right\}$ has terms of the form $a_n = (-1)^{n+1} b_n$, where

$$b_n = \frac{1}{\sqrt{n}}, \quad n = 1, 2, \ldots.$$

Now:

1. $b_n = \frac{1}{\sqrt{n}} \geq 0$, for $n = 1, 2, \ldots$;

2. $\{b_n\} = \{\frac{1}{\sqrt{n}}\}$ is a basic null sequence;

3. $\{b_n\} = \{\frac{1}{\sqrt{n}}\}$ is decreasing, because $\{\frac{1}{b_n}\} = \{\sqrt{n}\}$ is increasing.

Hence, by the Alternating Test, $\sum_{n=1}^{\infty} \frac{(-1)^{n+1}}{\sqrt{n}}$ is convergent.

(b) The sequence $\left\{\frac{(-1)^{n+1}}{n^4}\right\}$ has terms of the form $a_n = (-1)^{n+1} b_n$, where

Alternatively, we can show that this series is convergent by the Absolute Convergence Test.

$$b_n = \frac{1}{n^4}, \quad n = 1, 2, \ldots.$$

Now:

1. $b_n = \frac{1}{n^4} \geq 0$, for $n = 1, 2, \ldots$;
2. $\{b_n\} = \{\frac{1}{n^4}\}$ is a basic null sequence;
3. $\{b_n\} = \{\frac{1}{n^4}\}$ is decreasing, because $\{\frac{1}{b_n}\} = \{n^4\}$ is increasing.

Hence, by the Alternating Test, $\sum_{n=1}^{\infty} \frac{(-1)^{n+1}}{n^4}$ is convergent.

(c) The sequence $\{(-1)^{n+1}\}$ is not a null sequence. Hence, by the Non-null Test

Notice that the fact that the nth term includes a factor $(-1)^{n+1}$ does not imply at all that a series is convergent!

$$\sum_{n=1}^{\infty} (-1)^{n+1} \text{ is divergent.} \qquad \square$$

Problem 3 Determine which of the following series are convergent:

(a) $\sum_{n=1}^{\infty} \frac{(-1)^{n+1}}{n^3}$; (b) $\sum_{n=1}^{\infty} (-1)^{n+1} \frac{n}{n+2}$; (c) $\sum_{n=1}^{\infty} \frac{(-1)^{n+1}}{n^{1/3} + n^{1/2}}$.

3.3.3 Rearrangement of a series

In the last sub-section we saw that the series $1 - \frac{1}{2} + \frac{1}{3} - \frac{1}{4} + \frac{1}{5} - \frac{1}{6} + \cdots$ is convergent. Let us denote by ℓ the sum of this series.

If we temporarily ignore the need for careful mathematical proof and simply 'move terms about', look what we get

In fact this series is conditionally convergent, since the harmonic series $1 + \frac{1}{2} + \frac{1}{3} + \frac{1}{4} + \frac{1}{5} + \frac{1}{6} + \cdots$ is divergent.

$$\ell = 1 - \frac{1}{2} + \frac{1}{3} - \frac{1}{4} + \frac{1}{5} - \frac{1}{6} + \cdots \qquad (5)$$

This is the definition of ℓ.

$$= 1 - \frac{1}{2} - \frac{1}{4} + \frac{1}{3} - \frac{1}{6} - \frac{1}{8} + \frac{1}{5} - \frac{1}{10} - \frac{1}{12} + \frac{1}{7} - \frac{1}{14} - \frac{1}{16} + \cdots \qquad (6)$$

The pattern in (6) is to follow the next available positive term by the two next available negative terms.

$$= \left(1 - \frac{1}{2}\right) - \frac{1}{4} + \left(\frac{1}{3} - \frac{1}{6}\right) - \frac{1}{8} + \left(\frac{1}{5} - \frac{1}{10}\right) - \frac{1}{12}$$

$$+ \left(\frac{1}{7} - \frac{1}{14}\right) - \frac{1}{16} + \cdots$$

Here we insert some brackets.

$$= \frac{1}{2} - \frac{1}{4} + \frac{1}{6} - \frac{1}{8} + \frac{1}{10} - \frac{1}{12} + \frac{1}{14} - \frac{1}{16} + \cdots$$

Here we insert the value of each bracket.

$$= \frac{1}{2}\left(1 - \frac{1}{2} + \frac{1}{3} - \frac{1}{4} + \frac{1}{5} - \frac{1}{6} + \frac{1}{8} - \cdots\right)$$

Here we pull out a common factor of $\frac{1}{2}$.

$$= \frac{1}{2}\ell.$$

It follows from the fact that $\ell = \frac{1}{2}\ell$ that $\ell = 0$. However this is impossible, since the partial sum s_2 of the original series is $\frac{1}{2}$, and the even partial sums s_{2k} are increasing.

Recall that, since the series converges to ℓ, then s_{2k} converges to ℓ also.

What has gone wrong? We assumed that operations that are valid for sums of a finite number of terms also hold for sums of an infinite number of terms – we rearranged (infinitely many of) the terms in the series (5) to obtain the series (6) – without any justification. This rearrangement operation is not valid in general!

In fact the series (6) does converge; we ask you to supply a careful proof of this in Exercise 2 on Section 3.3.

We now give a precise definition of what we mean by a *rearrangement*. Loosely speaking, the series $\sum_{n=1}^{\infty} b_n = b_1 + b_2 + \cdots$ is a rearrangement of the series $\sum_{n=1}^{\infty} a_n = a_1 + a_2 + \cdots$ if precisely the same terms appear in the sequences $\{b_n\} = \{b_1, b_2, \ldots\}$ and $\{a_n\} = \{a_1, a_2, \ldots\}$, though they may occur in a different order.

Definition A series $\sum_{n=1}^{\infty} b_n = b_1 + b_2 + \cdots$ is a rearrangement of the series $\sum_{n=1}^{\infty} a_n = a_1 + a_2 + \cdots$ if there is a bijection f such that

$$f : \mathbb{N} \to \mathbb{N}$$
$$b_n \mapsto a_{f(n)}.$$

Recall that a bijection is a one–one onto mapping.

Example 2 Assuming that the sum of the series $\sum_{n=1}^{\infty} \frac{(-1)^{n+1}}{n} = 1 - \frac{1}{2} + \frac{1}{3} - \frac{1}{4} + \frac{1}{5} - \frac{1}{6} + \cdots$ is $\log_e 2$, prove that the series $\sum_{n=1}^{\infty} a_n = 1 + \frac{1}{3} - \frac{1}{2} + \frac{1}{5} + \frac{1}{7} - \frac{1}{4} + \frac{1}{9} + \frac{1}{11} - \frac{1}{6} + \cdots$ converges to $\frac{3}{2}\log_e 2$.

Solution Let s_n and t_n denote the nth partial sums of the series $1 + \frac{1}{3} - \frac{1}{2} + \frac{1}{5} + \frac{1}{7} - \frac{1}{4} + \frac{1}{9} + \frac{1}{11} - \frac{1}{6} + \cdots$ and $1 - \frac{1}{2} + \frac{1}{3} - \frac{1}{4} + \frac{1}{5} - \frac{1}{6} + \cdots$, respectively.

We now introduce an extra piece of notation, H_n; we denote by H_n the nth partial sum $\sum_{k=1}^{n} \frac{1}{k} = 1 + \frac{1}{2} + \cdots + \frac{1}{n}$ of the harmonic series.

This will enable us to tackle things much more easily!

Now, the terms of the series $\sum_{n=1}^{\infty} a_n$ come 'in a natural way' in threes, so it seems sensible to look at the partial sums s_{3n}

$$s_{3n} = 1 + \frac{1}{3} - \frac{1}{2} + \frac{1}{5} + \frac{1}{7} - \frac{1}{4} + \frac{1}{9} + \frac{1}{11} - \frac{1}{6} + \cdots$$
$$+ \frac{1}{4n-3} + \frac{1}{4n-1} - \frac{1}{2n}$$

From the definition of s_{3n}.

$$= \left(1 + \frac{1}{3} - \frac{1}{2}\right) + \left(\frac{1}{5} + \frac{1}{7} - \frac{1}{4}\right) + \left(\frac{1}{9} + \frac{1}{11} - \frac{1}{6}\right) + \cdots$$

We insert some brackets in a finite sum, for convenience.

$$+ \left(\frac{1}{4n-3} + \frac{1}{4n-1} - \frac{1}{2n}\right)$$

$$= \left(1 + \frac{1}{3} + \frac{1}{5} + \cdots + \frac{1}{4n-1}\right) - \left(\frac{1}{2} + \frac{1}{4} + \frac{1}{6} + \cdots + \frac{1}{2n}\right)$$

We separate out the positive and negative terms.

$$= \left[\left(1 + \frac{1}{2} + \frac{1}{3} + \cdots + \frac{1}{4n}\right) - \left(\frac{1}{2} + \frac{1}{4} + \frac{1}{6} + \cdots + \frac{1}{4n}\right)\right] - \frac{1}{2}H_n$$

We add some terms to the contents of the first bracket, then subtract them again; the last bracket is simply $\frac{1}{2}H_n$.

$$= \left[\left(1 + \frac{1}{2} + \frac{1}{3} + \cdots + \frac{1}{4n}\right) - \frac{1}{2}\left(1 + \frac{1}{2} + \frac{1}{3} + \cdots + \frac{1}{2n}\right)\right] - \frac{1}{2}H_n$$

$$= H_{4n} - \frac{1}{2}H_{2n} - \frac{1}{2}H_n. \qquad (7)$$

Furthermore

$$t_{2n} = 1 - \frac{1}{2} + \frac{1}{3} - \frac{1}{4} + \frac{1}{5} - \frac{1}{6} + \cdots + \frac{1}{2n-1} - \frac{1}{2n}$$

By the definition of t_{2n}.

$$= \left(1 + \frac{1}{2} + \frac{1}{3} + \cdots + \frac{1}{2n}\right) - 2\left(\frac{1}{2} + \frac{1}{4} + \frac{1}{6} + \cdots + \frac{1}{2n}\right)$$

We insert some positive terms, then remove them again.

$$= H_{2n} - \left(1 + \frac{1}{2} + \frac{1}{3} + \cdots + \frac{1}{n}\right)$$

By the definition of H_{2n}.

$$= H_{2n} - H_n. \qquad (8)$$

By the definition of H_n.

We now eliminate the H's from equations (7) and (8) as follows

$$s_{3n} = H_{4n} - \frac{1}{2}H_{2n} - \frac{1}{2}H_n$$

$$= (H_{4n} - H_{2n}) + \frac{1}{2}(H_{2n} - H_n)$$

$$= t_{4n} + \frac{1}{2}t_{2n}. \qquad (9)$$

But it follows from the hypotheses of the example that $t_n \to \log_e 2$, and so $t_{2n} \to \log_e 2$ and $t_{4n} \to \log_e 2$, as $n \to \infty$. Hence, letting $n \to \infty$ in equation (9), we see that

$$s_{3n} \to \log_e 2 + \frac{1}{2}\log_e 2$$

$$= \frac{3}{2}\log_e 2.$$

Next

$$s_{3n-1} = s_{3n} + \frac{1}{2n}$$

$$\to \frac{3}{2}\log_e 2$$

and

$$s_{3n-2} = s_{3n} + \frac{1}{2n} - \frac{1}{4n-1}$$

$$\to \frac{3}{2}\log_e 2 \quad \text{as } n \to \infty.$$

It follows from the above three results that $s_n \to \frac{3}{2}\log_e 2$ as $n \to \infty$. $\square$

Problem 4 Prove that the series $\sum\limits_{n=1}^{\infty} a_n = 1 - \frac{1}{2} - \frac{1}{4} + \frac{1}{3} - \frac{1}{6} - \frac{1}{8} + \frac{1}{5} - \frac{1}{10} - \frac{1}{12} + \frac{1}{7} - \frac{1}{14} - \frac{1}{16} + \cdots$ converges to $\frac{1}{2} \log_e 2$. [You may assume that the sum of the series $\sum\limits_{n=1}^{\infty} \frac{(-1)^{n+1}}{n} = 1 - \frac{1}{2} + \frac{1}{3} - \frac{1}{4} + \frac{1}{5} - \frac{1}{6} + \cdots$ is $\log_e 2$.]

In order to better understand the behaviour of series with positive and negative terms, we introduce some notation. For a conditionally convergent series

$$\sum_{n=1}^{\infty} a_n = a_1 + a_2 + \cdots,$$

we define the quantities a_n^+ and a_n^- as follows

$$a_n^+ = \begin{cases} a_n, & \text{if } a_n \geq 0, \\ 0, & \text{if } a_n < 0, \end{cases} \quad \text{and} \quad a_n^- = \begin{cases} 0, & \text{if } a_n \geq 0, \\ -a_n, & \text{if } a_n < 0. \end{cases}$$

In other words, the sequence $\{a_n^+\}$ picks out the non-negative terms and the sequence $\{a_n^-\}$ picks out the non-positive terms (and discards their sign).

In particular, $a_n = a_n^+ - a_n^-$, and $a_n^+ \geq 0$ and $a_n^- \geq 0$.

Next, since the series $\sum\limits_{n=1}^{\infty} a_n$ is convergent, $a_n \to 0$ as $n \to \infty$. It follows that

$$a_n^+ = \frac{1}{2}(|a_n| + a_n) \to 0 \quad \text{as } n \to \infty$$

and

$$a_n^- = \frac{1}{2}(|a_n| - a_n) \to 0 \quad \text{as } n \to \infty.$$

Now, since $\sum\limits_{n=1}^{\infty} a_n$ is conditionally convergent, it must contain infinitely many negative terms, since otherwise $\sum\limits_{n=1}^{\infty} a_n$ and $\sum\limits_{n=1}^{\infty} |a_n|$ would be the same apart from a finite number of terms. Since altering a finite number of terms does not affect the convergence of a series, it would then follow that the series $\sum\limits_{n=1}^{\infty} |a_n|$ would be convergent – which we know is not the case.

Similarly, there must be infinitely many positive terms in $\sum a_n$.

Finally, we show that, if the series $\sum\limits_{n=1}^{\infty} a_n$ is conditionally convergent, then the corresponding series $\sum\limits_{n=1}^{\infty} a_n^+$ (the 'positive part' of $\sum\limits_{n=1}^{\infty} a_n$) and $\sum\limits_{n=1}^{\infty} a_n^-$ (the 'negative part' of $\sum\limits_{n=1}^{\infty} a_n$) are both divergent.

For, suppose that $\sum\limits_{n=1}^{\infty} a_n^+$ were convergent. Then, since $a_n = a_n^+ - a_n^-$, we have $a_n^- = a_n^+ - a_n$; it follows that the series $\sum\limits_{n=1}^{\infty} a_n^-$ must also be convergent. And then, since $|a_n| = a_n^+ + a_n^-$, it would also follow that $\sum\limits_{n=1}^{\infty} |a_n|$ would be convergent – which we know is not the case.

A similar argument shows that $\sum\limits_{n=1}^{\infty} a_n^-$ cannot be convergent.

In particular, it follows from the divergence of the series $\sum\limits_{n=1}^{\infty} a_n^+$ and $\sum\limits_{n=1}^{\infty} a_n^-$, whose terms are non-negative, that their partial sums must tend to ∞ as $n \to \infty$.

For example, for the series

$$\sum_{n=1}^{\infty} a_n = 1 - \frac{1}{2} + \frac{1}{3} - \frac{1}{4} + \frac{1}{5} - \frac{1}{6} + \cdots,$$

we have

$$\sum_{n=1}^{\infty} a_n^+ = 1 - 0 + \frac{1}{3} - 0 + \frac{1}{5} - 0 + \cdots$$

and

$$\sum_{n=1}^{\infty} a_n^- = 0 + \frac{1}{2} + 0 + \frac{1}{4} + 0 + \frac{1}{6} \cdots.$$

These two expressions for a_n^+ and a_n^- are easily checked by considering the two cases $a_n \geq 0$, $a_n < 0$ separately.

For the series $\sum\limits_{n=1}^{\infty} a_n$ is assumed to be only conditionally convergent.

For the series $\sum\limits_{n=1}^{\infty} a_n$ is assumed to be only conditionally convergent.

By the Boundedness Theorem for series.

In general it is *not* true that *any* rearrangement of a conditionally convergent series converges. The situation is much more interesting!

> *Some rearrangements converge, some diverge.*

Theorem 3   Riemann's Rearrangement Theorem

Let the series $\sum\limits_{n=1}^{\infty} a_n = a_1 + a_2 + \cdots$ be conditionally convergent, with nth partial sum s_n. Then there are rearrangements of the series that have the following properties:

1. For any given number s, the rearranged series converges to s;

2. For any given numbers x and y, one subsequence of the s_n converges to x and another subsequence of the s_n converges to y;

3. The sequence s_n converges to ∞.

4. One subsequence of the s_n converges to ∞ and another subsequence of the s_n converges to $-\infty$.

> *This is not a complete list of all the possibilities that can occur!*

We will not prove Theorem 3. Instead, we illustrate case (1) using the facts that we have already discovered about conditionally convergent series. Using similar ideas, we can prove the various parts of the theorem.

Example 3   Find a rearrangement of the series $\sum\limits_{n=1}^{\infty} \frac{(-1)^{n+1}}{n} = 1 - \frac{1}{2} + \frac{1}{3} - \frac{1}{4} + \frac{1}{5} - \frac{1}{6} + \cdots$ that converges to the sum 3.

> *Recall that the sum of this series is $\log_e 2$ or approximately 0.69.*

Solution   Since the 'positive' part of the series $\sum\limits_{n=1}^{\infty} \frac{(-1)^{n+1}}{n}$ has partial sums that tend to ∞, we start to construct the desired rearranged series as follows. Take enough of the 'positive' terms $1, \frac{1}{3}, \frac{1}{5}, \ldots, \frac{1}{2N_1 - 1}$ so that the sum $1 + \frac{1}{3} + \frac{1}{5} + \cdots + \frac{1}{2N_1 - 1}$ is greater than 3, choosing N_1 so that it is the first integer such that this sum is greater than 3. Then these N_1 terms will form the first N_1 terms in our desired rearranged series.

> *This is possible since*
> $$1 + \frac{1}{3} + \frac{1}{5} + \cdots \to \infty,$$
> *as $n \to \infty$.*

Next, take enough of the 'negative' terms $\frac{1}{2}, \frac{1}{4}, \frac{1}{6}, \ldots, \frac{1}{2N_2}$ so that the expression

$$\left(1 + \frac{1}{3} + \frac{1}{5} + \cdots + \frac{1}{2N_1 - 1}\right) - \left(\frac{1}{2} + \frac{1}{4} + \frac{1}{6} + \cdots + \frac{1}{2N_2}\right)$$

is less than 3, choosing N_2 so that it is the first integer such that this sum is less than 3. These $N_1 + N_2$ terms, in this order, form the first $N_1 + N_2$ terms in our desired rearranged series.

Now add in some more 'positive' terms $\frac{1}{2N_1+1}, \frac{1}{2N_1+3}, \ldots, \frac{1}{2N_3-1}$ so that the sum

$$\left(1 + \frac{1}{3} + \frac{1}{5} + \cdots + \frac{1}{2N_1 - 1}\right) - \left(\frac{1}{2} + \frac{1}{4} + \frac{1}{6} + \cdots + \frac{1}{2N_2}\right)$$
$$+ \left(\frac{1}{2N_1 + 1} + \frac{1}{2N_1 + 3} + \cdots + \frac{1}{2N_3 - 1}\right)$$

is greater than 3, choosing N_3 so that it is the first available integer such that this sum is greater than 3. Then these $N_2 + N_3$ terms, in this order, will form the first $N_2 + N_3$ terms in our desired rearranged series.

We then add in just enough 'negative' terms to make the next sum, of $N_3 + N_4$ terms, less than 3, and so on indefinitely.

In each set of two steps in this process we must use at least one of the 'positive' terms and one of the 'negative' terms, so that eventually all the 'positive' terms and all the 'negative' terms of the original series will be taken exactly once in the new series, which we denote by $\sum_{n=1}^{\infty} b_n$. So certainly the series $\sum_{n=1}^{\infty} b_n$ is a rearrangement of the original series $\sum_{n=1}^{\infty} \frac{(-1)^{n+1}}{n}$.

> At each step there are always 'positive' terms or 'negative' terms left to choose, since there are infinitely many of each.

But how do we know that the sum of the rearranged series is 3? As we keep making the partial sums of $\sum_{n=1}^{\infty} b_n$ swing back and forth from one side of 3 to the other, we need the 'radius' of the swings to tend to 0. We achieve this by always changing the sign of the terms from $\sum_{n=1}^{\infty} \frac{(-1)^{n+1}}{n}$ that we choose next, once the partial sum of $\sum_{n=1}^{\infty} b_n$ has crossed *beyond* the value 3. For this ensures that (from $\frac{1}{2N_2}$ onwards) the difference between a partial sum and 3 will always be less than the absolute value of the last term $\frac{1}{2N_k-1}$ or $\frac{1}{2N_k}$ used, and we know that this last term tends to 0 as we go further out along the original series.

> For clearly $\frac{1}{N_k} \le \frac{1}{k}$, since we use at least 1 term in the original series at each step of the rearrangement process.

It follows that the rearranged series $\sum_{n=1}^{\infty} b_n$ converges to 3, as required. $\square$

Problem 5   Find a rearrangement of the series $\sum_{n=1}^{\infty} \frac{(-1)^{n+1}}{n} = 1 - \frac{1}{2} + \frac{1}{3} - \frac{1}{4} + \frac{1}{5} - \frac{1}{6} + \cdots$ whose partial sums tend to ∞.

The situation is very much simpler, though, in the case of absolutely convergent series.

Theorem 4   Let the series $\sum_{n=1}^{\infty} a_n = a_1 + a_2 + \cdots$ be absolutely convergent. Then any rearrangement $\sum_{n=1}^{\infty} b_n$ of the series also converges absolutely, and $\sum_{n=1}^{\infty} b_n = \sum_{n=1}^{\infty} a_n$.

For example, the series

$$1 - \frac{1}{2^2} + \frac{1}{3^2} - \frac{1}{4^2} + \frac{1}{5^2} - \frac{1}{6^2} + \frac{1}{7^2} - \frac{1}{8^2} + \frac{1}{9^2} - \cdots$$

> For the series
> $$\sum_{n=1}^{\infty} \left| \frac{(-1)^{n+1}}{n^2} \right| = \sum_{n=1}^{\infty} \frac{1}{n^2},$$ which is a basic convergent series.

is absolutely convergent. It follows from Theorem 4 that the following series is also absolutely convergent – and to the same sum

$$\underbrace{1}_{1 \text{ term}} \quad \underbrace{-\frac{1}{2^2} - \frac{1}{4^2}}_{2 \text{ terms}} \quad \underbrace{+\frac{1}{3^2} + \frac{1}{5^2} + \frac{1}{7^2}}_{3 \text{ terms}} \quad \underbrace{-\frac{1}{6^2} - \frac{1}{8^2} - \frac{1}{10^2} - \frac{1}{12^2}}_{4 \text{ terms}} + \cdots .$$

3.3.4   Multiplication of series

We now look at the question of how we might multiply together two infinite series $\sum_{n=0}^{\infty} a_n$ and $\sum_{n=0}^{\infty} b_n$ to obtain a single series $\sum_{n=0}^{\infty} c_n$ with the property that

> In this sub-section we sum our series from $n = 0$ rather than from $n = 1$, simply because in many applications of Theorem 5 (below) this is the version that we shall require.

$$\sum_{n=0}^{\infty} c_n = \left(\sum_{n=0}^{\infty} a_n\right) \times \left(\sum_{n=0}^{\infty} b_n\right).$$

Now, a first thought might be that $c_n = a_n b_n$, so that $\sum_{n=0}^{\infty} a_n b_n = \left(\sum_{n=0}^{\infty} a_n\right) \times \left(\sum_{n=0}^{\infty} b_n\right)$. However this is not the case!

For example, let $a_n = \left(\frac{1}{2}\right)^n$ and $b_n = \left(\frac{1}{3}\right)^n$. Then

$$\sum_{n=0}^{\infty} a_n = \sum_{n=0}^{\infty} \left(\frac{1}{2}\right)^n = 2 \quad \text{and} \quad \sum_{n=0}^{\infty} b_n = \sum_{n=0}^{\infty} \left(\frac{1}{3}\right)^n = \frac{3}{2},$$

These are both geometric series.

but

$$\sum_{n=0}^{\infty} a_n b_n = \sum_{n=0}^{\infty} \left(\frac{1}{6}\right)^n = \frac{6}{5}.$$

Note that $\frac{6}{5} \neq 2 \times \frac{3}{2}$!

We can get a clue to what the correct formula for c_n might be by, temporarily, abandoning our rigorous approach and simply 'pushing symbols about'! If we multiply out both brackets and collect terms in a convenient way, we get

$$\begin{aligned}
\left(\sum_{n=0}^{\infty} a_n\right) \times \left(\sum_{n=0}^{\infty} b_n\right) &= (a_0 + a_1 + a_2 + a_3 + \cdots) \times (b_0 + b_1 + b_2 + b_3 + \cdots) \\
&= a_0(b_0 + b_1 + b_2 + b_3 + \cdots) + a_1(b_0 + b_1 + b_2 + b_3 + \cdots) \\
&\quad + a_2(b_0 + b_1 + b_2 + b_3 + \cdots) + a_3(b_0 + b_1 + b_2 + b_3 + \cdots) + \cdots \\
&= a_0 b_0 + a_0 b_1 + a_0 b_2 + a_0 b_3 + \cdots + a_1 b_0 + a_1 b_1 + a_1 b_2 + a_1 b_3 + \cdots \\
&\quad + a_2 b_0 + a_2 b_1 + a_2 b_2 + a_2 b_3 + \cdots + a_3 b_0 + a_3 b_1 + a_3 b_2 + a_3 b_3 + \cdots \\
&= a_0 b_0 + (a_0 b_1 + a_1 b_0) + (a_0 b_2 + a_1 b_1 + a_2 b_0) \\
&\quad + (a_0 b_3 + a_1 b_2 + a_2 b_1 + a_3 b_0) + \cdots.
\end{aligned}$$

Notice that every term that we would expect to be in such a product appears exactly once in this final expression. We have grouped the terms in the final expression in a particularly convenient way, so that we can see a pattern emerging from this non-rigorous argument.

The pattern here is that in the kth bracket, the sum of the subscripts of each term is precisely k.

We can formalise this discussion as follows:

Theorem 5   Product Rule

Let the series $\sum_{n=0}^{\infty} a_n$ and $\sum_{n=0}^{\infty} b_n$ be absolutely convergent, and let

$$c_n = a_0 b_n + a_1 b_{n-1} + a_2 b_{n-2} + \cdots + a_n b_0 = \sum_{k=0}^{n} a_k b_{n-k}.$$

Then the series $\sum_{n=0}^{\infty} c_n$ is absolutely convergent, and

$$\sum_{n=0}^{\infty} c_n = \left(\sum_{n=0}^{\infty} a_n\right) \times \left(\sum_{n=0}^{\infty} b_n\right).$$

Let us now return to the two series $\sum_{n=0}^{\infty} a_n = \sum_{n=0}^{\infty}\left(\frac{1}{2}\right)^n$ and $\sum_{n=0}^{\infty} b_n = \sum_{n=0}^{\infty}\left(\frac{1}{3}\right)^n$ that we considered earlier, whose sums were 2 and $\frac{3}{2}$, respectively. With the formula in Theorem 5 for c_n, we have

$$c_n = \sum_{k=0}^{n} a_k b_{n-k} = \sum_{k=0}^{n} \frac{1}{2^k} \times \frac{1}{3^{n-k}}$$

$$= \frac{1}{3^n} \sum_{k=0}^{n} \left(\frac{3}{2}\right)^k$$

$$= \frac{1}{3^n} \times \frac{1 - \left(\frac{3}{2}\right)^{n+1}}{1 - \frac{3}{2}}$$

$$= \frac{1}{3^n} \times (-2) \times \left(1 - \left(\frac{3}{2}\right)^{n+1}\right) = \frac{3}{2^n} - \frac{2}{3^n}.$$

It follows that

By the Combination Rules for convergent series.

$$\sum_{n=0}^{\infty} c_n = 3 \times \sum_{n=0}^{\infty} \frac{1}{2^n} - 2 \times \sum_{n=0}^{\infty} \frac{1}{3^n}$$

$$= 3 \times 2 - 2 \times \left(\frac{3}{2}\right) = 6 - 3 = 3.$$

Since $2 \times \frac{3}{2} = 3$, we see that the sum of the series $\sum_{n=0}^{\infty} c_n$ is what Theorem 5 predicts it should be.

Remark

Notice that the theorem requires that the series $\sum_{n=0}^{\infty} a_n$ and $\sum_{n=0}^{\infty} b_n$ are absolutely convergent. Without this assumption, the result may be false.

For example, let $a_n = b_n = \frac{(-1)^{n+1}}{\sqrt{n+1}}$. Then both $\sum_{n=0}^{\infty} a_n$ and $\sum_{n=0}^{\infty} b_n$ are convergent, by the Alternating Test – but they are not absolutely convergent. Now, by the formula for c_n, we have

$$c_n = \sum_{k=0}^{n} a_k b_{n-k} = \sum_{k=0}^{n} \frac{(-1)^{k+1}}{\sqrt{k+1}} \times \frac{(-1)^{n-k+1}}{\sqrt{n-k+1}}$$

$$= (-1)^n \sum_{k=0}^{n} \frac{1}{\sqrt{k+1} \times \sqrt{n-k+1}},$$

so that

$$|c_n| = \sum_{k=0}^{n} \frac{1}{\sqrt{k+1} \times \sqrt{n-k+1}}$$

$$\geq \sum_{k=0}^{n} \frac{1}{\sqrt{n+1} \times \sqrt{n+1}} = 1.$$

Since we therefore do not have $c_n \to 0$ as $n \to \infty$, it follows that the series $\sum_{n=0}^{\infty} c_n$ is divergent.

Yet again, this demonstrates that the hypotheses of our theorems really do matter!

We shall use Theorem 5 in Section 3.4 to give a definition of the exponential function $x \mapsto e^x$ in terms of series and to examine its properties, and in Section 8.4 on power series.

3.3.5   Overall strategy for testing for convergence

We now give an overall strategy for testing a series $\sum a_n$ for convergence, as a flow chart.

We suggest that, given a series which you wish to test for convergence or absolute convergence, you use the tests in the order indicated in the chart. Naturally, you may be able to short-circuit this strategy in certain cases.

For example, if you wish to examine a series whose terms alternate in sign, it is best to go straight to the Alternating Test.

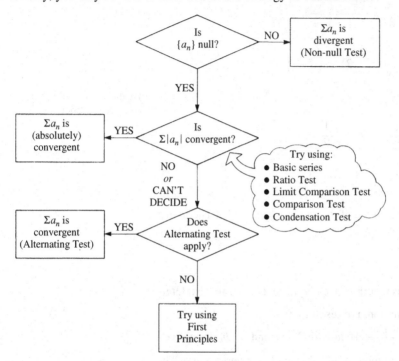

Remark

We have labelled the final box 'First Principles', to indicate that, if the various tests do not produce a result, then it may be possible to work directly with the sequence of partial sums $\{s_n\}$.

> **Problem 6**   Test the following series for convergence and absolute convergence:
>
> (a) $\sum_{n=1}^{\infty} \frac{1}{2}n$; (b) $\sum_{n=1}^{\infty} \frac{5n+2^n}{3^n}$; (c) $\sum_{n=1}^{\infty} \frac{3}{2n^3-1}$;
>
> (d) $\sum_{n=1}^{\infty} \frac{(-1)^{n+1}}{n^{\frac{1}{3}}}$; (e) $\sum_{n=1}^{\infty} \frac{(-1)^{n+1}n^2}{n^2+1}$; (f) $\sum_{n=1}^{\infty} \frac{(-1)^{n+1}n}{n^3+5}$;
>
> (g) $\sum_{n=1}^{\infty} \frac{2^n}{n^6}$; (h) $\sum_{n=1}^{\infty} \frac{(-1)^{n+1}n}{n^2+2}$; (i) $\sum_{n=2}^{\infty} \frac{1}{n(\log_e n)^{\frac{3}{4}}}$.

There is a variety of further tests for convergence or divergence which can be applied to series with non-negative terms. In this book we give only one further test, called the Integral Test. In particular it will enable us to prove that

You will meet the Integral Test in Section 7.4.

$$\sum_{n=1}^{\infty} \frac{1}{n^p} \text{ is convergent, } \quad \text{for all } p > 1.$$

3.3.6 Proofs

You may omit these proofs on a first reading.

We now supply the proofs omitted earlier in this section.

Theorem 1   Absolute Convergence Test

If $\sum\limits_{n=1}^{\infty} a_n$ is absolutely convergent, then $\sum\limits_{n=1}^{\infty} a_n$ is convergent.

Proof   We know that $\sum\limits_{n=1}^{\infty} |a_n|$ is convergent, and we want to prove that $\sum\limits_{n=1}^{\infty} a_n$ is convergent.

To do this, we define two new sequences

$$a_n^+ = \begin{cases} a_n, & \text{if } a_n \geq 0, \\ 0, & \text{if } a_n < 0, \end{cases} \quad \text{and} \quad a_n^- = \begin{cases} 0, & \text{if } a_n \geq 0, \\ -a_n, & \text{if } a_n < 0. \end{cases}$$

Both the sequences $\{a_n^+\}$ and $\{a_n^-\}$ are non-negative, and

$$a_n = a_n^+ - a_n^-, \quad \text{for } n = 1, 2, \ldots.$$

Also

$$a_n^+ \leq |a_n|, \quad \text{for } n = 1, 2, \ldots, \tag{10}$$

and

$$a_n^- \leq |a_n|, \quad \text{for } n = 1, 2, \ldots. \tag{11}$$

Since $\sum\limits_{n=1}^{\infty} |a_n|$ is convergent, we deduce from (10) and (11) that $\sum\limits_{n=1}^{\infty} a_n^+$ and $\sum\limits_{n=1}^{\infty} a_n^-$ are convergent, by the Comparison Test. Thus

$$\sum_{n=1}^{\infty} a_n = \sum_{n=1}^{\infty} \left(a_n^+ - a_n^-\right) = \sum_{n=1}^{\infty} a_n^+ - \sum_{n=1}^{\infty} a_n^- \tag{12}$$

is convergent, by the Combination Rules for series. ∎

Sub-section 3.1.4.

For example, if

$$\sum_{n=1}^{\infty} a_n = 1 - \frac{1}{2^2} + \frac{1}{3^2} - \frac{1}{4^2} + \cdots,$$

then

$$\sum_{n=1}^{\infty} a_n^+ = 1 + 0 + \frac{1}{3^2} + 0 + \cdots$$

and

$$\sum_{n=1}^{\infty} a_n^- = 0 + \frac{1}{2^2} + 0 + \frac{1}{4^2} + \cdots.$$

Triangle Inequality (infinite form)

If $\sum\limits_{n=1}^{\infty} a_n$ is absolutely convergent, then $\left| \sum\limits_{n=1}^{\infty} a_n \right| \leq \sum\limits_{n=1}^{\infty} |a_n|$.

Proof   Let

$$s_n = a_1 + a_2 + \cdots + a_n, \quad n = 1, 2, \ldots,$$

and

$$t_n = |a_1| + |a_2| + \cdots + |a_n|, \quad n = 1, 2, \ldots.$$

Then, by the Absolute Convergence Test

$$\lim_{n \to \infty} s_n = \sum_{n=1}^{\infty} a_n \text{ exists};$$

also

$$\lim_{n \to \infty} t_n = \sum_{n=1}^{\infty} |a_n|.$$

Alternatively, the infinite form of the Triangle Inequality can be deduced from statements (10), (11) and (12) in the proof of the Absolute Convergence Test.

Now, by the Triangle Inequality

$$|s_n| = |a_1 + a_2 + \cdots + a_n|$$
$$\leq |a_1| + |a_2| + \cdots + |a_n| = t_n,$$

and so

$$-t_n \leq s_n \leq t_n.$$

Thus, by the Limit Inequality Rule for sequences

$$-\lim_{n\to\infty} t_n \leq \lim_{n\to\infty} s_n \leq \lim_{n\to\infty} t_n;$$

that is

$$-\sum_{n=1}^{\infty} |a_n| \leq \sum_{n=1}^{\infty} a_n \leq \sum_{n=1}^{\infty} |a_n|,$$

and so

$$\left| \sum_{n=1}^{\infty} a_n \right| \leq \sum_{n=1}^{\infty} |a_n|. \qquad \blacksquare$$

In Remark 3 following the proof of the Triangle Inequality in Sub-section 1.3.1, we commented that the Triangle Inequality for two numbers, a and b, can be extended in the obvious way to any finite sum of numbers.

You met this in Sub-section 2.3.3, Theorem 3.

Theorem 2   Alternating Test

If

$$a_n = (-1)^{n+1} b_n, \quad n = 1, 2, \ldots,$$

where $\{b_n\}$ is a decreasing null sequence with positive terms, then

$$\sum_{n=1}^{\infty} a_n = b_1 - b_2 + b_3 - b_4 + \cdots \quad \text{is convergent.}$$

Proof   We can write the even partial sums s_{2k} of $\sum_{n=1}^{\infty} a_n$ as follows

$$s_{2k} = (b_1 - b_2) + (b_3 - b_4) + \cdots + (b_{2k-1} - b_{2k}).$$

Since $\{b_n\}$ is decreasing, all the brackets are non-negative, and so the even subsequence of partial sums, $\{s_{2k}\}$, is increasing.

We can also write the even partial sums s_{2k} as

$$s_{2k} = b_1 - (b_2 - b_3) - (b_4 - b_5) - \cdots - (b_{2k-2} - b_{2k-1}) - b_{2k}.$$

Again, all the brackets are non-negative, and so s_{2k} is bounded above, by b_1. Hence $\{s_{2k}\}$ is convergent, by the Monotone Convergence Theorem.
Now let

$$\lim_{k\to\infty} s_{2k} = s.$$

Since $s_{2k} = s_{2k-1} - b_{2k}$ so that $s_{2k-1} = s_{2k} + b_{2k}$, and $\{b_n\}$ is null, we have

$$\lim_{k\to\infty} s_{2k-1} = \lim_{k\to\infty} (s_{2k} + b_{2k})$$
$$= \lim_{k\to\infty} s_{2k} + \lim_{k\to\infty} b_{2k} = s,$$

by the Sum Rule for sequences. Thus the odd and even subsequences of $\{s_n\}$ both tend to the same limit s, and so $\{s_n\}$ tends to s. Hence

By Theorem 6, Subsection 2.4.4.

$$\sum_{n=1}^{\infty} a_n = b_1 - b_2 + b_3 - b_4 + \cdots \quad \text{is convergent, with sum } s. \quad \blacksquare$$

Theorem 4   Let the series $\sum_{n=1}^{\infty} a_n = a_1 + a_2 + \cdots$ be absolutely convergent. Then any rearrangement $\sum_{n=1}^{\infty} b_n$ of the series also converges absolutely, and $\sum_{n=1}^{\infty} b_n = \sum_{n=1}^{\infty} a_n$.

Proof   First of all, we prove the result in the special case that $a_k \geq 0$ for all k.

Choose any positive integer n. Then choose a positive integer N such that $b_1, b_2, \ldots, b_n$ all occur among the terms $a_1, a_2, \ldots, a_N$ of the original series. It follows that

In particular, $N \geq n$.

For the right-hand side is simply the left-hand side with possibly some more non-negative terms inserted.

$$\sum_{k=1}^{n} b_k \leq \sum_{k=1}^{N} a_k$$

$$\leq \sum_{k=1}^{\infty} a_k. \tag{13}$$

For the partial sums of a series of non-negative terms are always less than or equal the sum of the whole series.

Hence the partial sums $\sum_{k=1}^{n} b_k$ of the rearranged series $\sum_{k=1}^{\infty} b_k$ are bounded above, so that the rearranged series must be convergent. It also follows from (13) that $\sum_{k=1}^{\infty} b_k \leq \sum_{k=1}^{\infty} a_k$.

By reversing the roles of the terms $a_1, a_2, \ldots$ and $b_1, b_2, \ldots$ in the above argument, the same argument shows that $\sum_{k=1}^{\infty} a_k \leq \sum_{k=1}^{\infty} b_k$. It thus follows that the two series $\sum_{k=1}^{\infty} a_k$ and $\sum_{k=1}^{\infty} b_k$ must in fact have the *same* sum.

To complete our proof, we now drop the condition that $a_k \geq 0$ for all k. We then define the quantities a_k^+, a_k^-, b_k^+ and b_k^- as follows

$$a_k^+ = \begin{cases} a_k, & \text{if } a_k \geq 0, \\ 0, & \text{if } a_k < 0, \end{cases} \qquad a_k^- = \begin{cases} 0, & \text{if } a_k \geq 0, \\ -a_k, & \text{if } a_k < 0, \end{cases}$$

$$b_k^+ = \begin{cases} b_k, & \text{if } b_k \geq 0, \\ 0, & \text{if } b_k < 0, \end{cases} \qquad b_k^- = \begin{cases} 0, & \text{if } b_k \geq 0, \\ -a_k, & \text{if } b_k < 0. \end{cases}$$

Now, $\sum_{k=1}^{\infty} a_k^+$ is convergent, since $a_k^+ = \frac{1}{2}(|a_k| + a_k)$ and both $\sum_{k=1}^{\infty} |a_k|$ and $\sum_{k=1}^{\infty} a_k$ converge. $\sum_{k=1}^{\infty} b_k^+$ is a rearrangement of $\sum_{k=1}^{\infty} a_k^+$, and both series have non-negative terms. It follows from the first part of the proof that both series converge, and have the same sum.

By the Combination Rules for series.

Similarly, $\sum_{k=1}^{\infty} a_k^-$ and $\sum_{k=1}^{\infty} b_k^-$ both converge, and have the same sum.

But $|b_k| = b_k^+ + b_k^-$, and so the series $\sum_{k=1}^{\infty} |b_k|$ converges, by the Sum Rule for series. In other words, the rearranged series $\sum_{k=1}^{\infty} b_k$ is absolutely convergent.

And, in fact
$$\sum_{k=1}^{\infty} |b_k| = \sum_{k=1}^{\infty} b_k^+ + \sum_{k=1}^{\infty} b_k^-.$$

Finally

$$\sum_{k=1}^{\infty} b_k = \sum_{k=1}^{\infty} b_k^+ - \sum_{k=1}^{\infty} b_k^-$$

$$= \sum_{k=1}^{\infty} a_k^+ - \sum_{k=1}^{\infty} a_k^-$$

$$= \sum_{k=1}^{\infty} a_k.$$

∎

Theorem 5   Product Rule

Let the series $\sum_{n=0}^{\infty} a_n$ and $\sum_{n=0}^{\infty} b_n$ be absolutely convergent, and let

$$c_n = a_0 b_n + a_1 b_{n-1} + a_2 b_{n-2} + \cdots + a_n b_0 = \sum_{k=0}^{n} a_k b_{n-k}.$$

Then the series $\sum_{n=0}^{\infty} c_n$ is absolutely convergent, and

$$\sum_{n=0}^{\infty} c_n = \left(\sum_{n=0}^{\infty} a_n \right) \times \left(\sum_{n=0}^{\infty} b_n \right).$$

Proof   First, we introduce some notation, for $n \geq 0$

$$s_n = \sum_{k=0}^{n} a_k, \quad t_n = \sum_{k=0}^{n} b_k, \quad u_n = \sum_{k=0}^{n} c_k,$$

$$s_n' = \sum_{k=0}^{n} |a_k|, \quad t_n' = \sum_{k=0}^{n} |b_k|, \quad u_n' = \sum_{k=0}^{n} |c_k|,$$

$$s = \sum_{k=0}^{\infty} a_k, \quad t = \sum_{k=0}^{\infty} b_k,$$

$$s' = \sum_{k=0}^{\infty} |a_k|, \quad t' = \sum_{k=0}^{\infty} |b_k|.$$

In particular, we know that $s_n \to s$, $t_n \to t$, $s_n' \to s'$ and $t_n' \to t'$ as $n \to \infty$.

Also, by the Product Rule for sequences, $s_n t_n \to st$ and $s_n' t_n' \to s't'$, as $n \to \infty$. Hence, from the definition of limit, it follows that, for any positive number ε, there is some integer N for which

$$|s_n t_n - st| < \frac{1}{3}\varepsilon \quad \text{and} \quad |s_n' t_n' - s't'| < \frac{1}{3}\varepsilon, \quad \text{for all } n \geq N.$$

Here we use an integer N rather than a general number X and a weak inequality $n \geq N$ rather than a strict inequality; this simplifies the notation in the rest of the argument a little.

It follows that, for $n \geq N$

$$|s_n' t_n' - s_N' t_N'| = |(s_n' t_n' - s't') - (s_N' t_N' - s't')|$$

$$\leq |s_n' t_n' - s't'| + |s_N' t_N' - s't'|$$

$$< \frac{1}{3}\varepsilon + \frac{1}{3}\varepsilon$$

$$= \frac{2}{3}\varepsilon.$$

Now let n be such that $n \geq 2N$, and consider the terms $a_i\, b_j$ that occur in the expression $u_n - s_N t_N$. Every such term has $i + j \leq n$, since u_n consists of all such terms; but none of these terms has *both* $i \leq N$ *and* $j \leq N$.

For terms with both $i \leq N$ and $j \leq N$ are all in $s_N t_N$.

Hence, for every term $a_i\, b_j$ in $u_n - s_N t_N$, a corresponding term $|a_i b_j|$ will occur in the expression $s'_n t'_n - s'_N t'_N$ – for this last expression consists of all terms $|a_i b_j|$ with $i \leq n$ and $j \leq n$, but *not with both* $i \leq N$ *and* $j \leq N$. (Note that we made the requirement that $n \geq 2N$ in order that every term in $s_N t_N$ appears in u_n.)

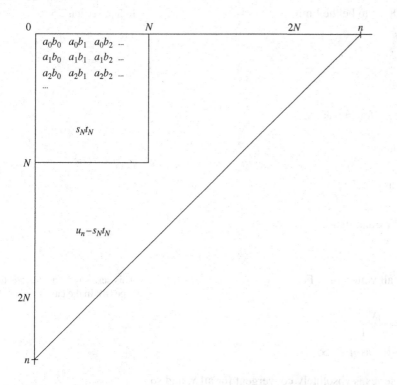

In the above diagram, we set out the terms $a_i\, b_j$ in rows and columns, where the term $a_i\, b_j$ occurs in the $(i+1)$th row and the $(j+1)$th column. Then the small square contains all the terms $a_i b_j$ with $0 \leq i \leq N$ and $0 \leq j \leq N$; these add up to $s_N t_N$.

Hence, for all $n \geq 2N$, we have

$$
\begin{aligned}
|u_n - st| &= |(u_n - s_N t_N) + (s_N t_N - st)| \\
&\leq |u_n - s_N t_N| + |s_N t_N - st| \\
&\leq |s'_n t'_n - s'_N t'_N| + |s_N t_N - st| \\
&< \frac{2}{3}\varepsilon + \frac{1}{3}\varepsilon \\
&= \varepsilon;
\end{aligned}
$$

By the Triangle Inequality.

By the discussions above concerning $u_n - s_N t_N$ and $s'_n t'_n - s'_N t'_N$.

it follows, from this inequality, that $u_n \to st$ as $n \to \infty$.

Finally, we have

$$
\begin{aligned}
u'_n &\leq s'_n t'_n \\
&\leq s' t'.
\end{aligned}
$$

For the product $s'_n t'_n$ contains more non-negative terms than does u'_n; and the sequences $\{s'_n\}$ and $\{t'_n\}$ are increasing.

Hence, the increasing sequence $\{u'_n\}$ is bounded above, and so tends to a limit as $n \to \infty$. Thus the series $\sum_{n=0}^{\infty} c_n$ is absolutely convergent. ■

3.4   The exponential function $x \mapsto e^x$

Earlier, we defined $e = 2.71828\ldots$ to be the limit

In Sub-section 2.5.3.

$$e = \lim_{n \to \infty} \left(1 + \frac{1}{n}\right)^n,$$

and we defined e^x to be the limit

$$e^x = \lim_{n \to \infty} \left(1 + \frac{x}{n}\right)^n, \quad \text{for any real } x.$$

Here we show that the formula

$$e^x = \sum_{n=0}^{\infty} \frac{x^n}{n!}, \quad \text{for any real } x,$$

is an equivalent definition of the quantity e^x.

Remark

The series $\sum_{n=0}^{\infty} \frac{x^n}{n!}$ converges for all values of x. For

This series is a basic series of type (d), in the case that $x \geq 0$.

$$\left| \frac{x^{n+1}}{(n+1)!} \right| \bigg/ \left| \frac{x^n}{n!} \right| = \frac{|x|}{n+1}$$
$$\to 0 \quad \text{as } n \to \infty,$$

so that, by the Ratio Test, the series is absolutely convergent for all x, and so is convergent for all x.

3.4.1   The definition of e^x as a power series, for $x > 0$

If we plot the partial sum functions of the infinite series of powers of x

$$\sum_{n=0}^{\infty} \frac{x^n}{n!} = 1 + x + \frac{x^2}{2!} + \frac{x^3}{3!} + \cdots, \tag{1}$$

the resulting graph appears to be that of e^x. (A series of multiples of increasing powers of x is called a *power series*.) In particular, when $x = 1$, the sum of the series

$$\sum_{n=0}^{\infty} \frac{1}{n!} = 1 + 1 + \frac{1}{2!} + \frac{1}{3!} + \cdots$$

is approximately $2.71828\ldots$.

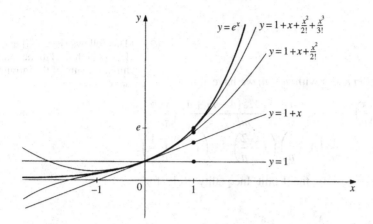

However the fact that the sequence of partial sums of the series (1) *appears* to converge to e^x does not constitute a proof of this fact. Our first aim is to supply this proof in the particular case that $x > 0$.

We shall deal with the case $x < 0$ in Sub-section 3.4.3.

Theorem 1   If $x > 0$, then $\displaystyle\sum_{n=0}^{\infty} \frac{x^n}{n!} = e^x$.

Hence, for $x > 0$, the series $\displaystyle\sum_{n=0}^{\infty} \frac{x^n}{n!}$ gives an equivalent definition of e^x.

Proof   We defined e^x to be $e^x = \lim\limits_{n \to \infty} \left(1 + \frac{x}{n}\right)^n$, and so we have to show that

You may omit this proof at a first reading.

$$\sum_{n=0}^{\infty} \frac{x^n}{n!} = \lim_{n \to \infty} \left(1 + \frac{x}{n}\right)^n, \quad \text{for } x > 0.$$

The $(n + 1)$th partial sum of the above series is

$$s_{n+1} = 1 + x + \frac{x^2}{2!} + \frac{x^3}{3!} + \cdots + \frac{x^n}{n!}.$$

By the Binomial Theorem

$$\left(1 + \frac{x}{n}\right)^n = 1 + n\left(\frac{x}{n}\right) + \frac{n(n-1)}{2!}\left(\frac{x}{n}\right)^2 + \cdots + \left(\frac{x}{n}\right)^n.$$

A typical term in this expansion is

$$\frac{n(n-1)\ldots(n-k+1)}{k!}\left(\frac{x}{n}\right)^k = \frac{x^k}{k!}\left(1 - \frac{1}{n}\right)\left(1 - \frac{2}{n}\right)\cdots\left(1 - \frac{k-1}{n}\right)$$
$$\leq \frac{x^k}{k!},$$

since each bracket is less than 1.
    Thus

$$\left(1 + \frac{x}{n}\right)^n \leq 1 + x + \frac{x^2}{2!} + \frac{x^3}{3!} + \cdots + \frac{x^n}{n!}$$
$$= s_{n+1},$$

and so, by the Limit Inequality Rule for sequences

$$\lim_{n \to \infty} \left(1 + \frac{x}{n}\right)^n \leq \lim_{n \to \infty} s_{n+1},$$

so that

$$e^x \leq \sum_{n=0}^{\infty} \frac{x^n}{n!}. \qquad (2)$$

This follows since $\{s_n\}$ and $\{s_{n+1}\}$ both tend to the same limit, the sum of the infinite series.

On the other hand, for any integers m and n with $m \leq n$, we have

$$\left(1 + \frac{x}{n}\right)^n \geq 1 + n\left(\frac{x}{n}\right) + \frac{n(n-1)}{2!}\left(\frac{x}{n}\right)^2 + \cdots + \frac{n(n-1)\ldots(n-m+1)}{m!}\left(\frac{x}{n}\right)^m$$

$$= 1 + x + \frac{x^2}{2!}\left(1 - \frac{1}{n}\right) + \cdots + \frac{x^m}{m!}\left(1 - \frac{1}{n}\right)\left(1 - \frac{2}{n}\right)\ldots\left(1 - \frac{m-1}{n}\right).$$

Now keep m fixed and let $n \to \infty$. By the Limit Inequality Rule for sequences, we obtain

$$\lim_{n \to \infty}\left(1 + \frac{x}{n}\right)^n \geq 1 + x + \frac{x^2}{2!} + \frac{x^3}{3!} + \cdots + \frac{x^m}{m!},$$

and so

$$e^x \geq s_{m+1}.$$

Now let $m \to \infty$; by the Limit Inequality Rule for sequences, we obtain

$$e^x \geq \lim_{m \to \infty} s_{m+1},$$

so that

$$e^x \geq \sum_{n=0}^{\infty} \frac{x^n}{n!} \qquad (3)$$

Combining inequalities (2) and (3), we obtain

$$e^x = \sum_{n=0}^{\infty} \frac{x^n}{n!}, \quad \text{for } x > 0. \qquad \blacksquare$$

Problem 1 Estimate e^2 (to 3 decimal places) by calculating the seventh partial sum of the series (1) when $x = 2$.

3.4.2 Calculating e

The representation of e by the infinite series

$$e = \sum_{n=0}^{\infty} \frac{1}{n!} = 1 + 1 + \frac{1}{2!} + \frac{1}{3!} + \cdots \qquad (4)$$

provides a much more efficient way of calculating approximate values for e than the equation $e = \lim_{n \to \infty}\left(1 + \frac{1}{n}\right)^n$. This is illustrated by the following table of approximate values:

n	1	2	3	4	5
$\left(1 + \frac{1}{n}\right)^n$	2	2.25	2.37	2.44	2.49
$\sum_{k=0}^{n} \frac{1}{k!}$	2	2.50	2.67	2.71	2.717

The calculation of e via the limit is a new calculation each time, whereas via the series involves only adding one extra term to the previous approximation.

We can estimate how quickly the sequence of partial sums

$$s_n = \sum_{k=0}^{n-1} \frac{1}{k!} = 1 + 1 + \frac{1}{2!} + \frac{1}{3!} + \cdots + \frac{1}{(n-1)!}, \quad n = 1, 2, \ldots,$$

converges to e as follows. The difference between e and s_n is given by

$$e - s_n = \sum_{k=n}^{\infty} \frac{1}{k!} = \frac{1}{n!} + \frac{1}{(n+1)!} + \frac{1}{(n+2)!} + \cdots$$

$$= \frac{1}{n!} \left[1 + \frac{1}{(n+1)} + \frac{1}{(n+1)(n+2)} + \cdots \right]$$

$$< \frac{1}{n!} \left[1 + \frac{1}{n} + \frac{1}{n^2} + \cdots \right].$$

Here we replace various terms by larger terms (because their denominators are smaller); so the new sum is greater.

The last expression in square brackets is a geometric series with first term 1 and common ratio $\frac{1}{n}$, and so its sum is $\frac{1}{1-\frac{1}{n}} = \frac{n}{n-1}$. Hence

$$0 < e - s_n < \frac{1}{n!} \times \frac{n}{n-1}$$

$$= \frac{1}{(n-1)!} \times \frac{1}{n-1}, \quad \text{for } n = 1, 2, \ldots. \tag{5}$$

For example, this estimate shows that

$$0 < e - s_6 < \frac{1}{5!} \times \frac{1}{5} = \frac{1}{600} = 0.001\overline{6}.$$

Since the partial sum

$$s_6 = 1 + 1 + \frac{1}{2!} + \frac{1}{3!} + \frac{1}{4!} + \frac{1}{5!}$$

$$= 1 + 1 + \frac{1}{2} + \frac{1}{6} + \frac{1}{24} + \frac{1}{120} = \frac{163}{60} = 2.71\overline{6},$$

we deduce very easily that

$$2.71\overline{6} < e < 2.71\overline{6} + 0.001\overline{6} = 2.718\overline{3}.$$

In fact,

$$e = 2.71828182845 \ldots.$$

Problem 2   Estimate how many terms in the series (4) are needed to determine e to ten decimal places.

The inequality (5) can also be used to show that e is irrational.

Theorem 2   The number e is irrational.

Proof   Suppose that $e = \frac{m}{n}$, where m and n are positive integers. Now, by the above estimate (5)

This is a proof by contradiction.

$$0 < e - s_{n+1} < \frac{1}{n!} \times \frac{1}{n},$$

and so

$$0 < n!(e - s_{n+1}) < \frac{1}{n}.$$

Since $e = \frac{m}{n}$, we have

$$0 < n! \left[\frac{m}{n} - \left(1 + 1 + \frac{1}{2!} + \frac{1}{3!} + \cdots + \frac{1}{n!} \right) \right] < \frac{1}{n}.$$

But $n! \left[\frac{m}{n} - \left(1 + 1 + \frac{1}{2!} + \frac{1}{3!} + \cdots + \frac{1}{n!} \right) \right]$ is an integer, since $n!$ times each expression in the square bracket is itself an integer. So we have found an integer which lies strictly between 0 and 1. This is clearly impossible, and so e cannot be rational after all. ∎

Here we have simply substituted $\frac{m}{n}$ for e in the previous expression.

3.4.3   The definition of e^x as a power series, for all real x

Earlier we saw that $\sum_{n=0}^{\infty} \frac{x^n}{n!} = e^x$, for all $x > 0$. We now show that this formula is valid for all real values of x, and we will also verify the *fundamental property of the exponential function*, namely that $e^x e^y = e^{x+y}$ for all real values of x and y. The Product Rule for series will be our crucial tool in this work.

First we check the following.

This was Theorem 1 in Sub-section 3.4.1.

This was Theorem 5 in Sub-section 3.3.4.

Lemma 1   For any $x \in \mathbb{R}$, $\sum_{n=0}^{\infty} \frac{x^n}{n!} \times \sum_{n=0}^{\infty} \frac{(-x)^n}{n!} = 1$.

Proof   By the Product Rule for series, $\sum_{n=0}^{\infty} \frac{x^n}{n!} \times \sum_{n=0}^{\infty} \frac{(-x)^n}{n!} = \sum_{n=0}^{\infty} c_n$, where, for $n \geq 1$, we have

$$c_n = \sum_{k=0}^{n} \frac{x^k}{k!} \times \frac{(-x)^{n-k}}{(n-k)!}$$

$$= \sum_{k=0}^{n} \frac{x^k (-x)^{n-k}}{k!(n-k)!}$$

$$= \frac{1}{n!} \sum_{k=0}^{n} \frac{n!}{k!(n-k)!} x^k (-x)^{n-k}$$

$$= \frac{1}{n!} (x + (-x))^n = 0.$$

Since $c_0 = 1 \times 1 = 1$, the result then follows. ∎

We may apply the Product Rule since the series $\sum_{n=0}^{\infty} \frac{x^n}{n!}$ and $\sum_{n=0}^{\infty} \frac{(-x)^n}{n!}$ are both absolutely convergent.

Here we apply the Binomial Theorem to $(x + (-x))^n$.

We can then use the result in Lemma 1 to prove that $\sum_{n=0}^{\infty} \frac{x^n}{n!} = e^x$, for $x < 0$ as well as $x > 0$. Since, obviously, $\sum_{n=0}^{\infty} \frac{x^n}{n!} = e^x$ when $x = 0$, this completes the proof that $\sum_{n=0}^{\infty} \frac{x^n}{n!} = e^x$, for all $x \in \mathbb{R}$.

In other words, the definition of e^x as a power series is equivalent to its definition as a limit.

Theorem 3   If $x < 0$, then $\sum_{n=0}^{\infty} \frac{x^n}{n!} = e^x$.

Proof   For $x < 0$, the following chain of equalities holds

$$\sum_{n=0}^{\infty} \frac{x^n}{n!} = 1 \left/ \sum_{n=0}^{\infty} \frac{(-x)^n}{n!} \right.$$ (by Lemma 1)

Here the definition of e^x on the right is that
$$e^x = \lim_{n \to \infty} \left(1 + \frac{x}{n} \right)^n.$$

$$= 1 \Big/ \lim_{n \to \infty} \Big(1 + \frac{-x}{n}\Big)^n \qquad \text{(by Theorem 1, since } -x > 0)$$

$$= 1/e^{-x} \qquad \text{(this is the definition of } e^{-x})$$

$$= e^x \qquad \text{(by the Inverse Property of } e^x).$$

This completes the proof.   ■

The Inverse Property of e^x was Theorem 4 of Sub-section 2.5.3.

In order to be crystal clear where we have reached, we now combine the results of Theorem 1 and Theorem 3 into the following:

Theorem 4   For all $x \in \mathbb{R}$, $\sum_{n=0}^{\infty} \frac{x^n}{n!} = \lim_{n \to \infty} \Big(1 + \frac{x}{n}\Big)^n = e^x$.

Finally, we verify the Fundamental Property of the exponential function, which was left as unfinished business at the end of Sub-section 2.5.3.

Theorem 5   Fundamental Property of the exponential function
For all $x, y \in \mathbb{R}$, $e^x e^y = e^{x+y}$.

Proof   By the Product Rule for series, $e^x \times e^y = \sum_{n=0}^{\infty} \frac{x^n}{n!} \times \sum_{n=0}^{\infty} \frac{y^n}{n!} = \sum_{n=0}^{\infty} c_n$,

where, for $n \geq 1$, we have

$$c_n = \sum_{k=0}^{n} \frac{x^k}{k!} \times \frac{y^{n-k}}{(n-k)!}$$

$$= \sum_{k=0}^{n} \frac{x^k y^{n-k}}{k!(n-k)!}$$

$$= \frac{1}{n!} \sum_{k=0}^{n} \frac{n!}{k!(n-k)!} x^k y^{n-k}$$

$$= \frac{(x+y)^n}{n!}.$$

Since $c_0 = 1 \times 1 = 1$, the result then follows.   ■

3.5   Exercises

Section 3.1

1. Prove that $\sum_{n=1}^{\infty} \Big(-\frac{3}{4}\Big)^n$ is convergent, and find its sum.

2. Interpret $0.\overline{12}$ as an infinite series, and hence find the value of $0.\overline{12}$ as a fraction.

3. Prove that

$$\frac{1}{4n^2 - 1} = \frac{1}{2}\left(\frac{1}{2n-1} - \frac{1}{2n+1}\right), \quad \text{for } n = 1, 2, \ldots,$$

and deduce that

$$\sum_{n=1}^{\infty} \frac{1}{4n^2 - 1} = \frac{1}{2}.$$

4. Determine whether the following series converge:

(a) $\sum_{n=1}^{\infty} \left(\left(\frac{4}{5}\right)^n + \frac{4}{n(n+2)}\right);$ (b) $\sum_{n=1}^{\infty} \left(1 + \left(\frac{1}{2}\right)^n\right);$ (c) $\sum_{n=1}^{\infty} \left(\frac{1}{\sqrt{n}} - \frac{1}{\sqrt{n+1}}\right).$

Section 3.2

1. Determine whether the following series converge:

(a) $\sum_{n=1}^{\infty} \frac{\cos(1/n)}{2n^2+3};$ (b) $\sum_{n=1}^{\infty} \frac{n^2}{2n^3-n};$ (c) $\sum_{n=1}^{\infty} \frac{\sqrt{2n}}{4n^3+n+2};$

(d) $\sum_{n=1}^{\infty} \frac{(n+1)^5}{2^n};$ (e) $\sum_{n=1}^{\infty} \frac{n^2 3^n}{n!};$ (f) $\sum_{n=1}^{\infty} \frac{(n!)^2}{(2n)!}.$

2. (a) Use the Ratio Test to prove that $\sum_{n=1}^{\infty} \frac{2^n n!}{n^n}$ converges, but that $\sum_{n=1}^{\infty} \frac{3^n n!}{n^n}$ diverges.

   (b) For which positive values of c can you use the Ratio Test to prove that $\sum_{n=1}^{\infty} \frac{c^n n!}{n^n}$ is convergent?

3. Prove that

$$\frac{1}{\sqrt{n}} - \frac{1}{\sqrt{n+1}} = \frac{1}{\sqrt{n}\sqrt{n+1}\left(\sqrt{n+1} + \sqrt{n}\right)}, \quad \text{for } n = 1, 2, \ldots,$$

and use Exercise 4(c) on Section 3.1 to deduce that $\sum_{n=1}^{\infty} \frac{1}{n^{\frac{3}{2}}}$ is convergent.

4. Determine whether the following series converge:

(a) $\sum_{n=3}^{\infty} \frac{1}{n(\log_e n)(\log_e(\log_e n))};$ (b) $\sum_{n=2}^{\infty} \frac{\sqrt{n+1}}{(2n^2-3n+1)\left(\log_e n + (\log_e n)^2\right)}.$

Section 3.3

1. Test the following series for convergence and absolute convergence:

(a) $\sum_{n=1}^{\infty} \frac{(-1)^{n+1}}{1+\sqrt{n}};$ (b) $\sum_{n=1}^{\infty} \frac{\sin n}{n^2};$ (c) $\sum_{n=1}^{\infty} \frac{(-1)^{n+1} n!}{n^4+3};$ (d) $\sum_{n=1}^{\infty} \frac{n+2^n}{3^n+5}.$

2. (a) Prove that

$$\frac{1}{3n-2} + \frac{1}{3n-1} - \frac{1}{3n} > \frac{1}{3n}, \quad \text{for } n = 1, 2, \ldots,$$

   and deduce that

$$1 + \frac{1}{2} - \frac{1}{3} + \frac{1}{4} + \frac{1}{5} - \frac{1}{6} + \cdots \quad \text{is divergent.}$$

(b) Prove that

$$\left|\frac{1}{n} - \frac{1}{2n} - \frac{1}{2n+2}\right| < \frac{1}{n^2}, \quad \text{for } n = 1, 2, \ldots,$$

and deduce that

$$1 - \frac{1}{2} - \frac{1}{4} + \frac{1}{3} - \frac{1}{6} - \frac{1}{8} + \frac{1}{5} - \frac{1}{10} - \frac{1}{12} + \frac{1}{7} - \frac{1}{14} - \frac{1}{16} + \cdots$$

is convergent.

(c) Prove that

$$\frac{1}{\sqrt{2n-1}} - \frac{1}{2n} \geq \frac{1}{2n}, \quad \text{for } n = 1, 2, \ldots,$$

and deduce that

$$1 - \frac{1}{2} + \frac{1}{\sqrt{3}} - \frac{1}{4} + \frac{1}{\sqrt{5}} - \frac{1}{6} + \cdots$$

is divergent.

3. Prove that there exists a rearrangement of the series $\sum_{n=1}^{\infty} \frac{(-1)^{n+1}}{n} = 1 - \frac{1}{2} + \frac{1}{3} - \frac{1}{4} + \frac{1}{5} - \frac{1}{6} + \cdots$ for which one subsequence of its partial sums tends to 1 and another subsequence of its partial sums tends to -1. Alternatively, prove that no such rearrangement exists.

4 Continuity

The graphs of many functions that we come across look to be smooth curves, without any *jumps*.

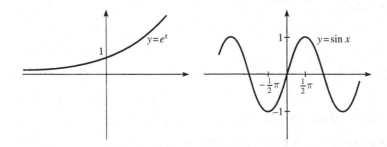

On the other hand, the graphs of some functions that arise naturally, or that we construct for various purposes, do not appear to be smooth, but to contain jumps.

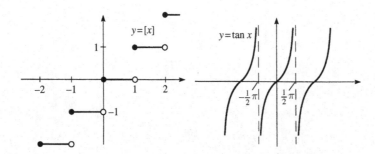

Indeed there are some functions where the graphs appear to have an unreasonable number of jumps!

How can we describe this complicated situation? The key underlying idea is that of *continuity*. Loosely speaking, a function is said to be *continuous* if we can draw its graph without taking our pencil off the page. This is the same as the graph 'having no jumps'. The concept is an important one in Analysis since, in many situations, the crucial step in proving that a function has some property is to prove that the function is continuous. This is the first of two chapters that study continuous functions.

In Section 4.1, we define the phrase:

the function f is *continuous* at the point c,

and we give a number of rules which state, for example, that various combinations and compositions of continuous functions are themselves continuous.

Using these rules, together with a list of basic continuous functions, we can deduce that many functions are continuous at each point of their domains. For example, the functions $x \mapsto x + \frac{1}{x}$ and $x \mapsto x \sin \frac{1}{x}$ are continuous at each point of $\mathbb{R} - \{0\}$, and the trigonometric and exponential functions are continuous at each point of their domains.

Section 4.2 is devoted to the *properties* of continuous functions. The two fundamental properties of continuous functions are the Intermediate Value Theorem and the Extreme Value Theorem. We shall see a useful application of the Intermediate Value Theorem to finding the zeros of various functions.

In particular, to finding solutions of polynomial equations.

In Section 4.3, we discuss the Inverse Function Rule; this rule gives conditions under which a continuous function f has a continuous *inverse function* f^{-1}. We then use the Inverse Function Rule to obtain the inverses of standard functions. Some of these inverse functions will be familiar to you already, but the Inverse Function Rule enables us to establish their properties by a rigorous argument.

Finally, in Section 4.4, we shall provide a rigorous definition of the *exponential function* $x \mapsto a^x$, where $x \in \mathbb{R}$ and $a > 0$.

4.1 Continuous functions

4.1.1 What is continuity?

To accord with our intuitive idea of what we should mean by '*the function f is continuous at the point c*', we wish to define this concept in such a way that the following two functions are continuous at the point c:

1.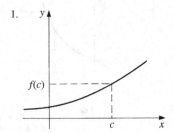

2.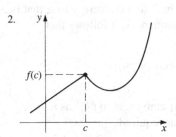

On the other hand, we wish to formulate our definition so that the following two functions are *not* continuous at the point c:

3.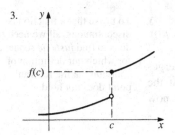

4.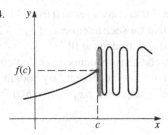

So, our definition must say, in a precise way, that:

if x tends to c, then $f(x)$ tends to $f(c)$.

There are several ways of making this idea precise. In this chapter we adopt a definition which involves the *convergence of sequences*, as this will enable us to use the results about sequences that we met in Chapter 2.

In Chapter 5 we shall meet a different definition, and see that the two definitions are in fact *equivalent*.

Definitions   A function f defined on an interval I that contains c as an interior point is **continuous at c** if:

for each sequence $\{x_n\}$ in I such that $x_n \to c$, then $f(x_n) \to f(c)$.

If f is not continuous at the point c in I, then it is **discontinuous** at c.

Thus, for example, in graph 3 above, if $\{x_n\}$ is a strictly decreasing sequence that tends to c as $n \to \infty$, then $\{f(x_n)\}$ is a strictly decreasing sequence that tends to $\{f(c)\}$. On the other hand, if $\{x_n\}$ is a strictly increasing sequence that tends to c as $n \to \infty$, then $\{f(x_n)\}$ is a strictly increasing sequence that tends to a limit as $n \to \infty$ – but that limit is some number less than $f(c)$.

However, in graph 4, if $\{x_n\}$ is a strictly increasing sequence that tends to c as $n \to \infty$, then $\{f(x_n)\}$ is a strictly increasing sequence that tends to $f(c)$; but, if $\{x_n\}$ is a strictly decreasing sequence that tends to c as $n \to \infty$, then the sequence $\{f(x_n)\}$ may not tend to any limit as $n \to \infty$.

On the other hand, in graphs 1 and 2, no matter how the sequence $\{x_n\}$ tends to c, then for sure we do have that $\{f(x_n)\}$ tends to $f(c)$.

For these examples, then, the definition agrees with what we believe the essence of continuity should be.

Example 1   Prove that the function $f(x) = x^3, x \in \mathbb{R}$, is continuous at the point $\tfrac{1}{2}$.

Solution   Let $\{x_n\}$ be any sequence in $\mathbb{R}$ that converges to $\tfrac{1}{2}$; that is, $x_n \to \tfrac{1}{2}$. Then, by the Combination Rules for sequences, it follows that

$$f(x_n) = x_n^3 \to \left(\frac{1}{2}\right)^3 = \frac{1}{8} \text{ as } n \to \infty,$$

while $f\left(\tfrac{1}{2}\right) = \tfrac{1}{8}$. In other words, $\{f(x_n)\}$ converges to $f\left(\tfrac{1}{2}\right)$ as $n \to \infty$.

It follows that f is continuous at $\tfrac{1}{2}$, as required.   □

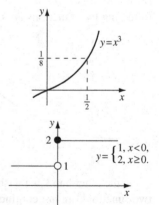

Example 2   Prove that the function $f(x) = \begin{cases} 1, & x < 0, \\ 2, & x \geq 0, \end{cases}$ is discontinuous at 0.

Solution   Let $\{x_n\}$ be any sequence in $\mathbb{R}$ that converges to 0 from the right. By looking at the graph $y = f(x)$, it is clear that for such a sequence $f(x_n) \to 2 = f(0)$. This will not help us to show that f is discontinuous at 0!

However, if we let $\{x_n\}$ be any non-constant sequence in $\mathbb{R}$ that converges to 0 from the left, the situation will be very different. By looking at the graph $y = f(x)$, it is clear that for such a sequence $f(x_n) \to 1 \neq f(0)$. We now make this precise.

To prove that a function is discontinuous, all we need to do is to find *just one* sequence for which our definition of continuity at the relevant point does *not* hold.

Choose the sequence $\{x_n\} = \{-\frac{1}{n}\}, n = 1, 2, \ldots$. For this sequence
$$f(x_n) = 1 \to 1 \neq f(0) \text{ as } n \to \infty.$$
It follows that the function f cannot be continuous at 0. □

Here we make a specific choice for the sequence $\{x_n\}$ to prove that f cannot be continuous at 0.

Problem 1

(a) Determine whether the function $f(x) = x^3 - 2x^2$, $x \in \mathbb{R}$, is continuous at 2.

(b) Determine whether the function $f(x) = [x]$, $x \in \mathbb{R}$, is continuous at 1.

Here $[\cdot]$ is the integer part function.

Problem 2

(a) Prove that the function $f(x) = 1$, $x \in \mathbb{R}$, is continuous on $\mathbb{R}$.

(b) Prove that the function $f(x) = x$, $x \in \mathbb{R}$, is continuous on $\mathbb{R}$.

By 'continuous on $\mathbb{R}$', we mean 'continuous at each point of $\mathbb{R}$'.

However, as we know, not every function is defined on the whole of $\mathbb{R}$ – but we still want to be able to discuss the continuity of such functions. For example:

- $f(x) = \sqrt{x}$, where the domain of f is the interval $[0, \infty)$;
- $f(x) = x^{-\frac{1}{2}} + (1 - x)^{\frac{1}{2}}$, where the domain of f is the half-closed interval $(0, 1]$;
- $f(x) = \frac{1}{x}$, where the domain of f is $\mathbb{R} - \{0\}$.

How can we deal with these various different situations?

The first two situations are dealt with by introducing the notion of *one-sided continuity*.

Suppose that f is defined on some set S that contains an interval $[c, c + r)$ for some $r > 0$, but where f is not necessarily defined on any interval $(c - r, c + r)$ that contains c as an interior point. Then it seems reasonable to say that f is *continuous* on one side of c. In particular, that f is *continuous on the right* at the point c of S.

Similarly, if f is defined on some set S that contains an interval $(c - r, c]$ for some $r > 0$, but f is not necessarily defined on any interval $(c - r, c + r)$ that contains c as an interior point, then it seems reasonable to say that f is *continuous* on one side of c. In particular, that f is *continuous on the left* at the point c of S.

The case of the function $f(x) = \frac{1}{x}$, where the domain of f is $\mathbb{R} - \{0\}$, is different. The domain of f is the whole of $\mathbb{R}$, with just one point omitted. That is, it is the union of two open intervals $(-\infty, 0)$ and $(0, \infty)$ – to both of which the definition of continuity applies in its originally stated form. So it makes sense to simply drop the original restriction that the domain of f must be an open interval in $\mathbb{R}$.

These considerations lead us to making the following less restrictive definition of continuity than our original definition:

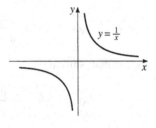

Definitions A function f defined on a set S in $\mathbb{R}$ that contains a point c is **continuous at c** if:

for each sequence $\{x_n\}$ in S such that $x_n \to c$, then $f(x_n) \to f(c)$.

If f is continuous on the whole of its domain, we often simply say that f is **continuous** (without explicit mention of the points of continuity).

If f is not continuous at the point c in S, then it is **discontinuous** at c.

---

*Here f is *continuous on S*.*

In addition, we have the following related definitions:

Definitions A function f whose domain contains an interval $[c, c + r)$ for some $r > 0$ is **continuous on the right at c** if:

for each sequence $\{x_n\}$ in $[c, c + r)$ such that $x_n \to c$, then $f(x_n) \to f(c)$.

A function f whose domain contains an interval $(c - r, c]$ for some $r > 0$ is **continuous on the left at c** if:

for each sequence $\{x_n\}$ in $(c - r, c]$ such that $x_n \to c$, then $f(x_n) \to f(c)$.

A function f whose domain contains an interval I is **continuous on I** if it is continuous at each interior point of I, continuous on the right at the left endpoint of I (if this belongs to I), and continuous on the left at the right endpoint of I (if this belongs to I).

The connection between the definitions of continuity and of one-sided continuity is rather obvious.

Theorem 1 A function f whose domain contains an interval I that contains c as an interior point is **continuous at c** if and only if f is both continuous on the left at c *and* continuous on the right at c.

We omit a proof of this straight-forward result.

Example 3 Determine whether the function f given by $f(x) = \sqrt{x}$, $x \geq 0$, is continuous on $\{x : x \in \mathbb{R}, x \geq 0\}$.

In fact, f is *continuous on the right* at 0, and *continuous* at all other points of its domain.

Solution The domain of f is the interval $I = \{x : x \geq 0\}$.

Thus, we have to show that for each c in I:

for each sequence $\{x_n\}$ in I such that $x_n \to c$, then $\sqrt{x_n} \to \sqrt{c}$.

First, if $c = 0$, then we know already that, for any null sequence $\{x_n\}$ in I, $\{\sqrt{x_n}\}$ is also a null sequence.

Next, let $c > 0$. We have to prove that if $\{x_n - c\}$ is a null sequence, then so is $\{\sqrt{x_n} - \sqrt{c}\}$. Now, from the identity

$$\sqrt{x_n} - \sqrt{c} = \frac{x_n - c}{\sqrt{x_n} + \sqrt{c}},$$

we see that, since $c \neq 0$

$$\sqrt{x_n} - \sqrt{c} \to \frac{0}{2\sqrt{c}} = 0 \quad \text{as } n \to \infty.$$

Recall that $c \neq 0$, so that $\frac{1}{\sqrt{c}}$ is indeed defined.

In other words, $\{\sqrt{x_n} - \sqrt{c}\}$ is indeed a null sequence. This completes the proof. □

Problem 3 Prove that the nth root function

$$f(x) = x^{\frac{1}{n}}, \text{ where } n \in \mathbb{N} \quad \text{and} \quad \begin{cases} x \geq 0, & \text{if } n \text{ is even}, \\ x \in \mathbb{R}, & \text{if } n \text{ is odd}, \end{cases}$$

is continuous.

You may omit this Problem if you are short of time.

We defined $a^{\frac{1}{n}}$ in Sub-section 1.5.3.

Hints: Use the result of Exercise 3(b) on Section 2.3, in Section 2.6, and the identity $a^q - b^q = (a - b)(a^{q-1} + a^{q-2}b + a^{q-3}b^2 + \cdots + b^{q-1})$ with suitable choices for *a, b* and *q*. In your solution, be careful not to use the letter *n* for two different purposes.

Example 3 above was the special case of Problem 3, where $n = 2$.

Example 4 Determine whether the function $f(x) = \frac{1}{x}$, $x \in \mathbb{R} - \{0\}$, is continuous.

Solution The domain of *f* is the set $\mathbb{R} - \{0\}$, the union of the two open intervals $(-\infty, 0)$ and $(0, \infty)$.

Let *c* be any point of $\mathbb{R} - \{0\}$, and let $\{x_n\}$ be any sequence in $\mathbb{R} - \{0\}$ that converges to *c*. Then, by the Quotient Rule for sequences, it follows that $\{f(x_n)\} = \{\frac{1}{x_n}\} \to \frac{1}{c}$ as $n \to \infty$; in other words, that $f(x_n) \to f(c) = \frac{1}{c}$ as $n \to \infty$. So *f* is continuous at *c*.

Sub-section 2.3.2.

Since *c* is an arbitrary point of $\mathbb{R} - \{0\}$, it follows that *f* is continuous on $\mathbb{R} - \{0\}$. ☐

Our work so far on continuity illustrates the following general strategy:

Strategy for continuity

- To prove that a function $f: S \to \mathbb{R}$ is *continuous* at a point *c* of *S*, prove that:

    for each sequence $\{x_n\}$ in *S* such that $x_n \to c$, then $f(x_n) \to f(c)$.

- To prove that a function $f: S \to \mathbb{R}$ is *discontinuous* at a point *c* of *S*:

    find one sequence $\{x_n\}$ in *S* such that $x_n \to c$ but $f(x_n) \not\to f(c)$.

Recall that *just one* such sequence suffices.

Problem 4 Prove that the following functions are continuous on $\mathbb{R}$:

(a) $f(x) = c$, $x \in \mathbb{R}$;

(b) $f(x) = x^n$, $x \in \mathbb{R}$, $n \in \mathbb{N}$;

(c) $f(x) = |x|$, $x \in \mathbb{R}$.

Problem 5 Determine the points of continuity and discontinuity of the *signum function*

$$f(x) = \begin{cases} -1, & x < 0, \\ 0, & x = 0, \\ 1, & x > 0. \end{cases}$$

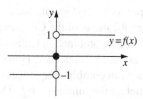

Remarks

We now look again closely at the definition of *continuity at a point*. Consider a function *f* (say) from a set in $\mathbb{R}$, the domain *A* (say), to another set in $\mathbb{R}$, the codomain *B* (say).

1. Let *f* be continuous at a point *c* in *A*, and assume that, for some set $A' \subseteq A$, $c \in A'$. Suppose that another function *g* has domain A' on which $g(x) = f(x)$. Technically *g* is a different function from *f*, for sure. However if *f* is continuous at *c*, then it is a simple matter of some definition checking to

You can think of the mapping *f* as being a formula $x \mapsto y$ that maps a point *x* of *A* onto to a point *y* of *B*, and we write $y = f(x)$ to indicate this.

verify that g too is continuous at c. So, restriction of a function to a smaller domain does not affect its continuity at a point.

2. Let f and g be functions defined on sets in $\mathbb{R}$ that contain an open interval I, and $c \in I$. Then, if $f(x) = g(x)$ for all $x \in I$, f is continuous at c if g is continuous at c, and f is discontinuous at c if g is discontinuous at c. Again, we may simply ignore any difference between the domains of the functions when we are studying continuity at a point.

3. The underlying point here is that *continuity at a point is a local property*. It is only the behaviour of the function near that point that determines whether it is continuous at the point.

4.1.2 Rules for continuous functions

We have seen how to recognise whether a given function is continuous at a point. However, it would be tedious to have to go back to first principles to determine on each occasion whether a complicated function is continuous. As usual, we avoid such problems by having a set of rules that enable us to construct continuous functions.

Combination Rules for continuous functions

If f and g are functions that are continuous at a point c, then so are:

Sum Rule　　　$f + g$;

Multiple Rule　λf,　for $\lambda \in \mathbb{R}$;

Product Rule　fg;

Quotient Rule　$\frac{f}{g}$, provided that $f(c) \neq 0$.

For example, any polynomial $p(x) = a_0 + a_1 x + \cdots + a_n x^n$, $x \in \mathbb{R}$, is continuous at all points of $\mathbb{R}$ since we can build up the expression for p by successive applications of the Combination Rules for continuous functions.

For instance, the function
$$p(x) = x^{17} - 34x^5 + 5x - 8$$
is continuous on $\mathbb{R}$.

Similarly, any rational function $r(x) = \frac{p(x)}{q(x)}$, where p and q are polynomials and the domain of r is $\mathbb{R}$ minus the points where q vanishes, is continuous at all points of its domain, since we can build up the expression for r by successive applications of the Combination Rules for continuous functions.

For instance, the function
$$r(x) = \frac{1 - 2x}{x^2 - 4},$$
$x \in \mathbb{R} - \{\pm 2\}$, is continuous on its domain.

The Combination Rules above are all natural analogues of the corresponding results for sequences. However, we can combine functions in more ways than we can combine sequences – for example, we can *compose* functions f and g to obtain the function $g \circ f$. This approach too will often enable us to obtain new continuous functions.

Composition Rule　Let f be continuous on a set S_1 that contains a point c, and g be continuous on a set S_2 that contains the point $f(c)$. Then $g \circ f$ is continuous at c.

Functions obtained in this way are called *composite functions*.

For example, we know that the function $f(x) = x^2 + 1$, $x \in \mathbb{R}$, is continuous on $\mathbb{R}$ and that the function $g(x) = \sqrt{x}$, $x \geq 0$, is continuous on $I = \{x : x \geq 0\}$. It follows from the Composition Rule that the function $g \circ f : x \mapsto \sqrt{x^2 + 1}$ is continuous at all points c of $\mathbb{R}$ (the domain of f) whose image $f(c)$ lies in I. But all such points $f(c)$ lie in I; so it follows that in fact the composite $g \circ f$ is continuous on $\mathbb{R}$.

Problem 6   Prove that the function f given by $f(x) = x^{\frac{3}{2}}$, $x \geq 0$, is continuous.

Finally, just as we had a Squeeze Rule for convergent sequences, we have a corresponding Squeeze Rule for continuous functions.

We state this rule only in the case of an interior point of an open interval. However more general versions also exist!

Theorem 1   Squeeze Rule

Let the functions f, g and h be defined on an open interval I, and $c \in I$. If:

1. $g(x) \leq f(x) \leq h(x)$,   for all $x \in I$,

2. $g(c) = f(c) = h(c)$,

3. g and h are continuous at c,

then f is also continuous at c.

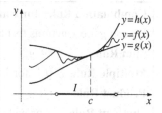

Example 5   Prove that the function f given by $f(x) = \begin{cases} x^2 \sin\left(\frac{1}{x}\right), & x \neq 0, \\ 0, & x = 0, \end{cases}$ is continuous at 0.

We shall study the trigonometric functions in detail in Sub-section 4.1.3. The only property that you need here is that, for any real number x, $|\sin x| \leq 1$.

Solution   The diagram in the margin suggests that we should find functions g and h that squeeze f near 0.

So, we define $g(x) = -x^2$, $x \in \mathbb{R}$, and $h(x) = x^2$, $x \in \mathbb{R}$. With these two chosen, we check the conditions of the Squeeze Rule.

The first condition is of course vital! Now we know that

$$-1 \leq \sin\left(\frac{1}{x}\right) \leq 1, \quad \text{for any } x \neq 0.$$

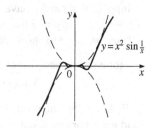

It follows that

$$-x^2 \leq x^2 \sin\left(\frac{1}{x}\right) \leq x^2, \quad \text{for any } x \neq 0,$$

so that

$$g(x)\left[= -x^2\right] \leq f(x) \leq \left[x^2 =\right] h(x), \quad \text{for any } x \in \mathbb{R}.$$

These inequalities are obviously true for $x = 0$, as well as for $x \neq 0$.

So condition 1 of the Squeeze Rule is satisfied.

Next, the functions f, g and h all take the value 0 at the point 0. Thus condition 2 of the Squeeze Rule is satisfied.

Finally, the functions g and h are polynomials, and so in particular they are continuous at 0. So condition 3 of the Squeeze Rule is satisfied.

Recall that all polynomials are continuous on $\mathbb{R}$.

It follows then from the Squeeze Rule that f is continuous at 0, as required. $\square$

To test your understanding of these techniques, try the following problems.

Problem 7   Prove that the following function is continuous on $\mathbb{R}$, stating each rule or fact about continuity that you are using

$$f(x) = \sqrt{x^2 + x + 1} - \frac{5x}{1 + x^2}, \quad x \in \mathbb{R}.$$

Problem 8   Using the elementary properties of the trigonometric functions, determine whether the following functions are continuous at 0:

(a) $f(x) = \begin{cases} x\sin(\frac{1}{x}), & x \neq 0, \\ 0, & x = 0; \end{cases}$ (b) $f(x) = \begin{cases} \sin(\frac{1}{x}), & x \neq 0, \\ 0, & x = 0. \end{cases}$

Proofs

We now give the proofs of the Combination, Composition and Squeeze Rules. First, recall the Combination Rules.

You may omit the rest of this Sub-section at a first reading.

Combination Rules for continuous functions

If f and g are functions that are continuous at a point c, then so are:

Sum Rule $f + g$;

Multiple Rule λf,   for $\lambda \in \mathbb{R}$;

Product Rule fg;

Quotient Rule $\frac{f}{g}$, provided that $f(c) \neq 0$.

The proofs of all these are very similar, and depend simply on the corresponding results for sequences. We prove only the Sum Rule.

Proof of the Sum Rule   We want to prove that $f + g$ is continuous at c.

Suppose that f and g have domains S_1 and S_2, respectively. Then the domain of $f + g$ is $S_1 \cap S_2$, and this set contains the point c.

Thus, we have to show that:

for each sequence $\{x_n\}$ in $S_1 \cap S_2$ such that $x_n \to c$, then
$$f(x_n) + g(x_n) \to f(c) + g(c).$$

We know that the sequence $\{x_n\}$ lies in S_1 and in S_2, and that both functions f and g are continuous at c. Hence
$$f(x_n) \to f(c) \quad \text{and} \quad g(x_n) \to g(c);$$

and it follows from the Sum Rule for Sequences that $f(x_n) + g(x_n) \to f(c) + g(c)$, as required.   ∎

Next, recall the Composition Rule.

Composition Rule   Let f be continuous on a set S_1 that contains a point c, and g be continuous on a set S_2 that contains the point $f(c)$. Then $g \circ f$ is continuous at c.

Proof   We want to prove that $g \circ f$ is continuous at c.

Now, we know that $g \circ f$ is certainly defined on the set $S = \{x : x \in S_1 \text{ and } f(x) \in S_2\}$, and this set contains the point c.

Thus, we have to show that:

for each sequence $\{x_n\}$ in S such that $x_n \to c$, then $g(f(x_n)) \to g(f(c))$.

We know that the sequence $\{x_n\}$ lies in S_1, and that f is continuous at c. Hence, we have that $f(x_n) \to f(c)$.

We also know that $\{f(x_n)\}$ lies in S_2, and that g is continuous at $f(c)$. It follows that $g(f(x_n)) \rightarrow g(f(c))$, as required. ■

Finally, recall the Squeeze Rule.

Theorem 1   Squeeze Rule

Let the functions f, g and h be defined on an open interval I, and $c \in I$. If:

1. $g(x) \leq f(x) \leq h(x)$, for all $x \in I$,
2. $g(c) = f(c) = h(c)$,
3. g and h are continuous at c,

then f is also continuous at c.

Proof We have to show that f is continuous at c.

Thus, we have to prove that:

> for each sequence $\{x_n\}$ in the domain of f such that $x_n \rightarrow c$, then $f(x_n) \rightarrow f(c)$.

Now, since $x_n \rightarrow c$ there is some number X such that $x_n \in I$, for all $n > X$. Hence, by condition 1 we have that

This follows from the definition of a sequence converging to c.

$$g(x_n) \leq f(x_n) \leq h(x_n), \quad \text{for all } n > X. \tag{1}$$

So, if we now let $n \rightarrow \infty$ and use conditions 2 and 3, we get that

$$\lim_{n\to\infty} g(x_n) = g(c) = f(c) \quad \text{and} \quad \lim_{n\to\infty} h(x_n) = h(c) = f(c). \tag{2}$$

It follows, from (1), (2) and the Squeeze Rule for sequences, that

Theorem 2, Sub-section 2.3.3.

$$\lim_{n\to\infty} f(x_n) = f(c),$$

as required. ■

4.1.3 Trigonometric functions and the exponential function

Trigonometric functions

For the moment we are assuming that you have a knowledge of the trigonometric functions arising from your study of trigonometry.

We will prove that the trigonometric functions are continuous on the whole of their domains. But, first, we need a basic inequality for the sine function.

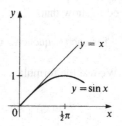

Lemma 1 $0 \leq \sin x \leq x, \quad \text{for } 0 \leq x \leq \frac{\pi}{2}.$

Proof If $x = 0$, then $\sin 0 = 0$; so there is an equality.

Suppose next that $0 < x \leq \frac{\pi}{2}$, and consider the following diagram, which represents a quarter circle, centred at the origin, with radius 1.

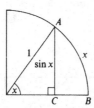

Since the circle has radius 1, the arc AB has length x and the perpendicular AC has length $\sin x$. Hence

$$0 < \sin x \le x, \quad \text{for } 0 < x \le \frac{\pi}{2}.$$ ∎

For the shortest distance from the point A to the line BC is the perpendicular from A to BC.

We can now extend the inequality in Lemma 1 to obtain a more general result.

Theorem 2   The Sine Inequality
$$|\sin x| \le |x|, \quad \text{for } x \in \mathbb{R}.$$

Proof We saw in Lemma 1 that the inequality holds for $0 \le x \le \frac{\pi}{2}$. For $x > \frac{\pi}{2}$, we have

$$|\sin x| \le 1 < \frac{1}{2}\pi < x = |x|,$$

and so the desired inequality is also true in this case.

Finally, the inequality also holds for $x < 0$, since

$$|\sin(-x)| = |\sin x| \quad \text{and} \quad |-x| = |x|.$$ ∎

This is the key tool that we need to prove the continuity of the trigonometric functions.

Theorem 3   The trigonometric functions (sine, cosine and tangent) are continuous.

Recall that this means they are continuous at each point of their domains.

Proof To prove that the sine function is continuous at each point $c \in \mathbb{R}$, we need to show that:

for each sequence $\{x_n\}$ in $\mathbb{R}$ such that $x_n \to c$, then $\sin x_n \to \sin c$.   (3)

We use the formula

$$\sin x_n - \sin c = 2\cos\left(\frac{1}{2}(x_n + c)\right)\sin\left(\frac{1}{2}(x_n - c)\right);$$

This is a standard trigonometric formula.

it follows that

$$|\sin x_n - \sin c| = 2\left|\cos\left(\frac{1}{2}(x_n + c)\right)\right| \times \left|\sin\left(\frac{1}{2}(x_n - c)\right)\right|$$

By taking moduli.

$$\leq 2\left|\sin\left(\frac{1}{2}(x_n - c)\right)\right|$$

For $\left|\cos\frac{1}{2}(x_n + c)\right| \leq 1$.

$$\leq 2\left|\frac{1}{2}(x_n - c)\right| = |x_n - c|.$$

Here we use the Sine Inequality, Theorem 2.

Hence, if $\{x_n - c\}$ is null, then $\{\sin x_n - \sin c\}$ is also null, by the Squeeze Rule for null sequences. In other words, the result (3) holds.

The continuity of the cosine and tangent functions now follows from the formulas

$$\cos x = \sin\left(x + \frac{1}{2}\pi\right) \quad \text{and} \quad \tan x = \frac{\sin x}{\cos x},$$

using the Combination Rules and the Composition Rule. ■

Problem 9 Prove that the following function is continuous on $\mathbb{R}$, stating each rule or fact about continuity that you are using

$$f(x) = x^2 + 1 + 3\sin(x^2 + 1), \quad \text{for } x \in \mathbb{R}.$$

Problem 10 Prove that the function $f(x) = \sin\left(\frac{\pi}{2}\cos x\right)$ is continuous on $\mathbb{R}$.

The exponential function $x \mapsto e^x$

We now prove that the exponential function is continuous on $\mathbb{R}$. But first we start with some inequalities that we will need to do this.

Theorem 4   The Exponential Inequalities

(a) $e^x > 1 + x$,   for $x > 0$;   and $e^x \geq 1 + x$,   for $x \geq 0$;

(b) $e^x \leq \frac{1}{1-x}$,   for $0 \leq x < 1$;

(c) $1 + x \leq e^x \leq \frac{1}{1-x}$,   for $|x| < 1$.

Proof We prove inequalities (a) and (b) using the exponential series

$$e^x = 1 + x + \frac{x^2}{2!} + \frac{x^3}{3!} + \cdots, \quad \text{for } x \geq 0.$$

You met this series in Sub-section 3.4.1.

(a) For $x > 0$, we have $\frac{x^2}{2!} > 0$, $\frac{x^3}{3!} > 0$, and so on. Hence

$$e^x > 1 + x, \quad \text{for } x > 0.$$

It follows from this that $e^x \geq 1 + x$, for $x \geq 0$, since $e^0 = 1$.

(b) For $x \geq 0$, we have $\frac{x^2}{2!} \leq x^2$, $\frac{x^3}{3!} \leq x^3$, and so on. Hence

$$e^x \leq 1 + x + x^2 + x^3 + \cdots.$$

The series on the right is a geometric series; it converges to the sum $\frac{1}{1-x}$, for $0 \leq x < 1$. Hence

$$e^x \leq \frac{1}{1-x}, \quad \text{for } 0 \leq x < 1.$$

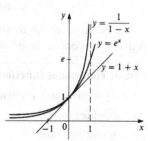

(c) We have just seen that these inequalities hold for $0 \leq x < 1$.
For $-1 < x < 0$, we have that $0 < -x < 1$, so that, by parts (a) and (b)

$$1 + (-x) \leq e^{-x} \leq \frac{1}{1 - (-x)}, \quad \text{for } -1 < x < 0.$$

By taking reciprocals and reversing the inequalities, we may reformulate this in the form

$$1 + x \leq e^x \leq \frac{1}{1 - x}, \quad \text{for } -1 < x < 0.$$

This completes the proof of the desired result. ∎

Remark

Notice that, when $x \neq 0$, the results of Theorem 4 hold with *strict* inequalities.

We are now able to prove the continuity of the exponential function.

Theorem 5 The exponential function $x \mapsto e^x, x \in \mathbb{R}$, is continuous.

Proof To prove that the exponential function is continuous at each point $c \in \mathbb{R}$, we need to show that:

for each sequence $\{x_n\}$ in $\mathbb{R}$ such that $x_n \to c$, then $e^{x_n} \to e^c$. (4)

We use the formula $e^{x_n} - e^c = e^c(e^{x_n-c} - 1)$. If we apply Theorem 4, part (c), with $x_n - c$ in place of x, we obtain

$$1 + (x_n - c) \leq e^{x_n-c} \leq \frac{1}{1 - (x_n - c)}, \quad \text{for } |x_n - c| < 1.$$

Thus, if $\{x_n - c\}$ is null, then $|x_n - c| < 1$ eventually, and so $e^{x_n-c} \to 1$ as $n \to \infty$, by the Squeeze Rule for sequences. Hence $e^{x_n} \to e^c$, so that the desired result (4) holds. ∎

Remark

Later we shall give a rigorous definition of the general exponential function $x \mapsto a^x$, for $x \in \mathbb{R}$ and $a > 0$, and we shall show that all exponential functions are continuous on $\mathbb{R}$.

Section 4.4.

Problem 11 Prove that the following function is continuous on $\mathbb{R}$, stating each rule or fact about continuity that you are using

$$f(x) = \cos(x^5 - 5x^2) + 7e^{-x^2}.$$

We end this section by listing the various types of functions that we have found to be continuous on their domains.

For example

$f(x) = 3x^7 - 4x^2 + 5;$
$f(x) = \dfrac{3x}{x^4 - 6};$
$f(x) = |x|;$
$f(x) = x^{\frac{1}{2}}, x \geq 0;$
$f(x) = x^{\frac{1}{3}}, x \in \mathbb{R};$
$f(x) = \sin x, \cos x, \tan x;$
$f(x) = e^x.$

Basic continuous functions The following functions are continuous:
- polynomials and rational functions;
- modulus function;
- nth root function;
- trigonometric functions (sine, cosine and tangent);
- the exponential function.

4.2 Properties of continuous functions

4.2.1 The Intermediate Value Theorem

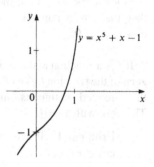

In Section 4.3 we shall prove that the function $f(x) = x^5 + x - 1$, $x \in \mathbb{R}$, is a one-one function on $\mathbb{R}$. From the graph of f it certainly *looks* as though f maps $\mathbb{R}$ *onto* $\mathbb{R}$; how can we *prove* this?

For example, is there a value of x for which $f(x) = 0$? In other words, is there a root of the equation $x^5 + x - 1 = 0$? The shape of the graph $y = x^5 + x - 1$ certainly suggests that such a number x exists; since $f(0) = -1$ and $f(1) = 1$, we would expect there to be some number x in the interval $(0, 1)$ such that $f(x) = 0$. However we do not have a formula for solving the equation to find x.

Now, we *introduced* the concept of continuity on the grounds that it would enable us to pin down precisely the idea that the graph of a 'well-behaved' function does not have 'gaps' or 'jumps', and this is the key to proving that such a number x exists.

Theorem 1   Intermediate Value Theorem

Let f be a continuous function on $[a, b]$, and let k be any number lying (strictly) between $f(a)$ and $f(b)$. Then there exists a number c in (a, b) such that $f(c) = k$.

This is one of the main existence theorems in Analysis.

This result is illustrated below in the two possible cases:

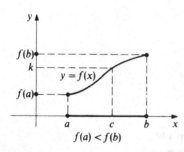

$f(a) < f(b)$

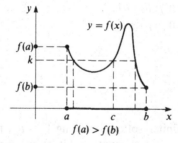

$f(a) > f(b)$

Note that $f(a) \neq f(b)$ and $a < c < b$.

As the graph above on the right shows, there may be more than one possible value of c such that $f(c) = k$. All that we claim is that there is *at least one* such point c.

The requirement in Theorem 1 that f be continuous at each point of $[a, b]$ is essential. For example, the function

$$f(x) = \begin{cases} \frac{1}{x}, & -1 \leq x < 0, \\ 0, & x = 0, \\ \frac{1}{x}, & 0 < x \leq 1, \end{cases}$$

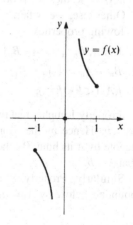

is continuous on $[-1, 1]$ except at 0 – where it is discontinuous. For this function, $f(-1) = -1$ and $f(1) = 1$, but there is no number c in $(-1, 1)$ such that $f(c) = \frac{1}{2}$ (for example).

The following example shows a typical application of the Intermediate Value Theorem.

Example 1   Prove that there is a number c in $(0, 1)$ such that $c^5 + c - 1 = 0$.

Solution   Consider the function $f(x) = x^5 + x - 1$ on the interval $[0, 1]$. Then f is continuous on $[0, 1]$ since it is a basic continuous function, and $f(0) = -1$ and $f(1) = 1$.   *For f is a polynomial.*

Since $f(0) < 0 < f(1)$, it then follows from the Intermediate Value Theorem that there is a number c in $(0, 1)$ such that $f(c) = 0$; that is, such that $c^5 + c - 1 = 0$.   □

The function f is strictly increasing on $[0, 1]$, and so the number c must be unique in this case.

If f is a function and c is a real number such that $f(c) = 0$, then c is called a **zero** of the function f. We often show that an equation has a solution by proving that a related continuous function has a zero (by using the Intermediate Value Theorem with $k = 0$).

We do not consider the possibility of complex zeros in this book.

> **Problem 1**   Prove that there is a real number c in $(0, 1)$ such that $\cos c = c$.

> **Problem 2**   Let the function $f: [0, 1] \to [0, 1]$ be continuous. Prove that there is a real number c in $[0, 1]$ such that $f(c) = c$.

For example
$$f(x) = \tfrac{1}{3} + \tfrac{1}{2}x\sin\left(\tfrac{\pi}{2}x\right).$$

Proof of Theorem 1   We use the method of repeated bisection.

We shall assume that $f(a) < f(b)$. If, in fact, $f(a) > f(b)$, the proof is very similar.

First, denote the closed interval $[a, b]$ as $[A_1, B_1]$. Then, denote by p the midpoint of the interval $[A_1, B_1]$. Notice that, if $f(p) = k$, then the proof is complete, since we can take $c = p$.

Otherwise, we define one of the two intervals $[A_1, p]$ and $[p, B_1]$ to be $[A_2, B_2]$ in the following way

$$[A_2, B_2] = \begin{cases} [A_1, p], & \text{if } f(p) > k, \\ [p, B_1], & \text{if } f(p) < k. \end{cases}$$

In either case, we obtain:

1. $[A_2, B_2] \subset [A_1, B_1]$;
2. $B_2 - A_2 = \tfrac{1}{2}(B_1 - A_1)$;
3. $f(A_2) < k < f(B_2)$.

You should at least skim this proof on a first reading, as the method is an important one.

$p = \tfrac{1}{2}(A_1 + B_1).$

We now repeat this process indefinitely often, bisecting $[A_2, B_2]$ to obtain $[A_3, B_3]$, and so on. If, at any stage, we encounter a bisection point p such that $f(p) = k$, then the proof is complete.

Otherwise, we obtain a sequence of closed intervals $\{[A_n, B_n]\}$ with the following properties:

1. $[A_{n+1}, B_{n+1}] \subset [A_n, B_n]$,   for each $n \in \mathbb{N}$;
2. $B_n - A_n = \left(\tfrac{1}{2}\right)^{n-1}(B_1 - A_1)$,   for each $n \in \mathbb{N}$;
3. $f(A_n) < k < f(B_n)$,   for each $n \in \mathbb{N}$.

Property 1 implies that the sequence $\{A_n\}$ is increasing and bounded above by $B_1 = b$. Hence by the Monotone Convergence Theorem, $\{A_n\}$ is convergent; denote by A its limit. By the Limit Inequality Rule for sequences, we must have that $A \leq b$.   *Theorem 3, Sub-section 2.3.3.*

Similarly, Property 1 implies that the sequence $\{B_n\}$ is decreasing and bounded below by $A_1 = a$. Hence by the Monotone Convergence Theorem,

$\{B_n\}$ is convergent; denote by B its limit. By the Limit Inequality Rule for sequences, we must have that $B \geq a$.

We may then deduce, by letting $n \to \infty$ in Property 2 and using the Combination Rules for sequences, that

$$B - A = \lim_{n\to\infty} B_n - \lim_{n\to\infty} A_n = \lim_{n\to\infty} (B_n - A_n)$$
$$= \lim_{n\to\infty} \left(\tfrac{1}{2}\right)^{n-1} (B_1 - A_1)$$
$$= (B_1 - A_1) \lim_{n\to\infty} \left(\tfrac{1}{2}\right)^{n-1} = 0.$$

In other words, the sequences $\{A_n\}$ and $\{B_n\}$ both converge to a common limit. Denote this limit by c; then we must have $a \leq c \leq b$.

That is, $c \in [a, b]$.

Now we use the fact that f is continuous at c. It follows that

$$\lim_{n\to\infty} f(A_n) = f(c) \quad \text{and} \quad \lim_{n\to\infty} f(B_n) = f(c).$$

But, by Property 3, $f(A_n) < k$, for $n = 1, 2, \ldots$; so that, by letting $n \to \infty$, we obtain $f(c) \leq k$ by the Limit Inequality Rule for sequences. Similarly, we can deduce from the fact that $k < f(B_n)$ that $f(c) \geq k$. It follows that $f(c) = k$.

Recall that taking limits 'flattens inequalities'.

Finally, notice that, since $f(a) < k < f(b)$, we cannot have either $c = a$ or $c = b$; so that, in fact, $a < c < b$ as required.   ∎

That is, $c \in (a, b)$.

The method of repeated bisection is very powerful, and of wide application in Mathematics.

Now, we saw above that the continuous function $f(x) = x^5 + x - 1$, $x \in [0, 1]$, has a zero in $(0, 1)$. Then, since $f\left(\tfrac{1}{2}\right) = \tfrac{1}{32} + \tfrac{1}{2} - 1 = -\tfrac{15}{32} < 0$ and $f(1) = 1 > 0$, we can make the stronger statement that f must have a zero in $\left(\tfrac{1}{2}, 1\right)$, by the Intermediate Value Theorem.

Problem 3   Use the method of repeated bisection to find an interval of length $\tfrac{1}{8}$ that contains a zero of the function $f(x) = x^5 + x - 1$, $x \in [0, 1]$.

Antipodal points

Two points on the surface of the Earth are called **antipodal points** if the line between them passes through the centre of the Earth. (In this sense, antipodal points are 'opposite each other'.)

The following result is a rather interesting application of the Intermediate Value Theorem! It makes the (physically reasonable) assumption that temperature is a continuous function of position on the Earth's surface.

Theorem 2   Antipodal Points Theorem
There is always a pair of antipodal points on the Equator of the Earth at which the temperature is the same.

In fact the same result holds for any 'Great Circle' on the Earth's surface – that is, the intersection of a plane through the Earth's centre with the Earth's surface.

To prove this, we must set up the situation as a mathematical problem. Let $f(\theta)$ denote the temperature at a point on the Equator at an angle θ radians East of Greenwich, for $0 \leq \theta < 2\pi$, and extend f to be defined on $[0, 2\pi]$ by

This process is often called *Mathematical modelling*.

requiring that $f(2\pi) = f(0)$. Then Theorem 2 can be rephrased in the following equivalent way:

Theorem 2′   Antipodal Points Theorem

Let $f: [0, 2\pi] \to \mathbb{R}$ be continuous, with $f(0) = f(2\pi)$. Then there exists a number c in $[0, \pi]$ such that $f(c) = f(c + \pi)$.

For, if $f(c) = f(c + \pi)$, then c and $c + \pi$ are antipodal points with the same temperature.

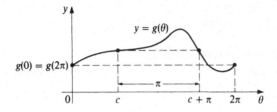

Proof of Theorem 2′   Notice, first, that if $f(0) = f(\pi)$ then we can take $c = 0$.

We now prove the result under the assumption that $f(0) < f(\pi)$. (The proof in the case that $f(0) > f(\pi)$ is very similar.)

So, define the function g as follows

$$g(\theta) = f(\theta) - f(\theta + \pi), \quad \text{for } \theta \in [0, \pi].$$

Then, since f is continuous on $[0, 2\pi]$, it follows that g is continuous on $[0, \pi]$, by the Combination Rules.

But

$$g(0) = f(0) - f(\pi) < 0$$

and

$$g(\pi) = f(\pi) - f(2\pi) = f(\pi) - f(0) > 0.$$

It then follows from the Intermediate Value Theorem, with $k = 0$, that there exists some number c in $(0, \pi)$ such that $g(c) = 0$; in other words, such that $f(c) - f(c + \pi) = 0$. This completes the proof. ∎

Quite often in mathematics we obtain our results by applying a standard result to some cunningly chosen *auxiliary* function!

4.2.2   Zeros of polynomials

You will have already met a standard method for solving a polynomial equation of degree 2, and possibly equations of degrees 3 and 4 too. However there exists no method of solving a general polynomial equation of degree 5 or higher by means of formulas. How many zeros can a polynomial equation have?

In fact, a polynomial equation of degree n can have at most n roots in $\mathbb{R}$.

These are called *quadratic*, *cubic* and *quartic* equations, respectively.

This is a quite difficult result to prove!

Theorem 3   Fundamental Theorem of Algebra

Let $p(x) = a_n x^n + a_{n-1} x^{n-1} + \cdots + a_1 x + a_0$, $x \in \mathbb{R}$, where $a_n \neq 0$. Then the equation $p(x) = 0$ has at most n roots in $\mathbb{R}$.

We do not prove this result, which requires methods beyond the scope of this book.

Remark

In fact, in its most general form the Fundamental Theorem of Algebra states that, if $p(x) = a_n x^n + a_{n-1} x^{n-1} + \cdots + a_1 x + a_0$, where the coefficients a_k may be real or complex numbers and $a_n \neq 0$, then the equation $p(x) = 0$ has exactly n roots in $\mathbb{C}$, the set of all complex numbers.

Now, it is sometimes straight-forward to locate zeros of a polynomial if we have some idea of where to look for them in the first place.

> **Problem 4** Let $p(x) = x^6 - 4x^4 + x + 1$, $x \in \mathbb{R}$. Prove that p has a zero in each of the intervals $(-1, 0)$, $(0, 1)$ and $(1, 2)$.

However to follow this approach for finding zeros (if any) of a given polynomial, we need first to have an inkling where to look for them. We usually start by applying the following result, which gives an interval in which the zeros *must* lie:

Theorem 4   Zeros Localisation Theorem

Let $p(x) = x^n + a_{n-1}x^{n-1} + \cdots + a_1 x + a_0$, $x \in \mathbb{R}$, be a polynomial. Then all the zeros of p (if there are any) lie in the open interval $(-M, M)$, where

$$M = 1 + \max\{|a_{n-1}|, \ldots, |a_1|, |a_0|\}.$$

Notice that the coefficient of x^n, the *leading coefficient* of p, has been set as 1.

The proof appears at the end of the sub-section.

Example 2 Prove that the polynomial $p(x) = x^4 - 2x^2 - x + 1$, $x \in \mathbb{R}$, has at least two zeros in $\mathbb{R}$.

Solution We will apply the Zeros Localisation Theorem to p, since its leading coefficient is 1. Since

$$M = 1 + \max\{|-2|, |-1|, |1|\}$$
$$= 3,$$

it follows that all the zeros of p lie in $(-3, 3)$.

We now compile a table of values of $p(x)$, for $x = -3, -2, -1, 0, 1, 2, 3$:

x	-3	-2	-1	0	1	2	3
$p(x)$	67	11	1	1	-1	7	61

We find that $p(0)$ and $p(1)$ have opposite signs, as do $p(1)$ and $p(2)$, so p must have a zero in each of the intervals $(0, 1)$ and $(1, 2)$, by the Intermediate Value Theorem.

Thus we have proved that p has *at least* two zeros in $\mathbb{R}$. □

In fact this polynomial has *exactly* two zeros in $\mathbb{R}$.

> **Problem 5** Prove that the polynomial $p(x) = x^5 + 3x^4 - x - 1$, $x \in \mathbb{R}$, has at least three zeros in $\mathbb{R}$.

Although we cannot at this stage prove the Fundamental Theorem of Algebra, nevertheless we can use the Zeros Localisation Theorem and the Intermediate Value Theorem to prove the following:

Theorem 5   Every real polynomial of odd degree has at least one zero in $\mathbb{R}$.

Thus, for example, for sufficiently large x the polynomial $p(x) = x^n + a_{n-1}x^{n-1} + \cdots + a_1 x + a_0$, $x \in \mathbb{R}$, is essentially dominated by its leading term x^n, and so is positive for large positive values of x and is negative for large negative values of x.

For example, $p(x) = x^5 + 3x^4 - x - 1$.

Since n is odd.

No corresponding result holds for polynomials of even degree. Thus, for instance, the polynomial $p(x) = x^2 + 1$ has no zeros in $\mathbb{R}$.

Proofs of Theorems 4 and 5

We supply these proofs which were omitted earlier, so as not to disturb the flow of the text.

You may omit these at a first reading.

Theorem 4   Zeros Localisation Theorem

Let $p(x) = x^n + a_{n-1}x^{n-1} + \cdots + a_1 x + a_0$, $x \in \mathbb{R}$, be a polynomial. Then all the zeros of p (if there are any) lie in the open interval $(-M, M)$, where
$$M = 1 + \max\{|a_{n-1}|, \ldots, |a_1|, |a_0|\}.$$

Recall that the coefficient of x^n, the *leading coefficient* of p, is 1.

Proof   In order to concentrate on the dominating term x^n, we define the function r as follows
$$r(x) = \frac{p(x)}{x^n} - 1 = \frac{a_{n-1}}{x} + \cdots + \frac{a_1}{x^{n-1}} + \frac{a_0}{x^n}, \quad x \in \mathbb{R} - \{0\}.$$
Then, by using the Triangle Inequality, we obtain that, for $|x| > 1$

$$
\begin{aligned}
|r(x)| &= \left| \frac{a_{n-1}}{x} + \cdots + \frac{a_1}{x^{n-1}} + \frac{a_0}{x^n} \right| \\
&\leq \left| \frac{a_{n-1}}{x} \right| + \cdots + \left| \frac{a_1}{x^{n-1}} \right| + \left| \frac{a_0}{x^n} \right| \\
&\leq \max\{|a_{n-1}|, \ldots, |a_1|, |a_0|\} \times \left(\frac{1}{|x|} + \cdots + \frac{1}{|x|^{n-1}} + \frac{1}{|x|^n} \right) \\
&< (M - 1) \times \left(\frac{1}{|x|} + \cdots + \frac{1}{|x|^{n-1}} + \frac{1}{|x|^n} + \cdots \right) \\
&= (M - 1) \times \frac{\frac{1}{|x|}}{1 - \frac{1}{|x|}} = \frac{M - 1}{|x| - 1}.
\end{aligned}
$$

Here the modulus of each coefficient is at most the maximum value over *all* coefficients.

$|x| > 1$, so that $\frac{1}{|x|} < 1$.

By summing the geometric series.

It follows that, if $|x| \geq M = 1 + \max\{|a_{n-1}|, \ldots, |a_1|, |a_0|\}$, then $|r(x)| < 1$. Now, from the definition of $r(x)$ we see that
$$p(x) = x^n(1 + r(x)), \quad \text{for } |x| \geq M.$$
But since $|r(x)| < 1$, we certainly have that $1 + r(x) > 0$. It follows from the above expression for $p(x)$ in terms of $r(x)$ that $p(x)$ must have the same sign as x^n, for $|x| \geq M$.

$|r(x)| < 1 \Leftrightarrow -1 < r(x) < 1$.

It follows that any zero of p must lie in $(-M, M)$.   ∎

Theorem 5   Every polynomial of odd degree has at least one zero in $\mathbb{R}$.

Proof   By dividing the polynomial by its leading coefficient, we may assume that the polynomial is of the form
$$p(x) = x^n + a_{n-1}x^{n-1} + \cdots + a_1 x + a_0, x \in \mathbb{R}, \text{ where } n \text{ is odd.}$$
We can then define M and $r(x)$ as in Theorem 4, and all the statements in the proof of Theorem 4 then hold here too.

Since n is odd, x^n is positive for $x \geq M$ and negative for $x \leq -M$. It follows from the arguments in the proof of Theorem 4 that $p(x)$ must be positive for $x \geq M$ and negative for $x \leq -M$. In particular, $p(M) > 0$ and $p(-M) < 0$.

Then, by the Intermediate Value Theorem, p must have a zero in $(-M, M)$.   ∎

So we have certainly got good value from the arguments in the previous proof!

4.2.3   The Extreme Values Theorem

We now look at whether continuous functions on closed intervals are bounded or unbounded. It turns out that they are bounded, and we use this fact a great deal throughout Analysis!

We discussed bounds, suprema, infima, maxima and minima of functions earlier, in Sub-section 1.4.2.

Theorem 6   The Extreme Values Theorem

A continuous function on a closed interval possesses a maximum value and a minimum value on the interval. In other words, if f is a function that is continuous on a closed interval $[a, b]$, then there exist points c and d in $[a, b]$ such that

$$f(c) \le f(x) \le f(d), \quad \text{for } x \in [a, b].$$

We prove this at the end of the sub-section.

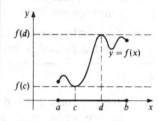

If f is continuous on an interval that is not closed, then it may not be bounded. For example, the function $f(x) = \frac{1}{x}, x \in (0, 1]$, is continuous on $(0,1]$ but is not bounded above (and so it is not bounded), but it is bounded below (by 1).

Similarly, if a function is not continuous on a closed interval, then it may not be bounded. For example, the function

$$f(x) = \begin{cases} 1, & x = 0, \\ \frac{1}{x}, & 0 < x \le 1, \end{cases}$$

is bounded below, by 0, but is not bounded above (and so it is not bounded).

Problem 6

(a) Determine the maximum and the minimum of the function $f(x) = x^2, x \in [-1, 2]$, on $[-1, 2]$. Specify all points in $[-1, 2]$ where these are attained.

(b) Determine the maximum and the minimum of the function $g(x) = \sin x, x \in [0, 2\pi]$, on $[0, 2\pi]$. Specify all points in $[0, 2\pi]$ where these are attained.

Since a function is bounded if and only it is both bounded above and bounded below, we sometimes use the following version of Theorem 6 in applications:

For it will often be sufficient for our purposes.

Corollary 1   The Boundedness Theorem

A continuous function on a closed interval is bounded. In other words, if f is a function that is continuous on the closed interval $[a, b]$, then there exist a number M such that

$$|f(x)| \le M, \quad \text{for } x \in [a, b].$$

On other occasions, we shall find the following consequence of Theorem 6 and the Intermediate Value Theorem useful:

Corollary 2   The Interval Image Theorem

The image of a closed interval under a continuous function is a closed interval.

Proof of Theorem 6

Without some systematic approach to studying continuous functions, we would never be able to prove this theorem, and would be reduced to much hand-waving arguments and loose assertions. So we shall use our earlier work on sequences, and the following related result:

Lemma 1   Let f be a function defined on an interval I.

(a) If f is bounded above on I and $\sup\{f(x): x \in I\} = M$, then there exists some sequence $\{x_n\}$ in I such that $f(x_n) \to M$ as $n \to \infty$.

(b) If f is not bounded above on I, then there exists some sequence $\{x_n\}$ in I such that $f(x_n) \to \infty$ as $n \to \infty$.

Notice that this is a result about sup for any function f on an interval; no assumption of continuity is involved.

A similar result holds for functions that are bounded below on I or are unbounded below on I.

Proof

(a) If there is some point, c say, in I for which $f(c) = M$, then the constant sequence $\{c\}$ has the desired property.

This is a proof by contradiction.

Now suppose that no such point c exists. Then, since $\sup\{f(x): x \in I\} = M$, it follows from the definition of supremum that there is some point, x_1 say, in I for which

$$f(x_1) > M - 1.$$

Next, since $f(x_1) < M$ (from our assumption that there is no point where f takes the value M), choose a point, x_2 say, in I for which

$$f(x_2) > \max\left\{f(x_1),\ M - \frac{1}{2}\right\};$$

notice, in particular, that this last inequality ensures that $x_2 \neq x_1$.

So $f(x_2) > f(x_1)$ and $f(x_2) > M - \frac{1}{2}$. Also, $f(x_2) < M$.

Continuing this process indefinitely, we obtain a sequence $\{x_n\}$ of distinct points in I for which

$$f(x_n) > \max\left\{f(x_1),\ f(x_2), \ldots,\ f(x_{n-1}),\ M - \frac{1}{n}\right\}.$$

Since the sequence $\{M - \frac{1}{n}\}$ converges to M, it follows, by the Squeeze Rule for sequences, that $f(x_n) \to M$ as $n \to \infty$.

For $M - \frac{1}{n} < f(x_n) < M$.

(b) It follows from the definition of 'unbounded above' that there is some point, x_1 say, in I for which $f(x_1) > 1$. Then, for each $n > 1$, we can construct a sequence $\{x_n\}$ of distinct points in I for which

$$f(x_n) > \max\{f(x_1),\ f(x_2), \ldots,\ f(x_{n-1}), n\}.$$

Since the sequence $\{n\}$ tends to ∞, it follows, by the Squeeze Rule for sequences, that $f(x_n) \to \infty$ as $n \to \infty$. $\square$

We are now in a position to prove Theorem 6.

Theorem 6   The Extreme Values Theorem

A continuous function on a closed interval possesses a maximum value and a minimum value on the interval. In other words, if f is a function that is continuous on the closed interval $[a, b]$, then there exist points c and d in $[a, b]$ such that

$$f(c) \leq f(x) \leq f(d), \quad \text{for } x \in [a, b].$$

Proof First, we shall assume that f is not bounded above, and verify that this assumption leads to a contradiction with known facts about f.

It follows from part (b) of Lemma 1 that there exists some sequence $\{x_n\}$ in $[a, b]$ such that $f(x_n) \to \infty$ as $n \to \infty$. Then, by the Bolzano–Weierstrass Theorem, $\{x_n\}$ must contain a convergent subsequence $\{x_{n_k}\}$; denote by d the limit of $\{x_{n_k}\}$. Since all the x_{n_k} lie in $[a, b]$, it follows, by the Limit Inequality Rule for sequences, that $d \in [a, b]$.

Since f is continuous at d and the sequence $\{x_{n_k}\}$ converges to d, it follows that $f(x_{n_k}) \to f(d)$. However, $\{f(x_{n_k})\}$ is a subsequence of the sequence $\{f(x_n)\}$ which tends to ∞, so that $f(x_{n_k}) \to \infty$.

This is a contradiction. So f must be bounded above on $[a, b]$ after all.

Next, denote by M the number $\sup\{f(x): x \in [a, b]\}$. It follows, from part (a) of Lemma 1, that there exists some sequence $\{x_n\}$ in $[a, b]$ such that $f(x_n) \to M$ as $n \to \infty$. Then, by the Bolzano–Weierstrass Theorem, $\{x_n\}$ must contain a convergent subsequence $\{x_{n_k}\}$; denote by d the limit of $\{x_{n_k}\}$. Since all the x_{n_k} lie in $[a, b]$, it follows, by the Limit Inequality Rule for sequences, that $d \in [a, b]$.

Since f is continuous at d and the sequence $\{x_{n_k}\}$ converges to d, it follows that $f(x_{n_k}) \to f(d)$. Therefore, since $\{f(x_{n_k})\}$ is a subsequence of the sequence $\{f(x_n)\}$ which tends to M, we have $M = f(d)$.

This complete the proof that there exists a point d in $[a, b]$ such that
$$f(x) \leq f(d), \quad \text{for all } x \in [a, b].$$

The proof of the existence of a point c in $[a, b]$ such that $f(c) \leq f(x)$, for all $x \in [a, b]$, is similar; we omit it. ∎

It is also possible to prove Theorem 6 by the method of repeated bisection.

Theorem 3, Sub-section 2.5.1.

$x_{n_k} \in [a, b] \Leftrightarrow a \leq x_{n_k} \leq b$.

Inheritance Property of Subsequences, Theorem 5, Sub-section 2.4.4.

M must exist, since we have shown that the set $\{f(x): x \in [a, b]\}$ is bounded.

4.3 Inverse functions

4.3.1 Existence of an inverse function

Let f be the function $f(x) = 2x$, $x \in \mathbb{R}$. Then, given any number y in $\mathbb{R}$, we can find a *unique* number $x = \frac{1}{2}y$ in the domain of f such that $y = f(x) = 2x$.

The inverse function f^{-1}, defined by $f^{-1}(y) = \frac{1}{2}y$, $y \in \mathbb{R}$, undoes the 'effect' of f; that is, $f^{-1}(f(x)) = x$, $x \in \mathbb{R}$.

Also, f undoes the effect of f^{-1}; that is, $f(f^{-1}(y)) = y$, $y \in \mathbb{R}$.

Not every function has an inverse function! For example, consider the function
$$g(x) = x^2, \quad x \in \mathbb{R}.$$

Since $g(2) = 4 = g(-2)$, we cannot define $g^{-1}(4)$ uniquely. Thus g fails to have an inverse function, because it is not one–one. However the function
$$h(x) = x^2, \quad x \in [0, \infty),$$
is one–one and has an inverse function
$$h^{-1}(y) = \sqrt{y}, \quad y \in [0, \infty).$$

In general, if $f: A \to \mathbb{R}$ is one–one, then, for each point y in the image $f(A)$, there is a *unique* point x in A such that $f(x) = y$. Thus f is a *one–one correspondence* between A and $f(A)$; and so we can define the inverse function f^{-1} by $f^{-1}(y) = x$, where $y = f(x)$.

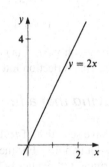

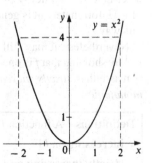

> **Definition** Let $f: A \to \mathbb{R}$ be a one–one function. Then the **inverse function** f^{-1} has domain $f(A)$ and is specified by
> $$f^{-1}(y) = x, \quad \text{where } y = f(x), x \in A.$$

For *some* functions f, we can find the inverse function f^{-1} directly, by solving the equation $y = f(x)$ algebraically to obtain x in terms of y.

Example 1 Prove that the function f defined by $f(x) = \frac{1}{1-x}$, $x \in (-\infty, 1)$, has an inverse function defined on $(0, \infty)$.

Solution First, we solve the equation $y = \frac{1}{1-x}$ to give x in terms of y. By simple manipulation, we obtain

$$y = \frac{1}{1-x} \Leftrightarrow x = 1 - \frac{1}{y}.$$

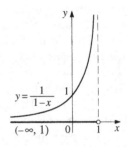

Now, for each $x \in (-\infty, 1)$, we have $x < 1$, and so $f(x) = \frac{1}{1-x} > 0$; thus $f((-\infty,1)) \subseteq (0,\infty)$. Also, for each $y \in (0, \infty)$, we have

$$x = 1 - \frac{1}{y} \in (-\infty, 1);$$

and so f is a one–one correspondence between $(-\infty, 1)$ and $(0, \infty)$. Hence

$$f^{-1}(y) = 1 - \frac{1}{y}, \quad y \in (0, \infty). \qquad \square$$

Remark

Usually, when defining a function we write x for the domain variable. To conform with this practice, we may rewrite the inverse function f^{-1} in Example 1 as follows

$$f^{-1}(x) = 1 - \frac{1}{x}, \quad x \in (0, \infty).$$

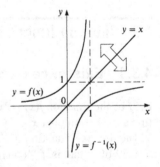

The graph $y = f^{-1}(x)$ is obtained by reflecting the graph $y = f(x)$ in the line $y = x$. This reflection interchanges the x- and y-axes.

Proving that a function f is one–one

We have seen that, if $f: A \to \mathbb{R}$ is one–one, then f has an inverse function f^{-1} with domain $f(A)$. For the function f considered in Example 1, it is possible to determine f^{-1} explicitly by solving the equation $y = f(x)$ to obtain x in terms of y. Unfortunately, it is generally *not* possible to solve the equation $y = f(x)$ in this way.

Nevertheless, it may still be possible to prove that f has an inverse function f^{-1} by showing that f is one–one in some other way. For example, f is one–one if it is either *strictly increasing* or *strictly decreasing*; that is, if f is *strictly monotonic*.

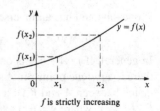

f is strictly increasing

> **Definitions** A function f defined on an interval I is:
> - **increasing** on I if $x_1 < x_2 \Rightarrow f(x_1) \le f(x_2)$, for $x_1, x_2 \in I$;
> - **strictly increasing** on I if $x_1 < x_2 \Rightarrow f(x_1) < f(x_2)$, for $x_1, x_2 \in I$;
> - **decreasing** on I if $x_1 < x_2 \Rightarrow f(x_1) \ge f(x_2)$, for $x_1, x_2 \in I$;

- **strictly decreasing** on I if $\quad x_1 < x_2 \Rightarrow f(x_1) > f(x_2)$, for $x_1, x_2 \in I$;
- **monotonic** on I if $\qquad f$ is *either* increasing on I *or* decreasing on I;
- **strictly monotonic** on I if $\quad f$ is *either* strictly increasing on I *or* strictly decreasing on I.

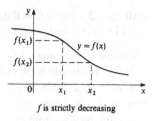

f is strictly decreasing

The most powerful technique for proving that a function f is strictly monotonic is to compute the derivative f' of f and examine the sign of $f'(x)$. We shall not study differentiation in detail until Chapter 6, so here we consider only functions which can be proved to be strictly monotonic by manipulating inequalities rather than by Calculus.

For example, if $n \in \mathbb{N}$, then the function $f(x) = x^n$, $x \in [0, \infty)$, is strictly increasing; and, if n is odd, the function $f(x) = x^n$, $x \in \mathbb{R}$, is strictly increasing.

Similarly, if $n \in \mathbb{N}$, then the function $f(x) = x^{-n}$, $x \in (0, \infty)$, is strictly decreasing.

Example 2 Prove that the function $f(x) = x^5 + x - 1, x \in \mathbb{R}$, is one–one.

Solution If $x_1 < x_2$, then $x_1^5 < x_2^5$. Hence
$$x_1^5 + x_1 - 1 < x_2^5 + x_2 - 1,$$
so f is strictly increasing, and thus one–one. $\qquad\square$

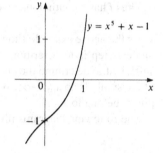

> **Problem 1** Prove that the following functions are one–one:
> (a) $f(x) = x^4 + 2x + 3, \quad x \in [0, \infty)$;
> (b) $f(x) = x^2 - \frac{1}{x}, \quad x \in (0, \infty)$.

4.3.2 The Inverse Function Rule

If the function $f: A \to \mathbb{R}$ is strictly monotonic, then f is one–one, and so f has an inverse function f^{-1} with domain $f(A)$. However, it is not always easy to determine $f(A)$. However, if f is known to be continuous on A, then the following result simplifies the problem immediately.

We prove the Inverse Function Rule in Sub-section 4.3.4.

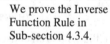

Theorem 1   Inverse Function Rule

Let $f: I \to J$, where I is an interval and J is the image $f(I)$, be a function such that:

1. f is strictly increasing on I;
2. f is continuous on I.

Then J is an interval, and f has an inverse function $f^{-1}: J \to I$ such that:

1'. f^{-1} is strictly increasing on J;
2'. f^{-1} is continuous on J.

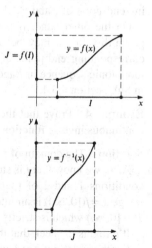

Remarks

1. The interval I may be any type of interval: open or closed, half-open, bounded or unbounded.
2. There is a similar version of the Inverse Function Rule with 'strictly increasing' replaced by 'strictly decreasing'.

Example 3 Prove that the function $f(x) = x^5 + x - 1$, $x \in \mathbb{R}$, has a continuous inverse function, with domain $\mathbb{R}$.

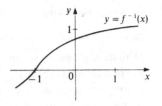

Solution The domain of f is $\mathbb{R}$, which is an interval.

We have already seen, in Example 2, that f is strictly increasing on $\mathbb{R}$ and so has an inverse function f^{-1}. Since f is a polynomial, it is a basic continuous function on $\mathbb{R}$. Thus f satisfies the hypotheses of the Inverse Function Rule.

Then it follows from the Rule that the image $J = f(\mathbb{R})$ is an interval, and that the inverse function $f^{-1}: J \to \mathbb{R}$ is strictly increasing and continuous on J.

It remains to check that the image J is the whole of $\mathbb{R}$. Notice that $J = f(\mathbb{R})$ contains each of the numbers

$$f(n) = n^5 + n - 1, \quad n \in \mathbb{Z}.$$

Now, J contains each of the intervals $[f(-n), f(n)]$; and $f(-n) = -n^5 - n - 1 \to -\infty$ as $n \to \infty$, while $f(n) = n^5 + n - 1 \to \infty$ as $n \to \infty$. It follows that, in fact, $J = f(\mathbb{R})$ must be $(-\infty, \infty) = \mathbb{R}$.

Thus f has a continuous inverse function $f^{-1}: \mathbb{R} \to \mathbb{R}$. $\qquad\square$

As the above example shows, when we apply the Inverse Function Rule the hardest step is to determine the image $J = f(I)$. Since J is an interval, it is sufficient to determine the *end-points* of J, which may be real numbers or one of the symbols ∞ and $-\infty$. We must also determine whether or not these end-points belong to J.

The following diagrams illustrate two examples:

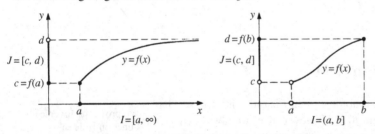

Notice that, if a is an end-point of I and $a \in I$, then $c = f(a)$ is the corresponding end-point of J and $c \in J$.

On the other hand, if a is an end-point of I and $a \notin I$ (this includes the possibility that a may be ∞ or $-\infty$), then it is a little harder to find the corresponding end-point of J. However, it can be shown that, if $\{x_n\}$ is a monotonic sequence in I and $x_n \to a$, then $f(x_n) \to c$, and $c \notin J$. (We prove this in Sub-section 4.3.4.)

Example 4 Prove that the function $f(x) = x^4 + 2x + 3$, $x \in [0, \infty)$, has a continuous inverse function with domain $[3, \infty)$.

Solution The domain of f is $[0, \infty)$, which is an interval.

Also, we know that f is strictly increasing and continuous on $[0, \infty)$, and so conditions 1 and 2 of the Inverse Function Rule hold. It follows that the image $J = f([0, \infty))$ is an interval, and f has a continuous inverse function $f^{-1}: J \to [0, \infty)$ which is strictly increasing on J.

It remains to check that the image J is $[3, \infty)$.

For the end-point 0 of I, we have $0 \in I$, so the corresponding end-point of J is $f(0) = 3$, and $3 \in J$.

For example:

$(0, 1]$ has end-points 0 and 1;

$[1, \infty)$ has end-points 1 and ∞.

Do not let this use of the *symbol* ∞ tempt you to think that ∞ is a real number.

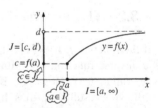

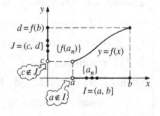

You saw this in Problem 1(a), above.

The other end-point of I is ∞, so to find the corresponding end-point of J we choose the monotonic sequence $\{n\}$, which lies in I and tends to infinity. Then $f(n) = n^4 + 2n + 3 \to \infty$ as $n \to \infty$. It follows that the corresponding end-point of J is ∞. Thus, $J = [3, \infty)$, as required.   □

We now summarise the strategy for establishing that a given continuous function f has a continuous inverse function.

Strategy   To prove that $f: I \to J$, where I is an interval with end-points a and b, has a continuous inverse $f^{-1}: J \to I$:

1. show that f is strictly increasing on I;

2. show that f is continuous on I;

3. determine the end-point c of J corresponding to the end-point a of I as follows:

   - if $a \in I$, then $f(a) = c$ (and $c \in J$);
   - if $a \notin I$, then $f(x_n) \to c$ (and $c \notin J$),

     where $\{x_n\}$ is a monotonic sequence in I such that $x_n \to a$;

4. determine the end-point d of J corresponding to the end-point b of I, similarly.

There is a corresponding version of this strategy for the case that f is *decreasing* on I.

Problem 2   Use the above strategy to prove that the function $f(x) = x^2 - \frac{1}{x}$, $x \in (0, \infty)$, has a continuous inverse function with domain $\mathbb{R}$.

Hint: Use the result of Problem 1(b) in Sub-section 4.3.1.

4.3.3   Inverses of standard functions

We now use the Inverse Function Rule and the above strategy to define continuous inverse functions for certain standard functions. Although you will be familiar with these inverse functions already, we can now *prove* that they exist and are continuous. We also remind you of some properties of these inverse functions.

For each function, we give brief remarks on the four steps of the strategy. In each case, the continuity of the function f follows directly from the results of Section 4.1.

The n*th root function*

We asserted the existence of the nth root function in Sub-section 1.5.2, Theorem 1. We can at last provide the proof of that assertion!

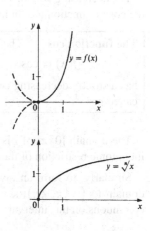

The nth root function   For any positive integer $n \geq 2$, the function
$$f(x) = x^n, \quad x \in [0, \infty),$$
has a strictly increasing continuous inverse function $f^{-1}(x) = \sqrt[n]{x}$ with domain $[0, \infty)$, called the **nth root function**.

In this case, the strategy is easy to apply:

1. f is strictly increasing on $[0, \infty)$;
2. f is continuous on $[0, \infty)$;
3. $f(0) = 0$;
4. $f(k) = k^n \to \infty$ as $k \to \infty$.

It follows, from the Inverse Function Rule, that f has a strictly increasing continuous inverse function $f^{-1}: [0, \infty) \to [0, \infty)$.

f is a basic continuous function.

We use $\{k\}$ rather than $\{n\}$ here, to avoid using n for two different purposes in the same expression.

Remark

If n is *odd*, then the nth root function can be extended to a continuous function whose domain is the whole of $\mathbb{R}$.

Inverse trigonometric functions

> **The function $\sin^{-1}$**   The function
> $$f(x) = \sin x, \quad x \in \left[-\frac{1}{2}\pi, \ \frac{1}{2}\pi \right],$$
> has a strictly increasing continuous inverse function with domain $[-1, 1]$, called $\sin^{-1}$.

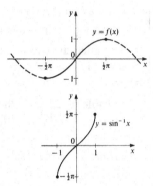

In this case:

1. the geometric definition of $f(x) = \sin x$ shows that f is strictly increasing on $\left[-\frac{1}{2}\pi, \frac{1}{2}\pi \right]$;
2. f is continuous on $\left[-\frac{1}{2}\pi, \frac{1}{2}\pi \right]$;
3. $\sin\left(-\frac{1}{2}\pi \right) = -1$;
4. $\sin\left(\frac{1}{2}\pi \right) = 1$.

It follows that $f\left(\left[-\frac{1}{2}\pi, \frac{1}{2}\pi \right] \right) = [-1, 1]$. Hence, by the Inverse Function Rule, f has a strictly increasing continuous inverse function $f^{-1}: [-1, 1] \to \left[-\frac{1}{2}\pi, \frac{1}{2}\pi \right]$.

The decreasing version of the strategy can be applied similarly to prove that the cosine function has an inverse, if we restrict its domain suitably.

> **The function $\cos^{-1}$**   The function
> $$f(x) = \cos x, \quad x \in [0, \ \pi],$$
> has a strictly decreasing continuous inverse function with domain $[-1, 1]$, called $\cos^{-1}$.

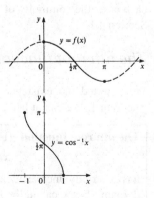

The domain $[0, \pi]$ of f is chosen here by convention, so that f is a strictly monotonic restriction of the cosine function.

Similarly, to form an inverse of the tangent function we must restrict its domain to $\left(-\frac{1}{2}\pi, \frac{1}{2}\pi \right)$, since the tangent function is strictly increasing and continuous on this interval.

> **The function $\tan^{-1}$** The function
> $$f(x) = \tan x, \quad x \in \left(-\frac{1}{2}\pi, \frac{1}{2}\pi\right),$$
> has a strictly increasing continuous inverse function with domain $\mathbb{R}$, called
> **$\tan^{-1}$**.

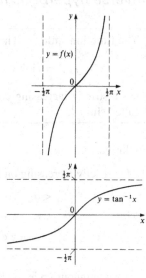

In this case, the image set $f\left(\left(-\frac{1}{2}\pi, \frac{1}{2}\pi\right)\right)$ is $\mathbb{R}$, because (for example) if $\{x_n\}$ is a monotonic sequence in $\left(-\frac{1}{2}\pi, \frac{1}{2}\pi\right)$ and $x_n \to \frac{1}{2}\pi$ as $n \to \infty$, then

$$f(x_n) = \tan x_n = \frac{\sin x_n}{\cos x_n} \to \infty \quad \text{as } n \to \infty.$$

Remark

Some texts use arc sin, arc cos and arc tan instead of $\sin^{-1}$, $\cos^{-1}$ and $\tan^{-1}$, respectively.

Problem 3
(a) Determine the values of $\sin^{-1}\left(\frac{1}{\sqrt{2}}\right)$, $\cos^{-1}\left(-\frac{1}{2}\right)$ and $\tan^{-1}\left(\sqrt{3}\right)$.
(b) Prove that $\cos\left(2\sin^{-1} x\right) = 1 - 2x^2$, for $x \in [-1, 1]$.
   Hint: Let $y = \sin^{-1} x$.

The function $\log_e$

We now discuss one of the most important inverse functions.

> **The function $\log_e$** The function
> $$f(x) = e^x, \quad x \in \mathbb{R},$$
> has a strictly increasing continuous inverse function f^{-1} with domain $(0, \infty)$, called **$\log_e$**.

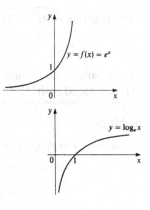

In this case:
1. f is strictly increasing on $\mathbb{R}$, since
$$x_1 < x_2 \Rightarrow x_2 - x_1 > 0$$
$$\Rightarrow e^{x_2 - x_1} > 1 \quad \text{(since } e^x \geq 1 + x > 1, \text{ for } x > 0\text{)}$$
$$\Rightarrow e^{x_2} > e^{x_1};$$
2. f is continuous on $\mathbb{R}$;
3. $f(n) = e^n \to \infty$ as $n \to \infty$;
4. $f(-n) = e^{-n} \to 0$ as $n \to \infty$.

It follows that the image of $\mathbb{R}$ under f is $f(\mathbb{R}) = (0, \infty)$. Hence, by the Inverse Function Rule, f has a strictly increasing continuous inverse function $f^{-1} \colon (0, \infty) \to \mathbb{R}$.

Problem 4 Prove that $\log_e x + \log_e y = \log_e(xy)$, for $x, y \in (0, \infty)$.
Hint: Let $a = \log_e x$ and $b = \log_e y$.

Inverse hyperbolic functions

The function $\sinh^{-1}$ The function
$$f(x) = \sinh x = \frac{1}{2}(e^x - e^{-x}), \quad x \in \mathbb{R},$$
has a strictly increasing continuous inverse function f^{-1} with domain $\mathbb{R}$, called $\sinh^{-1}$.

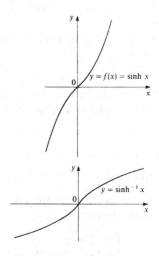

In this case:

1. f is strictly increasing on $\mathbb{R}$, since
$$x_1 < x_2 \Rightarrow e^{x_1} < e^{x_2}$$
$$\Rightarrow -e^{-x_1} < -e^{-x_2}$$
$$\Rightarrow e^{x_1} - e^{-x_1} < e^{x_2} - e^{-x_2}$$
$$\Rightarrow \sinh x_1 < \sinh x_2;$$
2. f is continuous on $\mathbb{R}$, by the Combination Rules;
3. $f(n) = \frac{1}{2}(e^n - e^{-n}) \to \infty$ as $n \to \infty$;
4. $f(-n) = \frac{1}{2}(e^{-n} - e^n) \to -\infty$ as $n \to \infty$.

It follows that the image of $\mathbb{R}$ under f is $f(\mathbb{R}) = \mathbb{R}$. Hence, by the Inverse Function Rule, f has a strictly increasing continuous inverse function $f^{-1}: \mathbb{R} \to \mathbb{R}$.

The function $\cosh^{-1}$ The function
$$f(x) = \cosh x = \frac{1}{2}(e^x + e^{-x}), \quad x \in [0, \infty),$$
has a strictly increasing continuous inverse function f^{-1} with domain $[1, \infty)$, called $\cosh^{-1}$.

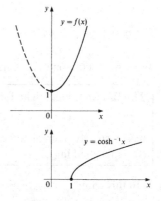

In this case:

1. f is strictly increasing on $[0, \infty)$, since
$$x_1 < x_2 \Rightarrow \sinh x_1 < \sinh x_2$$
$$\Rightarrow (1 + \sinh^2 x_1)^{\frac{1}{2}} < (1 + \sinh^2 x_2)^{\frac{1}{2}}$$
$$\Rightarrow \cosh x_1 < \cosh x_2, \text{ since } \cosh^2 x = 1 + \sinh^2 x;$$
2. f is continuous on $[0, \infty)$, by the Combination Rules;
3. $f(0) = 1$;
4. $f(n) = \frac{1}{2}(e^n + e^{-n}) \to \infty$ as $n \to \infty$.

It follows that the image of $[0, \infty)$ under f is $f([0, \infty)) = [1, \infty)$. Hence, by the Inverse Function Rule, f has a strictly increasing continuous inverse function $f^{-1}: [1, \infty) \to [0, \infty)$.

The strategy can be applied in a similar way to show that $f(x) = \tanh x$ is strictly increasing and continuous on $\mathbb{R}$, with $f(\mathbb{R}) = (-1, 1)$. We omit the details.

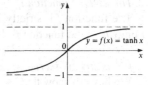

The function tanh⁻¹   The function
$$f(x) = \tanh x = \frac{\sinh x}{\cosh x}, \quad x \in \mathbb{R},$$
has a strictly increasing continuous inverse function f^{-1} with domain $(-1, 1)$, called **tanh⁻¹**.

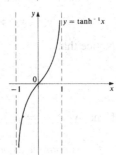

The inverse hyperbolic functions can be expressed in terms of $\log_e$, as the following example shows.

Example 5   Prove that $\sinh^{-1} x = \log_e\left(x + \sqrt{x^2 + 1}\right)$, for $x \in \mathbb{R}$.

Solution   Let $y = \sinh^{-1} x$, for $x \in \mathbb{R}$. Then
$$x = \sinh y = \frac{1}{2}(e^y - e^{-y}),$$
and so
$$e^{2y} - 2xe^y - 1 = 0.$$
This is a quadratic equation in e^y, so that
$$e^y = x \pm \sqrt{x^2 + 1}.$$
Since $e^y > 0$, we must choose the $+$ sign here, so that
$$y = \log_e\left(x + \sqrt{x^2 + 1}\right). \qquad \square$$

Problem 5   Prove that $\cosh^{-1} x = \log_e\left(x + \sqrt{x^2 - 1}\right)$, for $x \in [1, \infty)$.

4.3.4 Proof of the Inverse Function Rule

We now prove the Inverse Function Rule, and justify our strategy for determining the domains of inverse functions.

You may omit the proofs in this sub-section at a first reading, but you should at least read the statement of Theorem 2 below.

Theorem 1   Inverse Function Rule

Let $f: I \to J$, where I is an interval and J is the image $f(I)$, be a function such that:

1. f is strictly increasing on I;
2. f is continuous on I.

Then J is an interval, and f has an inverse function $f^{-1}: J \to I$ such that:

1′. f^{-1} is strictly increasing on J;
2′. f^{-1} is continuous on J.

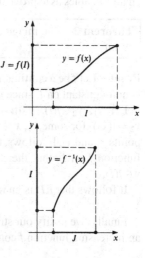

Proof   The proof is in four parts:
$J = f(I)$ is an interval.

Let $y_1, y_2 \in f(I)$, with $y_1 < y_2$, and let $y \in (y_1, y_2)$. Now $y_1 = f(x_1)$ and $y_2 = f(x_2)$, for some $x_1, x_2 \in I$, with $x_1 < x_2$ since f is strictly increasing on I. It follows, by the Intermediate Value Theorem, that there is some number $x \in (x_1, x_2)$ for which $f(x) = y$. Hence $y \in f(I)$.

It follows that $f(I)$ is an interval.

The inverse function $f^{-1}: J \to I$ exists.

The function f is strictly increasing and is therefore one–one; also, f maps I onto J, so the function $f^{-1}: J \to I$ exists, by definition.

f^{-1} is strictly increasing on J.

We have to show that

$$y_1 < y_2 \Rightarrow f^{-1}(y_1) < f^{-1}(y_2), \quad \text{for } y_1, y_2 \in J.$$

Notice that

$$f^{-1}(y_1) \geq f^{-1}(y_2) \Rightarrow f\bigl(f^{-1}(y_1)\bigr) \geq f\bigl(f^{-1}(y_2)\bigr)$$
$$\Rightarrow y_1 \geq y_2.$$

Hence $y_1 < y_2 \Rightarrow f^{-1}(y_1) < f^{-1}(y_2)$, as required.

f^{-1} is continuous on J.

Let $y \in J$, and (for simplicity) assume that y is not an end-point of J. Then $y = f(x)$, for some $x \in I$, and we want to prove that

$$y_n \to y \Rightarrow f^{-1}(y_n) \to f^{-1}(y) = x.$$

Thus, we want to deduce that:

for each $\varepsilon > 0$, there is some number X such that
$$x - \varepsilon < f^{-1}(y_n) < x + \varepsilon, \quad \text{for all } n > X. \tag{1}$$

Since f is strictly increasing, we know that
$$f(x - \varepsilon) < f(x) < f(x + \varepsilon);$$
also, since $y_n \to y = f(x)$, there is some number X such that
$$f(x - \varepsilon) < y_n < f(x + \varepsilon), \quad \text{for all } n > X.$$

If we then apply the strictly increasing function f^{-1} to these inequalities, we obtain (1), as required.

We assume, for convenience, that ε is sufficiently small that $(x - \varepsilon, x + \varepsilon) \subseteq I$.

This completes the proof of the Inverse Function Rule. ∎

In fact, a careful check of the first part of the above proof shows that a slight change enables us to prove the following result that is of interest in its own right.

Theorem 2 The image of an interval under a continuous function is also an interval.

Proof Let f be a continuous function on an interval I. We shall assume that f is non-constant on I, since otherwise the result is trivial.

Let $y_1, y_2 \in f(I)$, with $y_1 < y_2$, and let $y \in (y_1, y_2)$. Now $y_1 = f(x_1)$ and $y_2 = f(x_2)$, for some $x_1, x_2 \in I$, and $x_1 \neq x_2$. Let I' denote the interval with end-points x_1 and x_2. It follows, by applying the Intermediate Value Theorem to the function f on I', that there is some number $x \in I'$ for which $f(x) = y$. Hence $y \in f(I)$.

It follows that $f(I)$ is an interval. ∎

In Corollary 2, Sub-section 4.2.3, we showed that the image of a *closed* interval under a continuous function is a *closed* interval.

We must introduce the symbol I' since we do not know in general whether $x_1 < x_2$ or $x_2 < x_1$.

Finally, we justify our strategy for determining the end-points of $J = f(I)$ for an increasing function f continuous on an interval I.

> **Strategy for finding the end-points of J** If a is an end-point of the interval I, then we can find the corresponding end-point c of J as follows:
>
> - if $a \in I$, then $f(a) = c$ (and $c \in J$);
> - if $a \notin I$, then $f(x_n) \to c$ (and $c \notin J$),
>
>   where $\{x_n\}$ is a monotonic sequence in I such that $x_n \to a$.

Here we consider only the end-point a of I; a similar result holds for the other end-point b of I.

Proof For simplicity, we shall suppose that a is the left end-point of I.

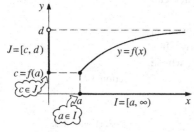 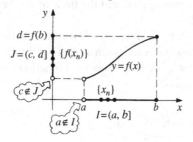

If $a \in I$, then $c = f(a) \in J$ and

$$f(x) \geq f(a) = c, \quad \text{for } x \in I,$$

and so c is the corresponding left end-point of J.

On the other hand, if $a \notin I$, then we select any decreasing sequence $\{x_n\}$ in I such that $x_n \to a$ as $n \to \infty$. Then $\{f(x_n)\}$ is also decreasing; it therefore follows, by the Monotonic Sequence Theorem for sequences, that

$$f(x_n) \to c \quad \text{as } n \to \infty, \tag{2}$$

where c is a real number or $-\infty$.

This was Theorem 2 in Sub-section 2.5.1: if the sequence $\{a_n\}$ is monotonic, then either $\{a_n\}$ is convergent or $a_n \to \pm\infty$.

Now, $f(x_n) \in J$ for $n = 1, 2, \ldots$; therefore, since J is an interval, it follows that

$$[f(x_n), f(x_1)] \subseteq J, \quad \text{for } n = 1, 2, \ldots.$$

Hence, by (2), we have that

$$(c, f(x_1)] = \bigcup_{n=1}^{\infty} [f(x_n), f(x_1)] \subseteq J.$$

Here $\bigcup_{n=1}^{\infty} [f(x_n), f(x_1)]$ denotes the set of points that belong to all the intervals $[f(x_n), f(x_1)]$; that is, the *union* of the intervals.

To deduce, finally, that c is the left end-point of J, we need to show that $c \notin J$. Suppose that in fact $c \in J$. Then $c = f(x)$, for some $x \in I$; it follows that

$$\begin{aligned} f(x) = c &< f(x_n), & \text{for } n = 1, 2, \ldots \\ \Rightarrow x &< x_n, & \text{for } n = 1, 2, \ldots \\ \Rightarrow x &\leq a. \end{aligned}$$

Thus $x \notin I$. This contradiction completes the proof. ∎

Taking limits 'flattens' inequalities.

4.4 Defining exponential functions

4.4.1 The definition of a^x

Earlier, we looked at the definition of the irrational number $\sqrt{2} = 1.4142\ldots$. We have also defined a^x for $a > 0$ when x is rational, but we have not yet defined a^x when x is irrational.

Sub-section 1.1.1.
Sub-section 1.5.3.

One possible method for defining the irrational power $2^{\sqrt{2}}$ involves the decimal representation of $\sqrt{2}$. Each of the truncations of $\sqrt{2} = 1.4142\ldots$ is a rational number, and the corresponding rational numbers

$$2^1, 2^{1.4}, 2^{1.41}, 2^{1.414}, 2^{1.4142}, \ldots \tag{1}$$

are defined, and form an increasing sequence which is bounded above by $2^2 = 4$ (for example). Hence, by the Monotone Convergence Theorem for sequences, the sequence (1) is convergent, and the limit of this sequence can be taken as the definition of $2^{\sqrt{2}}$.

Sub-section 2.5.1.

We can define a^x for $a > 0$ and $x \in \mathbb{R}$ similarly. However, with this definition it is difficult to establish the properties of a^x, such as the Exponential Laws. It is more convenient to define a^x by using the exponential function $x \mapsto e^x$, whose properties we have already discussed.

Section 3.4.

Recall that

$$e^x = \lim_{n \to \infty} \left(1 + \frac{x}{n}\right)^n = \sum_{n=0}^{\infty} \frac{x^n}{n!}, \quad \text{for } x \in \mathbb{R},$$

Sub-section 3.4.3.

$$e^x = \left(e^{-x}\right)^{-1}, \quad \text{for } x \in \mathbb{R},$$

and

$$e^{x+y} = e^x \times e^y, \quad \text{for } x, y \in \mathbb{R}. \tag{2}$$

Recall too that the function $x \mapsto e^x$ is strictly increasing and continuous; and hence, by the Inverse Function Rule, it has a strictly increasing continuous inverse function $x \mapsto \log_e x$, $x \in (0, \infty)$. Thus we have

$$\log_e(e^x) = x, \quad \text{for } x \in \mathbb{R}, \tag{3}$$

and

$$e^{\log_e x} = x, \quad \text{for } x \in (0, \infty).$$

Now let $a = e^x$ and $b = e^y$. Then, by equation (2), we have

$$ab = e^x e^y = e^{x+y};$$

so that, by equation (3), we obtain $\log_e(ab) = \log_e a + \log_e b$, for $a, b \in (0, \infty)$.

We deduce that, if $a > 0$ and $n \in \mathbb{N}$, then

$$\log_e(a^n) = n \log_e a,$$

and so

$$a^n = e^{n \log_e a}.$$

With a little more manipulation, we can show that the equation

$$a^x = e^{x \log_e a}$$

is true for each *rational* number x. This suggests that we *define* a^x, for $a > 0$ and x *irrational*, by means of this equation.

Definition If $a > 0$, then $a^x = e^{x \log_e a}$, for $x \in \mathbb{R}$.

For example, $2^x = e^{x \log_e 2}$ for $x \in \mathbb{R}$, so that the graph $y = 2^x$ is obtained from the graph $y = e^x$ by a scaling in the x-direction with scale factor $\frac{1}{\log_e 2}$.

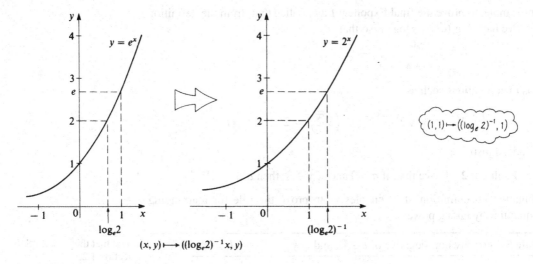

$$(x, y) \longmapsto ((\log_e 2)^{-1}x, y)$$

This relationship between the graphs $y = e^x$ and $y = 2^x$ suggests that the function $x \mapsto 2^x$ must also be continuous. In fact, we can deduce the continuity of the function $x \mapsto 2^x = e^{x \log_e 2}, x \in \mathbb{R}$, from the continuity of the function $x \mapsto e^x$, by using the Multiple Rule and the Composition Rule for continuity.

Remark

Since $x \mapsto 2^x$ is continuous, we deduce that the sequence

$$2^1, 2^{1.4}, 2^{1.41}, 2^{1.414}, 2^{1.4142}, \ldots,$$

where 1, 1.4, 1.41, 1.414, 1.4142, ... are truncations of $\sqrt{2} = 1.4142\ldots$, does converge to $2^{\sqrt{2}}$, and so both definitions of $2^{\sqrt{2}}$ agree.

In general, we have the following result.

> **Theorem 1** If $a > 0$, then the function $x \mapsto a^x = e^{x \log_e a}$, $x \in \mathbb{R}$, is continuous.

Problem 1 Prove that the following functions are continuous:

(a) $f(x) = x^\alpha$, where $x \in (0, \infty)$ and $\alpha \in \mathbb{R}$;

(b) $f(x) = x^x$, where $x \in (0, \infty)$.

4.4.2 Further properties of exponentials

Our definition of a^x enables us to give straight-forward proofs of the following Exponent Laws.

> **Exponent Laws**
> - If $a, b > 0$ and $x \in \mathbb{R}$, then $a^x b^x = (ab)^x$.
> - If $a > 0$ and $x, y \in \mathbb{R}$, then $a^x a^y = a^{x+y}$.
> - If $a > 0$ and $x, y \in \mathbb{R}$, then $(a^x)^y = a^{xy}$.

You first met these laws in Sub-section 1.5.3, but there $x, y \in \mathbb{Q}$.

For example, to prove the final Exponent Law, notice that, from the definition of a^x, we have $\log_e(a^x) = x \log_e a$; so that

$$(a^x)^y = e^{y \log_e(a^x)}$$
$$= e^{xy \log_e a} = a^{xy}.$$

Thus manipulations such as

$$\left(\sqrt{2}^{\sqrt{2}}\right)^{\sqrt{2}} = \sqrt{2}^{\left(\sqrt{2} \times \sqrt{2}\right)} = \left(\sqrt{2}\right)^2 = 2$$

are indeed justified.

Problem 2 Prove that, if $a > 0$ and $x, y \in \mathbb{R}$, then $a^x a^y = a^{x+y}$.

Finally, our definition of a^x enables us to prove the rule for rearranging inequalities by taking powers.

Rule 5 For any non-negative $a, b \in \mathbb{R}$, and any $p > 0$, $a < b \Leftrightarrow a^p < b^p$.

You met this Rule in Sub-section 1.2.1.

If $a = 0$, the result is obvious. In general, since the functions $x \mapsto \log_e x$ and $x \mapsto e^x$ are strictly increasing, we have

$$a < b \Leftrightarrow \log_e a < \log_e b$$
$$\Leftrightarrow p \log_e a < p \log_e b \quad \text{(since } p > 0)$$
$$\Leftrightarrow e^{p \log_e a} < e^{p \log_e b}$$
$$\Leftrightarrow a^p < b^p.$$

The following example illustrates the use of Rule 5.

Example 1 Determine which of the numbers e^π and π^e is greater.

Solution We use the inequality

$$e^x > 1 + x, \quad \text{for } x > 0.$$

Applying this inequality with $x = \left(\frac{\pi}{e}\right) - 1$, we obtain

$$e^{\left(\frac{\pi}{e}\right)-1} > 1 + \left(\left(\frac{\pi}{e}\right) - 1\right)$$
$$= \frac{\pi}{e} = \pi e^{-1};$$

then, by multiplying through by the positive factor e, we obtain that $e^{\frac{\pi}{e}} > \pi$.

Finally, by applying Rule 5 with $p = e$, we obtain $e^\pi > \pi^e$. $\square$

This was one of the Exponential Inequalities: part (a) of Theorem 4 in Sub-section 4.1.3.

Problem 3 Prove that $e^x > x^e$, for $x > e$.

4.5 Exercises

Section 4.1

1. Use the appropriate rules, together with the list of basic continuous functions, to prove that the following functions are continuous:

(a) $f(x) = \exp(\sin(x^2 + 1))$, $x \in \mathbb{R}$;

(b) $f(x) = e^{\sqrt{x}} + x^5$, $\qquad x \in [0, \infty)$.

2. Determine whether the following functions are continuous at 0:

(a) $f(x) = \begin{cases} \sin x \sin\left(\frac{1}{x}\right), & x \neq 0, \\ 0, & x = 0; \end{cases}$

(b) $f(x) = \begin{cases} \frac{1}{x}\sin\left(\frac{1}{x}\right), & x \neq 0, \\ 0, & x = 0. \end{cases}$

3. Prove that each of the following sequences is convergent, and determine its limit:

(a) $\left\{\sin\left(e^{\frac{1}{n}} - 1\right)\right\}$; (b) $\left\{\left(\cos\left(\frac{1}{2^n}\right)\right)^{\frac{1}{2}}\right\}$.

4. Let f be defined on an open interval I, and $c \in I$. Prove that, if f is continuous at c and $f(c) \neq 0$, then there is an open interval $J \subseteq I$ such that $c \in J$ and $f(x) \neq 0$, for any $x \in J$.

We shall use this result in Chapter 6.

5. Determine the points where the function $f(x) = \begin{cases} 1, & x = 0, 1, \\ x + [2x], & 0 < x < 1, \end{cases}$ is

(a) continuous on the left; (b) continuous on the right;

(c) continuous.

6. Prove that the following function is continuous on $\mathbb{R}$

$$f(x) = \begin{cases} -1, & x \leq -\frac{1}{2}\pi, \\ \sin x, & -\frac{1}{2}\pi < x < \frac{1}{2}\pi, \\ 1, & x \geq \frac{1}{2}\pi. \end{cases}$$

7. Determine at which points the following function is continuous

$$f(x) = \begin{cases} x, & -1 \leq x < 0, \\ e^x, & 0 \leq x \leq 1. \end{cases}$$

8. Write down examples of functions with the following properties:

(a) f and g are discontinuous on $\mathbb{R}$, but $g \circ f$ is continuous on $\mathbb{R}$;

(b) f and g are discontinuous on $\mathbb{R}$, but $f + g$ and fg are continuous on $\mathbb{R}$;

(c) f is continuous on $\mathbb{R} - \left\{1, \frac{1}{2}, \frac{1}{3}, \frac{1}{4}, \ldots\right\}$ but discontinuous at $\left\{1, \frac{1}{2}, \frac{1}{3}, \frac{1}{4}, \ldots\right\}$.

Section 4.2

1. The function f is continuous on $(0, 1)$, and takes every real value at most once. Use the Intermediate Value Theorem to prove that f is strictly monotonic on $(0, 1)$.

2. Give examples of functions f continuous on the half-open interval $[0, 1)$ in $\mathbb{R}$, to show that $f([0,1))$ can be open, closed or half-open.

3. Prove that each of the following polynomials has the stated number of (real) zeros:

(a) $p(x) = x^4 - 4x^3 + 3x^2 + 2x - 1$, 4 zeros;

(b) $p(x) = 3x^3 - 8x^2 + x + 3$, 3 zeros.

4. Prove that the function $f(x) = x - \sin x - \frac{2}{3}\pi$, $x \in \mathbb{R}$, has a zero in $\left(\frac{2}{3}\pi, \frac{5}{6}\pi\right)$.

5. Using the Zeros Localisation Theorem and the Extreme Values Theorem, prove that every polynomial of even degree n

$$p(x) = a_n x^n + a_{n-1}x^{n-1} + \cdots + a_1 x + a_0, \quad x \in \mathbb{R},$$

where $a_n \neq 0$, has a minimum value on $\mathbb{R}$.

6. Write down an example of a function $f(x)$, $x \in (0, 1)$, that is continuous on $(0, 1)$ and for which $f((0, 1)) = (0, 1)$, but such that there is no point c in $(0, 1)$ for which $f(c) = c$.

This result is closely related to Problem 2 in Sub-section 4.2.1, so you might like to look back at that problem and compare the two.

Section 4.3

1. For each of the following functions, prove that it has a continuous inverse function and determine the domain of that inverse function:

(a) $f(x) = x^3 + 1 - \frac{1}{x^2}$, $x \in (0, \infty)$;

(b) $f(x) = \frac{1}{(1+x^3)^2}$, $x \in (-1, \infty)$.

2. Determine whether each of the following statements is true:

(a) $\sin(\sin^{-1} x) = x$, for $x \in [-1, 1]$;

(b) $\sin^{-1}(\sin x) = x$, for $x \in \mathbb{R}$.

3. (a) Prove that $\tan^{-1} x + \tan^{-1} y = \tan^{-1}\left(\frac{x+y}{1-xy}\right)$, provided that $\tan^{-1} x + \tan^{-1} y$ lies in $\left(-\frac{1}{2}\pi, \frac{1}{2}\pi\right)$.

(b) Use the result in part (a) to evaluate $\tan^{-1}\left(\frac{1}{2}\right) + \tan^{-1}\left(\frac{1}{3}\right)$.

This is known as the *Addition Formula* for $\tan^{-1}$.

5 Limits and continuity

In Chapter 4 we made *some* progress in pinning down the idea of 'a well-behaved function' in precise terms. Using our earlier work on sequences we defined what is meant by a function being *continuous*: roughly speaking, its graph has no jumps or gaps.

However, a number of functions arise quite naturally in mathematics where we need to handle functions that are already defined near some particular point, but that are not defined at the point itself.

For example, the function

$$f(x) = \frac{\sin x}{x}, \ x \neq 0,$$

which arises when we examine the question of whether the sine function is differentiable. Clearly from their graphs, the behaviour of f differs significantly from the behaviour of the function

$$g(x) = \sin\left(\frac{1}{x}\right), \ x \neq 0,$$

that you have already met. It is possible to assign a value (namely, 1) to f at 0 so that this extension of the domain of f makes f continuous on a domain that is an interval. On the other hand, it is not possible to assign a value to g at 0 so that this extension of the domain of g makes g continuous on a domain that is an interval.

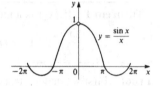

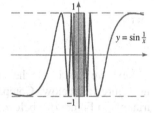

In Section 5.1, we discuss *limits of functions*, and show that the existence of this helpful value 1 for f can be stated as $\lim_{x \to 0} \frac{\sin x}{x} = 1$. The concept of a limit of a function is closely related to that of a continuous function, and many of the rules for calculating limits are similar to those for continuity. We also discuss one-sided limits.

In Section 5.2, we discuss the behaviour of functions near asymptotes of their graphs. In particular, we prove that, if $n \in \mathbb{Z}$, then $\lim_{x \to \infty} x^n e^{-x} = 0$.

Next, in Section 5.3, we introduce a slightly different definition of the limit $\lim_{x \to c} f(x)$ of a function f at a point c. Instead of using sequences tending to c to define the limit of f at c, we define $\lim_{x \to c} f(x)$ directly in terms of inequalities involving x and $f(x)$, and we verify that this new definition is completely equivalent to the earlier definition. We also illustrate the changes to proofs of results about limits that the new definition involves.

At first sight this new definition will seem more complicated. However throughout the rest of the book we shall see just how powerful this new definition turns out to be!

In Section 5.4, we introduce a definition for the continuity of a function f at a point c in terms of inequalities involving x and $f(x)$, rather than in terms of the behaviour of f on sequences that tend to c. We verify that this new definition is completely equivalent to the earlier definition of continuity, and illustrate the changes to proofs of results about continuity that the new definition involves.

Finally, in Section 5.5, we introduce a concept that will be extremely powerful in your further study of Analysis; namely, that of *uniform continuity*.

Before starting on Sections 5.3 and 5.4, you may find it useful to quickly revise the material in Sections 1.2 and 1.3 on inequalities.

'may' means 'will'!

5.1 Limits of functions

5.1.1 What is a limit of a function?

The graph of the function

$$f(x) = \frac{\sin x}{x}, \quad x \neq 0,$$

shows that, if x takes values which are 'close to' but distinct from 0, then $f(x)$ takes values which are 'close to' 1. The closer that x gets to 0, the closer $f(x)$ gets to 1. We now pin down this idea precisely.

First, we need the following fact.

Theorem 1   If $\{x_n\}$ is a null sequence whose terms are non-zero, then

$$\frac{\sin x_n}{x_n} \to 1 \quad \text{as } n \to \infty.$$

That is, $\displaystyle\lim_{n\to\infty} \frac{\sin x_n}{x_n} = 1$.

Proof   First we deduce from the inequality

$$\sin x \leq x, \quad \text{for } 0 < x < \frac{\pi}{2},$$

that

$$\frac{\sin x}{x} \leq 1, \quad \text{for } 0 < x < \frac{\pi}{2}.$$

Lemma 1, Sub-section 4.1.3.

Next, we need to use the formula for the area of a sector of a disc of radius 1 in Figure (a), below. Compare the area of this sector with the area of the triangle in Figure (b), below.

We discussed the area and perimeter of a disc of radius 1 in Sub-section 2.5.4 and in Exercise 4 on Section 2.5 in Section 2.6.

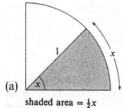

(a)

shaded area = $\frac{1}{2}x$

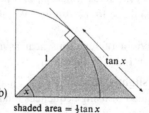

(b)

shaded area = $\frac{1}{2}\tan x$

We find that

$$x \leq \tan x, \quad \text{for } 0 < x < \frac{\pi}{2}.$$

Since $\tan x = \frac{\sin x}{\cos x}$ and $\cos x > 0$ for $0 < x < \frac{\pi}{2}$, we obtain

$$\cos x \leq \frac{\sin x}{x}, \quad \text{for } 0 < x < \frac{\pi}{2}.$$

Combining this inequality with our earlier upper estimate for $\frac{\sin x}{x}$, we find that

$$\cos x \leq \frac{\sin x}{x} \leq 1, \quad \text{for } 0 < x < \frac{\pi}{2}.$$

In fact, this inequality holds for $0 < |x| < \frac{\pi}{2}$, since

$$\cos(-x) = \cos x \quad \text{and} \quad \frac{\sin(-x)}{(-x)} = \frac{\sin x}{x}.$$

Now, if $\{x_n\}$ is any null sequence with non-zero terms, then the terms x_n must eventually satisfy the inequality $|x_n| < \frac{\pi}{2}$, and so there is some number X such that

Take $\varepsilon = \frac{\pi}{2}$ in the definition of null sequence.

$$\cos x_n \leq \frac{\sin x_n}{x_n} \leq 1, \quad \text{for } n > X.$$

But $\cos x_n \to 1$ as $n \to \infty$, since the cosine function is continuous at 0 and $\cos 0 = 1$. Hence, by the Squeeze Rule for sequences

$$\frac{\sin x_n}{x_n} \to 1. \qquad \blacksquare$$

This behaviour of $\frac{\sin x}{\cos x}$ near 0 is an example of a function f tending to a limit as x tends to a point c.

To define this concept, we need to ensure that the function f is defined *near* the point c, but *not necessarily at* the point c itself. We first introduce the idea of a punctured neighbourhood of c.

Definitions A **neighbourhood** of a point c of $\mathbb{R}$ is an open interval that contains the point c, and a **punctured neighbourhood** of a point c of $\mathbb{R}$ is a neighbourhood of c from which the point c itself has been deleted.

So the 'puncture' is at c.

For example, the sets $(0, 9)$, $(1, \infty)$ and $\mathbb{R}$ are neighbourhoods of the point 2, and $(1, 2) \cup (2, 5)$ and $(-\infty, 2) \cup (2, 4)$ are punctured neighbourhoods of the point 2. In general, a neighbourhood of c is an interval of the form $(c - r, c + s)$, for some $r, s > 0$, and a punctured neighbourhood of c is the union of a pair of intervals $(c - r, c) \cup (c, c + s)$, for some $r, s > 0$. In practice, we often choose as a neighbourhood of c an open interval $(c - r, c + r)$, $r > 0$, with centre at c; and as a punctured neighbourhood of c the union $(c - r, c) \cup (c, c + r)$ of two open intervals of equal length.

These choices are simply matters of convenience!

We now define the limit of a function in terms of limits of sequences.

Definition Let the function f be defined on a punctured neighbourhood N of a point c. Then $f(x)$ **tends to the limit ℓ as x tends to c** if:

for each sequence $\{x_n\}$ in N such that $x_n \to c$, then $f(x_n) \to \ell$. (1)

We write this as either '$\lim_{x \to c} f(x) = \ell$' or '$f(x) \to \ell$ as $x \to c$'.

Note that $x_n \neq c$, for any n.

Let us check that this definition holds for $f(x) = \frac{\sin x}{x}$ at 0. This function f is defined on the domain $\mathbb{R} - \{0\}$, and so in particular on every punctured neighbourhood of 0. We have just seen in Theorem 1 that the statement (1) holds; it follows then that

For example, f is defined on $(-1, 0) \cup (0, 1)$.

$$\lim_{x \to 0} \frac{\sin x}{x} = 1.$$

Since the definition of the limit of a function involves the limit of sequences, we can use our various Combination Rules for sequences to determine the limits of many functions.

Example 1   Prove that each of the following functions tends to a limit as x tends to 2, and determine these limits:

(a) $f(x) = \frac{x^2 - 4}{x - 2}$;   (b) $f(x) = \frac{x^3 - 3x - 2}{x^2 - 3x + 2}$.

Solution

(a) First, notice that f is defined on every punctured neighbourhood of 2.

   Next, notice that

$$f(x) = \frac{x^2 - 4}{x - 2} = x + 2, \quad \text{for } x \neq 2.$$

We can cancel $x - 2$, since $x \neq 2$.

   Thus, if $\{x_n\}$ is any sequence that lies in some punctured neighbourhood N of 2 and $x_n \to 2$, then

$$f(x_n) = x_n + 2 \to 4 \quad \text{as } n \to \infty,$$

For example, $N = (1, 2) \cup (2, 3)$; any *other* punctured neighbourhood of 2 would serve our purpose equally well.

   by the Sum Rule for sequences. It follows that

$$\lim_{x \to 2} \frac{x^2 - 4}{x - 2} = 4.$$

(b) The function

$$f(x) = \frac{x^3 - 3x - 2}{x^2 - 3x + 2} = \frac{(x - 2)(x^2 + 2x + 1)}{(x - 2)(x - 1)}$$

   has domain $\mathbb{R} - \{1, 2\}$, and so f is defined on the punctured neighbourhood $N = (1, 2) \cup (2, 3)$ of 2.

   Next, notice that

$$f(x) = \frac{x^3 - 3x - 2}{x^2 - 3x + 2} = \frac{x^2 + 2x + 1}{x - 1},$$

Of course any *smaller* punctured neighbourhood of 2 would serve our purpose equally well.

   for $x \in N = (1, 2) \cup (2, 3)$.

   Thus, if $\{x_n\}$ is any sequence that lies in $N = (1, 2) \cup (2, 3)$ and $x_n \to 2$, then

$$f(x_n) = \frac{x_n^2 + 2x_n + 1}{x_n - 1} \to \frac{4 + 4 + 1}{2 - 1} = 9 \quad \text{as } n \to \infty,$$

   by the Combination Rules for sequences. It follows that

$$\lim_{x \to 2} \frac{x^3 - 3x - 2}{x^2 - 3x + 2} = 9. \qquad \square$$

Later in this section we give several further techniques for calculating limits. Our next example illustrates how to prove that a limit does *not* exist.

Sub-sections 5.1.2 and 5.1.3.

Example 2   Prove that each of the following functions does not tend to a limit as x tends to 0:

(a) $f(x) = \frac{1}{x}$, $x \neq 0$;   (b) $f(x) = \sin\left(\frac{1}{x}\right)$, $x \neq 0$;   (c) $f(x) = \sqrt{x}$, $x \geq 0$.

Solution

(a) The function f is defined on the punctured neighbourhood $N = (-2, 0) \cup (0, 2)$ of 0. The null sequence $\left\{\frac{1}{n}\right\}$ lies in N and tends to 0, but $f\left(\frac{1}{n}\right) = n \to \infty$. Hence f does not tend to a limit as x tends to 0.

Any other punctured neighbourhood of 0 would serve equally well here, of course.

(b) The function f is defined on the punctured neighbourhood $N = (-2, 0) \cup (0, 2)$ of 0.

To prove that $f(x)$ does not tend to a limit as x tends to 0, we choose two null sequences $\{x_n\}$ and $\{x'_n\}$ that lie in N, such that

$$f(x_n) \to 1 \quad \text{and} \quad f(x'_n) \to -1.$$

Since $\sin\left(2n\pi + \frac{1}{2}\pi\right) = 1$ and $\sin\left(2n\pi + \frac{3}{2}\pi\right) = -1$ for $n \in \mathbb{Z}$, we can choose

$$x_n = \frac{1}{2n\pi + \frac{1}{2}\pi} \quad \text{and} \quad x'_n = \frac{1}{2n\pi + \frac{3}{2}\pi}, \quad \text{for } n = 0, 1, 2, \ldots.$$

Hence f does not tend to a limit as x tends to 0.

(c) The function $f(x) = \sqrt{x}$ has domain $[0, \infty)$, and so f is not defined on any punctured neighbourhood of zero. Hence f does not tend to a limit as x tends to zero. □

In particular, the terms x_n and x'_n will be non-zero.

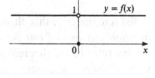

We collect these techniques together in the form of a strategy.

Strategy   To show that $\lim\limits_{x\to c} f(x)$ does *not* exist:
1. Show that there is no punctured neighbourhood N of c on which f is defined;

OR

2. Find two sequences $\{x_n\}$ and $\{x'_n\}$ (in some punctured neighbourhood N of c) which tend to c, such that $\{f(x_n)\}$ and $\{f(x'_n)\}$ have different limits;

OR

3. Find a sequence $\{x_n\}$ (in some punctured neighbourhood N of c) which tends to c, such that $f(x_n) \to \infty$ or $f(x_n) \to -\infty$.

In fact *at least one* of these possibilities MUST occur if $\lim\limits_{x\to c} f(x)$ does not exist; we omit a proof of this useful fact.

Problem 1   Determine whether the following limits exist:
(a) $\lim\limits_{x\to 0} \frac{x^2+x}{x}$;   (b) $\lim\limits_{x\to 0} \frac{|x|}{x}$.

5.1.2 Limits and continuity

Consider the function

$$f(x) = \begin{cases} 1, & x \neq 0, \\ 0, & x = 0. \end{cases}$$

Does this function tend to a limit as x tends to zero; and, if it does, what is the limit?

Certainly, f is defined on any punctured neighbourhood of zero, since the domain of f is $\mathbb{R}$. Also, if $\{x_n\}$ is any null sequence with non-zero terms, then $f(x_n) = 1$, for $n = 1, 2, \ldots$, so that $f(x_n) \to 1$ as $n \to \infty$. It follows that $\lim\limits_{x\to 0} f(x) = 1$.

This example serves to emphasise that the value of a limit $\lim\limits_{x\to c} f(x)$ has nothing to do with the value of $f(c)$ – even if f happens to be defined at the point c.

However, if f is defined at c, and f is also continuous at c, then the only possible value for $\lim\limits_{x\to c} f(x)$ is $f(c)$.

We put these observations together in the following result.

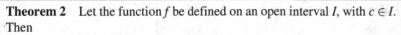

Theorem 2   Let the function f be defined on an open interval I, with $c \in I$. Then

$$f \text{ is continuous at } c \Leftrightarrow \lim_{x\to c} f(x) = f(c).$$

Notice that I is a neighbourhood of c.

Using Theorem 2 and our knowledge of continuous functions, we can evaluate many limits rather easily.

For example, to determine $\lim_{x \to 2}(3x^5 - 5x^2 + 1)$ notice that the function $f(x) = 3x^5 - 5x^2 + 1$ is defined on $\mathbb{R}$ and is continuous at 2, since f is a polynomial. Hence, by Theorem 2

> Polynomials are basic continuous functions: see Sub-section 4.1.3.

$$\lim_{x \to 2}(3x^5 - 5x^2 + 1) = f(2) = 77.$$

We saw earlier that the following functions are continuous at 0

$$f(x) = \begin{cases} x \sin\left(\frac{1}{x}\right), & x \neq 0, \\ 0, & x = 0; \end{cases} \quad \text{and} \quad f(x) = \begin{cases} x^2 \sin\left(\frac{1}{x}\right), & x \neq 0, \\ 0, & x = 0. \end{cases}$$

> Problem 8 and Example 5, Sub-section 4.1.2, respectively.

It therefore follows from Theorem 2 that

$$\lim_{x \to 0} x \sin\left(\frac{1}{x}\right) = 0 \quad \text{and} \quad \lim_{x \to 0} x^2 \sin\left(\frac{1}{x}\right) = 0.$$

On the other hand, we saw in part (b) of Example 2 that

$$\lim_{x \to 0} \sin\left(\frac{1}{x}\right) = 0 \text{ does } not \text{ exist.}$$

> Sub-section 5.1.1.

It follows from Theorem 2 that, no matter how we try to extend the domain of $f(x) = \sin\left(\frac{1}{x}\right)$ to include $x = 0$, we can *never* obtain a continuous function.

Remark

If f is defined on an open interval I, with $c \in I$, and $\lim_{x \to c} f(x)$ exists but $\lim_{x \to c} f(x) \neq f(c)$, then f is said to have a **removable discontinuity** at c. For example, the function $f(x) = \begin{cases} x \sin\left(\frac{1}{x}\right), x \neq 0, \\ 3, \qquad x = 0, \end{cases}$ has a removable discontinuity at 0.

> This means that, if we redefine the value of f just at c itself, the resulting function is then continuous at c.

Problem 2 Use Theorem 1 to determine the following limits:

(a) $\lim_{x \to 2} \sqrt{x}$; (b) $\lim_{x \to \frac{\pi}{2}} \sqrt{\sin x}$; (c) $\lim_{x \to 1} \frac{e^x}{1+x}$

In the remainder of this chapter we shall frequently use Theorem 2. *When the function f is one of our basic continuous functions*, however, we shall not always refer to the theorem explicitly. For example, $f(x) = x^2 + 1$ and $g(x) = \sin x$ are basic continuous functions, and so we can write $\lim_{x \to 2}(x^2 + 1) = 5$ and $\lim_{x \to 0} \sin x = 0$ without further explanation.

> *Basic continuous functions:*
> - polynomials and rational functions;
> - modulus function;
> - nth root function;
> - trigonometric functions (sine, cosine and tangent);
> - the exponential function.

5.1.3 Rules for limits

As you might expect from your experience with sequences, series and continuous functions, we often find limits by using various rules. First, we state the Combination Rules.

Theorem 3 **Combination Rules**
If $\lim_{x \to c} f(x) = \ell$ and $\lim_{x \to c} g(x) = m$, then:
Sum Rule $\lim_{x \to c}(f(x) + g(x)) = \ell + m$;
Multiple Rule $\lim_{x \to c} \lambda f(x) = \lambda \ell$, for $\lambda \in \mathbb{R}$;

Product Rule $\quad \lim\limits_{x \to c} f(x)g(x) = \ell m;$

Quotient Rule $\quad \lim\limits_{x \to c} \frac{f(x)}{g(x)} = \frac{\ell}{m}$, provided that $m \neq 0$.

For example, since $\lim\limits_{x \to 0} \frac{\sin x}{x} = 1$ and $\lim\limits_{x \to 0}(x^2 + 1) = 1$, we have

$$\lim_{x \to 0}\left(\frac{\sin x}{x} + 2(x^2 + 1)\right) = 1 + 2 \times 1 = 3.$$

The proofs of these rules are all simple consequences of the corresponding results for sequences. We illustrate this by proving just one rule.

Proof of the Sum Rule Let the functions f and g be defined on a punctured neighbourhood N of a point c. We want to show that:

> for each sequence $\{x_n\}$ in N such that $x_n \to c$, then $f(x_n) + g(x_n) \to \ell + m$.

Now, since $\lim\limits_{x \to c} f(x) = \ell$, we know that, for any sequence $\{x_n\}$ in N for which $x_n \to c$, then $f(x_n) \to \ell$. Also, since $\lim\limits_{x \to c} g(x) = m$, we know that, as $x_n \to c$, then $g(x_n) \to m$.

It follows by the Sum Rule for sequences that, as $x_n \to c$, then $f(x_n) + g(x_n) \to \ell + m$. This completes the proof. ∎

There are no new ideas in the proof. All that we have to do is to set up things so that we can use the Sum Rule for sequences.

We can also use the following Composition Rule.

Theorem 4 Composition Rule

If $\lim\limits_{x \to c} f(x) = \ell$ and $\lim\limits_{x \to \ell} g(x) = L$, then $\lim\limits_{x \to c} g(f(x)) = L$, provided that:

EITHER $\quad f(x) \neq \ell$ in some punctured neighbourhood of c;

OR $\quad\quad$ g is defined at ℓ and is continuous at ℓ.

Note that the second limit is a limit as $x \to \ell$, not as $x \to c$.

Remarks on the Composition Rule

(a) In any particular case, the Composition Rule may be FALSE if we do not ensure that one or other of the two provisos holds! For example, if

$$f(x) = 0, x \in \mathbb{R}, \quad \text{and} \quad g(x) = \begin{cases} \frac{\sin x}{x}, & x \neq 0, \\ 0, & x = 0, \end{cases}$$

then

$$\lim_{x \to 0} f(x) = 0 \quad \text{and} \quad \lim_{x \to 0} g(x) = 1, \quad \text{but} \quad \lim_{x \to 0} g(f(x)) = \lim_{x \to 0}(0) = 0.$$

Here we take $c = 0$, $\ell = 0$ and $L = 1$.

So, in this case, $\lim\limits_{x \to 0} g(f(x)) \neq L$.

(b) Suppose the first proviso '$f(x) \neq \ell$ in some punctured neighbourhood of c' in Theorem 4 holds. Then we know that, for each sequence $\{x_n\}$ in a punctured neighbourhood N of c for which $x_n \to c$, $f(x_n)$ does not *take* the value ℓ but must lie in a *punctured neighbourhood* of ℓ. In this case, the desired result will follow from the facts that (*i*) the sequence $\{f(x_n)\}$ converges to ℓ, (*ii*) $f(x_n)$ lies in a punctured neighbourhood of ℓ, and (*iii*) $\lim\limits_{x \to \ell} g(x) = L$.

We omit the details of the proof in this case, which are now straight-forward to write down.

(c) Suppose the second proviso 'g is defined at ℓ and is continuous at ℓ' in Theorem 4 holds. In this case, the desired result will follow from the facts that (*i*) the sequence $\{f(x_n)\}$ converges to ℓ, and (*ii*) g is continuous at ℓ.

The following example illustrates the use of the Composition Rule.

Example 3   Determine the following limits:

(a) $\lim\limits_{x \to 0} \dfrac{\sin(\sin x)}{\sin x}$;   (b) $\lim\limits_{x \to 0} \left(1 + \left(\dfrac{\sin x}{x}\right)^2\right)$.

Solution

(a) Let $f(x) = \sin x$, $x \in \mathbb{R}$, and $g(x) = \dfrac{\sin x}{x}$, $x \neq 0$. Then

$$\lim_{x \to 0} f(x) = \lim_{x \to 0} \sin x = 0 \quad \text{and} \quad \lim_{x \to 0} g(x) = \lim_{x \to 0} \frac{\sin x}{x} = 1.$$

Here we have $c = 0$, $\ell = 0$ and $L = 1$.

Also, $f(x) = \sin x \neq 0$ in the punctured neighbourhood $(-\pi, 0) \cup (0, \pi)$ of 0 (for example). It follows, by the Composition Rule, that

$$\lim_{x \to 0} g(f(x)) = \lim_{x \to 0} \frac{\sin(\sin x)}{\sin x} = 1.$$

(b) Let $f(x) = \dfrac{\sin x}{x}$, $x \neq 0$, and $g(x) = 1 + x^2$, $x \in \mathbb{R}$. Then

$$\lim_{x \to 0} \frac{\sin x}{x} = 1 \quad \text{and} \quad \lim_{x \to 1}(1 + x^2) = 2.$$

Also, g is defined and continuous at 1. It follows, by the Composition Rule, that

$$\lim_{x \to 0} g(f(x)) = \lim_{x \to 0}\left(1 + \left(\frac{\sin x}{x}\right)^2\right) = 2. \qquad \Box$$

Here we have $c = 0$, $\ell = 1$ and $L = 2$.

Problem 3   Use the Combination Rules and the Composition Rule to determine the following limits:

(a) $\lim\limits_{x \to 0} \dfrac{\sin x}{2x + x^2}$;   (b) $\lim\limits_{x \to 0} \dfrac{\sin(x^2)}{x^2}$;   (c) $\lim\limits_{x \to 0}\left(\dfrac{x}{\sin x}\right)^{\frac{1}{2}}$.

Problem 4   For the functions

$$f(x) = \begin{cases} 0, & x = 0, \\ -2 & x = 1, \\ 2 + x, & x \neq 0, 1, \end{cases} \quad \text{and} \quad g(x) = \begin{cases} 0, & x = 0, \\ 1 + x, & x \neq 0, \end{cases}$$

determine $(f \circ g)(x)$, $(f \circ g)(0)$, $f\left(\lim\limits_{x \to 0} g(x)\right)$ and $\lim\limits_{x \to 0} f(g(x))$.

There is also a Squeeze Rule for limits.

Theorem 5   Squeeze Rule

Let the functions f, g and h be defined on a punctured neighbourhood N of a point c. If:
1. $g(x) \leq f(x) \leq h(x)$, for $x \in N$, and
2. $\lim\limits_{x \to c} g(x) = \lim\limits_{x \to c} h(x) = \ell$,
then $\lim\limits_{x \to c} f(x) = \ell$.

The proof of Theorem 5 is a straight-forward application of the Squeeze Rule for sequences.

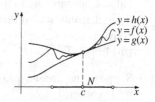

Problem 5   Use the Squeeze Rule for limits to prove that:

(a) $\lim\limits_{x \to 0} x^2 \sin\left(\frac{1}{x}\right) = 0$;   (b) $\lim\limits_{x \to 0} x \cos\left(\frac{1}{x}\right) = 0$.

The Limit Inequality Rule is another result for limits of functions that is an analogue of the corresponding result for sequences.

Theorem 3, Sub-section 2.3.3.

Theorem 6 Limit Inequality Rule

If $\lim_{x \to c} f(x) = \ell$ and $\lim_{x \to c} g(x) = m$, and also

$$f(x) \leq g(x), \text{ on some punctured neighbourhood of } c,$$

then

$$\ell \leq m.$$

5.1.4 One-sided limits

In Example 2(c) above, we saw that $\lim_{x \to 0} \sqrt{x}$ does not exist, because the function $f(x) = \sqrt{x}$, $x \geq 0$, is not defined on any punctured neighbourhood of zero. Nevertheless, if $\{x_n\}$ is any non-zero sequence in the domain $[0, \infty)$ of f for which $x_n \to 0$, then $\sqrt{x_n} \to 0$. Thus $f(x) = \sqrt{x}$ tends to $f(0) = 0$ as x tends to 0 *from the right*.

Definition   Let f be defined on $(c, c+r)$, for some $r > 0$. Then $f(x)$ **tends to the limit ℓ as x tends to c from the right** if:

for each sequence $\{x_n\}$ in $(c, c+r)$ such that $x_n \to c$, then $f(x_n) \to \ell$.

We write this either as '$\lim_{x \to c^+} f(x) = \ell$' or as '$f(x) \to \ell$ as $x \to c^+$'.

Note that f need not be defined
at the point c.

For example, $\lim_{x \to 0^+} \sqrt{x} = 0$.

There is a corresponding definition for limits as x tends to c from the left.

Definition   Let f be defined on $(c-r, c)$, for some $r > 0$. Then $f(x)$ **tends to the limit ℓ as x tends to c from the left** if:

for each sequence $\{x_n\}$ in $(c-r, c)$ such that $x_n \to c$, then $f(x_n) \to \ell$.

We write this either as '$\lim_{x \to c^-} f(x) = \ell$' or as '$f(x) \to \ell$ as $x \to c^-$'.

Again, f need not be defined at
the point c.

Sometimes both left and right limits exist. Even when this happens, the two values need not be equal, as the following example shows.

Example 4   Prove that the function $f(x) = \frac{|x|}{x}$, $x \neq 0$, tends to different limits as x tends to 0 from the right and from the left.

Solution   The function f is defined on $(0, 1)$, and $f(x) = 1$ on this interval. Thus, if $\{x_n\}$ is a null sequence in $(0, 1)$, then

$$\lim_{n \to \infty} f(x_n) = \lim_{n \to \infty} (1) = 1.$$

So $\lim_{x \to 0^+} f(x) = 1$.

Similarly, f is defined on $(-1, 0)$, and $f(x) = -1$ on this interval. Thus, if $\{x_n\}$ is a null sequence in $(-1, 0)$, then

$$\lim_{n \to \infty} f(x_n) = \lim_{n \to \infty} (-1) = -1.$$

So $\lim_{x \to 0^-} f(x) = -1.$ □

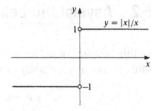

Problem 6   Write down a function f defined on the interval $[-1, 3]$ for which $\lim_{x \to 0^-} f(x)$ does not exist but $\lim_{x \to 0^+} f(x) = 1$. Verify that f has the specified properties.

The relationship between one-sided limits and 'ordinary' limits is given by the following result.

> **Theorem 7**   Let f be defined on a punctured neighbourhood of the point c. Then
> $$\lim_{x \to c} f(x) = \ell$$
> $$\Leftrightarrow \lim_{x \to c^+} f(x) \quad \text{and} \quad \lim_{x \to c^-} f(x) \quad \text{both exist and equal } \ell.$$

We omit a proof of this result.

Remark

If f is defined on an open interval I that contains the point c, and
$$\lim_{x \to c^+} f(x) \text{ and } \lim_{x \to c^-} f(x) \text{ both exist, but } \lim_{x \to c^+} f(x) \neq \lim_{x \to c^-} f(x),$$
then f is said to have a **jump discontinuity** at c.

Analogues of the Combination Rules, Composition Rule and Squeeze Rule can also be used to determine one-sided limits. In the statements of all these rules, we simply replace $\lim_{x \to c}$ by $\lim_{x \to c^+}$ or $\lim_{x \to c^-}$, and replace the open interval I containing c by $(c, c+r)$ or $(c-r, c)$, as appropriate.

There are strategies such as that at the end of Sub-section 5.1.1 for proving that one-sided limits do *not* exist.

Problem 7   Prove that:
(a) $\lim_{x \to 0^+} \left(\frac{\sin x}{x} + \sqrt{x} \right) = 1$;   (b) $\lim_{x \to 0^+} \frac{\sin(\sqrt{x})}{\sqrt{x}} = 1$.

Problem 8   Use the inequalities $1 + x \leq e^x \leq \frac{1}{1-x}$, for $|x| < 1$, to prove that
$$\lim_{x \to 0^+} \frac{e^x - 1}{x} = \lim_{x \to 0^-} \frac{e^x - 1}{x} = 1.$$

We verified these inequalities as part (c) of Theorem 4 ('The Exponential Inequalities') in Sub-section 4.1.3.

Deduce that
$$\lim_{x \to 0} \frac{e^x - 1}{x} = 1.$$

5.2   Asymptotic behaviour of functions

In this section we define *formally* a number of statements that you will have already met in your study of Calculus, such as
$$\frac{1}{x} \to \infty \text{ as } x \to 0^+ \quad \text{and} \quad e^x \to \infty \text{ as } x \to \infty,$$
and describe the relationship between them.

5.2.1   Functions which tend to infinity

Just as we defined the statement $\lim_{x \to c} f(x) = \ell$ in terms of the convergence of sequences, so we can define the statement $f(x) \to \infty$ as $x \to c$.

Definition   Let f be defined on a punctured neighbourhood N of the point c. Then $f(x) \to \infty$ as $x \to c$ if:

> for each sequence $\{x_n\}$ in N such that $x_n \to c$, then $f(x_n) \to \infty$.

The statement '$f(x) \to -\infty$ as $x \to c$' is defined similarly, with ∞ replaced by $-\infty$.

As with sequences, there is a version of the Reciprocal Rule which relates the behaviour of functions which tend to infinity and functions which tend to 0.

Theorem 2, Sub-section 2.4.3.

Theorem 1   Reciprocal Rule
(a) If the function f satisfies the following two conditions:
 1. $f(x) > 0$ for all x in some punctured neighbourhood of c, and
 2. $f(x) \to 0$ as $x \to c$,
 then $\frac{1}{f(x)} \to \infty$ as $x \to c$.
(b) If $f(x) \to \infty$ as $x \to c$, then $\frac{1}{f(x)} \to 0$ as $x \to c$.

For example, $\frac{1}{x^2} \to \infty$ as $x \to 0$, since $f(x) = x^2 > 0$, for $x \in \mathbb{R} - \{0\}$, and $\lim_{x \to 0} x^2 = 0$.

The statements

$$\text{`}f(x) \to \infty \text{ or } -\infty \text{ as } x \to c^+\text{'} \quad \text{and} \quad \text{`}f(x) \to \infty \text{ or } -\infty \text{ as } x \to c^-\text{'}$$

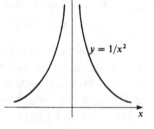

are defined similarly, with the punctured neighbourhood of c being replaced by open intervals $(c, c+r)$ or $(c-r, c)$, as appropriate, for some $r > 0$.

The Reciprocal Rule can also be applied with '$x \to c$' replaced by '$x \to c^+$' or '$x \to c^-$', and with the punctured neighbourhood of c being replaced by open intervals $(c, c+r)$ or $(c-r, c)$, as appropriate, for some $r > 0$.

For example, $\frac{1}{x} \to \infty$ as $x \to 0^+$, since $f(x) = x > 0$, for $x \in (0, \infty)$, and $\lim_{x \to 0^+} x = 0$.

> **Problem 1**   Prove the following:
> (a) $\frac{1}{|x|} \to \infty$ as $x \to 0$;   (b) $\frac{1}{1-x^3} \to -\infty$ as $x \to 1^+$;
>
> (c) $\frac{\sin x}{x^3} \to \infty$ as $x \to 0$.

Remark

There are also versions of the Combination Rules and Squeeze Rule for functions which tend to ∞ or $-\infty$ as x tends to c, c^+ or c^-; here we state only the Combination Rules for functions which tend to ∞ as x tends to c.

These are similar to results stated for sequences in Theorems 3 and 4, Sub-section 2.4.3.

Theorem 2   Combination Rules
If $f(x) \to \infty$ and $g(x) \to \infty$ as $x \to c$, then:

Sum Rule $f(x) + g(x) \to \infty$;
Multiple Rule $\lambda f(x) \to \infty$, for $\lambda > 0$;
Product Rule $f(x)\, g(x) \to \infty$.

Problem 2 Prove that the following statements are false:

(a) If $f(x) \to \infty$ and $g(x) \to \infty$ as $x \to 0$, then $f(x) - g(x) \to \infty$;

(b) If $f(x) \to \infty$ and $g(x) \to 0$ as $x \to 0$, then $f(x)g(x) \to 0$ or $f(x)g(x) \to \infty$.

5.2.2 Behaviour of *f(x)* as *x* tends to ∞ or −∞

Finally, we define various types of behaviour of real functions $f(x)$ as x tends to ∞ or $-\infty$. To avoid repetition, here we allow ℓ to denote either a real number or one of the symbols ∞ and $-\infty$.

We adopt this convention for simplicity ONLY in this sub-section!

> **Definition** Let f be defined on an interval (R, ∞), for some real number R. Then $f(x) \to \ell$ as $x \to \infty$ if:
>
> for each sequence $\{x_n\}$ in (R, ∞) such that $x_n \to \infty$, then $f(x_n) \to \ell$.

The statement '$f(x) \to \ell$ as $x \to -\infty$' is defined similarly, with ∞ replaced by $-\infty$, and (R, ∞) by $(-\infty, R)$. In practice, we usually prove that $f(x) \to \ell$ as $x \to -\infty$ by showing that $f(-x) \to \ell$ as $x \to \infty$.

When ℓ is a real number, we also use the notation

$$\lim_{x \to \infty} f(x) = \ell \quad \text{and} \quad \lim_{x \to -\infty} f(x) = \ell.$$

Once again, we can use versions of the Combination Rules and Reciprocal Rule to obtain statements about the behaviour of given functions as x tends to ∞ or $-\infty$. In the statements of these rules, we need only replace c by ∞ or $-\infty$, and the punctured neighbourhood of c by (R, ∞) or $(-\infty, R)$, as appropriate.

For example, for any positive integer n

$$x^n \to \infty \quad \text{as } x \to \infty \quad \text{and} \quad \lim_{x \to \infty} x^{-n} = 0.$$

More generally, we have the following result for the behaviour of polynomials as $x \to \infty$.

> **Theorem 3** If $a_0, a_1, \ldots, a_{n-1}$ are real numbers and
> $$p(x) = x^n + a_{n-1}x^{n-1} + \cdots + a_1 x + a_0, \quad x \in \mathbb{R},$$
> then
> $$p(x) \to \infty \text{ as } x \to \infty \quad \text{and} \quad \frac{1}{p(x)} \to 0 \text{ as } x \to \infty.$$

We ask you to prove the first part of this result in Section 5.6. The second part then follows at once.

There are also versions of the Squeeze Rule for functions as x tends to ∞, which have some important applications.

> **Theorem 4 Squeeze Rule**
> Let the functions f, g and h be defined on some interval (R, ∞).
> (a) If f, g and h satisfy the following two conditions:
>
>   1. $g(x) \leq f(x) \leq h(x)$, for all x in (R, ∞), and
>   2. $\lim_{x \to \infty} g(x) = \lim_{x \to \infty} h(x) = \ell$,
>
>   then $\lim_{x \to \infty} f(x) = \ell$.

We omit the proof of this theorem, which is straightforward.

(b) If f and g satisfy the following two conditions:
   1. $f(x) \geq g(x)$, for all x in (R, ∞), and
   2. $g(x) \to \infty$ as $x \to \infty$,
   then $f(x) \to \infty$ as $x \to \infty$.

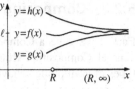

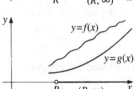

An important application of this Squeeze Rule is to show that e^x tends to ∞ faster than any power of x, as x tends to ∞. We prove this in the next example.

Example 1   Prove that for each $n = 0, 1, 2, \ldots$, $\dfrac{e^x}{x^n} \to \infty$ as $x \to \infty$.

Solution   We use the power series expansion

$$e^x = 1 + x + \frac{x^2}{2!} + \cdots + \frac{x^n}{n!} + \frac{x^{n+1}}{(n+1)!} + \cdots$$

See Section 3.4.

for $x \geq 0$. Since $x \geq 0$, all the terms on the right here are non-negative, and so

$$e^x \geq \frac{x^{n+1}}{(n+1)!}, \quad \text{for } x \geq 0.$$

It follows that

$$\frac{e^x}{x^n} \geq \frac{x}{(n+1)!}, \quad \text{for } x > 0.$$

Since $\frac{x}{(n+1)!} \to \infty$ as $x \to \infty$, it follows from part (b) of the Squeeze Rule that $\frac{e^x}{x^n} \to \infty$ as $x \to \infty$.   □

Remark

We can deduce from Example 1 and the Product Rule that, for ANY integer n, $x^n e^x \to \infty$ as $x \to \infty$. Thus, by the Reciprocal Rule, for any integer n, $x^n e^{-x} \to 0$ as $x \to \infty$.

> **Problem 3**   Determine the behaviour of the following functions as $x \to \infty$:
> (a) $f(x) = \frac{2x^3 + x}{x^3}$;   (b) $f(x) = \frac{\sin x}{x}$.

In our next example, we describe the behaviour of $\log_e x$ as $x \to \infty$.

Example 2   Prove that $\log_e x \to \infty$ as $x \to \infty$.

Solution   We prove this from first principles.
   First, note that the function $f(x) = \log_e x$ is defined on $(0, \infty)$.
   Next, we have to prove that:

   for each sequence $\{x_n\}$ in $(0, \infty)$ such that $x_n \to \infty$, then $\log_e x_n \to \infty$.

To prove that $\log_e x_n \to \infty$, we need to show that:

   for each positive number K, there is a number X such that
   $$\log_e x_n > K, \quad \text{for all } n > X. \tag{1}$$

However, since we know that $x_n \to \infty$, we can choose X such that
   $$x_n > e^K, \quad \text{for all } n > X.$$

Since the function $\log_e$ is strictly increasing, the statement (1) is therefore true. It follows that $\log_e x \to \infty$ as $x \to \infty$, as required.   □

There is no easy way to prove this result by using the Squeeze Rule, since $\log_e x$ tends to ∞ 'rather reluctantly'; that is, more slowly than any function that we have considered so far.

5.2.3 Composing asymptotic behaviour

Earlier we gave a Composition Rule for limits. We now describe a more general Composition Rule which permits the composition of the different types of asymptotic behaviour that we have now met.

Sub-section 5.1.3, Theorem 4.

For example, since

$$f(x) = \frac{1}{x} \to \infty \text{ as } x \to 0^+ \quad \text{and} \quad g(x) = e^x \to \infty \text{ as } x \to \infty,$$

we should expect that

$$g(f(x)) = e^{\frac{1}{x}} \to \infty \quad \text{as } x \to 0^+.$$

This result is true, as the following Composition Rule shows. To avoid repetition, we again allow ℓ and L to denote either a real number or one of the symbols ∞ and $-\infty$.

We again adopt this convention for simplicity ONLY in this sub-section!

Theorem 5   Composition Rule

If:

1. $f(x) \to \ell$ as $x \to c$ (or c^+, c^-, ∞ or $-\infty$), and

2. $g(x) \to L$ as $x \to \ell$,

then

$$g(f(x)) \to L \quad \text{as } x \to c \text{ (or } c^+, c^-, \infty \text{ or } -\infty, \text{ respectively)},$$

provided that:

EITHER $f(x) \neq \ell$ in some punctured neighbourhood of c (or in $(c, c+r)$, $(c-r, c)$, (R, ∞) or $(-\infty, R)$, respectively, for some $r > 0$)

OR ℓ is finite, and g is defined at ℓ and is continuous at ℓ.

We omit the proof of Theorem 5.

If ℓ denotes ∞ or $-\infty$, then the first proviso is automatically satisfied. Note that we *must* have conditions 1 and 2 and the first proviso satisfied *or* conditions 1 and 2 and the second proviso satisfied if we are to make any application of this result.

Example 3   Prove that

(a) $\frac{e^{\frac{x}{2}}}{x} \to \infty$ as $x \to \infty$;   (b) $xe^{\frac{1}{x}} \to \infty$ as $x \to 0^+$;

(c) $x\sin\left(\frac{1}{x}\right) \to 1$ as $x \to \infty$.

Solution

(a) Let $f(x) = \frac{x}{2}$, $x \in \mathbb{R}$, and $g(x) = \frac{e^x}{x}$, $x \in \mathbb{R} - \{0\}$, so that

$$g(f(x)) = \frac{e^{\frac{x}{2}}}{\frac{x}{2}} = \frac{2e^{\frac{x}{2}}}{x}, \quad \text{for } x \in \mathbb{R} - \{0\}, \text{ and so in}$$

particular for $x \in (0, \infty)$.

Now, by the Multiple Rule, $f(x) \to \infty$ as $x \to \infty$; and, by Example 1, $g(x) \to \infty$ as $x \to \infty$. It follows, by the Composition Rule, that

$$g(f(x)) = \frac{2e^{\frac{x}{2}}}{x} \to \infty \quad \text{as } x \to \infty,$$

so that, by the Multiple Rule, we have $\frac{e^{\frac{x}{2}}}{x} \to \infty$ as $x \to \infty$.

(b) Let $f(x) = \frac{1}{x}$, $x \in \mathbb{R} - \{0\}$, and $g(x) = \frac{e^x}{x}$, $x \in \mathbb{R} - \{0\}$, so that

$$g(f(x)) = \frac{e^{\frac{1}{x}}}{\frac{1}{x}} = xe^{\frac{1}{x}}, \quad \text{for } x \in \mathbb{R} - \{0\}, \text{ and so in}$$

particular for $x \in (0, \infty)$.

Now, $f(x) \to \infty$, as $x \to 0^+$; and, by Example 1, $g(x) \to \infty$ as $x \to \infty$. It follows, by the Composition Rule, that

$$g(f(x)) = xe^{\frac{1}{x}} \to \infty \quad \text{as } x \to 0^+.$$

(c) Let $f(x) = \frac{1}{x}$, $x \in \mathbb{R} - \{0\}$, and $g(x) = \frac{\sin x}{x}$, $x \in \mathbb{R} - \{0\}$, so that

$$g(f(x)) = \frac{\sin\left(\frac{1}{x}\right)}{\frac{1}{x}} = x \sin\left(\frac{1}{x}\right), \quad \text{for } x \in \mathbb{R} - \{0\}, \text{ and so}$$

for $x \in (0, \infty)$.

In Theorem 5, we have $\ell = 0$, $L = 1$.

Now

$$f(x) \to 0 \quad \text{as } x \to \infty,$$
$$g(x) \to 1 \quad \text{as } x \to 0, \quad \text{and}$$
$$f(x) \neq 0, \quad \text{for } x \in (0, \infty).$$

This is condition 1.

This is condition 2.

So the first proviso is satisfied.

It follows, by the Composition Rule, that

$$g(f(x)) = x \sin\left(\frac{1}{x}\right) \to 1 \quad \text{as } x \to \infty. \qquad \square$$

Problem 4 Prove that:
(a) $\log_e(\log_e x) \to \infty$ as $x \to \infty$; (b) $xe^{\frac{1}{x}} \to 0$ as $x \to 0^-$.
Hint for part (b): Use the fact that $\frac{1}{x} \to -\infty$ as $x \to 0^-$.

Problem 5 Give examples of functions f and g (and a specific value for each of ℓ and m) for which
1. $f(x) \to \ell$ as $x \to \infty$ and
2. $g(x) \to m$ as $x \to \ell$,
but for which $g(f(x)) \not\to m$ as $x \to \infty$.

5.3 Limits of functions – using ε and δ

Earlier we gave definitions of the limit of a sequence, continuity of a function and limit of a function, and strategies for using these definitions. In each case, the strategy was in two parts:

GUESS that the definition HOLDS: prove that it holds *for all* cases.

GUESS that the definition FAILS: find ONE counter-example.

Each definition is particularly convenient if we wish to prove that the definition FAILS, but it is not always easy to work with it when we wish to prove that the definition HOLDS.

For example, when we wish to prove that a given function f is *discontinuous* at a point c, then we have to find ONE sequence in a punctured neighbourhood of c such that

$$x_n \to c \quad \text{BUT} \quad f(x_n) \not\to f(c).$$

On the other hand, if we wish to use the definition to prove that f is continuous at c, then we have to show that:

> for each sequence $\{x_n\}$ in a punctured neighbourhood of c such that $x_n \to c$, then $f(x_n) \to f(c)$.

As it is sometimes tricky to determine the behaviour of every such sequence $\{x_n\}$, this definition can be very inconvenient to use.

In this section we introduce a definition of the limit of a function that is equivalent to our earlier definition, but which does not use sequences.

In Section 5.4 we shall introduce a definition of continuity which does not use sequences.

5.3.1   The $\varepsilon - \delta$ definition of limit of a sequence

We motivate our discussion by returning to the definition of a convergent sequence.

Definition   The sequence $\{a_n\}$ is **convergent** with **limit** ℓ, or **converges** to the **limit** ℓ, if $\{a_n - \ell\}$ is a null sequence. In other words, if:

> for each positive number ε, there is a number X such that
> $$|a_n - \ell| < \varepsilon, \quad \text{for all } n > X. \tag{1}$$

Sub-section 2.3.1.

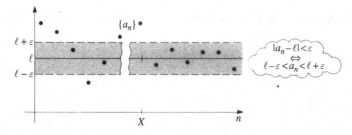

The condition (1) means that, from some point on, the terms of the sequence all lie in the shaded strip between $\ell - \varepsilon$ and $\ell + \varepsilon$, and thus lie 'close to' ℓ. If we choose a smaller number ε, then the shaded strip in the diagram becomes narrower, and we may need to choose a larger number X in order to ensure that the inequality (1) still holds. But, whatever positive number ε we choose, we can always find a number X for which (1) holds.

> **Problem 1**   Let $a_n = \frac{(-1)^n}{n^2}$, $n = 1, 2, \ldots$. How large must we take X in order that:
> (a) $|a_n - 0| < 0.1$,   for all $n > X$?
> (b) $|a_n - 0| < 0.01$,   for all $n > X$?
> (c) $|a_n - 0| < \varepsilon$,   for all $n > X$, where ε is a given positive number?

Earlier we described a 'game', based on the above definition, in which player A chooses a positive number ε and challenges player B to find a number X such that (1) holds. If the sequence in question converges, then such a number X always exists, and player B can always win. If the sequence does *not* converge, then, for SOME choice of ε, NO number X exists such that (1) holds, and so player A can always win.

Sub-section 2.2.1.

For example, consider the null sequence $a_n = \frac{1}{\sqrt{n}}$, $n = 1, 2, \ldots$. In this case, the game might proceed as follows:

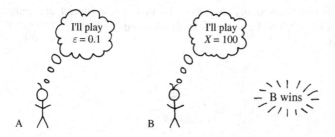

Here player B wins, since, for $n > 100$, we have

Here $X = 100$.

$$\left| \frac{1}{\sqrt{n}} - 0 \right| = \frac{1}{\sqrt{n}} < \frac{1}{\sqrt{100}} = 0.1.$$

Player A tries again:

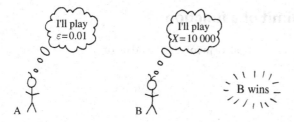

Player B again wins, since, for $n > 10\,000$, we have

Here $X = 10\,000$.

$$\left| \frac{1}{\sqrt{n}} - 0 \right| = \frac{1}{\sqrt{n}} < \frac{1}{\sqrt{10\,000}} = 0.01.$$

But now player B has figured out a winning strategy, and challenges player A to do his worst!

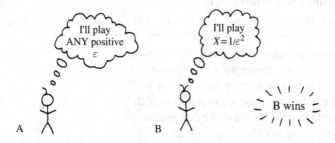

This is indeed a winning strategy, since

Here $X = \frac{1}{\varepsilon^2}$.

$$\text{for } n > \frac{1}{\varepsilon^2}, \text{ we have } \sqrt{n} > \frac{1}{\varepsilon}, \text{ and so} \left| \frac{1}{\sqrt{n}} - 0 \right| = \frac{1}{\sqrt{n}} < \varepsilon.$$

The reason for introducing ε in the definition of convergent sequence is to formalise the idea of 'closeness'. The statement (1) means that we can make the terms a_n of the sequence as close as we please to ℓ by choosing X large enough.

Put another way, we can think of a sequence $\{a_n\}$ as a function with domain $\mathbb{N}$

$$n \mapsto (n, a_n).$$

The number ε formalises the idea of closeness of a_n to ℓ (that is, in the codomain). The condition 'for all $n > X$' restricts the values n in the domain to values for which condition (1) holds.

In general, the smaller the value of ε, the larger the X that we have to choose.

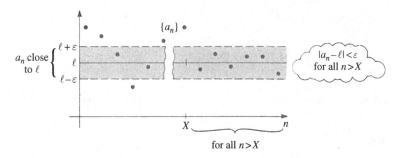

5.3.2   The $\varepsilon - \delta$ definition of limit of a function

The concept of the limit of a function as $x \to c$ also involves the idea of closeness. To define

$$f(x) \to \ell \quad \text{as } x \to c, \tag{2}$$

we require the statement

$$|f(x) - \ell| < \varepsilon;$$

this means that the values of $f(x)$ and ℓ are within a distance ε of each other.

Earlier, we formalised the statement (2) by defining the limit of a function in terms of limits of sequences.

Sub-section 5.1.1.

> **Definition**   Let the function f be defined on a punctured neighbourhood N of a point c. Then $f(x)$ **tends to the limit ℓ as x tends to c** if:
>
>   for each sequence $\{x_n\}$ in N such that $x_n \to c$, then $f(x_n) \to \ell$.   (3)

Note that $x_n \neq c$, for any n.

Intuitively, this means that, as x_n gets close to c, so $f(x_n)$ gets close to ℓ. More precisely, we can make $f(x_n)$ as close as we please to ℓ by choosing x_n sufficiently close to c; we can ensure this by considering only x_ns for sufficiently large n. But the sequence $\{x_n\}$ can be *any* sequence of points in N that converges to c, so what we are really saying is that we can make $f(x)$ as close as we please to ℓ by choosing x sufficiently close to c (but not equal to c).

Thus we now need to specify not only *closeness in the codomain*

$$|f(x) - \ell| < \varepsilon,$$

but also *closeness in the domain*. To do this, we introduce *another* small positive number δ (which depends on ε) and specify closeness in the domain by a statement of the form

$$0 < |x - c| < \delta.$$

Actually it is the requirement $|x - c| < \delta$ that specifies the closeness in the domain. The requirement $0 < |x - c|$ is made simply to ensure that $x \neq c$.

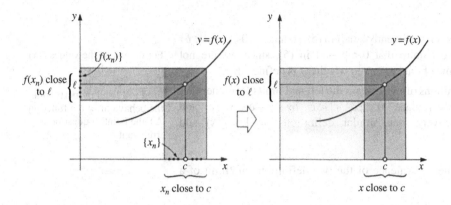

x_n close to c x close to c

Problem 2   Let $f(x) = 2x + 3$, $x \in \mathbb{R}$. How small must we choose δ in order that:

(a) $|f(x) - 5| < 0.1$,   for $0 < |x - 1| < \delta$?

(b) $|f(x) - 5| < 0.01$,   for $0 < |x - 1| < \delta$?

(c) $|f(x) - 5| < \varepsilon$,   for $0 < |x - 1| < \delta$, where ε is a given positive number?

In general, the statement '$f(x) \to \ell$ as $x \to c$' means the following:

for EACH positive number ε, there is a CORRESPONDING positive number δ such that

$$|f(x) - \ell| < \varepsilon, \quad \text{for all } x \text{ satisfying } 0 < |x - c| < \delta. \quad (4)$$

The following diagram illustrates how the numbers ε and δ are used to formalise the idea of closeness.

The same number δ may serve for various different values of ε; but, in general, the value of δ depends on the value of ε.

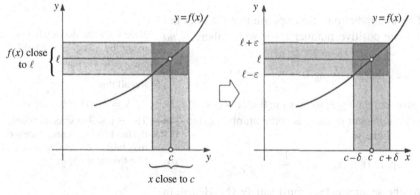

x close to c

Condition (4) means that, for all x with $0 < |x - c| < \delta$, the points $(x, f(x))$ of the graph lie in the heavily shaded rectangle, and thus lie 'close to' (c, ℓ).

This leads us to the following definition of the limit of a function.

Definition   Let the function f be defined on a punctured neighbourhood N of a point c. Then $f(x)$ **tends to the limit ℓ as x tends to c** if:

for each positive number ε, there is a positive number δ such that

$$|f(x) - \ell| < \varepsilon, \quad \text{for all } x \text{ satisfying} \quad 0 < |x - c| < \delta. \quad (5)$$

Thus

    $f(x)$ is 'close to' ℓ

whenever

    x is 'close to' c.

Remarks

1. We assume that δ is chosen sufficiently small in (5), so that $\{x: 0<|x-c|<\delta\}$ lies within N. We also require that $0<|x-c|$ in (5), since we are not concerned with the value of f at c, or even whether f is defined at c.

For otherwise the values $f(x)$ may be undefined.
The existence and value of a limit is a local property, but the behaviour of the function AT the point in question is irrelevant.

2. There are similar definitions of limits from the left and limits from the right at c, where we simply replace '$0<|x-c|<\delta$' by '$c-\delta<x<c$' and '$c<x<c+\delta$', respectively; and similar definitions of $\lim\limits_{x\to\infty} f(x)$ and $\lim\limits_{x\to-\infty} f(x)$.

We now formally state the equivalence of the two definitions of 'limit of a function'.

Theorem 1   The '$\varepsilon-\delta$ definition' and the 'sequential definition' of the statement $\lim\limits_{x\to c} f(x) = \ell$ are equivalent.

Proof   Let the function f be defined on a punctured neighbourhood N of a point c.

We have to show that:

We shall assume, for convenience, that
$$N = (c-r,c) \cup (c,c+r)$$
$$= \{x: 0<|x-c|<r\}.$$

for each sequence $\{x_n\}$ in N such that $x_n \to c$, then $f(x_n) \to \ell$ (6)

$\Leftrightarrow$ for each positive number ε, there is a positive number δ such that
$$|f(x) - \ell| < \varepsilon, \quad \text{for all } x \text{ satisfying } 0<|x-c|<\delta. \tag{7}$$

First, let us assume that (6) holds. Then, for *each* sequence $\{x_n\}$ in N and *each* positive number ε, there is a number X such that
$$|f(x_n) - \ell| < \varepsilon, \quad \text{for all } n > X. \tag{*}$$

Here we simply restate (6).

We will now use a 'proof by contradiction'. So, suppose that (7) does not hold. Then there must exist *some* positive number ε for which there is *no* positive number δ such that

Here we write down what is meant by '(7) does not hold'.

$$|f(x) - \ell| < \varepsilon, \quad \text{for all } x \text{ satisfying } 0<|x-c|<\delta. \tag{8}$$

In particular, we can always assume that $\delta \leq r$, so that $\{x: 0<|x-c|<\delta\} \subseteq N$.

It follows that, for each positive integer n, there is some number x_n (say) in N such that

Recall that
$$N = \{x: 0<|x-c|<r\}.$$
This is possible since (8) does not hold for any δ, and since we may take as successive choices of δ the values $1, \frac{1}{2}, \frac{1}{3}, \ldots$.

$$|f(x_n) - \ell| \geq \varepsilon, \quad \text{where } 0<|x_n-c|<\frac{1}{n}. \tag{9}$$

But, from our assumption (6), the sequence $\{x_n\}$ must satisfy (*). Hence, in addition, in particular, for the positive number ε that we have chosen for (*), there is a number X such that $|f(x_n) - \ell| < \varepsilon$, for all $n > X$. But this is inconsistent with (9), so that we have a contradiction. It follows, therefore, that (7) must hold after all!

For all $n > X, f(x_n)$ now has to satisfy the two inconsistent inequalities
$$|f(x_n) - \ell| \geq \varepsilon \text{ and}$$
$$|f(x_n) - \ell| < \varepsilon.$$

Next, let us assume that (7) holds. That is, for *each* positive number ε, there is a positive number δ such that
$$|f(x) - \ell| < \varepsilon, \quad \text{for all } x \text{ satisfying } 0<|x-c|<\delta. \tag{10}$$

Here we simply restate (7).

Now let $\{x_n\}$ be any sequence in N such that $x_n \to c$. In particular, for the positive number δ that we have chosen for (10), there is a number X such that $(0 <) |x_n - c| < \delta$ for all $n > X$. It follows from (10) that $|f(x_n) - \ell| < \varepsilon$ for all $n > X$.

We have $0 < |x_n - c|$ since x_n lies in a *punctured* neighbourhood of c.

Since this holds for *each* positive number ε, we deduce that $f(x_n) \to \ell$, so that (6) holds, as required. ∎

Remark

Depending on the particular circumstances, we can therefore use whichever definition of limit is the more convenient for our purposes.

We can think of the choice of δ in terms *of* ε in the definition of limit of a function as a game with players A and B, just as we did with the choice of X in the definition of convergent sequence.

Thus, player A chooses some value for ε, and then player B has to try to find a value of δ such that

$$|f(x) - \ell| < \varepsilon, \quad \text{for all } x \text{ satisfying } 0 < |x - c| < \delta.$$

For example, consider how we might show that $f(x) = x^2 \to 0$ as $x \to 0$. Here, for the function $f(x) = x^2, x \in \mathbb{R}$, and $c = 0$, player B has to find a number δ such that

$$|f(x) - 0| = \left|4x^2\right| < \varepsilon, \quad \text{for all } x \text{ satisfying } 0 < |x| < \delta.$$

The game might proceed as follows:

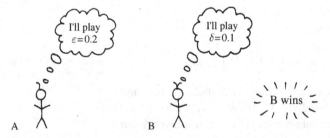

Here player B wins, since

for $0 < |x| < 0.1$, we have that $\left|4x^2\right| \leq 0.04 < 0.2$.

So player A tries again:

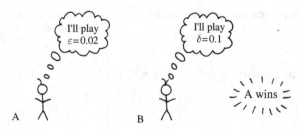

Here player A wins, since

$x = 0.09$ satisfies $0 < |x| < 0.1$ BUT $\left|4 \times 0.09^2\right| = 0.0324 \not< 0.02$.

Player B lost because he did not choose δ sufficiently small; so he tries again.

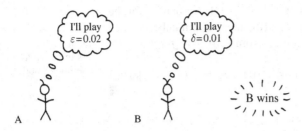

This time player B wins, since

for $0 < |x| < 0.01$, we have that $\left|4x^2\right| \le 0.0004 < 0.02$.

But now player B has figured out a winning strategy! He challenges player A to do his worst!

This is indeed a winning strategy, since

$$\text{for } 0 < |x| < \frac{1}{2}\sqrt{\varepsilon}, \text{ we have that } \left|4x^2\right| < 4 \times \left(\frac{1}{2}\sqrt{\varepsilon}\right)^2 = \varepsilon.$$

Whatever value of ε is chosen by player A, player B need only specify $\delta = \frac{1}{2}\sqrt{\varepsilon}$.

In general, if $f(x) \to \ell$ as $x \to c$, then such a number δ always exists, and player B can always win. On the other hand, if $f(x) \nrightarrow \ell$ as $x \to c$, then for SOME choices of ε no corresponding positive number δ exists, and so player A can always win.

Our strategy for using the '$\varepsilon - \delta$ definition' of $\lim\limits_{x \to c} f(x)$ in particular cases is as follows.

Strategy for using the '$\varepsilon - \delta$ definition' of limit of a function

1. To show that $f(x) \to \ell$ as $x \to c$, find an expression for δ in terms of ε such that:

    for each positive number ε
    $$|f(x) - \ell| < \varepsilon, \quad \text{for all } x \text{ satisfying } 0 < |x - c| < \delta. \text{ (11)}$$

2. To show that $f(x) \nrightarrow \ell$ as $x \to c$, find:

    ONE positive number ε for which there is NO positive number δ such that
    $$|f(x) - \ell| < \varepsilon, \quad \text{for all } x \text{ satisfying } 0 < |x - c| < \delta.$$

Of course the choice of δ may depend on c too.

Example 1   Using the strategy, prove that

(a) $\lim\limits_{x\to1}(2x+3)=5$;   (b) $\lim\limits_{x\to0}(2x^3)=0$;

(c) $\lim\limits_{x\to0}\left(x\sin\frac{1}{x}\right)=0$;   (d) $\lim\limits_{x\to0}\left(\sin\frac{1}{x}\right)$ does not exist.

Solution

(a) The function $f(x)=2x+3$ is defined on every punctured neighbourhood of 1.
We must prove that:

for each positive number ε, there is a positive number δ such that
$$|(2x+3)-5|<\varepsilon,\quad\text{for all }x\text{ satisfying }0<|x-1|<\delta;$$

that is

$$|2(x-1)|<\varepsilon,\quad\text{for all }x\text{ satisfying }0<|x-1|<\delta.\qquad(12)$$

We choose $\delta=\frac{1}{2}\varepsilon$; then the statement (12) holds, since

for $0<|x-1|<\frac{1}{2}\varepsilon$, we have $|2(x-1)|<\varepsilon$.

Hence, $\lim\limits_{x\to1}(2x+3)=5$.

> We could equally well choose δ to be any positive number less than $\frac{1}{2}\varepsilon$, since the statement (12) would still hold.

(b) The function $f(x)=2x^3$ is defined on every punctured neighbourhood of 0.
We must prove that:

for each positive number ε, there is a positive number δ such that
$$\left|2x^3-0\right|<\varepsilon,\quad\text{for all }x\text{ satisfying }0<|x-0|<\delta;$$

that is

$$\left|2x^3\right|<\varepsilon,\quad\text{for all }x\text{ satisfying }0<|x|<\delta.\qquad(13)$$

We choose $\delta=\frac{1}{2}\sqrt[3]{\varepsilon}$; then the statement (13) holds, since:

for $0<|x|<\frac{1}{2}\sqrt[3]{\varepsilon}$, we have that

$$\left|2x^3\right|=2|x|^3<2\times\left(\frac{1}{2}\sqrt[3]{\varepsilon}\right)^3=\frac{1}{4}\varepsilon<\varepsilon.$$

> Notice that many other choices of δ would also have served our purpose.

Hence, $\lim\limits_{x\to0}(2x^3)=0$.

(c) The function $f(x)=x\sin\frac{1}{x}$ is not defined at 0, but it is defined on every punctured neighbourhood of 0.
We must prove that:

for each positive number ε, there is a positive number δ such that
$$\left|x\sin\frac{1}{x}-0\right|<\varepsilon,\quad\text{for all }x\text{ satisfying }0<|x-0|<\delta;$$

that is

$$\left|x\sin\frac{1}{x}\right|<\varepsilon,\quad\text{for all }x\text{ satisfying }0<|x|<\delta.\qquad(14)$$

Now, $\left|\sin\frac{1}{x}\right|\le1$, for all non-zero x, so that

$$\left|x\sin\frac{1}{x}\right| \le |x|, \quad \text{for all non-zero } x.$$

We therefore choose $\delta = \varepsilon$; then the statement (14) holds.

Hence, $\lim\limits_{x\to 0}\left(x\sin\frac{1}{x}\right) = 0$.

Of course any choice for δ smaller than this particular value will also serve our purposes – such as $\frac{1}{7}\varepsilon$ or $\frac{1}{625}\varepsilon$.

(d) Suppose that the limit exists; call it ℓ. Take $\varepsilon = \frac{1}{2}$, say. Then there is some positive number δ, and points x_n and y_n of the form

$$x_n = \frac{1}{(2n+\frac{1}{2})\pi} \quad \text{and} \quad y_n = \frac{1}{(2n+\frac{3}{2})\pi}, \quad \text{where } n \in \mathbb{N},$$

such that

$$0 < |x_n| < \delta \quad \text{and} \quad |f(x_n) - \ell| < \frac{1}{2}, \quad \text{and}$$

$$0 < |y_n| < \delta \quad \text{and} \quad |f(y_n) - \ell| < \frac{1}{2}.$$

Hence, for sufficiently large n

$$|f(x_n) - f(y_n)| = |(f(x_n) - \ell) - (f(y_n) - \ell)|$$
$$\le |f(x_n) - \ell| + |f(y_n) - \ell| \quad < \frac{1}{2} + \frac{1}{2} = 1.$$

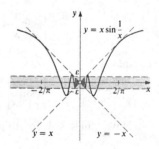

But $f(x_n) - f(y_n) = 1 - (-1) = 2$, which contradicts the previous inequality. It follows that no such numbers ℓ and δ exist, so that in fact $\lim\limits_{x\to 0}\left(\sin\frac{1}{x}\right)$ does not exist. $\square$

Remarks

1. In Example 1, parts (a) and (b), we can find an appropriate choice of δ by 'working backwards'.

For example, in part (b) we have

$$|2x^3| < \varepsilon \Leftrightarrow |x^3| < \frac{1}{2}\varepsilon$$
$$\Leftrightarrow |x| < \frac{1}{2^{\frac{1}{3}}}\sqrt[3]{\varepsilon},$$

so we can choose $\delta = \frac{1}{2^{\frac{1}{3}}}\sqrt[3]{\varepsilon}$. The choice that we made in Example 1 was to take $\delta = \frac{1}{2}\sqrt[3]{\varepsilon}$, which was a somewhat smaller (but equally valid) value for δ. Both possibilities are valid, since we only require to show that

$$0 < |x - 0| < \delta \Rightarrow |2x^3 - 0| < \varepsilon,$$

rather than

$$0 < |x - 0| < \delta \Leftrightarrow |2x^3 - 0| < \varepsilon.$$

On the other hand, in part (c) it is not possible to 'solve for δ' in this way – that is, to solve the inequality $\left|x\sin\frac{1}{x}\right| < \varepsilon$ to obtain a solution such that

$$\left|x\sin\frac{1}{x}\right| < \varepsilon, \quad \text{for } 0 < |x| < \delta.$$

For, if ε is sufficiently small, the solution set of the inequality $\left|x\sin\frac{1}{x}\right| < \varepsilon$ is not even an interval but a union of disjoint intervals.

2. In Example 1, part (d), the point is that, whatever the value of δ, there are always points in any punctured neighbourhood of 0 where f takes the values 1 and -1. Thus there cannot exist numbers ℓ and δ such that

$$|f(x) - \ell| < \frac{1}{2}, \text{ say, } \text{ for all } x \text{ satisfying } 0 < |x| < \delta.$$

Problem 3   Using the Strategy (except in part (d)), prove that:
(a) $\lim_{x \to 3}(5x - 2) = 13$;   (b) $\lim_{x \to 0}(1 - 7x^3) = 1$;
(c) $\lim_{x \to 0}\left(x^2 \cos\frac{1}{x^3}\right) = 0$;   (d) $\lim_{x \to 0}\frac{x + |x|}{x}$ does not exist.

Sometimes it is a little tricky to find an expression for δ in terms of ε such that

$$|f(x) - \ell| < \varepsilon, \text{ for all } x \text{ satisfying } 0 < |x - c| < \delta.$$

The following example illustrates one approach that is sometimes useful.

Example 2   Use the $\varepsilon - \delta$ definition to prove that $\lim_{x \to 2} x^2 = 4$.

Solution   According to the definition, we must show that for x 'close to 2' the following difference is small

$$\left|x^2 - 4\right| = |x + 2| \times |x - 2|.$$

For x close to 2, the term $|x + 2|$ is approximately 4, so that the product $|x + 2| \times |x - 2|$ is approximately equal to $4|x - 2|$. Certainly, if we restrict x to the punctured neighbourhood $(1, 2) \cup (2, 3)$ in which $|x + 2| < 5$, we know that

$$\left|x^2 - 4\right| < 5|x - 2|.$$

So, we proceed as follows.

The function $f(x) = x^2$ is defined on every punctured neighbourhood of 2. We must prove that

The discussion so far has been our motivation.

The formal proof now starts!

for each positive number ε, there is a positive number δ such that

$$\left|x^2 - 4\right| < \varepsilon, \text{ for all } x \text{ satisfying } 0 < |x - 2| < \delta;$$

that is

$$|x + 2| \times |x - 2| < \varepsilon, \text{ for all } x \text{ satisfying } 0 < |x - 2| < \delta.$$

We choose $\delta = \min\{1, \frac{1}{5}\varepsilon\}$, so that for $0 < |x - 2| < \delta$ we have

$$|x + 2| < 5 \text{ and } |x - 2| < \frac{1}{5}\varepsilon.$$

It follows that

$$|x + 2| \times |x - 2| < 5 \times \frac{1}{5}\varepsilon = \varepsilon, \text{ for all } x \text{ satisfying } 0 < |x - 2| < \delta.$$

Hence $\lim_{x \to 2} x^2 = 4$. $\square$

Problem 4   Use the $\varepsilon - \delta$ definition to prove that $\lim_{x \to 1} x^3 = 1$.
Hint: Try $\delta = \min\{1, \frac{1}{7}\varepsilon\}$.

Proofs

Since the definition of the limit of a function introduced in this section is equivalent to the earlier sequential definition, the rules for combining limits can also be proved using the $\varepsilon - \delta$ approach. To give you a flavour of what this involves, we prove here just one of the Combination Rules.

Example 3   Prove the Sum Rule:

$$\text{If } \lim_{x \to c} f(x) = \ell \text{ and } \lim_{x \to c} g(x) = m, \text{ then } \lim_{x \to c}(f(x) + g(x)) = \ell + m.$$

Solution   Since the limits for f and g exist, the functions f and g must be defined on some punctured neighbourhoods $(c - r_1, c) \cup (c, c + r_1)$ and $(c - r_2, c) \cup (c, c + r_2)$ of c. It follows that the function $f + g$ is certainly defined on the punctured neighbourhood $(c - r, c) \cup (c, c + r)$, where $r = \min\{r_1, r_2\}$.

Of course f and g may be defined on some larger sets too!

We want to prove that:

for each positive number ε, there is a positive number δ such that

$$|(f(x) + g(x)) - (\ell + m)| < \varepsilon, \quad \text{for all } x \text{ satisfying } 0 < |x - c| < \delta. \quad (15)$$

We know that, since $\lim_{x \to c} f(x) = \ell$, there is a positive number δ_1 such that

$$|f(x) - \ell| < \frac{1}{2}\varepsilon, \quad \text{for all } x \text{ satisfying } 0 < |x - c| < \delta_1; \quad (16)$$

similarly, since $\lim_{x \to c} g(x) = m$, there is a positive number δ_2 such that

$$|g(x) - m| < \frac{1}{2}\varepsilon, \quad \text{for all } x \text{ satisfying } 0 < |x - c| < \delta_2. \quad (17)$$

We now choose $\delta = \min\{\delta_1, \delta_2\}$. Then both statements (16) and (17) hold for all x satisfying $0 < |x - c| < \delta$, so that

$$\begin{aligned}
|(f(x) + g(x)) - (\ell + m)| &= |(f(x) - \ell) + (g(x) - m)| \\
&\leq |f(x) - \ell| + |g(x) - m| \\
&< \frac{1}{2}\varepsilon + \frac{1}{2}\varepsilon = \varepsilon,
\end{aligned}$$

so that the statement (15) holds.

Hence $\lim_{x \to c}(f(x) + g(x)) = \ell + m.$ $\square$

We now introduce an analogue of a useful tool for writing out proofs that you met earlier in your work on sequences, the so-called 'Kε Lemma'.

Sub-section 2.2.2.

In order to prove that a function f has a limit as $x \to c$, using the definition of limit, you need to prove that:

for each positive number ε, there is a positive number δ such that

$$|f(x) - \ell| < \varepsilon, \quad \text{for all } x \text{ satisfying } 0 < |x - c| < \delta.$$

It can be tedious in the process of the proof to make a complicated choice for δ in order to end up with a final inequality that says that some expression is '$<\varepsilon$'. While this means that we end up with an inequality that shows at once that the desired result holds, in fact it is not strictly necessary to end up with precisely '$< \varepsilon$'.

> **Lemma**   **The '$K\varepsilon$ Lemma'**   For the function f, suppose that:
>
> for each positive number ε, there is a positive number δ such that
> $$|f(x) - \ell| < K\varepsilon, \quad \text{for all } x \text{ satisfying } 0 < |x - c| < \delta,$$
> where K is a positive real number that does not depend on ε or x.
> Then $f(x) \to \ell$ as $x \to c$.

Loosely speaking, we may express this result as '$K\varepsilon$ is just as good as ε' in the definition of *limit*.

For example, K might be 2 or $\frac{\pi}{7}$ or 259, but it could not be $2x$ or $\frac{1}{\varepsilon}$.

We omit a proof of the Lemma, as it is essentially the same as the previous proof of the $K\varepsilon$ Lemma for sequences. There are analogues of the $K\varepsilon$ Lemma for the definition of continuity (in terms of ε and δ), differentiability and integrability, but we shall not always mention them explicitly every time an analogue could be stated.

From time to time we shall use this Lemma in order to avoid arithmetic complexity in proofs.

Problem 5   Use the $\varepsilon - \delta$ definition to prove the Product Rule:
$$\text{If } \lim_{x \to c} f(x) = \ell \text{ and } \lim_{x \to c} g(x) = m, \text{ then } \lim_{x \to c} f(x)g(x) = \ell m.$$

You may omit Problem 5 if you are short of time.

Problem 6   Use the $\varepsilon - \delta$ definition to prove that, if $\lim_{x \to c} f(x) = \ell$ and $\ell \neq 0$, then there exists some punctured neighbourhood of c on which $f(x)$ has the same sign as ℓ.

Hint: Take $\varepsilon = \frac{1}{2}|\ell|$ in the definition of limit.

5.4   Continuity – using ε and δ

In this section we introduce a definition of continuity of a function that is equivalent to our earlier definition, but which does not use sequences.

5.4.1   The ε–δ definition of continuity

Recall our earlier definitions of continuity. The first version was for continuity of a function at an interior point of an interval.

Sub-section 4.1.1.

> **Definition**   A function f defined on an interval I that contains c as an interior point is **continuous at c** if
>
> for each sequence $\{x_n\}$ in I such that $x_n \to c$, then $f(x_n) \to f(c)$.

We then saw how to extend this definition to include continuity of a function at an end-point of an interval.

Sub-section 4.1.1.

> **Definition**   A function f defined on a set S in $\mathbb{R}$ that contains a point c is **continuous at c** if:
>
> for each sequence $\{x_n\}$ in S such that $x_n \to c$, then $f(x_n) \to f(c)$.

Here c may be an interior point of an interval, or it may be an end-point of an interval; both possibilities are covered by the phrase 'each sequence $\{x_n\}$ in S'.

We start by reformulating the first version of this definition in terms of ε and δ. So, if we replace ℓ by $f(c)$ in the definition of limit of a function, we immediately obtain the following definition of continuity of a function at an interior point of an interval.

Sub-section 5.3.2.

> **Definition**   A function f defined on an interval I that contains c as an interior point is **continuous at c** if:
>
> for each positive number ε, there is a positive number δ such that
> $$|f(x) - f(c)| < \varepsilon, \quad \text{for all } x \text{ satisfying } |x - c| < \delta. \qquad (1)$$

Note that the inequality $|f(x) - f(c)| < \varepsilon$ specifies *closeness in the codomain*, whereas the inequality $|x - c| < \delta$ specifies *closeness in the domain*.

Remarks

1. Notice that in (1) we write $|x - c| < \delta$ rather than $0 < |x - c| < \delta$, since in fact (1) is always satisfied when $x = c$.

2. We may assume that δ has been chosen sufficiently small so that $(c - \delta, c + \delta)$ lies in I.

3. It is sufficient in (1) to have
$$|f(x) - f(c)| < K\varepsilon, \quad \text{for all } x \text{ satisfying } |x - c| < \delta,$$
   where K is some positive constant. This fact is often referred to as the '$K\varepsilon$ Lemma' for continuity.

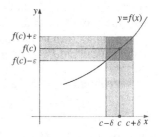

4. The condition (1) means that, for all x 'sufficiently close to' c, the values of $f(x)$ lie 'close to' $f(c)$. If we choose a smaller number ε, then we may need to choose a smaller number δ in order to ensure that (1) still holds. But, whatever positive number ε we choose, we can always find a number δ for which (1) holds.

5. We can describe the definition of continuity in terms of an '$\varepsilon - \delta$ game' in much the same way as we described the definition of limit. WHATEVER choice of ε player A makes, then player B can ALWAYS choose a value of δ such that the required condition (1) holds.

6. For a given value of the positive number ε, we *may* need to choose different values of δ for different points c.

We shall return to this point in Section 5.5.

7. It follows immediately from our definition of continuity in terms of limits that, if a function f is defined on an open interval containing c as an interior point, then
$$f \text{ is continuous at } c \Leftrightarrow \lim_{x \to c} f(x) = f(c).$$

8. We can define one-sided continuity in a similar fashion. Thus a function f is **continuous on the left at c** if:

   for each positive number ε, there is a positive number δ such that
   $$|f(x) - f(c)| < \varepsilon, \quad \text{for all } x \text{ satisfying } c - \delta < x < c;$$

   and is **continuous on the right at c** if:

   for each positive number ε, there is a positive number δ such that
   $$|f(x) - f(c)| < \varepsilon, \quad \text{for all } x \text{ satisfying } c < x < c + \delta.$$

   It follows from these definitions and the previous remark that, if a function f is defined on an open interval containing c as an interior point, then

   $$f \text{ is continuous at } c \Leftrightarrow f \text{ is continuous on the left and on the right at } c.$$

We shall not spend much time looking at one-sided continuity; however there is a wide range of results similar to the results for 'ordinary' continuity.

We can use the definitions in the last Remark to extend the definition of continuity of a function f from points of an open interval to all points of a general interval, in a natural way. We say that f is *continuous on an interval I* if f is continuous at all interior points of I, continuous on the right at the left endpoint of I if it belongs to I, and continuous on the left at the right end-point of I if it belongs to I.

This leads us to extend our initial definition of continuity, as follows.

Definition   A function f defined on an interval I that contains a point c is **continuous at c** if:

for each positive number ε, there is a positive number δ such that
$$|f(x) - f(c)| < \varepsilon, \quad \text{for all } x \text{ in } I \text{ satisfying } |x - c| < \delta.$$

f is said to be **continuous on I** if it is continuous at each point c of I.

Since the '$\varepsilon - \delta$ definition' and the 'sequential definition' of limits are equivalent, and the two definitions of continuity are simply reformulations of those definitions, it is obvious that the two definitions of continuity must also be equivalent. Depending on the particular circumstances, we can therefore use whichever definition of continuity is the more convenient for our purposes.

This was Theorem 1 in Subsection 5.3.2.

Theorem 1   The '$\varepsilon - \delta$ definition' and the 'sequential definition' of the statement 'f is continuous at c' are equivalent.

Our strategy for using the '$\varepsilon - \delta$ definition' of continuity in particular cases is also similar to the strategy for using the '$\varepsilon - \delta$ definition' of limits.

Strategy for using the '$\varepsilon - \delta$ definition' of continuity

1. To show that f is continuous at an interior point c of an interval in I, find an expression for δ in terms of ε such that:

    for each positive number ε
    $$|f(x) - f(c)| < \varepsilon, \quad \text{for all } x \text{ in } I \text{ satisfying } |x - c| < \delta. \quad (2)$$

2. To show that f is discontinuous at an interior point c of an interval in I, find ONE positive number ε for which there is NO positive number δ such that
    $$|f(x) - f(c)| < \varepsilon, \quad \text{for all } x \text{ in } I \text{ satisfying } |x - c| < \delta. \quad (3)$$

Example 1   Use the Strategy to prove that $f(x) = x^2$ is continuous at 2.

Solution   The function f is defined on $\mathbb{R}$.

We must prove that:

for each positive number ε, there is a positive number δ such that
$$|x^2 - 4| < \varepsilon, \quad \text{for all } x \text{ satisfying } |x - 2| < \delta;$$

that is

$$|x + 2||x - 2| < \varepsilon, \quad \text{for all } x \text{ satisfying } |x - 2| < \delta.$$

Note that, for x near to 2, $x + 2$ is near to 4 and $x - 2$ near to 0.

We choose $\delta = \min\{1, \frac{1}{5}\varepsilon\}$; then, for $|x - 2| < \delta$, we have

$$|x + 2| < 5 \quad \text{and} \quad |x - 2| < \frac{1}{5}\varepsilon.$$

Hence

$$|x^2 - 4| < 5 \times \frac{1}{5}\varepsilon = \varepsilon, \quad \text{for all } x \text{ satisfying } |x - 2| < \delta.$$

It follows that f is continuous at 2. □

We could equally well choose δ to be any positive number *smaller* than $\min\{1, \frac{1}{5}\varepsilon\}$, since the rest of the argument would still apply. We choose $\delta < 1$ to get a bound on the term $x + 2$, and $\delta < \frac{1}{5}\varepsilon$ as a small multiple of ε such that the final product is at most ε.

Problem 1 Use the Strategy to prove that $f(x) = x^3$ is continuous at 1.
Hint: Try $\delta = \min\{1, \frac{1}{7}\varepsilon\}$.

Problem 2 Use the Strategy to prove that $f(x) = \sqrt{x}$, $x \geq 0$, is continuous at 4.

Example 2 Use the Strategy to prove that the following function is discontinuous at 0

$$f(x) = \begin{cases} x, & \text{if } x \geq 0, \\ 1, & \text{if } x < 0. \end{cases}$$

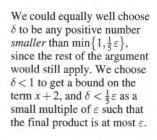

Solution The function f is defined on $\mathbb{R}$.

The graph of f suggests that, while f is 'well behaved' to the right of 0, we should examine the behaviour of f to the left of 0. If we choose ε to be any positive number less than 1, then there will always be points x (with $x < 0$) as close as we please to 0, where $f(x)$ is not within a distance ε of $f(0) = 0$.

So, take $\varepsilon = \frac{1}{2}$, say; and let $x_n = -\frac{1}{n}$, for $n = 1, 2, \ldots$. Then, for *any* positive number δ

$$x_n \left(= -\frac{1}{n} \right) \in (-\delta, 0), \quad \text{for all } n > X, \text{ where } X = \frac{1}{\delta},$$

but

$$\begin{aligned} |f(x_n) - f(0)| &= \left| f\left(-\frac{1}{n}\right) - f(0) \right| \\ &= 1 - 0 \\ &= 1 \not< \frac{1}{2} = \varepsilon. \end{aligned}$$

Thus, with this choice of ε, no value of δ exists such that

$$|f(x) - f(0)| < \varepsilon, \quad \text{for all } x \text{ satisfying } |x - 0| < \delta.$$

This proves that f is discontinuous at 0. □

For example, the points of the sequence $\{x_n\}$, where $x_n = -\frac{1}{n}$, for all $n \geq 1$, since then $f(x_n) = 1$.

We make a suitable choice of ε such that no corresponding δ exists that satisfies the requirement (3).

This was requirement (3).

Problem 3 Use the Strategy to prove that the following function is discontinuous at 2

$$f(x) = \begin{cases} x - \frac{1}{2}, & \text{if } x > 2, \\ 1, & \text{if } x = 2, \\ x - 1, & \text{if } x < 2. \end{cases}$$

Finally, we demonstrate a nice application of the $\varepsilon - \delta$ approach to continuity in order to obtain a result that will be useful later on.

> **Theorem 2** Let f be continuous at an interior point c of an interval I, and $f(c) \neq 0$. Then there exists a neighbourhood $N = (c - r, \ c + r)$ of c on which:
> (a) $f(x)$ has the same sign as $f(c)$, and
> (b) $|f(x)| > \frac{1}{2}|f(c)|$.

Here $r > 0$.

Proof Since $f(c) \neq 0$, we may take $\frac{1}{2}|f(c)|$ as the positive number ε in the definition of continuity at c. It follows that there exists some positive number r such that

$$|f(x) - f(c)| < \frac{1}{2}|f(c)|, \quad \text{for all } x \text{ satisfying } |x - c| < r. \quad (4)$$

We now rewrite (4) in the following equivalent form

$$-\frac{1}{2}|f(c)| < f(x) - f(c) < \frac{1}{2}|f(c)|, \quad \text{for all } x \in (c - r, \ c + r). \quad (5)$$

For convenience, here we use the symbol r rather than δ in the definition of continuity – this makes no real difference.

Suppose, first, that $f(c) > 0$. It follows from the left inequality in (5) that, for $x \in (c - r, \ c + r)$, we have $-\frac{1}{2}f(c) < f(x) - f(c)$ – in other words, that $\frac{1}{2}f(c) < f(x)$. Thus, in this case, both (a) and (b) hold for $x \in (c - r, c + r)$.

Suppose, next, that $f(c) < 0$. It follows from the right inequality in (5) that, for $x \in (c - r, \ c + r)$, we have $f(x) - f(c) < \frac{1}{2}|f(c)| = -\frac{1}{2}f(c)$ – in other words, that $f(x) < \frac{1}{2}f(c)$. Thus, in this case too, both (a) and (b) hold for $x \in (c - r, c + r)$.

This completes the proof of the theorem. ∎

5.4.2 The Dirichlet and Riemann functions

We can use our new $\varepsilon - \delta$ approach to continuity to tackle two very strange functions that were devised in the second half of the nineteenth century. We start with the Dirichlet function.

> **Definition** **Dirichlet's function** is defined on $\mathbb{R}$ by the formula
> $$f(x) = \begin{cases} 1, & \text{if } x \text{ is rational,} \\ 0, & \text{if } x \text{ is irrational.} \end{cases}$$

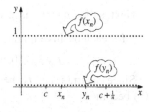

The graph of f looks rather like two parallel lines, but each line has 'infinitely many gaps in it'.

> **Theorem 3** The Dirichlet function is discontinuous at each point of $\mathbb{R}$.

Proof Let c be any point of $\mathbb{R}$. Then, for each $n = 1, 2, \ldots$, by the Density Property of $\mathbb{R}$, the open interval $\left(c, \ c + \frac{1}{n}\right)$ contains a rational number, x_n say, and an irrational number, y_n say. Then $x_n \to c$ and $y_n \to c$ as $n \to \infty$, but

$$f(x_n) = 1 \quad \text{and} \quad f(y_n) = 0.$$

Recall that the *Density Property* of $\mathbb{R}$ asserts that, between any two unequal real numbers, there exists at least one rational and one irrational number – Sub-section 1.1.4.

Thus $f(x)$ does not tend to a limit as x tends to c, whether c is rational or irrational.

It follows that f is discontinuous at each point c of $\mathbb{R}$. ∎

Our next function possesses even stranger behaviour.

Definition   **Riemann's function** is defined on $\mathbb{R}$ by the formula
$$f(x) = \begin{cases} \frac{1}{q}, & \text{if } x \text{ is a rational number } \frac{p}{q}, \text{ in lowest terms, with } q > 0, \\ 0, & \text{if } x \text{ is irrational.} \end{cases}$$

The expression 'in lowest terms' means that $|p|$ and q have no common factor.

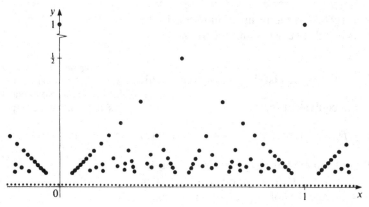

It is not clear from the graph of f whether f is continuous at any point of $\mathbb{R}$.

Theorem 4   Riemann's function is discontinuous at each rational point of $\mathbb{R}$, and continuous at each irrational point of $\mathbb{R}$.

Proof   First, let $c = \frac{p}{q}$ be any rational point of $\mathbb{R}$, where $q > 0$ and $\frac{p}{q}$ is in its lowest terms.

In this part of the proof, we use a sequential approach.

Then, by the Density Property of $\mathbb{R}$, for each $n = 1, 2, \ldots$, the open interval $\left(c - \frac{1}{n}, c + \frac{1}{n}\right)$ contains an irrational number x_n, for which $f(x_n) = 0 \neq \frac{1}{q} = f(c)$.

Note that $x_n \neq c$.

Hence as $n \to \infty$
$$x_n \to c \text{ but } f(x_n) \not\to f(c).$$

Thus f is discontinuous at any rational point c.

Next, suppose that c is any irrational point of $\mathbb{R}$. We must prove that:

In this part of the proof, we use an $\varepsilon - \delta$ approach.

for each positive number ε, there is a positive number δ such that
$$|f(x) - f(c)| < \varepsilon, \quad \text{for all } x \text{ satisfying } |x - c| < \delta. \tag{6}$$

Since c is irrational, $f(c) = 0$; also $f(x) \geq 0$ for all x. It follows that condition (6) becomes
$$f(x) < \varepsilon, \quad \text{for all } x \text{ satisfying } |x - c| < \delta. \tag{7}$$

Our task is therefore to find a value of δ such that (7) holds.

Now, let N be any positive integer such that $N > \frac{1}{\varepsilon}$, and let S_N be the set of rational numbers $\frac{p}{q}$ (in lowest terms) in the interval $(c - 1, c + 1)$ for which $0 < q < N$ – that is

Thus, in particular, $\frac{1}{N} < \varepsilon$. We choose $(c - 1, c + 1)$ simply so that we are *only* looking at points 'near to c'.

$$S_N = \left\{ \frac{p}{q} : \left| c - \frac{p}{q} \right| \leq 1, \, 0 < q < N \right\}.$$

Since there are only a finite number of points in S_N and $c \notin S_N$, we can define a positive number δ as follows

$c \notin S_N$ since c is irrational.

$$\delta = \min\{|x - c| : x \in S_N\}.$$

Thus there are NO rational numbers $\frac{p}{q}$, with $0 < q < N$, that lie in the interval $(c - \delta, c + \delta)$.

It follows that, if $|x - c| < \delta$, then:

EITHER x is irrational, so that $f(x) (= 0) < \varepsilon$;

OR x is rational, so that $x = \frac{p}{q}$ (in lowest terms), with $q \geq N$, and
$f(x)(=\frac{1}{q}) \leq \frac{1}{N} < \varepsilon.$

In either case, $f(x) < \varepsilon$.

Hence we have proved that (7) holds, and so that f is continuous at c. ■

> δ is the distance from c to the nearest point of S_N.

5.4.3 Proofs

Since the definition of continuity introduced in this section is equivalent to the earlier sequential definition, the rules for continuity can also be proved using the $\varepsilon - \delta$ approach. To give you a flavour of what this involves, we first prove one of the Combination Rules.

Example 3 (Sum Rule)   Let f and g be defined on an open interval I containing the point c. Then, if f and g are continuous at c, so is the sum function $f + g$.

Solution   The functions f, g and $f + g$ are certainly defined on some neighbourhood $(c - r, c + r) \subseteq I$ of c, where $r > 0$.

We want to prove that:

> for each positive number ε, there is a positive number δ such that
> $$|(f(x) + g(x)) - (f(c) + g(c))| < \varepsilon, \text{ for all } x \text{ satisfying } |x - c| < \delta. \quad (8)$$

We know that, since f is continuous at c, there is a positive number δ_1 (which we may assume is $\leq r$) such that

$$|f(x) - f(c)| < \frac{1}{2}\varepsilon, \quad \text{for all } x \text{ satisfying } |x - c| < \delta_1; \quad (9)$$

similarly, since g is continuous at c, there is a positive number δ_2 (which we may assume is also $\leq r$) such that

$$|g(x) - g(c)| < \frac{1}{2}\varepsilon, \quad \text{for all } x \text{ satisfying } |x - c| < \delta_2. \quad (10)$$

We now choose $\delta = \min\{\delta_1, \delta_2\}$. Then both statements (9) and (10) hold for all x satisfying $|x - c| < \delta$, so that

$$\begin{aligned} |(f(x) + g(x)) - (f(c) + g(c))| &= |(f(x) - f(c)) + (g(x) - g(c))| \\ &\leq |f(x) - f(c)| + |g(x) - g(c)| \\ &< \frac{1}{2}\varepsilon + \frac{1}{2}\varepsilon = \varepsilon, \end{aligned}$$

so that the statement (8) holds.

Hence $f + g$ is continuous at c. □

> You might like to compare this solution with that of Example 3 in Sub-section 5.3.2.

Problem 4   Prove that if f is defined on an open interval I containing the point c at which it is continuous, and $f(c) \neq 0$, then $\frac{1}{f}$ is defined on some neighbourhood of c and is also continuous at c.

Hint: Use the result of Theorem 2 in Sub-section 5.4.1.

Finally, we give a proof of the Composition Rule for continuity; this is a particularly pleasing example of the $\varepsilon - \delta$ approach.

Example 4   (Composition Rule) Prove that if f is continuous at c and g is continuous at $f(c)$, then $g \circ f$ is continuous at c.

We do not bother to mention appropriate neighbourhoods of c and $f(c)$ on which f and g are defined, respectively, just to simplify the statement of the Rule.

Solution   We must prove that:

for each positive number ε, there is a positive number δ such that

$$|g(f(x)) - g(f(c))| < \varepsilon, \quad \text{for all } x \text{ satisfying } |x - c| < \delta. \quad (11)$$

We know that, since g is continuous at $f(c)$, there is a positive number δ_1 such that:

$$|g(f(x)) - g(f(c))| < \varepsilon, \quad \text{for all } f(x) \text{ satisfying } |f(x) - f(c)| < \delta_1. \quad (12)$$

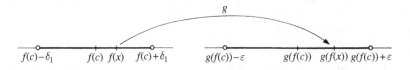

Also, since f is continuous at c, there is a positive number δ such that

$$|f(x) - f(c)| < \delta_1, \quad \text{for all } x \text{ satisfying } |x - c| < \delta. \quad (13)$$

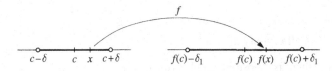

Combining (12) and (13), we deduce that:

for all x satisfying $|x - c| < \delta$, then $|f(x) - f(c)| < \delta_1$,

By (13).

so that

$$|g(f(x)) - g(f(c))| < \varepsilon.$$

By (12).

Hence $g \circ f$ is continuous at c.   □

5.5   Uniform continuity

We start by reminding you of the $\varepsilon - \delta$ definition of continuity at points of an interval I.

Definition   A function f defined on an interval I that contains a point c is **continuous at c** if:

for each positive number ε, there is a positive number δ such that

$$|f(x) - f(c)| < \varepsilon, \quad \text{for all } x \text{ in } I \text{ satisfying } |x - c| < \delta.$$

f is said to be **continuous on I** if it is continuous at each point c of I.

Here the choice of δ depends, in general, on both ε and c.

Earlier we pointed out that, for a given value of the positive number ε, we *may* need to choose different values of δ for different points c. This suggests the following related definition.

> **Definition**   A function f defined on an interval I is **uniformly continuous on I** if:
>
> for each positive number ε, there is a positive number δ such that
> $$|f(x) - f(c)| < \varepsilon, \text{ for all } x \text{ and } c \text{ in } I \text{ satisfying } |x - c| < \delta. \qquad (1)$$

Sub-section 5.4.1, Remark 6.

Here the choice of δ depends ONLY on ε, the same choice whatever c may be.

Since the definition of uniform continuity is more restrictive than that of continuity, it follows that uniform continuity implies continuity.

> **Theorem 1**   If a function f is uniformly continuous on an interval I, then it is continuous on I.

However the converse result is not true in general.

Example 1   Prove that the function $f(x) = \frac{1}{x}$, $0 < x \le 1$, is not uniformly continuous on $(0, 1]$.

Since f is a rational function, it is a basic continuous function, and so is continuous on $(0, 1]$.

Solution   To prove that f is NOT uniformly continuous on $(0,1]$, we need to

find ONE positive number ε for which there is NO positive number δ such that
$$|f(x) - f(c)| < \varepsilon, \quad \text{for all } x \text{ and } c \text{ in } (0, 1] \text{ satisfying} |x - c| < \delta. \qquad (2)$$

We will take as our choice of ε the number 1, and prove that there is NO corresponding positive number δ such that
$$|f(x) - f(c)| < 1, \quad \text{for all } x \text{ and } c \text{ in } (0, 1] \text{ satisfying } |x - c| < \delta. \qquad (3)$$

For the statement (2) asserts that statement (1) is not valid.

Suppose on the contrary that some number δ did exist such that (3) was valid. We will show that this assumption leads us to a contradiction.
  Choose a positive integer N with

This is what we now set out to do.

$$\frac{1}{N} < \delta.$$

Now each point of $(0,1]$ belongs to at least one interval of the form

$$\left(\frac{0}{2N}, \frac{2}{2N}\right), \left(\frac{1}{2N}, \frac{3}{2N}\right), \ldots, \left(\frac{k}{2N}, \frac{k+2}{2N}\right), \ldots, \left(\frac{2N-2}{2N}, \frac{2N}{2N}\right), \left(\frac{2N-1}{2N}, 1\right],$$
$$(4)$$

For, successive intervals in (4) overlap.

and, in addition, all points x and c in any given such interval satisfy the inequality

$$|x - c| < \left(\frac{2}{2N}=\right)\frac{1}{N} < \delta;$$

Hence, by (3)

all points x and c in $(0, 1]$ in any interval in (4) satisfy $|f(x) - f(c)| < 1$.
$$(5)$$

In particular, $f(x) < f(c) + 1$.

Now, from the definition of f, we have

$$f(x) < \frac{2N}{2N-1} = K, \text{ say}, \quad \text{for all } x \text{ in } \left(\frac{2N-1}{2N}, 1\right].$$

For, f is strictly decreasing and $f\left(\frac{2N-1}{2N}\right) = \frac{2N}{2N-1}$.

Then, since the intervals $\left(\frac{2N-1}{2N}, 1\right]$ and $\left(\frac{2N-2}{2N}, \frac{2N}{2N}\right)$ overlap, there is a point c in $\left(\frac{2N-2}{2N}, \frac{2N}{2N}\right)$ for which $f(c) < K$. It follows from (5) that in the second last interval $\left(\frac{2N-2}{2N}, \frac{2N}{2N}\right)$ in (4) we have

$$f(x) < K+1, \quad \text{for all } x \text{ in } \left(\frac{2N-2}{2N}, \frac{2N}{2N}\right).$$

Working backwards through the $2N$ intervals, we can check that

$$f(x) < K + (2N-1), \quad \text{for all } x \text{ in } \left(\frac{0}{2N}, \frac{2}{2N}\right). \qquad \text{For } f \text{ is strictly decreasing.}$$

It follows then, that in fact $f(x) < K + (2N-1)$, for all x in $(0,1]$. In other words, f is bounded in $(0,1]$. But this is not the case, since $f(x) \to \infty$ as $x \to 0^+$.

We have thus reached the desired contradiction. $\square$

> **Problem 1**   Let f be the function $f(x) = x^2$, $x \in [2, 3]$. For any given positive number ε, find a formula for δ in terms of ε such that (1) holds. This proves directly from the definition that f is uniformly continuous on $[2, 3]$.
>
> *Hint:*   Use the fact that $x^2 - c^2 = (x+c)(x-c)$.

Notice, though, that the function in Problem 1 is a continuous function on a closed interval, and recall that we saw earlier that continuous functions on closed intervals have 'nice properties'. It turns out that closed intervals are important too for uniform continuity.

Section 4.2.

Example 1 shows that, in general, f may NOT be uniformly continuous if its interval of continuity is a half-closed interval.

Theorem 2   If a function f is continuous on a closed interval $[a, b]$, then it is uniformly continuous on $[a, b]$.

Proof of Theorem 2   We use the Bolzano–Weierstrass Theorem, and a proof by contradiction.

So, suppose that, in fact, f is not uniformly continuous on $[a, b]$. Then for some choice of ε, there is NO corresponding positive number δ such that

$$|f(x) - f(y)| < \varepsilon, \quad \text{for all } x \text{ and } y \text{ in } [a, b] \text{ satisfying } |x - y| < \delta. \qquad (6)$$

We can reformulate this statement (6) in the following convenient way:

for any positive number δ, there are two points x, y in $[a, b]$ such that

$$|f(x) - f(y)| \geq \varepsilon \quad \text{and} \quad |x - y| < \delta. \qquad (7)$$

By applying (7) for the choices $\delta_n = \frac{1}{n}$, $n = 1, 2, \ldots$, in turn, we can then define two sequences $\{x_n\}$ and $\{y_n\}$ in $[a,b]$ such that

$$|f(x_n) - f(y_n)| \geq \varepsilon \quad \text{and} \quad |x_n - y_n| < \frac{1}{n}, \quad \text{for } n \in \mathbb{N}. \qquad (8)$$

Now, the sequence $\{x_n\}$ is bounded, so that by the Bolzano–Weierstrass Theorem it contains a subsequence $\{x_{n_k}\}$ that converges to some point c of $[a, b]$ as $k \to \infty$. Then, since $|x_{n_k} - y_{n_k}| < \frac{1}{n_k}$, it follows that the corresponding subsequence $\{y_{n_k}\}$ must also converge to c as $k \to \infty$.

Since f is continuous at c, we must have

$$f(x_{n_k}) \to f(c) \quad \text{and} \quad f(y_{n_k}) \to f(c), \text{ as } k \to \infty. \qquad (9)$$

You may omit this proof on a first reading.

ε is now fixed from here on in this proof.

We use y rather than c in (6), since it fits in better with the notation that we will use later in the proof.

The choice of x and y will depend, in general, on the choice of δ.

Sub-section 2.5.1, Theorem 3.

From (8).

Now, it follows from (8) that

$$|f(x_{n_k}) - f(y_{n_k})| \geq \varepsilon, \quad \text{for each } k \in \mathbb{N}. \tag{10}$$

So, if we now let $k \to \infty$ in (10) and use the limits in (9), we deduce that

$$0 \geq \varepsilon,$$

which is absurd.

Here we use the Limit Inequality Rule, Theorem 6 of Sub-section 5.1.3.

This contradiction shows that (7) and so (6) cannot be valid. It follows that f must be uniformly continuous on $[a, b]$ after all. ∎

We shall use Theorem 2 later, to prove that a function continuous on a closed interval $[a, b]$ is integrable on $[a, b]$.

In Theorem 3, Sub-section 7.2.2.

Theorem 2 is one of the *important* theorems in Analysis beyond the scope of this book.

5.6   Exercises

Section 5.1

1. Determine whether the following limits exist:

   (a) $\lim\limits_{x \to 1} \frac{x^3 - 1}{x - 1}$;   (b) $\lim\limits_{x \to 1} \frac{x^3 - 1}{|x - 1|}$;   (c) $\lim\limits_{x \to 0} e^{\sin x}$.

2. Determine the following limits:

   (a) $\lim\limits_{x \to 0} \left(\sin x + \frac{e^x - 1}{x} \right)$;   (b) $\lim\limits_{x \to 0} \frac{e^x - 1}{\sin x}$;

   (c) $\lim\limits_{x \to 0} \frac{e^{|x|} - 1}{|x|}$;   (d) $\lim\limits_{x \to 1^+} \frac{x^3 - 1}{|x - 1|}$.

3. Write out the proof of Theorem 4 (the Composition Rule) in Sub-section 5.1.3.

4. For the functions

$$f(x) = \begin{cases} 0, & x = 0, \\ 1, & x = 1, \\ 2, & x \neq 0, 1, \end{cases} \quad \text{and} \quad g(x) = \begin{cases} 0, & x = 0, \\ 1 + |x|, & x \neq 0, \end{cases}$$

   determine $f(g(0))$, $f\left(\lim\limits_{x \to 0} g(x) \right)$ and $\lim\limits_{x \to 0} f(g(x))$.

Section 5.2

1. Prove that
   (a) $\frac{1}{x^4} \to \infty$ as $x \to 0$;   (b) $\cot x \to \infty$ as $x \to 0^+$;
   (c) $\exp(e^x - x) \to \infty$ as $x \to \infty$;   (d) $\log_e x \to -\infty$ as $x \to 0^+$;
   (e) $x + \sin x \to \infty$ as $x \to \infty$;   (f) $x^x \to \infty$ as $x \to \infty$.

2. Let $p(x) = x^n + a_{n-1}x^{n-1} + \cdots + a_1 x + a_0$, where $a_0, a_1, \ldots, a_{n-1} \in \mathbb{R}$. Prove that $p(x) \to \infty$ as $x \to \infty$.

   Hint: Write $p(x) = x^n \left(1 + r\left(\frac{1}{x} \right) \right)$, for a suitable polynomial r.

3. For $a > 0$ and $n \in \mathbb{Z}$, prove that:
   (a) $e^{ax} x^n \to \infty$ as $x \to \infty$;   (b) $e^{ax} x^n \to 0$ as $x \to -\infty$.

4. For $a < 0$ and $n \in \mathbb{Z}$, prove that:

   (a) $e^{ax}x^n \to 0$ as $x \to \infty$;

   (b) $e^{ax}x^n \to \begin{cases} -\infty, & \text{as } x \to -\infty, & \text{if } n \text{ is odd}, \\ \infty, & \text{as } x \to -\infty, & \text{if } n \text{ is even}. \end{cases}$

Section 5.3

1. Use the strategy based on the $\varepsilon - \delta$ definition of limit to prove the following statements:

   (a) $\lim_{x \to 3}(3x - 4) = 5$;　　　(b) $\lim_{x \to 2^+} \sqrt{x^2 - 4} = 0$.

2. Use the strategy based on the $\varepsilon - \delta$ definition of limit to prove the following statements:

   (a) $\lim_{x \to 0} \frac{\sin x}{|x|}$ does not exist;　　　(b) $\lim_{x \to 0} \cos\left(\frac{1}{x}\right)$ does not exist.

3. Use the strategy based on the $\varepsilon - \delta$ definition of limit to prove that, if $\lim_{x \to c} f(x) = \ell$, where $\ell \neq 0$, then $\lim_{x \to c} \frac{1}{f(x)}$ exists and equals $\frac{1}{\ell}$.

Section 5.4

1. Use the strategy based on the $\varepsilon - \delta$ definition of continuity to prove that:

   (a) $f(x) = x^5$ is continuous at 0;　　　(b) $f(x) = \frac{1}{x}$ is continuous at 2.

2. Use the strategy based on the $\varepsilon - \delta$ definition of continuity to prove that the function

$$f(x) = \begin{cases} \sin\frac{1}{x}, & \text{if } x \neq 0, \\ 0, & \text{if } x = 0, \end{cases}$$

   is discontinuous at 0.

3. Use the $\varepsilon - \delta$ definition of continuity to prove that, if f and g are continuous at c, then so is the product fg.

Section 5.5

1. A function f is defined on a bounded interval I, on which it satisfies the following inequality for some given number $K \in \mathbb{R}$:

$$|f(x) - f(y)| \leq K|x - y|, \quad \text{for all } x, y \text{ in } I.$$

   Prove that f is (a) bounded on I, and (b) uniformly continuous on I.

2. Give an example of a function f that is continuous and bounded on $\mathbb{R}$ but is not uniformly continuous on $\mathbb{R}$, and verify your assertions.

3. Prove that, if a function f is continuous on a half-closed interval $(a, b]$, then it is uniformly continuous on $(a, b]$ if and only if $\lim_{x \to a^+} f(x)$ exists (and is finite).

6 Differentiation

The family of all functions is so large that there is really no possibility of finding many interesting properties that they all possess. In the last two chapters we concentrated our attention on the class of all *continuous* functions, and we found that continuous functions share some important properties – for example, they satisfy the Intermediate Value Theorem, the Extreme Values Theorem and the Boundedness Theorem. However, many of the most interesting and powerful properties of functions are obtained only when we further restrict our attention to the class of all *differentiable* functions.

You will have already met the idea of *differentiating* a given function f; that is, finding the slope of the tangent to the graph $y = f(x)$ at those points of the graph where a tangent exists. The slope of the tangent at the point $(c, f(c))$ is called the *derivative* of f at c, and is written as $f'(c)$. But when does a function have a derivative? Geometrically, the answer is: whenever the slope of the chord through the point $(c, f(c))$ and an arbitrary point $(x, f(x))$ of the graph approaches a limit as $x \to c$. In this chapter we investigate which functions are differentiable, and we discuss some of the important properties that all differentiable functions possess.

In Section 6.1 we give a strategy for determining whether a given function f is differentiable at a given point c. In particular, we prove that, if f is differentiable at c, then it is continuous at c; whereas a function f may be continuous at c, but not differentiable at c. We also consider functions that possess higher derivatives; that is, functions which can be differentiated more than once.

In fact there exist functions that are continuous everywhere on $\mathbb{R}$, but differentiable nowhere on $\mathbb{R}$. This discovery by Karl Weierstrass in 1872 caused a sensation in Analysis.

In Section 6.2 we obtain various standard derivatives and rules for differentiation, including the Inverse Function Rule.

In Sections 6.3 and 6.4 we study the properties of functions that are differentiable *on an interval*, and establish some powerful results about derivatives which are easy to describe geometrically.

In Section 6.5 we meet an important and useful result, called l'Hôpital's Rule, which enables us to find limits of the form

$$\lim_{x \to c} \frac{f(x)}{g(x)},$$

in some of the awkward cases when $f(c) = g(c) = 0$.

In this case, the Quotient Rule for limits of functions fails.

In Section 6.6 we construct the *Blancmange function*, a function that is continuous everywhere on $\mathbb{R}$, but differentiable nowhere on $\mathbb{R}$. It is related to certain types of *fractals*.

6.1 Differentiable functions

6.1.1 What is differentiability?

Differentiability arises from the geometric concept of the *tangent to a graph*. The tangent to the graph $y = f(x)$ at the point $(c, f(c))$ is the straight line through the point $(c, f(c))$ whose direction is the limiting direction of chords joining the points $(c, f(c))$ and $(x, f(x))$ on the graph, as $x \to c$.

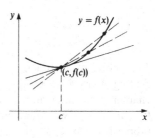

The following three examples illustrate some of the possibilities that can occur when we try to find tangents in particular instances.

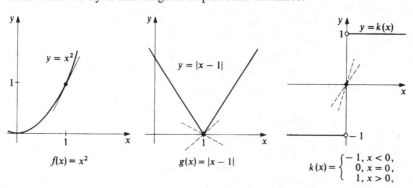

$$f(x) = x^2 \qquad g(x) = |x - 1| \qquad k(x) = \begin{cases} -1, & x < 0, \\ 0, & x = 0, \\ 1, & x > 0, \end{cases}$$

The function $f(x) = x^2$, $x \in \mathbb{R}$, is continuous on $\mathbb{R}$, and its graph has a tangent at each point; for example, the line $y = 2x - 1$ is the tangent to the graph at the point $(1, 1)$.

On the other hand, although the function $g(x) = |x - 1|$, $x \in \mathbb{R}$, is continuous on $\mathbb{R}$, its graph does not have a tangent at the point $(1, 0)$; no line through the point $(1, 0)$ is a tangent to the graph.

Finally, the *signum function*

$$k(x) = \begin{cases} -1, & x < 0, \\ 0, & x = 0, \\ 1, & x > 0, \end{cases}$$

is discontinuous at 0, and no line through the point $(0, 0)$ is a tangent to the graph.

We now make these ideas more precise, using the concept of *limit* to pin down what we meant above by 'limiting direction'. We define the slope of the graph at $(c, f(c))$ to be the limit, as x tends to c, of the slope of the chord through the points $(c, f(c))$ and $(x, f(x))$. The slope of this chord is

$$\frac{f(x) - f(c)}{x - c},$$

and this expression is called the **difference quotient for f at c**. Thus the slope of the graph of f at $(c, f(c))$ is

$$\lim_{x \to c} \frac{f(x) - f(c)}{x - c}, \tag{1}$$

provided that this limit exists.

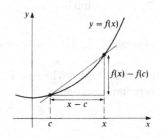

Sometimes it is more convenient to use an equivalent form of the difference quotient, particularly when we are examining a specific function for differentiability at a specific point. If we replace x by $c + h$, then '$x \to c$' in (1) is equivalent to '$h \to 0$'. Thus we can rewrite the difference quotient for f at c as

$$Q(h) = \frac{f(c+h) - f(c)}{h},$$

and the slope of the graph of f at $(c, f(c))$ is then

$$\lim_{h \to 0} Q(h), \tag{2}$$

provided that this limit exists.

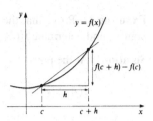

We say that f is *differentiable* at c if the graph $y = f(x)$ has a tangent at the point $(c, f(c))$, and that the *derivative* of f at c is the limit of the difference quotient given by expression (1) or expression (2). To formalise this concept, we need to ensure that f is defined near the point c, and so we assume that c belongs to some open interval I in the domain of f.

Definitions   Let f be defined on an open interval I, and $c \in I$. Then the **derivative of f at c** is

$$\lim_{x \to c} \frac{f(x) - f(c)}{x - c} \quad \text{or} \quad \lim_{h \to 0} Q(h),$$

provided that this limit exists. In this case, we say that f is **differentiable at c**.

The derivative of f at c is denoted by $f'(c)$, and the function $f': x \mapsto f'(x)$ is called the **derived function**. The operation of obtaining $f'(x)$ from $f(x)$ is called **differentiation**.

Note that the difference quotient depends on the choice of c; for difference choices of c, the difference quotient is generally different.

Sometimes f' is denoted by Df and $f'(x)$ by $Df(x)$.

For future reference, we reformulate the definition in terms of ε and δ as follows.

Definition   Let f be defined on an open interval I, and $c \in I$. Then f is **differentiable** at c with **derivative $f'(c)$** if:

for each positive number ε, there is a positive number δ such that
$$\left| \frac{f(x) - f(c)}{x - c} - f'(c) \right| < \varepsilon, \quad \text{for all } x \text{ satisfying } 0 < |x - c| < \delta.$$

This is a simple reformulation, with nothing else taking place. We shall use the definition occasionally in this form, especially when proving general results as distinct from looking at specific functions and specific points.

Remarks

1. In 'Leibniz notation', $f'(x)$ is written as $\frac{dy}{dx}$, where $y = f(x)$. As we shall see later, this notation is sometimes useful; however, it is important to recall that the symbol $\frac{dy}{dx}$ is purely notation and does NOT mean some quantity dy 'divided by' another quantity dx.

For instance, in the statement of the Composition Rule in Sub-section 6.2.2.

2. The existence of the derivative $f'(c)$ is not quite equivalent to the existence of a tangent to the graph $y = f(x)$ at the point $(c, f(c))$. For, if $\lim_{x \to c} \frac{f(x) - f(c)}{x - c}$ exists (and so is some real number), then the graph has a tangent at the point $(c, f(c))$, and the slope of the tangent is the value of this limit. On the other hand, the graph may have a *vertical* tangent at the point $(c, f(c))$. In this case, $\frac{f(x) - f(c)}{x - c} \to \infty$ or $-\infty$ as $x \to c$, so that $\lim_{x \to c} \frac{f(x) - f(c)}{x - c}$ does not exist; thus f is not differentiable at such points c.

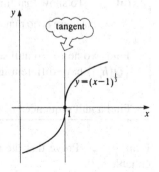

3. Since the concept of a derivative is defined in terms of a limit, we shall use many results obtained for limits of functions to prove analogous results for derivatives.

Example 1 Prove that the function $f(x) = x^3$, $x \in \mathbb{R}$, is differentiable at any point c, and determine $f'(c)$.

Solution At the point c

$$Q(h) = \frac{f(c+h) - f(c)}{h} = \frac{(c+h)^3 - c^3}{h}$$
$$= \frac{3c^2h + 3ch^2 + h^3}{h}$$
$$= 3c^2 + 3ch + h^2.$$

Recall that $h \neq 0$.

Thus, $Q(h) \to 3c^2$ as $h \to 0$. It follows that f is differentiable at c, and that $f'(c) = 3c^2$. □

Thus, the derived function of f is $f'(x) = 3x^2$, $x \in \mathbb{R}$.

For comparison, we now prove the same result using the $\varepsilon - \delta$ definition of differentiability; for simplicity, though, we shall assume that $c = 2$ and we shall simply check that $f'(2) = 12$. We think that you will see why the $Q(h)$ method is often preferred!

Solution We have to show that for each positive number ε, there is a positive number δ such that

$$\left| \frac{x^3 - 8}{x - 2} - 12 \right| < \varepsilon, \quad \text{for all } x \text{ satisfying } 0 < |x - 2| < \delta;$$

that is

$$|x^2 + 2x - 8| < \varepsilon, \quad \text{for all } x \text{ satisfying } 0 < |x - 2| < \delta.$$

Here we use the fact that $x^3 - 8 = (x-2)(x^2 + 2x + 4)$.

Now, $x^2 + 2x - 8 = (x+4)(x-2)$; so, for $0 < |x - 2| < 1$, we have $x \in (1, 3)$ and $|x + 4| < 7$.

Next, choose $\delta = \min\{1, \frac{1}{7}\varepsilon\}$. With this choice of δ, it follows that, for all x satisfying $0 < |x - 2| < \delta$, we have

$$|x^2 + 2x - 8| = |x + 4| \times |x - 2|$$
$$< 7 \times \frac{1}{7}\varepsilon = \varepsilon.$$

It follows that f is differentiable at 2, with derivative $f'(2) = 12$. □

We arrange for $x \in (1, 3)$, in order to concentrate simply on values of x near to 2. The bound 7 for $|x + 4|$ is simply used in order to keep some given bound (not necessarily a small bound) for this term.

It is now clear why we made that particular choice for δ; it arose from trying various values and then adjusting the choice appropriately in order to end up with a neat 'ε'. Alternatively, we could have used a different value for δ in terms of ε, and the $K\varepsilon$ Lemma.

To prove that a function is not differentiable at a point, we use the strategy for limits that applies in these situations.

Strategy To show that $\lim_{h \to 0} Q(h)$ does *not* exist:

1. Show that there is no punctured neighbourhood of c on which Q is defined;

OR

2. Find two non-zero null sequences $\{h_n\}$ and $\{h'_n\}$ such that $\{Q(h_n)\}$ and $\{Q(h'_n)\}$ have different limits;

OR

3. Find a null sequence $\{h_n\}$ such that $Q(h_n) \to \infty$ or $Q(h_n) \to -\infty$.

For convenience, we rephrase this strategy in terms of $\lim_{h \to 0} Q(h)$ rather than $\lim_{x \to c} f(x)$.

Example 2 Prove that the modulus function $f(x) = |x|$, $x \in \mathbb{R}$, is not differentiable at 0.

Solution The graph suggests that the slopes of chords joining the origin to points on the graph of f to the left and to the right do not tend to the same limit as these points approach the origin. This suggests that we investigate whether case 2 of the above strategy may be useful.

This is motivation for the approach to the solution rather than part of the solution itself.

At the point 0

This is the start of the solution.

$$Q(h) = \frac{f(0+h) - f(0)}{h} = \frac{|h| - |0|}{h}$$

$$= \frac{|h|}{h}.$$

Now let $\{h_n\}$ and $\{h'_n\}$ be two null sequences, where $h_n = \frac{1}{n}$ and $h'_n = -\frac{1}{n}$. Then

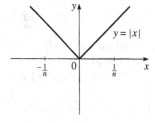

$y = |x|$

$$Q(h_n) = Q\left(\frac{1}{n}\right) = \frac{\left|\frac{1}{n}\right|}{\frac{1}{n}}$$

$$= \frac{\frac{1}{n}}{\frac{1}{n}} \to 1 \quad \text{as } n \to \infty,$$

and

$$Q(h'_n) = Q\left(-\frac{1}{n}\right) = \frac{\left|-\frac{1}{n}\right|}{-\frac{1}{n}}$$

$$= \frac{\frac{1}{n}}{-\frac{1}{n}} \to -1 \quad \text{as } n \to \infty.$$

It follows that f is not differentiable at 0. ☐

Problem 1

(a) Prove that the function $f(x) = x^n$, $x \in \mathbb{R}$, $n \in \mathbb{N}$, is differentiable at any point c, and determine $f'(c)$.

(b) Prove that the constant function on $\mathbb{R}$ is differentiable at any point c, with derivative zero.

In particular, the derivative of the identity function $f(x) = x$, $x \in \mathbb{R}$, is the constant function $f'(x) = 1$.

Problem 2 Prove that the function $f(x) = \frac{1}{x}$, $x \in \mathbb{R} - \{0\}$, is differentiable at any point $c \neq 0$, and determine $f'(c)$.

Problem 3 Use the $\varepsilon - \delta$ definition of differentiability to prove that the function $f(x) = x^4$ is differentiable at 1, with derivative $f'(1) = 4$.

Problem 4 Prove that the following functions f are not differentiable at the given point c:

(a) $f(x) = |x|^{\frac{1}{2}}$, $x \in \mathbb{R}$, $c = 0$; (b) $f(x) = [x]$, $x \in \mathbb{R}$, $c = 1$.

Here $[x]$ denotes the integer part of x.

Looking back at Example 2, it looks as though chords joining the origin to points $(h, f(h))$ have slopes that tend to a limit 1 if $h \to 0^+$, whereas they have slopes that tend to a limit -1 if $h \to 0^-$. This suggests the concept of one-sided derivatives that will be useful in our work later on.

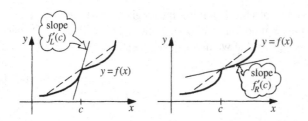

<table>
<tbody>
<tr><td>

Definitions Let f be defined on an interval I, and $c \in I$. Then the **left derivative of f at c** is

$$f'_L(c) = \lim_{x \to c^-} \frac{f(x) - f(c)}{x - c} \quad \text{or} \quad f'_L(c) = \lim_{h \to 0^-} Q(h),$$

provided that this limit exists. In this case, we say that f is **differentiable on the left at c**.

Similarly, the **right derivative of f at c** is

$$f'_R(c) = \lim_{x \to c^+} \frac{f(x) - f(c)}{x - c} \quad \text{or} \quad f'_R(c) = \lim_{h \to 0^+} Q(h),$$

provided that this limit exists. In this case, we say that f is **differentiable on the right at c**.

A function f whose domain contains an interval I is **differentiable on I** if it is differentiable at each interior point of I, differentiable on the right at the left end-point of I (if this belongs to I) and differentiable on the left at the right end-point of I (if this belongs to I).

</td></tr>
</tbody>
</table>

All these definitions are analogous to definitions for continuity that you met in Sub-section 4.1.1.

With this notation, the function f in Example 2 is differentiable on the left at 0 and $f'_L(0) = -1$; it is also differentiable on the right at 0, and $f'_R(0) = 1$.

The connection between the definitions of differentiability and of one-sided differentiability is rather obvious.

<table>
<tbody>
<tr><td>

Theorem 1 A function f whose domain contains an interval I that contains c as an interior point is **differentiable at c** if and only if f is both differentiable on the left at c *and* differentiable on the right at c AND

$$f'_L(c) = f'_R(c).$$

</td></tr>
</tbody>
</table>

We omit a proof of this straight-forward result.

The common value is simply $f'(c)$.

Example 3 Determine whether the function

$$f(x) = \begin{cases} x + x^2, & -1 \leq x < 0, \\ \sin x, & 0 \leq x \leq 2\pi, \end{cases}$$

is differentiable at the points $c = -1, 0$ and 2π, and determine the corresponding derivatives when they exist.

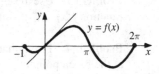

Solution We investigate the behaviour of the difference quotient $Q(h)$ at each point in turn.

At -1, the function is not defined to the left of -1; and, for $0 < h < 1$, we have

$$Q(h) = \frac{f(-1 + h) - f(-1)}{h}$$

$$= \frac{\left\{(-1+h)+(-1+h)^2\right\} - \left\{(-1)+(-1)^2\right\}}{h}$$

$$= \frac{-h+h^2}{h}$$

$$= -1+h \to -1 \quad \text{as } h \to 0^+.$$

It follows that f is differentiable on the right at -1, and $f_R'(-1) = -1$.

At 0, the function is defined on either side of 0, but by a different formula; we therefore examine each side separately.

At 0, for $0 < h < 2\pi$, we have

$$Q(h) = \frac{f(h)-f(0)}{h} = \frac{\sin h - \sin 0}{h}$$

$$= \frac{\sin h}{h} \to 1 \quad \text{as } h \to 0^+.$$

Also at 0, for $-1 < h < 0$, we have

$$Q(h) = \frac{f(h)-f(0)}{h} = \frac{\left\{h+h^2\right\} - \left\{\sin 0\right\}}{h}$$

$$= \frac{h+h^2}{h}$$

$$= 1+h \to 1 \quad \text{as } h \to 0^-.$$

It follows that f is differentiable on the right at 0 and $f_R'(0) = 1$, and that f is differentiable on the left at 0 and $f_L'(0) = 1$. Since the left and right derivatives at 0 are equal, it follows that f is differentiable at 0 and $f'(0) = 1$.

Finally, at 2π, the function is not defined to the right of 2π; and, for $-2\pi < h < 0$, we have

$$Q(h) = \frac{f(2\pi+h)-f(2\pi)}{h} = \frac{\sin(2\pi+h) - \sin 2\pi}{h}$$

$$= \frac{\sin h}{h} \to 1 \quad \text{as } h \to 0^-.$$

It follows that f is differentiable on the left at 2π, and $f_L'(2\pi) = 1$. $\qquad \square$

Problem 5   Determine whether the function

$$f(x) = \begin{cases} -x^2, & -2 \leq x < 0, \\ x^4, & 0 \leq x < 1, \\ x^3, & 1 \leq x \leq 2, \\ 0, & x > 2, \end{cases}$$

is differentiable at the points $c = -2$, 0, 1 and 2, and determine the corresponding derivatives when they exist.

Hint: Sketch the graph $y = f(x)$ first.

Remarks

Just as with continuity, the definition of differentiability involves a function f defined on a set in $\mathbb{R}$, the domain A (say), that f maps to another set in $\mathbb{R}$, the codomain B (say).

These remarks are analogous to similar remarks for continuity in Sub-section 4.1.1.

1. Let f and g be functions defined on open intervals I and J, respectively, where $J \subseteq I$; and let $f(x) = g(x)$ on J. Technically g is a different function from f, if $I \neq J$. However, if f is differentiable at an interior point c of J, it is a simple matter of some definition checking to verify that g too is differentiable at c. Similarly, if f is non-differentiable at c, g too is non-differentiable at c.

2. The underlying point here is that *differentiability at a point is a local property*. It is only the behaviour of the function near that point that determines whether it is differentiable at the point.

6.1.2 Differentiability and continuity

All the functions that we have met so far that are differentiable at a particular point c are also continuous at that point. In fact, this property holds in general.

differentiable $\Rightarrow$ continuous

Theorem 2   Let f be defined on an open interval I, and $c \in I$. If f is differentiable at c, then f is also continuous at c.

There is a similar result for one-sided derivatives.

Proof   If f is differentiable at c, then there is some number $f'(c)$ such that

$$\lim_{x \to c} \frac{f(x) - f(c)}{x - c} = f'(c).$$

It follows that

$$\lim_{x \to c}\{f(x) - f(c)\} = \lim_{x \to c}\left\{\frac{f(x) - f(c)}{x - c} \times (x - c)\right\}$$

$$= \left\{\lim_{x \to c}\frac{f(x) - f(c)}{x - c}\right\} \times \left\{\lim_{x \to c}(x - c)\right\}$$

$$= f'(c) \times 0 = 0.$$

By the Product Rule for Limits, Sub-section 5.1.3.

Hence, by the Sum and Multiple Rules for limits

$$\lim_{x \to c} f(x) = f(c).$$

Sub-section 5.1.3

Thus f is continuous at c.   ∎

In fact this gives us a useful test for non-differentiability.

Corollary 1   If f is discontinuous at c, then f is not differentiable at c.

For example, the *signum function*

$$f(x) = \begin{cases} -1, & x < 0, \\ 0, & x = 0, \\ 1, & x > 0, \end{cases}$$

Problem 5, Sub-section 4.1.1.

is discontinuous at 0, and so cannot be differentiable at 0.

It is often worth checking whether a function is even continuous at a point before setting out on a complicated investigation of difference quotients to determine whether it is differentiable at that point.

Problem 6 Prove that the function

$$f(x) = \begin{cases} \sin\frac{1}{x}, & x \neq 0, \\ 0, & x = 0, \end{cases}$$

is not differentiable at 0.

Example 4 Show that the function

$$f(x) = \begin{cases} x\sin\frac{1}{x}, & x \neq 0, \\ 0, & x = 0, \end{cases}$$

is continuous at 0, but not differentiable at 0.

Solution We proved earlier that f is continuous at 0.
At 0

$$Q(h) = \frac{f(0+h) - f(0)}{h} = \frac{h\sin\left(\frac{1}{h}\right) - 0}{h}$$

$$= \sin\left(\frac{1}{h}\right).$$

Sub-section 4.1.2, Problem 8 (a).

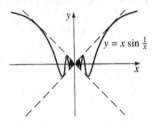

$y = x\sin\frac{1}{x}$

Thus, if we define two null sequences $\{x_n\} = \left\{\frac{1}{n\pi}\right\}$ and $\{x_n'\} = \left\{\frac{1}{(2n+\frac{1}{2})\pi}\right\}$, $n \in \mathbb{N}$, we have

$$Q(x_n) = \sin(n\pi) = 0 \quad \text{and} \quad Q(x_n') = \sin\left(\left(2n + \frac{1}{2}\right)\pi\right) = \sin\left(\frac{1}{2}\pi\right) = 1,$$

so that $\{Q(x_n)\}$ and $\{Q(x_n')\}$ tend to different limits as $n \to \infty$. It follows that f cannot be differentiable at 0. $\square$

Problem 7 Prove that the function

$$f(x) = \begin{cases} x^2\sin\frac{1}{x}, & x \neq 0, \\ 0, & x = 0, \end{cases}$$

is differentiable at 0. Given that, for $x \neq 0$, $f'(x) = 2x\sin\frac{1}{x} - \cos\frac{1}{x}$, is f' continuous at 0?

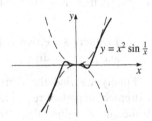

$y = x^2\sin\frac{1}{x}$

6.1.3 The sine, cosine and exponential functions

In order to study the differentiability of each of the sine, cosine and exponential functions, we need to use the following three standard limits.

Theorem 3   Three standard limits

(a) $\lim\limits_{x\to 0}\frac{\sin x}{x} = 1$;   (b) $\lim\limits_{x\to 0}\frac{1-\cos x}{x} = 0$;   (c) $\lim\limits_{x\to 0}\frac{e^x-1}{x} = 1$.

Proof The limits (a) and (c) have been verified earlier, so we have only to verify (b) now.

Using the half-angle formula for cosine, we have

$$\lim_{x\to 0}\frac{1-\cos x}{x} = \lim_{x\to 0}\frac{2\sin^2\left(\frac{1}{2}x\right)}{x}$$

(a) was proved in Sub-section 5.1.1; (c) in Sub-section 5.1.4, Problem 8.

For $\cos x = 1 - 2\sin^2\left(\frac{1}{2}x\right)$.

$$= \lim_{x\to 0}\left(\frac{\sin\left(\frac{1}{2}x\right)}{\frac{1}{2}x}\times\left(\sin\left(\frac{1}{2}x\right)\right)\right)$$

$$= \lim_{x\to 0}\left(\frac{\sin\left(\frac{1}{2}x\right)}{\frac{1}{2}x}\right)\times\lim_{x\to 0}\left(\sin\left(\frac{1}{2}x\right)\right)$$

By the Product Rule for limits.

$$= 1\times 0 = 0.$$

■ Since sine is continuous at 0.

With this tool, we can now tackle the various functions.

Theorem 4 The function $f(x)=\sin x$ is differentiable on $\mathbb{R}$, and $f'(x)=\cos x$.

Proof Let c be any point in $\mathbb{R}$. At c, the difference quotient for f is

$$Q(h)=\frac{\sin(c+h)-\sin c}{h}$$

$$=\frac{\sin c\cos h+\cos c\sin h-\sin c}{h}$$

$$=\cos c\times\frac{\sin h}{h}-\sin c\times\frac{1-\cos h}{h},$$

so that

$$\lim_{h\to 0}Q(h)=\cos c\times\lim_{h\to 0}\frac{\sin h}{h}-\sin c\times\lim_{h\to 0}\frac{1-\cos h}{h}$$

$$=\cos c\times 1-\sin c\times 0$$

$$=\cos c.$$

It follows that f is differentiable at c, and $f'(c)=\cos c$. ■

Problem 8 Prove that the function $f(x)=\cos x$ is differentiable on $\mathbb{R}$, and $f'(x)=-\sin x$.

Finally we find the derivative of the exponential function. This is an extremely important result, as the function $f(x)=\lambda e^x$, λ an arbitrary constant, is the only function f that satisfies the differential equation $f'(x)=f(x)$ on $\mathbb{R}$.

That is, f is its own derived function.

Theorem 5 The function $f(x)=e^x$ is differentiable on $\mathbb{R}$, and $f'(x)=e^x$.

Proof Let c be any point in $\mathbb{R}$. At c, the difference quotient for f is

$$Q(h)=\frac{e^{c+h}-e^c}{h}$$

$$=e^c\times\frac{e^h-1}{h},$$

so that

$$\lim_{h\to 0}Q(h)=e^c\times\lim_{h\to 0}\frac{e^h-1}{h}$$

$$=e^c\times 1$$

$$=e^c.$$

It follows that f is differentiable at c, and $f'(c)=e^c$. ■

6.1.4 Higher-order derivatives

In the previous sub-section, we saw that $\sin' = \cos$ and $\cos' = -\sin$. An exactly similar argument shows that $(-\sin)' = -\cos$ and $(-\cos)' = \sin$. Thus $\pm\sin$ and $\pm\cos$ are all differentiable on $\mathbb{R}$, and so clearly we can differentiate the functions $\sin$ and $\cos$ as many times as we please on $\mathbb{R}$.

In general, when we differentiate any given differentiable function f, we obtain a new function f' (whose domain may be smaller than that of f). The notion of differentiability can then be applied to the function f', just as before, yielding another function $f'' = (f')'$, whose domain consists of those points where f' is differentiable.

Definitions   Let f be differentiable on an open interval I, and $c \in I$. If f' is differentiable at c, then f is called **twice differentiable at c**, and the number $f''(c)$ is called the **second derivative of f at c**. The function f'' is called the **second derived function** of f.

   Provided that the derivatives exist, we can define f''' or $f^{(3)}$, $f^{(4)}$, ..., $f^{(n)}$, ... The functions f'', $f^{(3)}$, $f^{(4)}$, ..., $f^{(n)}$, ... are called the **higher-order derived functions of f**, whose values $f''(x)$, $f^{(3)}(x)$, $f^{(4)}(x)$, ..., $f^{(n)}(x)$, ... are called the **higher-order derivatives of f**.

$f'' = (f')'$ is sometimes written as $f^{(2)}$.

However, not all derived functions are themselves differentiable. You have already seen, for example, that the function

$$f(x) = \begin{cases} x^2 \sin\frac{1}{x}, & x \neq 0, \\ 0, & x = 0, \end{cases}$$

is differentiable at 0. In fact, it has as its derived function

$$f'(x) = \begin{cases} 2x\sin\frac{1}{x} - \cos\frac{1}{x}, & x \neq 0, \\ 0, & x = 0, \end{cases}$$

which is not even continuous at 0.

Example 5   Prove that the function

$$f(x) = \begin{cases} -\frac{1}{2}x^2, & x < 0, \\ \frac{1}{2}x^2, & x \geq 0, \end{cases}$$

is differentiable on $\mathbb{R}$, but that its derived function is not differentiable at 0.

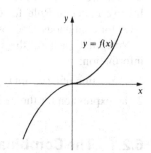

Solution   For $c > 0$, the difference quotient for f at c is

$$Q(h) = \frac{\frac{1}{2}(c+h)^2 - \frac{1}{2}c^2}{h}$$

$$= \frac{1}{2}(2c + h) \to c \quad \text{as } h \to 0,$$

so that f is differentiable at c and $f'(c) = c$.

   When $c = 0$ and $h > 0$, a similar argument shows that f is differentiable on the right at 0 and $f'_R(0) = 0$.

   For $c < 0$, the difference quotient for f at c is

$$Q(h) = \frac{-\frac{1}{2}(c+h)^2 + \frac{1}{2}c^2}{h}$$

$$= -\frac{1}{2}(2c + h) \to -c \quad \text{as } h \to 0,$$

We assume that h is sufficiently small that $|h| < c$, so that $c + h > 0$; and hence that we are using the correct value for f at $c + h$.

We assume that h is sufficiently small that $|h| < -c$, so that $c + h < 0$; and hence that we are using the correct value for f at $c + h$.

so that f is differentiable at c and $f'(c) = -c$.

When $c = 0$ and $h < 0$, a similar argument shows that f is differentiable on the left at 0 and $f'_L(0) = 0$.

Since f has $f'_L(0) = f'_R(0) = 0$, it follows that f is differentiable at 0 and $f(0) = 0$.

Thus the derived function for f is given by

$$f'(x) = |x|, \quad x \in \mathbb{R}.$$

This is the modulus function, which as we showed earlier is not differentiable at 0. It follows that f' is not differentiable at 0. □

Example 2, Sub-section 6.1.1.

Problem 9 Prove that the function

$$f(x) = \begin{cases} x^2, & x < 0, \\ x^3, & x \geq 0, \end{cases}$$

is differentiable on $\mathbb{R}$. Is f' differentiable at 0?

6.2   Rules for differentiation

In the last section we showed that the functions sin, cos and exp are differentiable on $\mathbb{R}$, by appealing directly to the definition of differentiability.

However, it would be very tedious if, every time that we wished to prove that a given function is differentiable and to determine its derived function, we had to use the difference quotient method described in Section 6.1. Sometimes we do need to use that method, but usually we can avoid the algebra involved in the difference quotient method by using various rules for differentiation. In this section, we introduce the Combination Rules, the Composition Rule and the Inverse Function Rule for differentiable functions. These are similar to the rules for continuous functions that you met earlier.

In Sections 4.1 and 4.3.

Note that each of the rules for differentiation supplies two pieces of information:

1. a function of a certain type is differentiable;

2. an expression for the derivative.

6.2.1   The Combination Rules

The Combination Rules for differentiable functions are a consequence of the Combination Rules for limits.

Theorem 1   Combination Rules
Let f and g be defined on an open interval I, and $c \in I$. Then, if f and g are differentiable at c, so are:

Sum Rule	$f + g$,	and $(f+g)'(c) = f'(c) + g'(c);$
Multiple Rule	λf, for $\lambda \in \mathbb{R}$,	and $(\lambda f)'(c) = \lambda f'(c);$
Product Rule	fg,	and $(fg)'(c) = f'(c)g(c) + f(c)g'(c);$

We differentiate each term in turn.

We differentiate one term at a time.

Quotient Rule $\frac{f}{g}$, provided that $g(c) \neq 0$; and

$$\left(\frac{f}{g}\right)'(c) = \frac{g(c)f'(c) - f(c)g'(c)}{(g(c))^2}.$$

In particular,

$$\left(\frac{1}{g}\right)'(c) = -\frac{g'(c)}{(g(c))^2}.$$

Now, we saw in Section 6.1 that, for any $n \in \mathbb{N}$, the function

$$x \mapsto x^n, \quad x \in \mathbb{R},$$

is differentiable on $\mathbb{R}$, and that its derived function is

$$x \mapsto nx^{n-1}, \quad x \in \mathbb{R}.$$

We can use this fact, together with the Combination Rules, to prove that any polynomial function is differentiable on $\mathbb{R}$, and that its derivative can be obtained by differentiating the polynomial term-by-term.

Corollary 1 Let $p(x) = a_0 + a_1 x + a_2 x^2 + \cdots + a_n x^n$, $x \in \mathbb{R}$, where a_0, a_1, a_2, ..., $a_n \in \mathbb{R}$. Then p is differentiable on $\mathbb{R}$, and its derivative is

$$p'(x) = a_1 + 2a_2 x + \cdots + n a_n x^{n-1}, \quad x \in \mathbb{R}.$$

Since a rational function is a quotient of two polynomials, it follows from Corollary 1 and the Quotient Rule that *a rational function is differentiable at all points where the denominator does not vanish (that is, the denominator does not take the value 0).*

Example 1 Prove that the function $f(x) = \frac{x^3}{x^2-1}$, $x \in \mathbb{R} - \{\pm 1\}$, is differentiable on its domain, and find its derivative.

Solution The function f is a rational function whose denominator does not vanish on $\mathbb{R} - \{\pm 1\}$; hence f is differentiable on this set (the whole of its domain).

By Corollary 1, the derived function of $x \mapsto x^3$ is $x \mapsto 3x^2$, and the derived function of $x \mapsto x^2 - 1$ is $x \mapsto 2x$. It follows from the Quotient Rule that the derivative of f is

$$f'(x) = \frac{(x^2 - 1) \times 3x^2 - x^3 \times (2x)}{(x^2 - 1)^2}$$

$$= \frac{x^4 - 3x^2}{(x^2 - 1)^2}. \qquad \square$$

Problem 1 Find the derivative of each of the following functions:

(a) $f(x) = x^7 - 2x^4 + 3x^3 - 5x + 1$, $x \in \mathbb{R}$;

(b) $f(x) = \frac{x^2 + 1}{x^3 - 1}$, $x \in \mathbb{R} - \{1\}$;

(c) $f(x) = 2 \sin x \cos x$, $x \in \mathbb{R}$;

(d) $f(x) = \frac{e^x}{3 + \sin x - 2 \cos x}$, $x \in \mathbb{R}$.

Problem 2 Find the third order derivative of the function $f(x) = xe^{2x}$, $x \in \mathbb{R}$.

In the last section we found the derived functions for sin, cos and exp. We now ask you to find the derived functions for the remaining trigonometric functions and the three most common hyperbolic functions.

Problem 3 Find the derivative of each of the following functions:

(a) $f(x) = \tan x, \quad x \in \mathbb{R} - \left\{ \pm \frac{1}{2}\pi, \pm \frac{3}{2}\pi, \pm \frac{5}{2}\pi, \ldots \right\}$;

(b) $f(x) = \operatorname{cosec} x, \quad x \in \mathbb{R} - \{0, \pm\pi, \pm 2\pi, \ldots\}$;

(c) $f(x) = \sec x, \quad x \in \mathbb{R} - \left\{ \pm \frac{1}{2}\pi, \pm \frac{3}{2}\pi, \pm \frac{5}{2}\pi, \ldots \right\}$;

(d) $f(x) = \cot x, \quad x \in \mathbb{R} - \{0, \pm\pi, \pm 2\pi, \ldots\}$.

Problem 4 Find the derivative of each of the following functions

(a) $f(x) = \sinh x, x \in \mathbb{R}$;   (b) $f(x) = \cosh x, x \in \mathbb{R}$;

(c) $f(x) = \tanh x, x \in \mathbb{R}$.

6.2.2 The Composition Rule

In the last sub-section we extended our stock of differentiable functions to include all rational, trigonometric and hyperbolic functions. However, to differentiate many other functions we need to differentiate composite functions, such as the function $f(x) = \sin(\cos x), x \in \mathbb{R}$, which is the composite of two differentiable functions – namely, $f = \sin \circ \cos$. The Composition Rule tells us that the composite of two differentiable functions is itself differentiable.

Theorem 2   Composition Rule

Let g and f be defined on open intervals I and J, respectively, and let $c \in I$ and $g(I) \subseteq J$. If g is differentiable at c and f is differentiable at $g(c)$, then $f \circ g$ is differentiable at c and

$$(f \circ g)'(c) = f'(g(c))g'(c).$$

This rule is sometimes known as the **Chain Rule**.

We give the proof of the Rule in Sub-section 6.2.4.

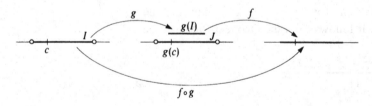

Remarks

1. When written in Leibniz notation, the Composition Rule has a form that is easy to remember: if we put

   $$u = g(x) \quad \text{and} \quad y = f(u) = f(g(x))$$

   then

   $$\frac{dy}{dx} = \frac{dy}{du} \times \frac{du}{dx}.$$

2. We can extend the Composition Rule to a composite of three (or more) functions; for example

   $$(f \circ g \circ h)'(x) = f'[g(h(x))]g'(h(x))h'(x).$$

   In Leibniz notation, if we put

   $$v = h(x), u = g(v) \quad \text{and} \quad y = f(u) = f[g(h(x))],$$

We frequently use this extension of the Composition Rule without mentioning it explicitly.

then we obtain the chain

$$\frac{dy}{dx} = \frac{dy}{du} \times \frac{du}{dv} \times \frac{dv}{dx}.$$

Example 2   Prove that each of the following composite functions is differentiable on its domain, and find its derivative:

(a) $k(x) = \sin(\cos x)$, $x \in \mathbb{R}$;   (b) $k(x) = \cosh(e^{2x})$, $x \in \mathbb{R}$;
(c) $k(x) = \tan\frac{1}{4}\left(e^{x^2} + \sin \pi x\right)$, $x \in (-1, 1)$.

Solution

(a) Here $k(x) = \sin(\cos x)$, so let

$$g(x) = \cos x \quad \text{and} \quad f(x) = \sin x, \quad \text{for } x \in \mathbb{R};$$

then

$$g'(x) = -\sin x \quad \text{and} \quad f'(x) = \cos x, \quad \text{for } x \in \mathbb{R}.$$

By the Composition Rule, $k = f \circ g$ is differentiable on $\mathbb{R}$, and

$$\begin{aligned} k'(x) &= f'(g(x))g'(x) \\ &= \cos(\cos x) \times (-\sin x) \\ &= -\cos(\cos x) \times \sin x. \end{aligned}$$

(b) Here $k(x) = \cosh(e^{2x})$, so let

$$h(x) = 2x, \ g(x) = e^x \quad \text{and} \quad f(x) = \cosh x, \quad \text{for } x \in \mathbb{R};$$

then h, g and f are differentiable on $\mathbb{R}$, and

$$h'(x) = 2, \ g'(x) = e^x \quad \text{and} \quad f'(x) = \sinh x, \quad \text{for } x \in \mathbb{R}.$$

By the (extended form of the) Composition Rule, $k = f \circ g \circ h$ is differentiable on $\mathbb{R}$, and

$$\begin{aligned} k'(x) &= f'[g(h(x))]g'(h(x))h'(x) \\ &= \left(\sinh e^{2x}\right) \times e^{2x} \times 2 \\ &= 2e^{2x} \sinh e^{2x}. \end{aligned}$$

(c) Here $k(x) = \tan\frac{1}{4}\left(e^{x^2} + \sin \pi x\right)$, $x \in (-1, 1)$, so let

$$g(x) = \frac{1}{4}\left(e^{x^2} + \sin \pi x\right), \ x \in (-1, 1), \quad \text{and}$$

$$f(x) = \tan x, \ x \in \left(-\frac{1}{2}\pi, \frac{1}{2}\pi\right).$$

Now, when $x \in (-1, 1)$, we have

$$\begin{aligned} |g(x)| &\le \frac{1}{4}\left(e^{x^2} + |\sin \pi x|\right) \\ &\le \frac{1}{4}(e + 1) \\ &< 1 \\ &< \frac{1}{2}\pi, \end{aligned}$$

so that $g(x)$ lies in $(-\frac{1}{2}\pi, \frac{1}{2}\pi)$, the domain of the differentiable function tan.

Now, by the Composition and Combination Rules, g is differentiable on $(-1, 1)$, and

$$g'(x) = \frac{1}{4}\left(2xe^{x^2} + \pi\cos\pi x\right),\ x \in (-1,1).$$

Also

$$f'(x) = \sec^2 x,\ x \in \left(-\frac{1}{2}\pi, \frac{1}{2}\pi\right).$$

Finally, it follows, by the Composition Rule, that $k = f \circ g$ is differentiable on $(-1, 1)$, and

$$k'(x) = f'(g(x))g'(x)$$
$$= \sec^2\left(\frac{1}{4}\left(e^{x^2} + \sin\pi x\right)\right) \times \frac{1}{4}\left(2xe^{x^2} + \pi\cos\pi x\right),\ x \in (-1,1).\quad \square$$

Problem 5   Find the derivative of each of the following functions:
(a) $f(x) = \sinh(x^2),\ x \in \mathbb{R}$;   (b) $f(x) = \sin(\sinh 2x),\ x \in \mathbb{R}$;
(c) $f(x) = \sin\left(\frac{\cos 2x}{x^2}\right),\ x \in (0, \infty)$.

6.2.3   The Inverse Function Rule

Earlier we showed that, if a function f with domain some interval I and image J is strictly monotonic on I, then f possesses a strictly monotonic and continuous inverse function f^{-1} on J. In particular, the power function, the trigonometric functions, the exponential function and the hyperbolic functions all have inverse functions, provided that we restrict the domains, where necessary.

Section 4.3.
Recall that strictly monotonic means that f is either strictly increasing or strictly decreasing.

These standard functions are all differentiable on their domains. Do their inverse functions also have this property, as their graphs suggest?

As you saw earlier, we obtain the graph $y = f^{-1}(x)$ by reflecting the graph $y = f(x)$ in the line $y = x$. This reflection maps a typical point $P(c, d)$ to the point $Q(d, c)$.

It follows that, if the slope of the tangent to the graph $y = f(x)$ at the point P is $f'(c) = m$, then the slope of the tangent to the graph $y = f^{-1}(x)$ at Q is $(f^{-1})'(d) = \frac{1}{m}$.

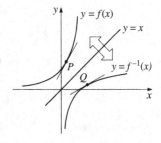

However, if the graph of f has a horizontal tangent ($m = 0$) at a point P, then the graph of f^{-1} has a vertical tangent at the corresponding point Q; in this case, f^{-1} is not differentiable at Q, since $\frac{1}{m}$ is not defined when $m = 0$. We therefore require the condition '$f'(x)$ is non-zero' in our statement of the Rule.

Thus, in the Inverse Function Rule we require:

(a) f is strictly monotonic and continuous on an open interval I, so that f has a strictly monotonic and continuous inverse function on the open interval $J = f(I)$;
(b) f is differentiable on I and $f'(x) \neq 0$ on I, so that $\frac{1}{f'}(x)$ is defined.

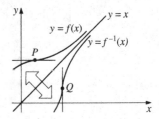

The Rule tells us that, under these conditions, f^{-1} is also differentiable, and there is a simple formula for $(f^{-1})'$.

Theorem 3   Inverse Function Rule

Let $f : I \to J$, where I is an interval and J is the image $f(I)$, be a function such that:

1. f is strictly monotonic on I;
2. f is differentiable on I;
3. $f'(x) \neq 0$ on I.

Then f^{-1} is differentiable on J. Further, if $c \in I$ and $d = f(c)$, then

$$(f^{-1})'(d) = \frac{1}{f'(c)}.$$

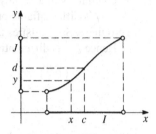

Remark

The Leibniz notation for derivatives can be used to express the Inverse Function Rule in a form that is easy to remember: if

$$y = f(x) \quad \text{and} \quad x = f^{-1}(y),$$

and we write $\frac{dy}{dx}$ for $f'(x)$, and $\frac{dx}{dy}$ for $(f^{-1})'(y)$, then

$$\frac{dx}{dy} = \frac{1}{\frac{dy}{dx}}.$$

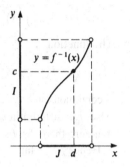

Proof   Let $F = f^{-1}$, and let $y = f(x)$, where $x \in I$, so that $x = F(y)$, where $y \in J$. Then the difference quotient for F at d is

$$\frac{F(y) - F(d)}{y - d} = \frac{x - c}{f(x) - f(c)}$$

$$= \frac{1}{\frac{f(x) - f(c)}{x - c}}.$$

Since f is one–one and differentiable on I, it is necessarily one–one and continuous on I. Thus F is one–one and continuous on J. It follows that, for $y \neq d$, we must have $x \neq c$, since $x = f^{-1}(y)$ and $c = f^{-1}(d)$.

Also, since f^{-1} is continuous, $x \to c$ as $y \to d$. It follows that

$$\lim_{y \to d} \frac{F(y) - F(d)}{y - d} = \lim_{x \to c} \left(\frac{1}{\frac{f(x) - f(c)}{x - c}} \right)$$

$$= \frac{1}{\lim_{x \to c} \left(\frac{f(x) - f(c)}{x - c} \right)}$$

$$= \frac{1}{f'(c)}.$$

Thus F is differentiable at c, with derivative $\frac{1}{f'(c)}$. ∎

Example 3   For each of the following functions, show that f^{-1} is differentiable and determine its derivative:

(a) $f(x) = x^n$, $x > 0$, $n \in \mathbb{N}$;   (b) $f(x) = \tan x$, $x \in \left(-\frac{1}{2}\pi, \frac{1}{2}\pi \right)$;
(c) $f(x) = e^x$, $x \in \mathbb{R}$.

Solution

(a) The function

$$f(x) = x^n, \quad x > 0,$$

is continuous and strictly increasing on $(0, \infty)$, and $f((0, \infty)) = (0, \infty)$. Also, f is differentiable on $(0, \infty)$, and its derivative $f'(x) = nx^{n-1}$ is non-zero there. So f satisfies the conditions of the Inverse Function Rule.

Hence f^{-1} is differentiable on $(0, \infty)$; and, if $y = f(x)$, then

$$(f^{-1})'(y) = \frac{1}{f'(x)} = \frac{1}{nx^{n-1}}$$

$$= \frac{1}{n}x^{1-n} = \frac{1}{n}y^{\frac{1-n}{n}}. \qquad\qquad y = x^n, \text{ so that } x = y^{\frac{1}{n}}.$$

If we now replace the domain variable y by x, we obtain

$$(f^{-1})'(x) = \frac{1}{n}x^{\frac{1}{n}-1}, \quad x \in (0, \infty).$$

(b) The function

$$f(x) = \tan x, \quad x \in \left(-\frac{1}{2}\pi, \frac{1}{2}\pi\right),$$

is continuous and strictly increasing on $\left(-\frac{1}{2}\pi, \frac{1}{2}\pi\right)$, and $f\left(\left(-\frac{1}{2}\pi, \frac{1}{2}\pi\right)\right) = \mathbb{R}$. Also, f is differentiable on $\left(-\frac{1}{2}\pi, \frac{1}{2}\pi\right)$, and its derivative $f'(x) = \sec^2 x$ is non-zero there. So f satisfies the conditions of the Inverse Function Rule.

Hence $f^{-1} = \tan^{-1}$ is differentiable on $\mathbb{R}$; and, if $y = f(x)$, then

$$(f^{-1})'(y) = \frac{1}{f'(x)} = \frac{1}{\sec^2 x}$$

$$= \frac{1}{1 + \tan^2 x} = \frac{1}{1 + y^2}.$$

If we now replace the domain variable y by x, we obtain

$$(\tan^{-1})'(x) = \frac{1}{1 + x^2}, \quad x \in \mathbb{R}.$$

(c) The function

$$f(x) = e^x, \quad x \in \mathbb{R},$$

is continuous and strictly increasing on $\mathbb{R}$, and $f(\mathbb{R}) = (0, \infty)$. Also, f is differentiable on $\mathbb{R}$, and its derivative $f'(x) = e^x$ is non-zero there. So f satisfies the conditions of the Inverse Function Rule.

Hence $f^{-1} = \log_e$ is differentiable on $(0, \infty)$; and, if $y = f(x)$, then

$$(f^{-1})'(y) = \frac{1}{f'(x)} = \frac{1}{e^x}$$

$$= \frac{1}{y}.$$

If we now replace the domain variable y by x, we obtain

$$(\log_e)'(x) = \frac{1}{x}, \quad x \in (0, \infty). \qquad\qquad \square$$

Problem 6 For each of the following functions f, show that f^{-1} is differentiable and determine its derivative:

(a) $f(x) = \cos x, x \in (0, \pi)$;   (b) $f(x) = \sinh x, x \in \mathbb{R}$.

Most equations of the form $y = f(x)$ cannot be solved explicitly to give x as some formula involving y alone. The Inverse Function Rule is often used in such situations to solve problems that would otherwise be intractable. The following problem illustrates this type of application.

Problem 7 Prove that the function $f(x) = x^5 + x - 1$, $x \in \mathbb{R}$, has an inverse function f^{-1} which is differentiable on $\mathbb{R}$. Find the values of $(f^{-1})'(d)$ at those points d corresponding to the points $c = 0$, 1 and -1, where $d = f(c)$.

Exponential functions

Earlier we defined the number a^x, for $a > 0$, by the formula

Section 4.4.

$$a^x = \exp(x \log_e a).$$

Since the functions exp and log are differentiable on $\mathbb{R}$ and $(0, \infty)$, respectively, Section 6.1. it follows that we can use this formula to determine the derivatives of functions such as $x \mapsto x^\alpha, x \mapsto a^x$ and $x \mapsto x^x$.

Example 4 Prove that, for $\alpha \in \mathbb{R}$, the power function $f(x) = x^\alpha$, $x \in (0, \infty)$, is differentiable on its domain, and that $f'(x) = \alpha x^{\alpha-1}$.

Solution By definition

$$f(x) = \exp(\alpha \log_e x).$$

The function $x \mapsto \alpha \log_e x$ is differentiable on $(0, \infty)$, with derivative $\frac{\alpha}{x}$. By the Composition Rule, f is differentiable on $(0, \infty)$ with derivative

$$f'(x) = \exp(\alpha \log_e x) \times \left(\frac{\alpha}{x}\right)$$
$$= x^\alpha \times \left(\frac{\alpha}{x}\right) = \alpha x^{\alpha-1}, \quad x \in (0, \infty). \qquad \square$$

Remark

In the case when α is a rational number, $\alpha = \frac{m}{n}$ say, the result of Example 4 can also be proved by applying the Composition Rule to the functions $x \mapsto x^m$ and $x \mapsto x^{\frac{1}{n}}$. The function $x \mapsto x^{\frac{1}{n}}$ is differentiable on $(0, \infty)$, by the Inverse Function Rule, as it is the inverse of the function $x \mapsto x^n$.

Example 5 Prove that, for $a > 0$, the function $f(x) = a^x$, $x \in \mathbb{R}$, is differentiable on its domain, and that $f'(x) = a^x \log_e a$.

Solution By definition

$$f(x) = \exp(x \log_e a).$$

The function $x \mapsto x \log_e a$ is differentiable on $\mathbb{R}$, with derivative $\log_e a$. By the Composition Rule, f is differentiable on $\mathbb{R}$, with derivative

$$f'(x) = \exp(x \log_e a) \times \log_e a$$
$$= a^x \log_e a, \quad x \in \mathbb{R}. \qquad \square$$

Problem 8 Prove that the function $f(x) = x^x$, $x \in (0, \infty)$, is differentiable on its domain, and find its derivative.

We end this sub-section with a list of basic differentiable functions.

Basic differentiable functions   The following functions are differentiable on their domains:

- polynomials and rational functions
- nth root function
- trigonometric functions (sine, cosine and tangent)
- exponential function
- hyperbolic functions (sinh, cosh and tanh).

6.2.4   Proofs

You may omit these proofs at a first reading.

We now supply the proofs of the Combination Rules and the Composition Rule, which we omitted earlier. We also illustrate how to prove such results using the $\varepsilon - \delta$ method.

Theorem 1   Combination Rules

Let f and g be defined on an open interval I, and $c \in I$. Then, if f and g are differentiable at c, so are:

Sum Rule $f + g$,   and $(f+g)'(c) = f'(c) + g'(c)$;

Multiple Rule λf, for $\lambda \in \mathbb{R}$, and $(\lambda f)'(c) = \lambda f'(c)$;

Product Rule fg,   and $(fg)'(c) = f'(c)\,g(c) + f(c)\,g'(c)$;

Quotient Rule $\frac{f}{g}$,   provided   that g

$(c) \neq 0$; and $\left(\dfrac{f}{g}\right)'(c) = \dfrac{g(c)f'(c) - f(c)g'(c)}{(g(c))^2}$.

Proof

Sum Rule

In this proof we use the Combination Rules for limits given in Sub-section 5.1.2.

Let $F = f + g$. Then

$$\lim_{x \to c} \frac{F(x) - F(c)}{x - c} = \lim_{x \to c} \frac{\{f(x) + g(x)\} - \{f(c) + g(c)\}}{x - c}$$

$$= \lim_{x \to c} \frac{f(x) - f(c)}{x - c} + \lim_{x \to c} \frac{g(x) - g(c)}{x - c}$$

$$= f'(c) + g'(c).$$

Thus F is differentiable at c, with derivative $f'(c) + g'(c)$.

Multiple Rule

This is just a special case of the Product Rule, with $g(x) = \lambda$.

Product Rule

Let $F = fg$. Then

$$\lim_{x \to c} \frac{F(x) - F(c)}{x - c} = \lim_{x \to c} \frac{f(x)g(x) - f(c)g(c)}{x - c}$$

$$= \lim_{x \to c} \frac{\{f(x) - f(c)\}g(x) + f(c)\{g(x) - g(c)\}}{x - c}$$

$$= \lim_{x \to c} \frac{f(x) - f(c)}{x - c} g(x) + \lim_{x \to c} f(c) \frac{g(x) - g(c)}{x - c}$$

$$= \lim_{x \to c} \frac{f(x) - f(c)}{x - c} \times \lim_{x \to c} g(x) + f(c) \times \lim_{x \to c} \frac{g(x) - g(c)}{x - c}$$

$$= f'(c)g(c) + f(c)g'(c),$$

since f and g are differentiable at c, and since g is continuous at c, so that $g(x) \to g(c)$ as $x \to c$.

For, differentiable $\Rightarrow$ continuous.

Thus F is differentiable at c, with derivative $f'(c)g(c) + f(c)g'(c)$.

Quotient Rule

Let $F = \frac{f}{g}$.

Recall, first, that, since g is continuous at c (as it is differentiable there) and $g(c) \neq 0$, there exists an open interval J containing c on which $g(x) \neq 0$. Thus the domain of F contains the open interval J, and $c \in J$.

Then

$$\lim_{x \to c} \frac{F(x) - F(c)}{x - c} = \lim_{x \to c} \frac{\frac{f(x)}{g(x)} - \frac{f(c)}{g(c)}}{x - c}$$

$$= \lim_{x \to c} \frac{f(x)g(c) - f(c)g(x)}{(x - c) \times g(x)g(c)}$$

$$= \lim_{x \to c} \frac{\{f(x) - f(c)\}g(c) - f(c)\{g(x) - g(c)\}}{(x - c) \times g(x)g(c)}$$

$$= \lim_{x \to c} \frac{f(x) - f(c)}{(x - c) \times g(x)} - \lim_{x \to c} \frac{f(c)}{g(c)} \times \frac{g(x) - g(c)}{(x - c) \times g(x)}$$

$$= \lim_{x \to c} \frac{f(x) - f(c)}{x - c} \times \lim_{x \to c} \frac{1}{g(x)} - \frac{f(c)}{g(c)} \times \lim_{x \to c} \frac{g(x) - g(c)}{(x - c)} \times \lim_{x \to c} \frac{1}{g(x)}$$

$$= \frac{f'(c)}{g(c)} - \frac{f(c)g'(c)}{g^2(c)} = \frac{f'(c)g(c) - f(c)g'(c)}{g^2(c)},$$

since f and g are differentiable at c, and g is continuous at c.

Thus F is differentiable at c, with derivative $\frac{f'(c)g(c) - f(c)g'(c)}{g^2(c)}$. ∎

As an illustration of how such results may be proved using the $\varepsilon - \delta$ method that we introduced in Section 5.4, we prove just the Sum Rule. Notice that in our proofs we avoid many complications by judicious use of the '$K\varepsilon$ Lemma'.

The '$K\varepsilon$ Lemma' first appeared in Sub-section 2.2.2.

Proof of the Sum Rule   Let $F = f + g$.

In view of the $K\varepsilon$ Lemma, we must prove that:

As you work through this proof, compare it with the earlier proof using limits.

for each positive number ε, there is a positive number δ such that

$$\left| \frac{F(x) - F(c)}{x - c} - \{f'(c) + g'(c)\} \right| < K\varepsilon \quad \text{for all } x \text{ satisfying}$$

Recall that K must NOT depend on x or ε.

$$0 < |x - c| < \delta. \tag{1}$$

First, we write the expression on the left-hand side of (1) in a convenient form as

$$\frac{\{f(x)+g(x)\}-\{f(c)+g(c)\}}{x-c}-\{f'(c)+g'(c)\}$$

$$=\frac{\{f(x)-f(c)\}+\{g(x)-g(c)\}}{x-c}-\{f'(c)+g'(c)\}$$

$$=\left\{\frac{f(x)-f(c)}{x-c}-f'(c)\right\}+\left\{\frac{g(x)-g(c)}{x-c}-g'(c)\right\}. \qquad (2)$$

We now use the information that we already have to tackle each of the terms in (2) in turn.

Since f is differentiable at c, there exists some positive number δ_1 such that

$$\left|\frac{f(x)-f(c)}{x-c}-f'(c)\right|<\varepsilon, \quad \text{for all } x \text{ satisfying } 0<|x-c|<\delta_1. \qquad (3)$$

Also, since g is differentiable at c, there exists some positive number δ_2 such that

$$\left|\frac{g(x)-g(c)}{x-c}-g'(c)\right|<\varepsilon, \quad \text{for all } x \text{ satisfying } 0<|x-c|<\delta_2. \qquad (4)$$

Now let $\delta=\min\{\delta_1,\delta_2\}$. It follows that, for all x satisfying $0<|x-c|<\delta$, both (3) and (4) hold. Hence, if we apply the Triangle Inequality to (2) and then use both (3) and (4), we find that, for all x satisfying $0<|x-c|<\delta$

$$\left|\frac{\{f(x)+g(x)\}-\{f(c)+g(c)\}}{x-c}-\{f'(c)+g'(c)\}\right|$$

This expression appears in (2).

$$\leq\left|\frac{f(x)-f(c)}{x-c}-f'(c)\right|+\left|\frac{g(x)-g(c)}{x-c}-g'(c)\right|$$

By the Triangle Inequality.

$$<\varepsilon+\varepsilon=2\varepsilon.$$

By (3) and (4).

But this is just the statement (1) that we set out to prove, with $K=2$. ∎

Remark

Note that our use of the '$K\varepsilon$ Lemma' meant that we did not need to know at the steps (3) and (4) to use expressions like $\frac{1}{2}\varepsilon$ rather than ε in order to end up with a final conclusion that 'some expression is $<\varepsilon$'! For 'some expression is $<K\varepsilon$' is sufficient.

We would only discover the need for such a choice once we had made a first attempt at the proof.

We end with the proof of the Composition Rule.

Theorem 2   Composition Rule
Let g and f be defined on open intervals I and J, respectively, and let $c\in I$ and $g(I)\subseteq J$. If g is differentiable at c and f is differentiable at $g(c)$, then $f\circ g$ is differentiable at c and

$$(f\circ g)'(c)=f'(g(c))g'(c).$$

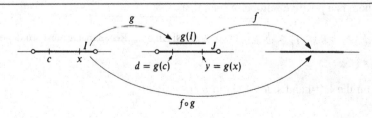

Proof   Let $F = f \circ g$, and let $y = g(x)$ and $d = g(c)$.

The difference quotient for F at c is

$$\frac{F(x) - F(c)}{x - c} = \frac{f(g(x)) - f(g(c))}{x - c}. \qquad (5)$$

Now, the right-hand side of (5) is equal to

$$\frac{f(y) - f(d)}{y - d} \times \frac{g(x) - g(c)}{x - c}, \qquad (6)$$

provided that $y \neq d$.

Unfortunately, for some choices of g and c it *is* possible for $y = d$ to occur when $y = g(x)$ and x is arbitrarily close to c but not actually equal to c. In such situations the expression (6) will not exist.

To get round this problem, we introduce a carefully chosen auxiliary function

Some texts ignore this possibility, and so give incomplete proofs of Theorem 2! In the Remark below, we describe one such situation.

$$h(y) = \begin{cases} \frac{f(y) - f(d)}{y - d}, & y \neq d, \\ f'(d), & y = d. \end{cases}$$

Since f is differentiable at d, $h(y) \to f'(d)$ as $y \to d$; and, since $h(d) = f'(d)$, it follows that h is continuous at d.

We then apply the Composition Rule for continuous functions, to deduce that the composite function

$$h \circ g(x) = \begin{cases} \frac{f(g(x)) - f(g(c))}{g(x) - g(c)}, & g(x) \neq g(c), \\ f'(g(c)), & g(x) = g(c), \end{cases}$$

is continuous at c.

Next, notice that, if $g(x) \neq g(c)$, it follows from (5) and (6) that

$$\frac{F(x) - F(c)}{x - c} = h \circ g(x) \times \frac{g(x) - g(c)}{x - c}; \qquad (7)$$

and also that this last statement (7) is also valid when $g(x) = g(c)$, since then both sides of (7) are zero.

Hence, if we let $x \to c$ in (7) and use the continuity of the function $h \circ g$, we obtain

$$\lim_{x \to c} \frac{F(x) - F(c)}{x - c} = h \circ g(c) \times g'(c)$$
$$= f'(g(c))g'(c).$$

Thus F is differentiable at c, with derivative $f'(g(c))g'(c)$.   ∎

Remark

Let

$$g(x) = \begin{cases} x^2 \sin\left(\frac{1}{x}\right), & x \neq 0, \\ 0, & x = 0, \end{cases} \quad c = 0, \quad \text{and} \quad d = g(0) = 0.$$

Since $\sin(n\pi) = 0$ for all $n \in \mathbb{N}$, it follows that $y = g(x)$ takes the value $d = 0$ at the points $\frac{1}{n\pi}$ that are arbitrarily close to $c = 0$ but not equal to 0.

This example shows that the paragraph 'Unfortunately ...' after equation (6) above does describe a situation that can occur.

6.3   Rolle's Theorem

In the next two sections we describe some of the fundamental properties of functions that are differentiable, not just *at a particular point* but *on an interval*. Our results are motivated by the geometric significance of differentiability in terms of tangents, and explain why the graphs of *differentiable* functions possess certain geometric properties.

6.3.1   The Local Extremum Theorem

Earlier we described some of the fundamental properties of functions which are continuous on a closed interval. In particular, we proved the Extreme Values Theorem, which states that, if a function f is continuous on a closed interval $[a, b]$, then there are points c, d in $[a, b]$ such that

$$f(c) \leq f(x) \leq f(d), \quad \text{for all } x \in [a,b];$$

we called $f(c)$ the **minimum** of f on $[a, b]$, and $f(d)$ the **maximum** of f on $[a, b]$.

We used the term **extremum** to denote either a maximum or a minimum.

But how do we determine the points c, d where these extrema occur? In general, unfortunately, this is not easy. However, if the function f is *differentiable*, then we can, in principle, determine c and d by first finding any *local maxima* or *local minima* of the function f on the interval $[a, b]$.

Roughly speaking, for a point c in $[a, b]$, the value $f(c)$ is a *local maximum* of f on $[a, b]$ if

$$f(x) \leq f(c), \quad \text{for all } x \text{ in } [a, b] \text{ sufficiently close to } c,$$

and a *local minimum* of f on $[a, b]$ if

$$f(x) \geq f(c), \quad \text{for all } x \text{ in } [a, b] \text{ sufficiently close to } c.$$

To make this idea more precise, we use the concept of *neighbourhood* that you met earlier.

Section 4.2.

Sub-section 4.2.3.

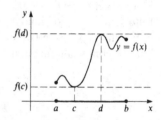

From this point onwards, we shall commonly use the letter c to denote an *extremum* – that is EITHER a maximum OR a minimum.

Sub-section 5.1.1.

Definitions   Let f be defined on an interval I, and let $c \in I$. Then:

- f has a **local maximum** $f(c)$ at c if there exists a neighbourhood N of c such that

$$f(x) \leq f(c), \quad \text{for } x \in N \cap I;$$

- f has a **local minimum** $f(c)$ at c if there exists a neighbourhood N of c such that

$$f(x) \geq f(c), \quad \text{for } x \in N \cap I;$$

- f has a **local extremum** at c if $f(c)$ is either a local maximum or a local minimum.

Notice that $N \cap I$ necessarily includes an interval of the form $(c - r, c]$ or $[c, c + s)$, where $r, s > 0$.

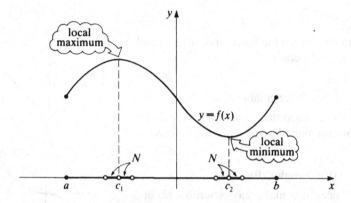

When we wish to locate local extrema of a differentiable function f, instead of using the above definition we usually use the following result, which gives a connection between local extrema of a function f and the points where f' vanishes.

Theorem 1   Local Extremum Theorem

Let f be defined on an interval $[a, b]$. If f has a local extremum at c, where $a < c < b$, and if f is differentiable at c, then
$$f'(c) = 0.$$

Proof   Suppose that f has a local maximum at c.

Since $a < c < b$, it follows from the definition of local maximum that there exists a neighbourhood N of c with $N \subset [a, b]$ such that
$$f(x) \leq f(c), \quad \text{for } x \in N.$$
We now choose numbers $r, s > 0$ such that $N = (c - r, c + s)$.

First, looking to the left of c, we have
$$f(x) - f(c) \leq 0 \quad \text{and} \quad x - c < 0, \quad \text{for } c - r < x < c,$$
so that
$$\frac{f(x) - f(c)}{x - c} \geq 0, \quad \text{for } c - r < x < c.$$
Thus
$$f_L'(c) = \lim_{x \to c^-} \frac{f(x) - f(c)}{x - c} \geq 0. \tag{1}$$
Next, looking to the right of c, we have
$$f(x) - f(c) \leq 0 \quad \text{and} \quad x - c > 0, \quad \text{for } c < x < c + s,$$
so that
$$\frac{f(x) - f(c)}{x - c} \leq 0, \quad \text{for } c < x < c + s.$$
Thus
$$f_R'(c) = \lim_{x \to c^+} \frac{f(x) - f(c)}{x - c} \leq 0. \tag{2}$$

Since f is differentiable at c, the left and right derivatives at c exist and are equal. Hence, in view of inequalities (1) and (2), their common value, $f'(c)$, must be 0. ∎

We prove only the local maximum version: the proof of the local minimum version is similar.

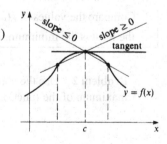

Remarks

1. The Local Extremum Theorem applies *only* if the function is differentiable at a local extremum. For example, the function
$$f(x) = |x|, \quad x \in [-1, 1],$$
has a local minimum 0 at 0, but f is not differentiable at 0.

2. The Local Extremum Theorem does *not* assert that a point where the derivative vanishes is necessarily a local extremum. For example, the function
$$f(x) = x^3, \quad x \in [-1, 1],$$
does not have a local extremum at 0, although $f'(0) = 0$.

3. The Local Extremum Theorem does *not* make any assertion about a local extremum that occurs at a point c that is one of the end-points of the interval $[a, b]$.

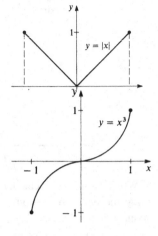

> **Problem 1** Find the local extrema of the function
> $$f(x) = \frac{1}{4}x^4 - \frac{1}{3}x^3, \quad x \in [-1, 2].$$

Clearly any extremum of f on $[a, b]$ that occurs at a point other than a or b must be a local extremum. It follows from Theorem 1 that such a point c must be a point where $f'(c) = 0$. Thus an immediate consequence of the Local Extremum Theorem is the following criterion for finding all the extrema of 'well-behaved functions' on closed intervals.

Corollary 1 Let f be continuous on the closed interval $[a, b]$ and differentiable on the open interval (a, b). Then the extrema of f on $[a, b]$ can occur only at a, at b, or at points c in (a, b) where $f'(c) = 0$.

We now reformulate Corollary 1 as a strategy for locating minima and maxima.

Strategy To determine the maximum and the minimum of a function f which is continuous on $[a, b]$ and differentiable on (a, b):

1. Determine the points $c_1, c_2, \ldots$ in (a, b) where f' vanishes;
2. Compare the values of $f(a), f(b), f(c_1), f(c_2), \ldots$;

   the least is the minimum, and the greatest is the maximum.

> **Problem 2** Use the above Strategy to determine the minimum and the maximum of the function $f(x) = \sin^2 x + \cos x$, for $x \in \left[0, \frac{1}{2}\pi\right]$.

6.3.2 Rolle's Theorem

In the previous sub-section we saw that, if a function f is continuous on the closed interval $[a, b]$ and differentiable on the open interval (a, b), then the extrema of f can occur only at a, at b, or at points c in (a, b) where $f'(c) = 0$.

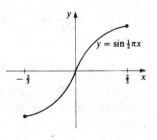

Now the function

$$f(x) = \sin\left(\frac{1}{2}\pi x\right), \quad x \in \left[-\frac{2}{3}, \frac{2}{3}\right],$$

shows that it can happen that no interior point of (a, b) corresponds to a maximum or a minimum of f, and that f' need not vanish at *any* interior point at all.

That is, the extrema may not occur at interior points of (a, b).

However, the situation is quite different for the function

$$f(x) = \sin\left(\frac{1}{2}\pi x\right), \quad x \in [-2, 2].$$

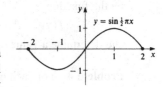

Here $f(-2) = f(2) = 0$; on $[-2, 2]$ the function f has a maximum at 1 and a minimum at -1, and at both these interior points f' vanishes. This is a special case of *Rolle's Theorem* which asserts that, if $f(a) = f(b)$, then there is at least one point c *strictly* between a and b at which f' vanishes.

Theorem 2   Rolle's Theorem

Let f be defined on the closed interval $[a, b]$ and differentiable on the open interval (a, b). If $f(a) = f(b)$, then there exists some point c, with $a < c < b$, for which $f'(c) = 0$.

This is an *existence* theorem. Often it is difficult to evaluate c explicitly.

Remarks

1. This apparently simple result is one of the most important results in Analysis.

2. In geometric terms, Rolle's Theorem means that, if the line joining the points $(a, f(a))$ and $(b, f(b))$ on the graph of f is horizontal, then so is the tangent to the graph at some point c in (a, b).

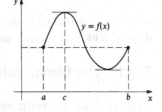

3. There may be more than one point c in (a, b) at which f' vanishes (as in the diagram in the margin). Rolle's Theorem simply asserts that *at least one* such point c exists.

Proof If f is constant on $[a, b]$, then $f'(x) = 0$ everywhere in (a, b); in this case, we may take c to be any point of (a, b).

If f is non-constant on $[a, b]$, then either the maximum or the minimum (or both) of f on $[a, b]$ is different from the common value $f(a) = f(b)$. Since one of the extrema occurs at some point c with $a < c < b$, the Local Extremum Theorem applied to the point c shows that $f'(c)$ must be zero. ∎

Since f is continuous on $[a, b]$, f must have both a maximum and a minimum on $[a, b]$, by the Extreme Values Theorem.

Example 1 Verify that the conditions of Rolle's Theorem are satisfied by the function

$$f(x) = 3x^4 - 2x^3 - 2x^2 + 2x, \quad x \in [-1, 1],$$

and determine a value of c in $(-1, 1)$ for which $f'(c) = 0$.

Solution Since f is a polynomial function, f is continuous on $[-1, 1]$ and differentiable on $(-1, 1)$. Also, $f(-1) = f(1) = 1$. Thus f satisfies the conditions of Rolle's Theorem on $[-1, 1]$.

It follows that there exists a number $c \in (-1, 1)$ for which $f'(c) = 0$. Now

$$f'(x) = 12x^3 - 6x^2 - 4x + 2$$
$$= 12\left(x^2 - \frac{1}{3}\right)\left(x - \frac{1}{2}\right),$$

so that f' vanishes at the points $\frac{1}{\sqrt{3}}$, $-\frac{1}{\sqrt{3}}$ and $\frac{1}{2}$ in $(-1, 1)$. Any of these three numbers will serve for c.   □

Problem 3 Verify that the conditions of Rolle's Theorem are satisfied by the function
$$f(x) = x^4 - 4x^3 + 3x^2 + 2, \quad x \in [1,3],$$
and determine a value of c in $(1,3)$ for which $f'(c) = 0$.

Problem 4 For each of the following functions, state whether Rolle's Theorem applies for the given interval:

(a) $f(x) = \tan x, \quad x \in [0, \pi]$;

(b) $f(x) = x + 3|x - 1|, \quad x \in [0, 2]$;

(c) $f(x) = x - 9x^{17} + 8x^{18}, \quad x \in [0, 1]$;

(d) $f(x) = \sin x + \tan^{-1} x, \quad x \in \left[0, \frac{1}{2}\pi\right]$.

6.4 The Mean Value Theorem

Here we continue to study the geometric properties of functions that are differentiable on intervals, and describe some of their applications.

6.4.1 The Mean Value Theorem

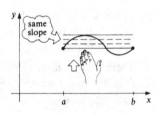

First, recall the geometric interpretation of Rolle's Theorem: Under suitable conditions, if the chord joining the points $(a, f(a))$ and $(b, f(b))$ of the graph of f is horizontal, then so is the tangent at some point c of (a, b).

If you imagine pushing the chord (as shown in the margin), always parallel to itself, until it is just about to lose contact with the graph of f, it looks as though at this point the chord becomes a tangent to the graph. Similarly, the 'chord-pushing' approach suggests that, even if the original chord is not horizontal (that is, if $f(a) \neq f(b)$), there must still be some point c of (a, b) at which the tangent is parallel to the chord.

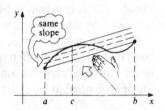

Example 1 Consider the function
$$f(x) = 3 - 3x + x^3, \quad x \in [1, 2].$$

Find a point c of $(1, 2)$ such that the tangent to the graph of f is parallel to the chord joining $(1, f(1))$ to $(2, f(2))$.

Solution Since $f(1) = 3 - 3 + 1 = 1$ and $f(2) = 3 - 6 + 8 = 5$, the slope of the chord joining the endpoints of the graph is

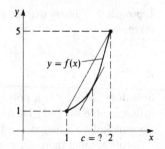

$$\frac{f(2) - f(1)}{2 - 1} = \frac{5 - 1}{2 - 1} = 4.$$

Now, since f is a polynomial, it is differentiable on $(1, 2)$ and its derivative is $f'(x) = -3 + 3x^2$; hence $f'(c) = 4$ when $3c^2 = 7$, or $c = \sqrt{\frac{7}{3}} \simeq 1.53$. Thus, at the point $(c, f(c))$ the tangent to the curve is parallel to the chord joining the end-points. □

We now generalise Rolle's Theorem and assert that there is always a point where the tangent to the graph is parallel to the chord joining the end-points. This result is known as the *Mean Value Theorem*, so-called since

$$\frac{f(b) - f(a)}{b - a}$$

can be thought of as the *mean value* of the derivative between a and b.

Theorem 1   Mean Value Theorem

Let f be continuous on the closed interval $[a, b]$ and differentiable on the open interval (a, b). Then there exists a point c in (a, b) such that

$$f'(c) = \frac{f(b) - f(a)}{b - a}.$$

Again, this is an *existence* theorem.

Note that when $f(a) = f(b)$, the Mean Value Theorem simply reduces to Rolle's Theorem.

The idea of the proof is as follows. We define $h(x)$ to be the vertical distance from the chord to the curve; then $h(a)$ and $h(b)$ are both 0; in fact, h satisfies all the conditions of Rolle's Theorem. Applying Rolle's Theorem to h, we obtain the desired result.

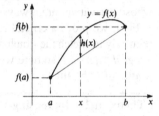

Proof   The slope of the chord joining the points $(a, f(a))$ and $(b, f(b))$ is

$$m = \frac{f(b) - f(a)}{b - a},$$

and so the equation of the chord is

$$y = m(x - a) + f(a).$$

It follows that the vertical height, $h(x)$, between points with ordinate x on the graph and those on the chord is given by

$$h(x) = f(x) - [m(x - a) + f(a)].$$

Now $h(a) = h(b) = 0$, and h is continuous on $[a, b]$ and differentiable on (a, b). Thus h satisfies all the conditions of Rolle's Theorem.

It follows from Rolle's Theorem that there exists some point c in (a, b) for which $h'(c) = 0$. But, since $h'(c) = f'(c) - m$, it follows that

$$f'(c) = m = \frac{f(b) - f(a)}{b - a}. \quad ■$$

Example 2   Verify that the conditions of the Mean Value Theorem are satisfied by the function $f(x) = \frac{x-1}{x+1}$, $x \in \left[1, \frac{7}{2}\right]$; and find a value for c that satisfies the conclusion of the theorem.

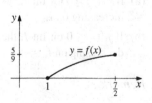

Solution   The function f is a rational function whose denominator is non-zero on $\left[1, \frac{7}{2}\right]$, so f is continuous on $\left[1, \frac{7}{2}\right]$ and differentiable on $\left(1, \frac{7}{2}\right)$. Thus f satisfies the conditions of the Mean Value Theorem.

Now

$$\frac{f\left(\frac{7}{2}\right) - f(1)}{\frac{7}{2} - 1} = \frac{\frac{5}{9} - 0}{\frac{5}{2}} = \frac{2}{9},$$

and
$$f'(x) = \frac{2}{(x+1)^2},$$
so the Mean Value Theorem asserts that there exists some point c in $\left(1, \frac{7}{2}\right)$ for which $f'(c) = \frac{2}{9}$. Thus
$$\frac{2}{(c+1)^2} = \frac{2}{9},$$
and so $(c+1)^2 = 9$. It follows that $c = 2$.   ☐

This is a quadratic equation with roots $c = 2$ and -4; we ignore the solution $c = -4$, since it lies outside $\left(1, \frac{7}{2}\right)$.

> **Problem 1**   For each of the following functions, verify that the conditions of the Mean Value Theorem are satisfied, and find a value for c that satisfies the conclusion of the theorem:
>
> (a) $f(x) = x^3 + 2x$, $x \in [-2, 2]$;   (b) $f(x) = e^x$, $x \in [0, 3]$.

6.4.2   Positive, negative and zero derivatives

We now study some consequences of the Mean Value Theorem for functions whose derivatives are always positive, always negative, or always zero.

First, we prove a crucial result about monotonic functions which you use regularly to sketch the graph of a function f. It concerns the behaviour of f on a *general* interval I, so here we denote the **interior** of I (the set of all **interior points** of I) by **Int I**.

For example, if $I = [0, 1)$, then Int $I = (0, 1)$.

Theorem 2   Increasing-Decreasing Theorem

Let f be continuous on an interval I and differentiable on Int I.

(a) If $f'(x) \geq 0$ on Int I, then f is increasing on I.

(b) If $f'(x) \leq 0$ on Int I, then f is decreasing on I.

Proof   Choose any two points x_1 and x_2 in I, with $x_1 < x_2$. The function f satisfies the conditions of the Mean Value Theorem on the interval $[x_1, x_2]$, so there exists a point c in (x_1, x_2) such that
$$\frac{f(x_2) - f(x_1)}{x_2 - x_1} = f'(c).$$
It follows that $f(x_2) - f(x_1)$ must have the same sign as $f'(c)$.

(a) If $f'(x) \geq 0$ on Int I, then $f(x_2) - f(x_1) \geq 0$, so that $f(x_2) \geq f(x_1)$. Thus f is increasing on I.

(b) If $f'(x) \leq 0$ on Int I, then $f(x_2) - f(x_1) \leq 0$, so that $f(x_2) \leq f(x_1)$. Thus f is decreasing on I.   ∎

Remark

If the inequalities in the statement of Theorem 2 are replaced by strict inequalities, the conclusions of the Theorem become the following:

(a) If $f'(x) > 0$ on Int I, then f is strictly increasing on I;

(b) If $f'(x) < 0$ on Int I, then f is strictly decreasing on I.

The proofs of these assertions are similar to the proofs in Theorem 2.

Problem 2 For each of the following functions f, determine whether f is increasing, strictly increasing, decreasing or strictly decreasing:

(a) $f(x) = 3x^{\frac{4}{3}} - 4x$, $x \in [1, \infty)$;   (b) $f(x) = x - \log_e x$, $x \in (0, 1]$.

Two useful consequences of Theorem 2 are the following corollaries.

Corollary 1   Zero Derivative Theorem

Let f be continuous on an interval I and differentiable on Int I. If $f'(x) = 0$ for all x in Int I, then f is constant on I.

Proof Cases (a) and (b) of Theorem 2 both apply, so that f is both increasing and decreasing on I.

Hence f is constant on I. ∎

As an illustration of the use of this important result, we can now prove the claim made earlier that the function $f(x) = \lambda e^x$, λ an arbitrary constant, is the only function f that satisfies the differential equation $f'(x) = f(x)$ on $\mathbb{R}$. For, if $f'(x) = f(x)$, then

$$\frac{d}{dx}\left(e^{-x}f(x)\right) = e^{-x}f'(x) - e^{-x}f(x)$$
$$= e^{-x}(f'(x) - f(x)) = 0;$$

Sub-section 6.1.3.

it then follows from Corollary 1 that $e^{-x}f(x)$ is just some constant λ, say, so that $f(x) = \lambda e^x$.

Corollary 2   Let f and g be continuous on an interval I and differentiable on Int I. If $f'(x) = g'(x)$ for all x in Int I, then

$$f(x) = g(x) + c \text{ for all } x \text{ in } I, \text{ for some constant } c.$$

Proof This follows immediately by applying Corollary 1 to the function $h = f - g$, since $h'(x) = 0$ for all x in the interior of I. ∎

For h is continuous on I and differentiable on Int I.

Example 3 Prove that $\sinh^{-1} x = \log_e\left(x + \sqrt{x^2 + 1}\right)$, for all $x \in \mathbb{R}$.

Solution Let

$$f(x) = \sinh^{-1} x, \quad x \in \mathbb{R},$$
$$g(x) = \log_e\left(x + \sqrt{x^2 + 1}\right), \quad x \in \mathbb{R}.$$

Then f and g are continuous on $\mathbb{R}$ and differentiable on $\mathbb{R}$.

Now

$$f'(x) = \frac{1}{\sqrt{x^2 + 1}}, \quad x \in \mathbb{R};$$

also, by the Composition Rule for Derivatives

$$g'(x) = \frac{1}{x + \sqrt{x^2 + 1}} \times \left[1 + \frac{1}{2} \times 2x(x^2 + 1)^{-\frac{1}{2}}\right]$$
$$= \frac{1 + \frac{x}{\sqrt{x^2+1}}}{x + \sqrt{x^2 + 1}}$$
$$= \frac{1}{\sqrt{x^2 + 1}}, \quad x \in \mathbb{R}.$$

Hence $f'(x) = g'(x)$ for all x in $\mathbb{R}$.

It follows from Corollary 2 that

$$\sinh^{-1} x = \log_e\left(x + \sqrt{x^2 + 1}\right) + c,$$

for some constant c. Putting $x = 0$ in the above identity, we obtain $0 = \log_e(1) + c$, so that $c = 0$. $\square$

Problem 3 Use Corollary 1 to prove the following identities:

(a) $\sin^{-1} x + \cos^{-1} x = \frac{1}{2}\pi$, for $x \in [-1, 1]$;

(b) $\tan^{-1} x + \tan^{-1}\left(\frac{1}{x}\right) = \frac{1}{2}\pi$, for $x > 0$.

Next, note that second derivatives can often be used to identify whether a point c where $f'(c) = 0$ is a local maximum or a local minimum of a function f.

Suppose that f is defined on a neighbourhood of c, and that

$$f'(c) = 0 \quad \text{and} \quad f''(c) > 0.$$

It can be shown that there is some punctured neighbourhood $N = (c - r, c) \cup (c, c + s)$ of c such that

$$\frac{f'(x) - f'(c)}{x - c} > 0, \quad \text{for } x \in N;$$

In Exercise 6 on this sub-section, in Section 6.7, we ask you to verify a result that implies this assertion.

that is

$$\frac{f'(x)}{x - c} > 0, \quad \text{for } x \in N.$$

We can rewrite this inequality in the form

$$f'(x) < 0, \quad \text{for } x \in (c - r, \ c),$$
$$f'(x) > 0, \quad \text{for } x \in (c, \ c + s).$$

It follows from the Increasing–Decreasing Theorem that f has a local minimum at c.

A similar argument shows that, if

$$f'(c) = 0 \quad \text{and} \quad f''(c) < 0,$$

then f has a local maximum at c.

Thus we have proved the following result.

Theorem 3   Second Derivative Test

Let f be defined on a neighbourhood of c, and $f'(c) = 0$.

(a) If $f''(c) > 0$, then $f(c)$ is a local minimum of f.

(b) If $f''(c) < 0$, then $f(c)$ is a local maximum of f.

The following diagrams are helpful in remembering this result.

Remark

The theorem gives us no information in the case that $f''(c) = 0$.

Problem 4 For the function $f(x) = x^3 - 3x^2 + 1, x \in \mathbb{R}$, determine those points c where $f'(c) = 0$. Using the Second Derivative Test, determine whether these correspond to local maxima, local minima or neither.

Inequalities

We now demonstrate how the Increasing–Decreasing Theorem can be used to prove certain inequalities involving *differentiable* functions.

Example 4 Prove that, for $\alpha > 1$,
$$(1+x)^\alpha \geq 1 + \alpha x, \quad \text{for } x \geq -1.$$

This is a generalisation of Bernoulli's Inequality
$$(1+x)^n \geq 1 + nx,$$
for $x \geq -1$, $n \in \mathbb{N}$, that you met in Sub-section 1.3.3.

Solution Let
$$f(x) = (1+x)^\alpha - (1 + \alpha x), \quad \text{for } x \in [-1, \infty).$$
The function f is continuous on $[-1, \infty)$ and differentiable on $(-1, \infty)$, and
$$f'(x) = \alpha(1+x)^{\alpha-1} - \alpha$$
$$= \alpha\left[(1+x)^{\alpha-1} - 1\right].$$

Firstly, if $-1 < x < 0$, then
$$0 < 1 + x < 1.$$

Since $\alpha > 1$, we can then take the $(\alpha - 1)$th power of each side of the inequality $1 + x < 1$ to obtain

Since $\alpha - 1 > 0$, we can use the Power Rule for inequalities, which you met in Sub-section 1.2.1.

$$(1+x)^{\alpha-1} < 1, \quad \text{for } -1 < x < 0,$$
so that
$$f'(x) < 0, \quad \text{for } -1 < x < 0.$$

Next, if $x > 0$, then
$$1 + x > 1,$$
so that
$$(1+x)^{\alpha-1} > 1, \quad \text{for } x > 0.$$

Since $\alpha - 1 > 0$, we can again use the Power Rule for inequalities.

Hence
$$f'(x) > 0, \quad \text{for } x > 0.$$

Bringing together these two arguments, we have
$$f'(x) < 0, \quad \text{for } -1 < x < 0,$$
$$f'(x) > 0, \quad \text{for } x > 0.$$

Also, $f(0) = 0$.

Hence, by the Increasing–Decreasing Theorem
$$f \text{ is decreasing on } [-1, 0],$$
$$f \text{ is increasing on } [0, \infty);$$
so we have
$$f(x) \geq f(0) = 0, \quad \text{for } x \in [-1, 0],$$
$$f(x) \geq f(0) = 0, \quad \text{for } x \in [0, \infty).$$

It follows that
$$(1+x)^\alpha - (1 + \alpha x) \geq 0, \quad \text{for } x \in [-1, \infty),$$
as required. $\quad\square$

Example 4 illustrates the following general strategy.

Strategy   To prove that $g(x) \geq h(x)$ on $[a, b]$:

1. Let
$$f(x) = g(x) - h(x),$$
and show that f is continuous on $[a, b]$ and differentiable on (a, b);

2. Prove that:

EITHER $f(a) \geq 0,$ and $f'(x) \geq 0$ on (a, b),

OR $f(b) \geq 0,$ and $f'(x) \leq 0$ on (a, b),

There is a corresponding version of the Strategy in which the weak inequalities are replaced by strict inequalities.

This also works if b is ∞.

This also works if a is $-\infty$.

Problem 5   Prove the following inequalities:

(a) $x \geq \sin x$, for $x \in \left[0, \frac{1}{2}\pi\right]$;

(b) $\frac{2}{3}x + \frac{1}{3} \geq x^{\frac{2}{3}}$, for $x \in [0, 1]$.

6.5   L'Hôpital's Rule

In order to differentiate the functions sin and exp, we needed to use the following results

$$\lim_{x \to 0} \frac{\sin x}{x} = 1 \quad \text{and} \quad \lim_{x \to 0} \frac{e^x - 1}{x} = 1.$$

Sub-section 6.1.3.

Each of these limits is the limit as $x \to 0$ of a quotient in which the numerator and denominator take the value 0 when x is 0. The evaluation of these limits was not trivial and required considerable care.

In Analysis and in Mathematical Physics we often need to evaluate limits of the form

$$\lim_{x \to c} \frac{f(x)}{g(x)}, \quad \text{where } f(c) = g(c) = 0.$$

See Sub-section 5.1.1 and Problem 8 in Sub-section 5.1.4, respectively.

Such limits cannot be evaluated by the Quotient Rule for limits of functions, because it does not apply in this situation.

For example, do the limits

$$\lim_{x \to \frac{\pi}{2}} \frac{\cos 3x}{\sin x - e^{\cos x}} \quad \text{and} \quad \lim_{x \to 0} \frac{x^2}{\cosh x - 1}$$

exist? If they do, what are their values?

Do we have to evaluate all such limits by a direct argument, or is there a handy rule which we can apply? We shall find that there is, called l'Hôpital's Rule.

The Quotient Rule for limits of functions states that

$$\lim_{x \to c} \frac{f(x)}{g(x)} = \frac{\lim_{x \to c} f(x)}{\lim_{x \to c} g(x)},$$

provided that these last two limits exist.

6.5.1   Cauchy's Mean Value Theorem

Recall that the Mean Value Theorem asserts that, under certain conditions, the graph of a function $f(x)$, for $x \in [a, b]$, has the property that at some intermediate point c the tangent to the graph is parallel to the chord joining the endpoints of the graph. In other words, there exists a point c in (a, b) such that

Theorem 1, Sub-section 6.4.1.

$$f'(c) = \frac{f(b) - f(a)}{b - a}. \tag{1}$$

The key tool that we shall need in Sub-section 6.5.2 in the proof of l'Hôpital's Rule is the following result.

Theorem 1   Cauchy's Mean Value Theorem

Let f and g be continuous on $[a, b]$ and differentiable on (a, b). Then there is some point c in (a, b) for which

$$f'(c)\{g(b) - g(a)\} = g'(c)\{f(b) - f(a)\}. \tag{2}$$

In particular, if $g(b) \neq g(a)$ and $g'(c) \neq 0$, this equation can be written in the form

$$\frac{f'(c)}{g'(c)} = \frac{f(b) - f(a)}{g(b) - g(a)}. \tag{3}$$

This form is easier to remember.

Notice that the Mean Value Theorem is simply the special case when $g(x) = x$. For then $g'(c) = 1$ and $g(b) - g(a) = b - a$, and equation (3) reduces to equation (1).

However, Theorem 1 is NOT a simple consequence of the Mean Value Theorem. For, if we apply the Mean Value Theorem separately to the functions f and g, we establish the existence of two points c_1 and c_2 in (a, b) for which

$$f'(c_1) = \frac{f(b) - f(a)}{b - a} \quad \text{and} \quad g'(c_2) = \frac{g(b) - g(a)}{b - a}.$$

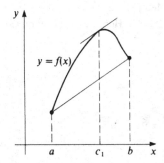 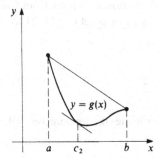

However, since c_1 and c_2 are usually unequal, we cannot deduce the existence of a *single* point c satisfying the statement (2).

Our proof of Theorem 1 is similar to that of the Mean Value Theorem: we choose a suitable 'auxiliary' function h on $[a, b]$ for which the conditions of Rolle's Theorem are satisfied.

Proof   Consider the function

$$h(x) = f(x)\{g(b) - g(a)\} - g(x)\{f(b) - f(a)\}, \quad \text{for } x \in [a, b].$$

By the Combination Rules for continuity and differentiability, h is continuous on $[a, b]$ and differentiable on (a, b). Also

$$h(a) = f(a)\{g(b) - g(a)\} - g(a)\{f(b) - f(a)\}$$
$$= f(a)g(b) - g(a)f(b)$$

Here two terms cancel.

and

$$h(b) = f(b)\{g(b) - g(a)\} - g(b)\{f(b) - f(a)\}$$
$$= f(a)g(b) - g(a)f(b),$$

Again two terms cancel.

so that $h(a) = h(b)$.

Thus h satisfies the conditions of Rolle's Theorem on $[a, b]$. Therefore there exists a point c in (a, b) for which

The common value of $h(a)$ and $h(b)$ does not matter; the important thing is that the two are equal.

$$h'(c) = 0;$$

that is, these exists a point c in (a, b) for which

$$f'(c)\{g(b) - g(a)\} = g'(c)\{f(b) - f(a)\}.$$

This is precisely equation (2).

Equation (3) follows immediately from equation (2).   ∎

Example 1   By applying Cauchy's Mean Value Theorem to the functions

$$f(x) = x^6 - 1 \quad \text{and} \quad g(x) = \frac{1}{2}x^4 + 3x^3 + 3x - 3 \text{ on } [1, 2],$$

prove that the equation $3x^5 - 2x^3 - 9x^2 = 3$ has at least one root in $(1, 2)$.

Solution   Since f and g are both polynomials, they are continuous on $[1, 2]$ and differentiable on $(1, 2)$. It follows that Cauchy's Mean Value Theorem applies to the functions f and g on $[1, 2]$.

Now, $g(2) = \frac{1}{2} \times 16 + 3 \times 8 + 3 \times 2 - 3 = 8 + 24 + 6 - 3 = 35$ and $g(1) = \frac{1}{2} + 3 + 3 - 3 = 3\frac{1}{2}$, so that $g(2) \neq g(1)$. Also, $f'(x) = 6x^5$ and $g'(x) = 2x^3 + 9x^2 + 3 \neq 0$ on $(1, 2)$. It therefore follows from Cauchy's Mean Value Theorem that there exists at least one point c in $(1, 2)$ for which

Here we use equation (3).

$$\frac{6c^5}{2c^3 + 9c^2 + 3} = \frac{63 - 0}{35 - 3\frac{1}{2}}$$
$$= \frac{63}{31\frac{1}{2}} = 2.$$

By cross-multiplying, we see that c satisfies the equation $6c^5 = 4c^3 + 18c^2 + 6$, or

$$3c^5 - 2c^3 - 9c^2 = 3.$$

In other words, the equation $3x^5 - 2x^3 - 9x^2 = 3$ has at least one root in $(1, 2)$.   □

Problem 1   By applying Cauchy's Mean Value Theorem to the functions

$$f(x) = x^3 + x^2 \sin x \quad \text{and} \quad g(x) = x \cos x - \sin x \text{ on } [0, \pi],$$

prove that the equation $3x = (\pi^2 - 2) \sin x - x \cos x$ has at least one root in $(0, \pi)$.

6.5.2   l'Hôpital's Rule

We are now in a position to state the key result that we need to evaluate certain types of limits in a fairly routine way.

Theorem 2   l'Hôpital's Rule

Let f and g be differentiable on a neighbourhood of the point c, at which $f(c) = g(c) = 0$. Then

$$\lim_{x \to c} \frac{f(x)}{g(x)} \text{ exists and equals } \lim_{x \to c} \frac{f'(x)}{g'(x)},$$

provided that this last limit exists.

Proof   We assume that

$$\lim_{x \to c} \frac{f'(x)}{g'(x)} \text{ exists and equals } \ell. \tag{4}$$

You may omit this proof at a first reading.

Hence there is some punctured neighbourhood $N = (c - r, c) \cup (c, c + s)$ of c on which $\frac{f'}{g'}$ is defined and $g'(x) \neq 0$.

We now prove that our assumption (4) implies that

$$\lim_{x \to c} \frac{f(x)}{g(x)} \text{ exists and equals } \ell. \tag{5}$$

Let y be any *specific* point in N for which $y > c$. The functions f and g are continuous on $[c, y]$ and differentiable on (c, y), so they satisfy the conditions of Cauchy's Mean Value Theorem on $[c, y]$.

Now $g(y) - g(c) \neq 0$; for otherwise we would have $g(c) = 0$ (from our assumptions) and so $g(y) = 0$; and this would imply, by Rolle's Theorem, that g' would vanish somewhere in (c, y), which it does not.

It follows from the conclusion (3) of Cauchy's Mean Value Theorem that there exists some point z in (c, y) for which

Theorem 1, Sub-section 6.5.1.

$$\frac{f'(z)}{g'(z)} = \frac{f(y) - f(c)}{g(y) - g(c)}$$

$$= \frac{f(y)}{g(y)}, \quad \text{since } f(c) = g(c) = 0.$$

Now let $y \to c^+$. Since $c < z < y$, it follows that $z \to c^+$ too. But we know that

$$\lim_{z \to c^+} \frac{f'(z)}{g'(z)} \text{ exists and has the value } \ell.$$

This is just a special case of (4), with z in place of x and $z \to c^+$ in place of $x \to c$.

It follows that

$$\lim_{y \to c^+} \frac{f(y)}{g(y)} \text{ exists and has the value } \ell.$$

This is exactly the same as the statement that

$$\lim_{x \to c^+} \frac{f(x)}{g(x)} \text{ exists and has the value } \ell.$$

Here x is simply a 'dummy variable': it does not matter what letter we assign to the variable in the limit.

A similar argument shows that

$$\lim_{x \to c^-} \frac{f(x)}{g(x)} \text{ exists and has the value } \ell.$$

Combining the last two statements, we obtain the desired result (5).   ∎

We simply repeat the whole argument, starting with a *specific* point y in N for which $y < c$.

We now show how we can use l'Hôpital's Rule to evaluate the two limits that we mentioned at the start of this section.

Example 2 Prove that $\lim\limits_{x \to \frac{\pi}{2}} \dfrac{\cos 3x}{\sin x - e^{\cos x}}$ (6)
exists and determine its value.

Solution Let

$$f(x) = \cos 3x \quad \text{and} \quad g(x) = \sin x - e^{\cos x}, \quad \text{for } x \in \mathbb{R}.$$

Then f and g are differentiable on $\mathbb{R}$, and

$$f\left(\frac{\pi}{2}\right) = g\left(\frac{\pi}{2}\right) = 0;$$

so that f and g satisfy the conditions of l'Hôpital's Rule at $\frac{\pi}{2}$.

Now

$$\frac{f'(x)}{g'(x)} = \frac{-3 \sin 3x}{\cos x + \sin x \times e^{\cos x}}.$$

Then, by l'Hôpital's Rule, the limit (6) exists and equals

$$\lim_{x \to \frac{\pi}{2}} \frac{f'(x)}{g'(x)} = \lim_{x \to \frac{\pi}{2}} \frac{-3 \sin 3x}{\cos x + \sin x \times e^{\cos x}},$$

provided that this last limit exists.

By the Quotient Rule for continuous functions, we know that $\frac{f'}{g'}$ is continuous at $\frac{\pi}{2}$, so that

$$\lim_{x \to \frac{\pi}{2}} \frac{f'(x)}{g'(x)} = \frac{f'\left(\frac{\pi}{2}\right)}{g'\left(\frac{\pi}{2}\right)}$$
$$= \frac{3}{1} = 3.$$

Here we are using the fact (see Remark 7, Sub-section 5.4.1) that, if f is continuous at c, then

$$\lim_{x \to c} f(x) = f(c).$$

(In future, we shall not mention this fact explicitly in this particular connection.)

It follows, from l'Hôpital's Rule, that the original limit (6) must also exist, and that its value is 3.   □

Example 3 Prove that $\lim\limits_{x \to 0} \dfrac{x^2}{\cosh x - 1}$ (7)
exists and determine its value.

Solution Let

$$f(x) = x^2 \quad \text{and} \quad g(x) = \cosh x - 1, \quad \text{for } x \in \mathbb{R}.$$

Then f and g are differentiable on $\mathbb{R}$, and

$$f(0) = g(0) = 0;$$

so that f and g satisfy the conditions of l'Hôpital's Rule at 0.

Now, the derivatives of f and g are

$$f'(x) = 2x \quad \text{and} \quad g'(x) = \sinh x, \quad \text{for } x \in \mathbb{R}.$$

It follows from l'Hôpital's Rule that the limit (7) exists and equals

$$\lim_{x \to 0} \frac{2x}{\sinh x}$$ (8)

provided that this last limit exists.

Now, both f' and g' are differentiable on $\mathbb{R}$, and $f'(0) = g'(0) = 0$; thus f' and g' satisfy the conditions of l'Hôpital's Rule at 0.

Since

$$f''(x) = 2 \quad \text{and} \quad g''(x) = \cosh x, \quad \text{for } x \in \mathbb{R},$$

We cannot assert that

$$\lim_{x \to 0} \frac{f'(x)}{g'(x)} = \frac{f'(0)}{g'(0)},$$

since $f'(0) = g'(0) = 0$.

it follows from l'Hôpital's Rule that the limit (8) exists and equals

$$\lim_{x \to 0} \frac{2}{\cosh x}, \tag{9}$$

provided that this last limit exists.

The function cosh is continuous on $\mathbb{R}$, and $\cosh 0 = 1$, so that $\lim_{x \to 0} \cosh x = 1$. It follows, from the Quotient Rule for limits, that the limit (9) does exist and that its value is

$$\frac{2}{\cosh 0} = \frac{2}{1} = 2.$$

Working backwards, we conclude that the limit (8) exists, and equals 2. Working further backwards, we conclude that the limit (7) also exists and equals 2. $\qquad\qquad\qquad\qquad\qquad\qquad\qquad\qquad\quad\square$

> Note that the carefully set out logic of the argument here is essential, since it is a consequence of what we know from Theorem 2.

Before applying a theorem, it is important to check that its conditions are satisfied; for if they are not satisfied you cannot use the theorem! For instance, a thoughtless application of l'Hôpital's Rule can give an incorrect answer!

For example, consider the problem of evaluating

$$\lim_{x \to 1} \frac{2x^2 - x - 1}{x^2 - x}. \tag{10}$$

If we put $f(x) = 2x^2 - x - 1$ and $g(x) = x^2 - x$, we might be tempted to evaluate (10) as follows

> Note that $f(1) = g(1) = 0$.

$$\lim_{x \to 1} \frac{f(x)}{g(x)} = \lim_{x \to 1} \frac{f'(x)}{g'(x)} \left(\text{which} = \lim_{x \to 1} \frac{4x - 1}{2x - 1} \right) \tag{11}$$
$$= \lim_{x \to 1} \frac{f''(x)}{g''(x)} = \lim_{x \to 1} \frac{4}{2}$$
$$= 2.$$

> Here the careful arguments that we gave in Examples 2 and 3 have been abandoned in favour of thoughtless 'formula-pushing'!

In fact, the value of the limit (10) is 3; so let us review the above argument carefully!

The line (11) in the calculation is valid, since f and g are differentiable on $\mathbb{R}$ and $f(1) = g(1) = 0$; so the conditions of l'Hôpital's Rule are satisfied for the *first* application of the Rule. However, $f'(1) = 3$ and $g'(1) = 1$, so the conditions are *not* satisfied for the *second* application of l'Hôpital's Rule!

In fact, we should have concluded directly from line (11) that

$$\lim_{x \to 1} \frac{f'(x)}{g'(x)} = \lim_{x \to 1} \frac{4x - 1}{2x - 1} = 3.$$

The moral is that you must apply l'Hôpital's Rule with care, particularly to make the proviso 'provided that this last limit exists' at the appropriate points. At the end, you then work backwards through the chain of applications of the Rule to reach your conclusion about the limit that you set out originally to examine.

Remark

Notice that we *cannot* apply l'Hôpital's Rule to evaluate the limits

$$\lim_{x \to 0} \frac{\sin x}{x} = 1 \quad \text{and} \quad \lim_{x \to 0} \frac{e^x - 1}{x} = 1,$$

because we *used* these limits to *find* the derivatives of $\sin x$ and e^x, respectively!

Problem 2 Prove that the following limits exist, and evaluate them.

(a) $\lim\limits_{x \to 0} \frac{\sinh 2x}{\sin 3x}$;

(b) $\lim\limits_{x \to 0} \frac{(1+x)^{\frac{1}{3}}-(1-x)^{\frac{1}{3}}}{(1+2x)^{\frac{2}{3}}-(1-2x)^{\frac{2}{3}}}$;

(c) $\lim\limits_{x \to 0} \frac{\sin\left(x^2+\sin x^2\right)}{1-\cos 4x}$;

(d) $\lim\limits_{x \to 0} \frac{\sin x - x\cos x}{x^3}$.

6.6 The Blancmange function

You may omit this section, apart from the next two paragraphs, at a first reading.

We now meet a function, called the *Blancmange function*, which is continuous at each point of $\mathbb{R}$ but is *differentiable nowhere* on $\mathbb{R}$. The construction of the first function with these properties by Karl Weierstrass caused a huge excitement among mathematicians!

You saw earlier that, if a function is differentiable at a point c, then it is also continuous at c. The Blancmange function shows, in a very striking way, that the converse result is false!

Sub-section 6.1.2.

6.6.1 What is the Blancmange function?

We give the formal definition first, and then look at the underlying geometry of the graph of the function.

> **Definition**   Let f be the function
> $$f(x) = \begin{cases} x - [x], & 0 \le x - [x] \le \frac{1}{2}, \\ 1 - (x - [x]), & \frac{1}{2} < x - [x] < 1, \end{cases}$$
> where $[x]$ denotes the integer part of x. Then the **Blancmange function** is defined on $\mathbb{R}$ by the formula
> $$B(x) = f(x) + \frac{1}{2}f(2x) + \frac{1}{4}f(4x) + \frac{1}{8}f(8x) + \cdots$$
> $$= \sum_{n=0}^{\infty} \frac{1}{2^n}f(2^n x).$$

This series converges by the Comparison Test, since $|f(2^n x)| \le \frac{1}{2}$, for all $x \in \mathbb{R}$.

So, what does the graph of B look like? And can we obtain any idea why the function has its strange properties?

The function f is continuous on $\mathbb{R}$, but is not differentiable at the points $\frac{1}{2}n$, for any integer n.

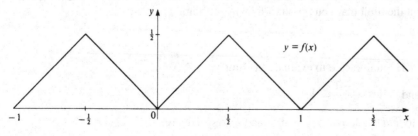

Next, we construct the graphs

$$y = \tfrac{1}{2}f(2x), \quad y = \tfrac{1}{4}f(4x), \quad y = \tfrac{1}{8}f(8x), \ldots;$$

we obtain each graph by scaling the previous graph by a factor of $\tfrac{1}{2}$ in both the x-direction and the y-direction.

Each graph has twice as many peaks in any interval as its predecessor, and each of these peaks is half the height of the peaks in its predecessor. Thus the number of points in the interval where the function is not differentiable doubles (approximately) at each stage.

The first few of these graphs are included in the diagrams below.

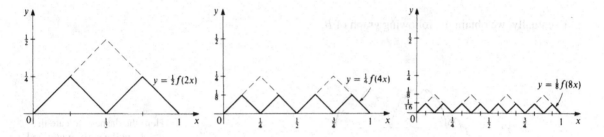

We obtain the Blancmange function B by adding together all these functions to form an infinite series $B(x) = f(x) + \tfrac{1}{2}f(2x) + \tfrac{1}{4}f(4x) + \tfrac{1}{8}f(8x) + \cdots$. Thus, for example

$$\begin{aligned}
B\left(\tfrac{1}{2}\right) &= f\left(\tfrac{1}{2}\right) + \tfrac{1}{2}f(1) + \tfrac{1}{4}f(2) + \tfrac{1}{8}f(4) + \cdots \\
&= \tfrac{1}{2} + \left(\tfrac{1}{2} \times 0\right) + \left(\tfrac{1}{4} \times 0\right) + \left(\tfrac{1}{8} \times 0\right) + \cdots \\
&= \tfrac{1}{2},
\end{aligned}$$

and

$$\begin{aligned}
B\left(\tfrac{1}{4}\right) &= f\left(\tfrac{1}{4}\right) + \tfrac{1}{2}f\left(\tfrac{1}{2}\right) + \tfrac{1}{4}f(1) + \tfrac{1}{8}f(2) + \cdots \\
&= \tfrac{1}{4} + \left(\tfrac{1}{2} \times \tfrac{1}{2}\right) + \left(\tfrac{1}{4} \times 0\right) + \left(\tfrac{1}{8} \times 0\right) + \cdots \\
&= \tfrac{1}{4} + \tfrac{1}{4} = \tfrac{1}{2}.
\end{aligned}$$

To get an idea of the shape of the graph of B, we look at the graphs of successive partial sum functions of B. To help you follow the construction, we also draw in the graph at the previous stage (in light dashes) and the function being added to it (in heavy dashes).

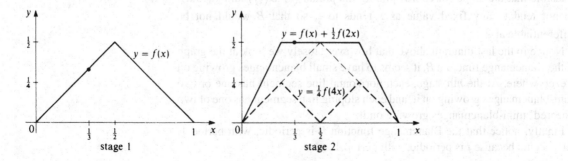

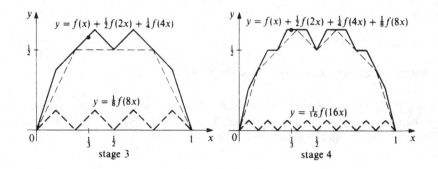

Eventually, we obtain the following graph of B:

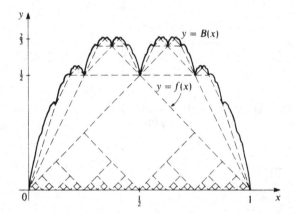

Here the dashes indicate the graphs at the early stages and also the graphs of functions occurring in the summation.

It *looks* as though B is continuous. However, it is NOT true in general that the sum of infinitely many continuous function is itself continuous, so a proof of the continuity of B is necessary.

We give this in Sub-section 6.6.2.

Similarly, it *looks* as though B might not be differentiable.

We prove this in Sub-section 6.6.3.

For example, we have marked the points on the above graphs corresponding to $x = \frac{1}{3}$. At the first stage, $\frac{1}{3}$ lies in the interval $\left[0, \frac{1}{2}\right]$, and so the point $\left(\frac{1}{3}, f\left(\frac{1}{3}\right)\right)$ lies on a line segment of slope 1. At the next stage, $\frac{1}{3}$ lies in the interval $\left[\frac{1}{4}, \frac{1}{2}\right]$, and so the point $\left(\frac{1}{3}, f\left(\frac{1}{3}\right) + \frac{1}{2}f\left(\frac{2}{3}\right)\right)$ lies on a line segment of slope 0. And so on.

At successive stages in the construction of the graph of B, the point corresponding to $\frac{1}{3}$ lies alternately on line segments of slope 0 and 1. Thus it seems plausible that the slopes of chords joining the points $\left(\frac{1}{3}, B\left(\frac{1}{3}\right)\right)$ and $(x, B(x))$ do not tend to any fixed value as x tends to $\frac{1}{3}$, so that B would not be differentiable at $\frac{1}{3}$.

Notice in the last diagram above that however closely we look at the graph of the Blancmange function B, it seems to have small blancmanges growing on it everywhere. At the nth stage, each horizontal line segment has one or two mini-blancmanges growing on it, and each sloping line segment has one or two 'sheared' mini-blancmanges growing on it.

Finally, notice that the Blancmange function B is periodic, with period 1. This occurs because f is periodic, with period 1.

6.6.2 Continuity of the Blancmange function

To verify that B is continuous on $\mathbb{R}$, we must use the power of the $\varepsilon - \delta$ definition of continuity.

Theorem 1   The Blancmange function B is continuous at each point $c \in \mathbb{R}$.

Proof   We must show that

for each positive number ε, there is a positive number δ such that
$$|B(x) - B(c)| < \varepsilon, \quad \text{for all } x \text{ satisfying } |x - c| < \delta. \tag{1}$$

Now, it follows, from the definition of B, that $B(x) - B(c) = \sum_{n=0}^{\infty} \frac{1}{2^n} (f(2^n x) - f(2^n c))$; hence, by the infinite form of the Triangle Inequality   Sub-section 3.3.1.

$$|B(x) - B(c)| \leq \sum_{n=0}^{\infty} \frac{1}{2^n} |f(2^n x) - f(2^n c)|. \tag{2}$$

For all x and c, both of the numbers $f(2^n x)$ and $f(2^n c)$ lie in $\left[0, \frac{1}{2}\right]$, so that the modulus of their difference is at most $\frac{1}{2}$; that is

$$|f(2^n x) - f(2^n c)| \leq \frac{1}{2}.$$

Now we choose an integer N such that $\frac{1}{2^N} < \frac{1}{2}\varepsilon$. (This choice is possible, since the sequence $\left\{\frac{1}{2^n}\right\}$ is null.) It follows that

$$\sum_{n=N}^{\infty} \frac{1}{2^n} |f(2^n x) - f(2^n c)| \leq \sum_{n=N}^{\infty} \frac{1}{2^n} \times \frac{1}{2}$$
$$= \frac{1}{2^N}$$
$$< \frac{1}{2}\varepsilon. \tag{3}$$

Next, since each of the functions $x \mapsto f(2^n x)$, $n = 0, 1, \ldots, N-1$, is continuous, it follows that

We apply the definition of continuity to each function in turn.

for each $n = 0, 1, \ldots, N-1$, there is a positive number δ_n such that
$$|f(2^n x) - f(2^n c)| < \tfrac{1}{4}\varepsilon, \quad \text{for all } x \text{ satisfying } |x - c| < \delta_n.$$

We use $\frac{1}{4}\varepsilon$ here, in order to obtain an ε in our final result (1).

Now we choose $\delta = \min\{\delta_0, \delta_1, \ldots, \delta_{N-1}\}$. Thus, for each $n = 0, 1, \ldots, N-1$, we must have

$$|f(2^n x) - f(2^n c)| < \tfrac{1}{4}\varepsilon, \quad \text{for all } x \text{ satisfying } |x - c| < \delta.$$

It follows that

$$\sum_{n=0}^{N-1} \frac{1}{2^n} |f(2^n x) - f(2^n c)| < \sum_{n=0}^{N-1} \frac{1}{2^n} \times \frac{1}{4}\varepsilon$$
$$< \sum_{n=0}^{\infty} \frac{1}{2^{n+2}} \times \varepsilon$$
$$= \frac{1}{2}\varepsilon. \tag{4}$$

Substituting the upper bounds from (3) and (4) into inequality (2), we deduce that, for all x satisfying $|x - c| < \delta$, we have

$$|B(x) - B(c)| \leq \sum_{n=0}^{N-1} \frac{1}{2^n} |f(2^n x) - f(2^n c)| + \sum_{n=N}^{\infty} \frac{1}{2^n} |f(2^n x) - f(2^n c)|$$
$$< \frac{1}{2}\varepsilon + \frac{1}{2}\varepsilon$$
$$= \varepsilon.$$

This is the desired result (1). Hence B is continuous at c. ∎

6.6.3 The Blancmange function is differentiable nowhere

The proof that B is nowhere differentiable on $\mathbb{R}$ is more tricky, and the following lemma about difference quotients in general (which is of interest in its own right) plays a crucial role.

Lemma Let f be defined on an open interval I and differentiable at the point $c \in I$. Let $\{x_n\}$ and $\{y_n\}$ be sequences in I converging to c such that

$$x_n \leq c \leq y_n \quad \text{and} \quad x_n < y_n, \quad \text{for } n = 0, 1, 2, \ldots.$$

Then

$$\lim_{n \to \infty} \frac{f(y_n) - f(x_n)}{y_n - x_n} \text{ exists and equals } f'(c).$$

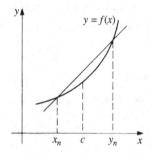

Proof We must prove that

for each positive number ε, there exists some number X such that
$$\left| \frac{f(y_n) - f(x_n)}{y_n - x_n} - f'(c) \right| < \varepsilon, \quad \text{for all } n > X. \quad (5)$$

Since f is differentiable at c, we know that, for each positive number ε, there exists some positive number δ such that

$$\left| \frac{f(x) - f(c)}{x - c} - f'(c) \right| < \frac{1}{4}\varepsilon, \quad \text{for all } x \text{ satisfying } 0 < |x - c| < \delta,$$

so that

$$|f(x) - f(c) - f'(c)(x - c)| \leq \frac{1}{4}\varepsilon|x - c|, \quad \text{for all } x \text{ satisfying } |x - c| < \delta. \quad (6)$$

Now, since $x_n \to c$ and $y_n \to c$ as $n \to \infty$, it follows that there are numbers X_1 and X_2 such that

$$|x_n - c| < \delta \text{ for all } n > X_1, \quad \text{and} \quad |y_n - c| < \delta \text{ for all } n > X_2;$$

so, if we set $X = \max\{X_1, X_2\}$, we certainly have

$$|x_n - c| < \delta \quad \text{and} \quad |y_n - c| < \delta \quad \text{for all } n > X.$$

It now follows from (6) that, for all $n > X$, we have

$$|f(x_n) - f(c) - f'(c)(x_n - c)| \leq \frac{1}{4}\varepsilon|x_n - c| \quad \text{and}$$

$$|f(y_n) - f(c) - f'(c)(y_n - c)| \leq \frac{1}{4}\varepsilon|y_n - c|.$$

This is the definition of differentiability at c, but with $\frac{1}{4}\varepsilon$ in place of ε, in order to obtain an ε in our final result (5).

In fact, this is a strict inequality if $x \neq c$. However later in the argument we will need to allow the possibility that $x = c$, so we must write (6) with a weak inequality sign.

We put $x = x_n$ and then $x = y_n$ into (6).

Hence, we can apply the Triangle Inequality to obtain

$$|f(y_n) - f(x_n) - f'(c)(y_n - x_n)|$$

$$= |\{f(y_n) - f(c) - f'(c)(y_n - c)\} - \{f(x_n) - f(c) - f'(c)(x_n - c)\}|$$

$$\le |f(y_n) - f(c) - f'(c)(y_n - c)| + |f(x_n) - f(c) - f'(c)(x_n - c)|$$

$$\le \frac{1}{4}\varepsilon|y_n - c| + \frac{1}{4}\varepsilon|x_n - c|$$

$$\le \frac{1}{4}\varepsilon|y_n - x_n| + \frac{1}{4}\varepsilon|y_n - x_n|$$

$$= \frac{1}{2}\varepsilon|y_n - x_n|, \quad \text{for all } n > X.$$

We first insert some terms and take them away again.

We use the two inequalities that we have just verified for $n > X$.

Since $x_n \le c \le y_n$, both $|y_n - c|$ and $|x_n - c|$ are $\le |y_n - x_n|$.

Since we know that $x_n \ne y_n$, we can divide this inequality by the non-zero term $y_n - x_n$ to obtain, for $n > X$

$$\left| \frac{f(y_n) - f(x_n)}{y_n - x_n} - f'(c) \right| \le \frac{1}{2}\varepsilon < \varepsilon.$$

This is the result (5) that we set out to prove.   ■

Remark

The hypothesis that x_n and y_n must lie on opposite sides of c cannot generally be omitted. For example, consider the function

$$f(x) = \begin{cases} x^2 \sin\left(\frac{1}{x}\right), & x \ne 0, \\ 0, & x = 0, \end{cases}$$

that you saw earlier is differentiable at 0, with $f'(0) = 0$. If we set

Problem 7, Sub-section 6.1.2.

$$x_n = \frac{1}{(2n+1)\pi} \quad \text{and} \quad y_n = \frac{1}{\left(2n+\frac{1}{2}\right)\pi}, \quad \text{for } n = 0, 1, 2, \ldots,$$

then

$$\frac{f(y_n) - f(x_n)}{y_n - x_n} = \frac{\left(\frac{1}{\left(2n+\frac{1}{2}\right)^2 \pi^2}\right) - (0)}{\left(\frac{1}{\left(2n+\frac{1}{2}\right)\pi}\right) - \left(\frac{1}{(2n+1)\pi}\right)}$$

$$= \frac{2(2n+1)}{\left(2n+\frac{1}{2}\right)\pi}$$

$$\to \frac{2}{\pi} \quad \text{as } n \to \infty.$$

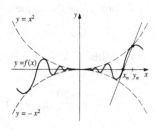

However, the value of this limit is not 0, the value of $f'(0)$.

We are now ready to prove the principal result in this sub-section.

Theorem 2   The Blancmange function B is not differentiable at any point $c \in \mathbb{R}$.

Proof   In order to apply the lemma, we construct two sequences $\{x_n\}$ and $\{y_n\}$ converging to c such that the corresponding sequence of difference quotients is not convergent. We use the method of repeated bisection to construct $\{x_n\}$ and $\{y_n\}$.

You may omit this proof at a first reading.

Since B is periodic with period 1, we assume for simplicity that $c \in [0, 1]$.

We start by defining $[x_0, y_0] = [0, 1]$. Then, since c lies in one of the intervals $\left[0, \frac{1}{2}\right]$ and $\left(\frac{1}{2}, 1\right]$, we can define

$$[x_1, y_1] = \begin{cases} \left[0, \frac{1}{2}\right], & \text{if } c \in \left[0, \frac{1}{2}\right], \\ \left[\frac{1}{2}, 1\right], & \text{if } c \in \left(\frac{1}{2}, 1\right]. \end{cases}$$

With this definition of $[x_1, y_1]$, we have that:

1. $[x_1, y_1] \subseteq [x_0, y_0]$;
2. $y_1 - x_1 = \frac{1}{2}$;
3. $x_1 \leq c \leq y_1$;
4. $x_1 = \frac{1}{2}p_1$ and $y_1 = \frac{1}{2}(p_1 + 1)$, for some integer p_1.

We can then repeat this process, bisecting the interval $[x_1, y_1]$ to obtain $[x_2, y_2]$, and so on. In this way, we obtain a sequence of closed intervals $[x_n, y_n]$, for $n = 1, 2, \ldots$, such that:

1. $[x_{n+1}, y_{n+1}] \subseteq [x_n, y_n]$;
2. $y_n - x_n = \left(\frac{1}{2}\right)^n$;
3. $x_n \leq c \leq y_n$;
4. $x_n = \frac{1}{2^n}p_n$ and $y_n = \frac{1}{2^n}(p_n + 1)$, for some integer p_n.

Properties 2 and 3 imply that both the sequences $\{x_n\}$ and $\{y_n\}$ converge to c, but we shall show that the sequence of difference quotients $\{Q_n\}$, where

$$Q_n = \frac{B(y_n) - B(x_n)}{y_n - x_n}$$
$$= 2^n(B(y_n) - B(x_n)),$$

We use Property 2 for the value of $y_n - x_n$.

is not convergent. It will then follow, from the lemma, that B is not differentiable at c.

To prove that the sequence $\{Q_n\}$ is divergent, it is sufficient to prove that

$$Q_{n+1} = Q_n \pm 1, \quad \text{for } n = 0, 1, 2, \ldots, \tag{7}$$

since it then follows that $\{Q_n\}$ cannot converge.

To prove (7), we put

$$z_n = \frac{1}{2}(x_n + y_n) = \frac{1}{2^{n+1}}(2p_n + 1).$$

Note that

$$[x_{n+1}, y_{n+1}] = \begin{cases} [x_n, z_n], & \text{if } c \in [x_n, z_n], \\ [z_n, y_n], & \text{if } c \in (z_n, y_n). \end{cases}$$

We now claim that, for $n = 0, 1, 2, \ldots$

$$B(z_n) = \frac{1}{2}(B(x_n) + B(y_n)) + \frac{1}{2^{n+1}}. \tag{8}$$

The structure of the proof is as follows
$(8) \Rightarrow (7) \Rightarrow$ Theorem 2.

It would then follow from (8) that, if $[x_{n+1}, y_{n+1}] = [x_n, z_n]$, then

$$Q_{n+1} = 2^{n+1}(B(z_n) - B(x_n))$$
$$= 2^n(B(y_n) - B(x_n)) + 1 = Q_n + 1;$$

whereas, if $[x_{n+1}, y_{n+1}] = [z_n, y_n]$, then

$$Q_{n+1} = 2^{n+1}(B(y_n) - B(z_n))$$
$$= 2^n(B(y_n) - B(x_n)) - 1 = Q_n - 1.$$

In either case, the required result (7) would then hold. Thus, in order to complete the proof that (7) holds, it is sufficient to prove that (8) holds.

To verify (8), note that, for $k = 0, 1, 2, \ldots$, we have

$$\left.\begin{array}{l} 2^k x_n = \dfrac{1}{2^{n-k}} p_n, \\[2mm] 2^k y_n = \dfrac{1}{2^{n-k}}(p_n + 1), \\[2mm] 2^k z_n = \dfrac{1}{2^{n-k+1}}(2p_n + 1), \end{array}\right\} \qquad (9)$$

We shall shortly use the expression $\sum_{k=0}^{\infty} \frac{1}{2^k} f(2^k x)$ for $B(x)$, hence some ks now appear.

and $f(p) = 0$ for any integer p. It then follows from the definition of B as an infinite series that

$$B(x_n) + B(y_n) = \sum_{k=0}^{\infty} \frac{1}{2^k}\left(f(2^k x_n) + f(2^k y_n)\right)$$

$$= \sum_{k=0}^{n-1} \frac{1}{2^k}\left(f(2^k x_n) + f(2^k y_n)\right), \quad \text{and}$$

For, $f(2^k x_n) = f(2^k y_n) = 0$ if $k > n - 1$.

$$B(z_n) = \sum_{k=0}^{\infty} \frac{1}{2^k} f(2^k z_n)$$

$$= \sum_{k=0}^{n} \frac{1}{2^k} f(2^k z_n).$$

For, $f(2^k z_n) = 0$ if $k > n$.

We deduce from (9) that, for $k = 0, 1, 2, \ldots, n - 1$, the terms $2^k x_n$, $2^k y_n$ and $2^k z_n$ all lie in the interval

$$[2^k x_n, \ 2^k y_n] = \left[\frac{1}{2^{n-k}} p_n, \ \frac{1}{2^{n-k}}(p_n + 1)\right],$$

These intervals are among the non-overlapping 'end-point to end-point' intervals along $\mathbb{R}$ of length $\frac{1}{2^{n-k}}$ that include $\left[0, \frac{1}{2^{n-k}}\right]$.

and so in some interval of the form $\left[\frac{1}{2} q_k, \frac{1}{2}(q_k + 1)\right]$, where q_k is an integer.

Now, the restriction of f to such an interval is linear, so that we have

$$f(2^k z_n) = \tfrac{1}{2}\left(f(2^k x_n) + f(2^k y_n)\right), \quad k = 0, 1, 2, \ldots, n - 1;$$

hence

$$\sum_{k=0}^{n-1} \frac{1}{2^k} f(2^k z_n) = \frac{1}{2} \sum_{k=0}^{n-1} \frac{1}{2^k}\left(f(2^k x_n) + f(2^k y_n)\right). \qquad (10)$$

Finally

$$\frac{1}{2^n} f(2^n z_n) = \frac{1}{2^n} f\left(p_n + \frac{1}{2}\right)$$

$$= \frac{1}{2^{n+1}}. \qquad (11)$$

If we then add (10) and (11), we obtain (8), as required.

This completes the proof of Theorem 2.   ∎

6.7   Exercises

Section 6.1

1. Determine whether each of the following functions f is differentiable at the specified point c; if it is, evaluate the derivative $f'(c)$.

(a) $f(x) = \begin{cases} -x^2, & x \le 0, \\ x^2, & x > 0, \end{cases}$ $c = 0;$

(b) $f(x) = \begin{cases} 1, & x < 1, \\ x^2, & x \ge 1, \end{cases}$ $c = 1;$

(c) $f(x) = \begin{cases} \tan x, & -\frac{1}{2}\pi < x < \frac{1}{3}\pi, \\ x^2, & x \ge \frac{1}{3}\pi, \end{cases}$ $c = \frac{1}{3}\pi;$

(d) $f(x) = \begin{cases} 1 - x, & x < 1, \\ x - x^2, & x \ge 1, \end{cases}$ $c = 1;$

(e) $f(x) = \begin{cases} x \sin\left(\frac{1}{x^2}\right), & x \ne 0, \\ 0, & x = 0, \end{cases}$ $c = 0;$

(f) $f(x) = \begin{cases} \sin x \sin\left(\frac{1}{x}\right), & x \ne 0, \\ 0, & x = 0, \end{cases}$ $c = 0.$

2. Write down an expression for a function f with domain $(-1, 2]$ and the following properties:

(a) f'_L exists at 1, and $f'_L(1) = 1;$

(b) f'_R exists at 1, and $f'_R(1) = 2.$

Verify that f has the properties (a) and (b).

Section 6.2

1. Use the rules for differentiation to verify that the function $f(x) = \log_e(1+x) + e^{x^2}$, $x \in (-1, \infty)$, is differentiable on its domain, and determine its derivative.

2. Write down (without justification) the derivatives of the following functions:

(a) $f(x) = \frac{x^2+1}{x-1}$, $x \in (1, \infty);$

(b) $f(x) = \log_e(\sin x)$, $x \in (0, \pi);$

(c) $f(x) = \log_e(\sec x + \tan x)$, $x \in \left(-\frac{1}{2}\pi, \frac{1}{2}\pi\right);$

(d) $f(x) = \frac{\cos x + \sin x}{\cos x - \sin x}$, $x \in \left(0, \frac{1}{4}\pi\right).$

3. Prove that the function $f(x) = \tanh x$, $x \in \mathbb{R}$, has an inverse function f^{-1} that is differentiable on $(-1, 1)$, and find an expression for $(f^{-1})'(x)$.

4. Prove that the function $f(x) = \tan x + 3x$, $x \in \left(-\frac{1}{2}\pi, \frac{1}{2}\pi\right)$, has an inverse function f^{-1} that is differentiable on $\mathbb{R}$, and find the value of $(f^{-1})'(0)$.

5. Determine the derivatives of the following functions, *assuming* that they are differentiable on their domains:

(a) $f(x) = \coth x$, $x \in \mathbb{R} - \{0\};$

(b) $f(x) = (\log_e x)^{\log_e x}$, $x \in (1, \infty).$

Section 6.3

1. The function f has domain $[-2, 2]$ and its graph consists of four line segments, as shown.

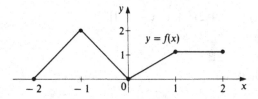

Identify (a) the local minima of f, (b) the minima of f, (c) the local maxima of f, and (d) the maxima of f, and state where these occur,

2. Let f be the function
$$f(x) = 2x^3 - 3x^2, \quad x \in [0, 2].$$
Find (a) the local minima and minima of f, (b) the local maxima and maxima of f, and state where these occur.

3. Prove that, of all rectangles with given perimeter, the square has the greatest area.

4. Verify that the conditions of Rolle's Theorem are satisfied by the function
$$f(x) = 1 + 2x - x^2, \quad x \in [0, 2],$$
and determine a value of c in $(0, 2)$ for which $f'(c) = 0$.

5. Use Rolle's Theorem to prove that, if p is a polynomial and $\lambda_1, \lambda_2, \ldots, \lambda_n$ are distinct zeros of p, then p' has at least $(n-1)$ zeros.

6. Use Rolle's Theorem to prove that, for *any* real number λ, the function
$$f(x) = x^3 - \frac{3}{2}x^2 + \lambda, \quad x \in \mathbb{R},$$
never has two distinct zeros in $[0, 1]$.

   Hint: Assume that f has two distinct zeros in $[0, 1]$, and obtain a contradiction.

7. The function f is twice differentiable on an interval $[a, b]$; and, for some point c in (a, b), $f(a) = f(b) = f(c)$. Prove that there exists some point d in (a, b) with $f''(d) = 0$.

8. The function f is differentiable on a neighbourhood N of a point c; f' is continuous at c and $f'(c) > 0$. Prove that f is increasing on some neighbourhood of c.

   Harder: Give an example of a function f that is differentiable on $\mathbb{R}$ with the properties that (a) $f'(0) > 0$ and (b) there is NO open neighbourhood of 0 on which f is increasing.

9. Let the function $f : [0, 1] \mapsto [0, 1]$ be continuous on $[0, 1]$ and differentiable on $(0, 1)$. We know that there is *at least one* point c in $[0, 1]$ for which $f(c) = c$. Use Rolle's Theorem to prove that, if $f'(x) \neq 1$ in $(0, 1)$, then there is *exactly one* such point c. You saw this in Problem 2 of Sub-section 4.2.1.

   Hint: Consider the function $h(x) = f(x) - x$, $x \in [0, 1]$; assume that two such points c exist, and obtain a contradiction.

Section 6.4

1. For each of the following functions, verify that the conditions of the Mean Value Theorem are satisfied, and determine a value of c that satisfies the conclusion of the theorem:

    (a) $f(x) = \frac{2x-1}{x-2}$, $x \in [-1, 1]$;   (b) $f(x) = x^3 + 2x^2 + x$, $x \in [0, 1]$.

2. Use the Mean Value Theorem to prove that
$$|\sin b - \sin a| \le |b - a|, \quad \text{for } a, b \in \mathbb{R}.$$

3. Use the Zero Derivative Theorem to prove that
$$\cosh^{-1} x = \log_e \left(x + \sqrt{x^2 - 1} \right), \quad \text{for } x \ge 1.$$

4. For the function $f(x) = x^3 - 2x^2 + x$, $x \in \mathbb{R}$, determine those points c where $f'(c) = 0$. Using the Second Derivative Test, determine whether these correspond to local maxima, local minima, or neither.

5. Prove the following inequalities:

    (a) $\log_e x \ge 1 - \frac{1}{x}$,   for $x \in [1, \infty)$;

    (b) $4x^{\frac{1}{4}} \le x + 3$,   for $x \in [0, 1]$;

    (c) $\log_e (1 + x) > x - \frac{1}{2}x^2$,   for $x \in (0, \infty)$.

6. Let f be defined on a neighbourhood of c, with $f'(c) > 0$. Prove that there is some punctured neighbourhood N of c such that
$$\frac{f(x) - f(c)}{x - c} > 0, \quad \text{for } x \in N.$$

7. By applying the Mean Value Theorem to the function $f(x) = \sqrt{x}$ on the interval [100, 102], prove that $10\frac{1}{11} < \sqrt{102} < 10\frac{1}{10}$.

8. Let f be differentiable on a closed interval $[a, b]$.

    (a) Prove that, if the minimum of f on $[a, b]$ occurs at a, then $f'_R(a) \ge 0$; and that, if the minimum of f on $[a, b]$ occurs at b, then $f'_L(b) \le 0$.

    (b) Prove that, if $f'_R(a) < 0$ and $f'_L(b) > 0$, then $f'(x) = 0$ for some $x \in (a, b)$.
    Hint: Check that the minimum of f on $[a, b]$ occurs at some point in (a, b).

    (c) By considering the function $g(x) = f(x) - kx$, $x \in [a, b]$, prove that, if $f'_R(a) < k < f'_L(b)$, then $f'(x) = k$ for some $x \in (a, b)$.

    This result is known as *Darboux's Theorem*, and is an intermediate value theorem for f'.

Section 6.5

1. Verify that the following functions satisfy the conditions of Cauchy's Mean Value Theorem on [0, 2], and determine a value of c that satisfies the conclusion of the theorem
$$f(x) = x^4 + 2x^2 \quad \text{and} \quad g(x) = x^3 + 3x, \quad \text{for } x \in [0, 2].$$

2. Use l'Hôpital's Rule to prove that the following limits exist, and evaluate the limits:

    (a) $\lim\limits_{x \to 0} \frac{\sinh(x + \sin x)}{\sin x}$;   (b) $\lim\limits_{x \to 1} \frac{(5x+3)^{\frac{1}{3}} - (x+3)^{\frac{1}{2}}}{x-1}$;

    (c) $\lim\limits_{x \to 0} \frac{1 - \cos x}{x^2}$;   (d) $\lim\limits_{x \to 0} \frac{\sinh x - x}{\sin(3x^3)}$.

7 Integration

We have already used the idea of the *area* of a set in the plane; for example, to define the number π and to prove that $\frac{\sin x}{x} \to 1$ as $x \to 0$. However, our earlier discussions begged the question of what exactly we mean by 'area'.

For simplicity, we shall restrict our attention in this book to defining only the area between a graph and the x-axis.

In Section 7.1 we use the idea of lower and upper estimates to give a rigorous definition of such an area. This is done by trapping the desired area between underestimates and overestimates, each of which is the sum of the areas of suitably chosen rectangles.

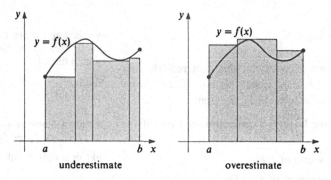

underestimate   overestimate

Then the area between the graph $y = f(x)$, $x \in [a, b]$, and the segment $[a, b]$ of the x-axis is defined to be A if:

the least upper bound of the underestimates $= A$,

and:

the greatest lower bound of the overestimates $= A$.

We call A the integral of f over $[a, b]$, and denote it by

$$\int_a^b f \quad \text{or} \quad \int_a^b f(x)dx.$$

In practice, it would be inconvenient if we had to revert to the definition in order to tell whether a given function was integrable or not, and we shall verify several criteria for integrability.

In Section 7.2, we identify some large classes of integrable functions, and verify the standard rules for integrable functions.

In Section 7.3 we meet the Fundamental Theorem of Calculus which enables us to avoid use of the definition of integrability in order to evaluate many integrals. For instance, we verify the usual methods for integration by parts and integration by substitution.

It follows from the Fundamental Theorem of Calculus that we can think of integration as the operation *inverse* to differentiation.

Often it is not possible to evaluate an integral explicitly, and the best that we can do is to obtain upper and lower estimates for its value. In Section 7.4, we meet a range of inequalities for integrals, including the *Triangle Inequality for integrals*

$$\left| \int_a^b f \right| \le \int_a^b |f|, \quad \text{where } a < b.$$

We use these inequalities to prove *Wallis's Formula* for π, namely that

$$\lim_{n \to \infty} \left(\frac{2}{1} \cdot \frac{2}{3} \cdot \frac{4}{3} \cdot \frac{4}{5} \cdot \frac{6}{5} \cdot \frac{6}{7} \cdot \cdots \cdot \frac{2n}{2n-1} \cdot \frac{2n}{2n+1} \right) = \frac{\pi}{2},$$

and to establish the *Maclaurin Integral Test*, which enables us to determine the convergence or divergence of series such as $\sum_{n=1}^{\infty} \frac{1}{n^p}$, for $p > 0$, and $\sum_{n=2}^{\infty} \frac{1}{n \log_e n}$.

Finally, in Section 7.5, we discuss the evaluation of $n!$. It is easy to evaluate $n!$ for values of n up to 10, say, by direct multiplication; but for $n = 100$ or $n = 200$, the number $n!$ cannot be evaluated by a standard scientific calculator. Surprisingly, we can use integration techniques to obtain an excellent estimate for $n!$, called *Stirling's Formula*

$$n! \sim \sqrt{2\pi n} \left(\frac{n}{e} \right)^n \quad \text{as } n \to \infty.$$

At least not in 2005, when these words are being written.

We shall also explain the *precise* meaning of the symbol tilda: '$\sim$'.

Before starting to read the chapter, we recommend that you refresh your memory of greatest lower bounds and least upper bounds of functions. Recall that, if f is a function defined on an interval $I \subseteq \mathbb{R}$, then:

You met sup and inf in Sub-section 1.4.2.

- A real number m is the **greatest lower bound**, or **infimum**, of f on I if:

  1. m is a lower bound of $f(I)$;
  2. if $m' > m$, then m' is not a lower bound of $f(I)$.

We often denote m by $\inf \{ f(x) : x \in I \}$, $\inf_{x \in I} f(x)$, $\inf_I f$ or simply by $\inf f$.

- M is the **least upper bound**, or **supremum**, of f on I if:

  1. M is an upper bound of $f(I)$;
  2. if $M' < M$, then M' is not an upper bound of $f(I)$.

We often denote M by $\sup \{ f(x) : x \in I \}$, $\sup_{x \in I} f(x)$, $\sup_I f$ or simply by $\sup f$.

In Sections 7.1 and 7.2 we shall discuss some new properties of infimum and supremum that are needed for our work on integrability.

Also, we shall recommend that you omit working your way through quite a number of detailed proofs during your first reading of this chapter. This is not necessarily because they are particularly difficult, but simply in order to guide you through the key ideas first before you return to study some of the 'gory details' later on once you have grasped the overall picture.

7.1 The Riemann integral

Earlier we defined the number π to be the area of the unit disc. We obtained lower and upper estimates for π by trapping the area of the disc between the area of a 3×2^n-sided inner polygon and the area of a 3×2^n-sided outer polygon, and letting $n \to \infty$.

Sub-section 2.5.4.

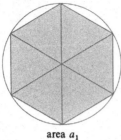

area a_1
underestimate

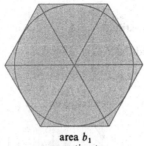

area b_1
overestimate

For simplicity, in this book we do not consider general areas in the plane, but restrict our attention to the area between a graph $y = f(x)$, $x \in [a, b]$, and the interval $[a, b]$ of the x-axis. For a continuous function f, we certainly require such a definition to agree with our intuitive notion of area!

However, it is not obvious that we can always say that the region between a graph and the x-axis *has* an area. For example, how can we define such an area for the following functions?

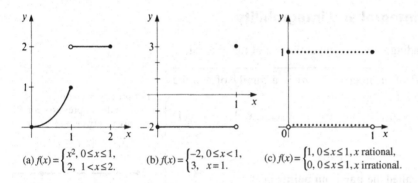

(a) $f(x) = \begin{cases} x^2, & 0 \leq x \leq 1, \\ 2, & 1 < x \leq 2. \end{cases}$ (b) $f(x) = \begin{cases} -2, & 0 \leq x < 1, \\ 3, & x = 1. \end{cases}$ (c) $f(x) = \begin{cases} 1, & 0 \leq x \leq 1, x \text{ rational}, \\ 0, & 0 \leq x \leq 1, x \text{ irrational}. \end{cases}$

Our approach is to find lower and upper estimates for the area, if such a thing exists, that we can assign to each region by splitting it up into smaller regions with areas which we can approximate by rectangles, and use the fact that

> Area of a rectangle = base × height.

We now illustrate the underlying idea of these estimates by considering the situation when the function f is positive on $[a, b]$ (see the diagrams below). First, we divide up the interval $[a, b]$ into a family of smaller intervals, called a *partition* of $[a, b]$. Then we approximate the area by finding two sequences of rectangles each with one of the subintervals as base. In one sequence, we choose rectangles as large as possible so that the sum of their individual areas forms an underestimate for the 'area' of the region; in the other, we choose rectangles as small as possible so that the sum of their individual areas forms an overestimate for the 'area'.

Then, if there is a real number A with the properties:

> the least upper bound of the underestimates $= A$,

When defining π, we used triangles rather than rectangles.

and:

the greatest lower bound of the overestimates $= A$,

we define the area between the graph and the x-axis to be A.

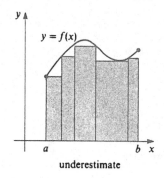
underestimate

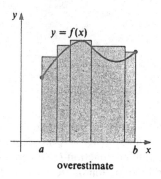
overestimate

We find that we can define such an area for the graphs (a) and (b) above, but not for the graph (c).

7.1.1 The Riemann integral and integrability

We now start on a process leading to the formal definition of the integral.

Definitions A **partition** P of an interval $[a, b]$ is a family of a finite number of subintervals of $[a, b]$

$$P = \{[x_0, x_1], [x_1, x_2], \ldots, [x_{i-1}, x_i], \ldots, [x_{n-1}, x_n]\}, \qquad (1)$$

where

$$a = x_0 < x_1 < x_2 < \cdots < x_{i-1} < x_i < \cdots < x_{n-1} < x_n = b.$$

The points x_i, $0 \le i \le n$, are called the **partition points** in P.

The **length of the ith subinterval** is denoted by $\delta x_i = x_i - x_{i-1}$, and the **mesh** of P is the quantity $\|P\| = \max\limits_{1 \le i \le n} \{\delta x_i\}$.

A **standard partition** is a partition with equal subintervals.

For brevity, we sometimes shorten this expression for P simply to $\{[x_{i-1}, x_i]\}_{i=1}^{n}$ or

$$\{[x_{i-1}, x_i] : 1 \le i \le n\}.$$

$$a = x_0 \; x_1 \cdots x_{i-1} \; x_i \; \cdots \; x_n = b$$

For example, consider the partition P of $[0, 1]$, where

$$P = \left\{ \left[0, \frac{1}{2}\right], \left[\frac{1}{2}, \frac{3}{5}\right], \left[\frac{3}{5}, \frac{3}{4}\right], \left[\frac{3}{4}, 1\right] \right\}.$$

Equivalently

$$P = \{[0, 0.5], [0.5, 0.6],$$
$$[0.6, 0.75], [0.75, 1]\}.$$

Here

$$\delta x_1 = \frac{1}{2} - 0 = \frac{1}{2}, \; \delta x_2 = \frac{3}{5} - \frac{1}{2} = \frac{1}{10}, \; \delta x_3 = \frac{3}{4} - \frac{3}{5} = \frac{3}{20} \quad \text{and}$$

$$\delta x_4 = 1 - \frac{3}{4} = \frac{1}{4},$$

and the mesh of P is

$$\|P\| = \max\left\{\frac{1}{2}, \frac{1}{10}, \frac{3}{20}, \frac{1}{4}\right\} = \frac{1}{2}.$$

P is not a standard partition of $[0, 1]$, since not all its subintervals are of equal length.

Next, we introduce some notation, m_i and M_i, associated with the height of the graph of a bounded function f on its subintervals. Since we are intending to develop a definition of integral that will apply to discontinuous functions as well as to continuous functions, we use the concepts of greatest lower bound and least upper bound of f on subintervals rather than minimum and maximum of f on the subintervals.

We also introduce some quantities, $L(f, P)$ and $U(f, P)$, that correspond to underestimates and overestimates of the possible area.

We need to do this, since, if f is not continuous on a typical subinterval $[x_{i-1}, x_i]$, it may not possess a minimum or a maximum on the subinterval.

> **Definitions** Let f be a bounded function on $[a, b]$, and P a partition of $[a, b]$ given by $P = \{[x_{i-1}, x_i] : 1 \le i \le n\}$. We denote by $\boldsymbol{m_i}$ and $\boldsymbol{M_i}$ the quantities
> $$m_i = \inf\{f(x) : x \in [x_{i-1}, x_i]\} \quad \text{and} \quad M_i = \sup\{f(x) : x \in [x_{i-1}, x_i]\}.$$
> Then the corresponding **lower** and **upper Riemann sums for** f on $[a, b]$ are
> $$L(f, P) = \sum_{i=1}^{n} m_i \delta x_i \quad \text{and} \quad U(f, P) = \sum_{i=1}^{n} M_i \delta x_i.$$

You met infimum and supremum earlier, in Sub-section 1.4.2. The quantities m_i and M_i exist, since f is bounded.

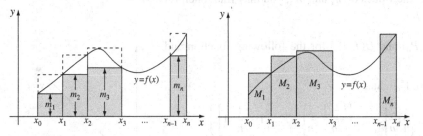

Example 1 Let $f(x) = x$, $x \in [0, 1]$, and let $P = \left\{\left[0, \frac{1}{5}\right], \left[\frac{1}{5}, \frac{1}{2}\right], \left[\frac{1}{2}, 1\right]\right\}$ be a partition of $[0, 1]$. Evaluate $L(f, P)$ and $U(f, P)$.

Solution In this case, the function f is increasing and continuous. Thus, on each subinterval in $[0, 1]$, the infimum of f is the value of f at the left end-point of the subinterval and the supremum of f is the value of f at the right end-point of the subinterval.

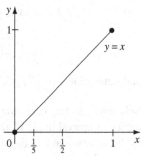

Hence, on the three subintervals in P, we have

$$m_1 = f(0) = 0, \qquad M_1 = f\left(\frac{1}{5}\right) = \frac{1}{5}, \qquad \delta x_1 = \frac{1}{5} - 0 = \frac{1}{5},$$

$$m_2 = f\left(\frac{1}{5}\right) = \frac{1}{5}, \qquad M_2 = f\left(\frac{1}{2}\right) = \frac{1}{2}, \qquad \delta x_2 = \frac{1}{2} - \frac{1}{5} = \frac{3}{10},$$

$$m_3 = f\left(\frac{1}{2}\right) = \frac{1}{2}, \qquad M_3 = f(1) = 1, \qquad \delta x_3 = 1 - \frac{1}{2} = \frac{1}{2}.$$

We use the fact that, if a function is continuous on a closed interval, then it *attains* its infimum and supremum there, by the Extreme Values Theorem in Sub-section 4.2.3.

It then follows, from the definitions of $L(f, P)$ and $U(f, P)$, that

$$L(f, P) = \sum_{i=1}^{3} m_i \delta x_i = m_1 \delta x_1 + m_2 \delta x_2 + m_3 \delta x_3$$

$$= 0 \times \frac{1}{5} + \frac{1}{5} \times \frac{3}{10} + \frac{1}{2} \times \frac{1}{2}$$

$$= 0 + \frac{3}{50} + \frac{1}{4} = \frac{31}{100},$$

$$U(f,P) = \sum_{i=1}^{3} M_i \delta x_i = M_1 \delta x_1 + M_2 \delta x_2 + M_3 \delta x_3$$

$$= \frac{1}{5} \times \frac{1}{5} + \frac{1}{2} \times \frac{3}{10} + 1 \times \frac{1}{2}$$

$$= \frac{1}{25} + \frac{3}{20} + \frac{1}{2} = \frac{69}{100}. \qquad \Box$$

Notice how a careful laying out of the calculation for the various terms makes it straight-forward to calculate these two sums.

Problem 1 Evaluate $L\,(f,\,P)$ and $U\,(f,\,P)$ for the following function and partition of $[0, 1]$:

$$f(x) = \begin{cases} 2x, & 0 \le x < \frac{1}{2}, \ \frac{1}{2} < x < 1, \\ 0, & x = \frac{1}{2}, \\ 1, & x = 1, \end{cases}$$

$x \in [0, 1]$, and $P = \left\{ \left[0, \frac{1}{3}\right], \left[\frac{1}{3}, \frac{3}{4}\right], \left[\frac{3}{4}, 1\right] \right\}$.

Hint: Be careful over the values of m_2 and M_3; you may find it helpful to sketch the graph of f.

It is quite important that you tackle this problem, to make sure that you understand the notation being used. You should also read its solution carefully.

Example 2 Evaluate $L(f, P_n)$ and $U(f, P_n)$ for the following function and standard partition of $[0, 1]$

$$f(x) = x, \ x \in [0, 1], \quad \text{and}$$

$$P_n = \left\{ \left[0, \frac{1}{n}\right], \left[\frac{1}{n}, \frac{2}{n}\right], \dots, \left[\frac{i-1}{n}, \frac{i}{n}\right], \dots, \left[1 - \frac{1}{n}, 1\right] \right\};$$

and determine $\lim_{n\to\infty} L(f, P_n)$ and $\lim_{n\to\infty} U(f, P_n)$, if these exist.

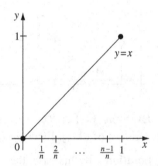

Solution In this case, the function f is increasing and continuous. Thus, on each subinterval in $[0, 1]$, the infimum of f is the value of f at the left end-point of the subinterval and the supremum of f is the value of f at the right end-point of the subinterval.

We follow the general structure of the solution to Example 1.

Hence, on the ith subinterval $\left[\frac{i-1}{n}, \frac{i}{n}\right]$ in P_n, for $1 \le i \le n$, we have

$$m_i = f\left(\frac{i-1}{n}\right) = \frac{i-1}{n}, \quad M_i = f\left(\frac{i}{n}\right) = \frac{i}{n}, \quad \text{and}$$

$$\delta x_i = \frac{i}{n} - \frac{i-1}{n} = \frac{1}{n}.$$

It then follows, from the definitions of $L(f, P_n)$ and $U(f, P_n)$, that

$$L(f, P_n) = \sum_{i=1}^{n} m_i \delta x_i = \sum_{i=1}^{n} \frac{i-1}{n} \times \frac{1}{n}$$

$$= \frac{1}{n^2} \left\{ \sum_{i=1}^{n} i - \sum_{i=1}^{n} 1 \right\}$$

$$= \frac{1}{n^2} \left\{ \frac{n(n+1)}{2} - n \right\} = \frac{n-1}{2n},$$

$$U(f, P_n) = \sum_{i=1}^{n} M_i \delta x_i = \sum_{i=1}^{n} \frac{i}{n} \times \frac{1}{n}$$

$$= \frac{1}{n^2} \sum_{i=1}^{n} i$$

$$= \frac{1}{n^2} \times \frac{n(n+1)}{2} = \frac{n+1}{2n}.$$

It follows that

$$\lim_{n \to \infty} L(f, P_n) = \lim_{n \to \infty} \frac{n-1}{2n} = \frac{1}{2} \quad \text{and}$$

$$\lim_{n \to \infty} U(f, P_n) = \lim_{n \to \infty} \frac{n+1}{2n} = \frac{1}{2}. \qquad \qquad \square$$

Problem 2 Evaluate $L(f, P_n)$ and $U(f, P_n)$ for the following function and standard partition of $[0, 1]$

$$f(x) = x^2, \quad x \in [0, 1], \quad \text{and}$$

$$P_n = \left\{ \left[0, \frac{1}{n}\right], \left[\frac{1}{n}, \frac{2}{n}\right], \ldots, \left[\frac{i-1}{n}, \frac{i}{n}\right], \ldots, \left[1 - \frac{1}{n}, 1\right] \right\};$$

and determine $\lim_{n \to \infty} L(f, P_n)$ and $\lim_{n \to \infty} U(f, P_n)$, if these exist.

Properties of Riemann sums

In all the examples that we have seen so far, the lower Riemann sum for f over an interval has been less than or equal to the corresponding upper Riemann sum. Also, in Example 2 we found that the value that 'we hope for' for $\int_0^1 x \, dx$, namely $\frac{1}{2}$, is greater than all the lower Riemann sums and less than all the upper Riemann sums; it seems reasonable to ask whether such a property holds in general.

So we now examine some of the key properties of Riemann sums. We shall need to use some properties of least upper bounds and greatest lower bounds.

> **Theorem 1** For any function f bounded on an interval $[a, b]$ and any partition P of $[a, b]$, $L(f, P) \leq U(f, P)$.

Proof Let $P = \{[x_{i-1}, x_i] : 1 \leq i \leq n\}$. Then, on each subinterval $[x_{i-1}, x_i]$, $1 \leq i \leq n$, we have $\inf\{f(x) : x \in [x_{i-1}, x_i]\} \leq \sup\{f(x) : x \in [x_{i-1}, x_i]\}$ – in other words, $m_i \leq M_i$.

It follows that

$$\sum_{i=1}^{n} m_i \delta x_i \leq \sum_{i=1}^{n} M_i \delta x_i;$$

in other words, $L(f, P) \leq U(f, P)$, as required. ∎

Next, we need some techniques for comparing the Riemann sums for different partitions on the interval $[a, b]$; this will enable us shortly to verify that, for two partitions P and P' of $[a, b]$, we must have $L(f, P) \leq U(f, P')$.

Here we are assuming that our rigorous treatment of integration will give the same answers as those that you obtained in your initial Calculus course!

We will set these out carefully as we need them.

For each term in the left-hand sum is less than or equal to the corresponding term in the right-hand sum.

Theorem 3, below.

Notice that this fact does NOT follow from Theorem 1, which deals only with *one* partition at a time.

Definition   Let $P = \{[x_0, x_1], [x_1, x_2], \ldots, [x_{i-1}, x_i], \ldots, [x_{n-1}, x_n]\}$ be a partition of an interval $[a, b]$. Then a partition P' of $[a, b]$ is a **refinement** of P if the partition points of P' include the partition points of P. A partition Q of $[a, b]$ is the **common refinement** of two partitions P and P' of $[a, b]$ if the partition points of Q comprise the partition points of P together with the partition points of P'.

For example, the partition $P' = \left\{[0, \frac{1}{2}], [\frac{1}{2}, \frac{3}{5}], [\frac{3}{5}, \frac{3}{4}], [\frac{3}{4}, 1]\right\}$ of $[0, 1]$ is a refinement of the partition $P = \left\{[0, \frac{1}{2}], [\frac{1}{2}, \frac{3}{4}], [\frac{3}{4}, 1]\right\}$, since it simply has one additional partition point $\frac{3}{5}$ as compared with P. Similarly, $P'' = \left\{[0, \frac{1}{3}], [\frac{1}{3}, \frac{1}{2}], [\frac{1}{2}, \frac{3}{5}], [\frac{3}{5}, \frac{3}{4}], [\frac{3}{4}, 1]\right\}$ is also a refinement of P. However the partition $Q = \left\{[0, \frac{1}{5}], [\frac{1}{5}, \frac{1}{2}], [\frac{1}{2}, 1]\right\}$ of $[0, 1]$ is not a refinement of P, since its partition points do not include all the partition points of P.

> The point $\frac{3}{4}$ is missing from Q.

Also, the common refinement of the partitions $P = \left\{[0, \frac{1}{2}], [\frac{1}{2}, \frac{3}{4}], [\frac{3}{4}, 1]\right\}$ and $P' = \left\{[0, \frac{1}{3}], [\frac{1}{3}, \frac{2}{3}], [\frac{2}{3}, 1]\right\}$ is

$$\left\{\left[0, \frac{1}{3}\right], \left[\frac{1}{3}, \frac{1}{2}\right], \left[\frac{1}{2}, \frac{2}{3}\right], \left[\frac{2}{3}, \frac{3}{4}\right], \left[\frac{3}{4}, 1\right]\right\}.$$

We shall need the following crucial result in our work on refinements.

Lemma 1   For any bounded function f defined on intervals I and J, with $I \subseteq J$, we have

$$\inf_{x \in J} f \leq \inf_{x \in I} f \quad \text{and} \quad \sup_{x \in I} f \leq \sup_{x \in J} f.$$

> Loosely speaking, the larger interval gives the function more space to get smaller and more space to get larger.

Proof   By definition of greatest lower bound, we know that $\inf\limits_{x \in J} f \leq f(x)$, for $x \in J$. It follows, from the fact that $I \subseteq J$, that

$$\inf_{x \in J} f \leq f(x), \quad \text{for } x \in I.$$

Hence $\inf\limits_{x \in J} f$ is a lower bound for f on I, so that $\inf\limits_{x \in J} f \leq \inf\limits_{x \in I} f$.

The proof that $\sup\limits_{x \in I} f \leq \sup\limits_{x \in J} f$ is similar, so we omit it.   ∎

> For $\inf\limits_{x \in I} f$ is the *greatest* lower bound of f on I.

Problem 3   Prove that, for any bounded function f defined on intervals I and J where $I \subseteq J$, $\sup\limits_{x \in I} f \leq \sup\limits_{x \in J} f$.

Lemma 2   Let f be a bounded function on an interval $[a, b]$. Let P and P' be partitions of $[a, b]$, where P' is a refinement of P that contains just one additional partition point. Then

$$L(f, P) \leq L(f, P') \quad \text{and} \quad U(f, P') \leq U(f, P).$$

> Loosely speaking, the addition of one partition point increases L and decreases U.

Proof   Let P be the partition

$$P = \{[x_0, x_1], [x_1, x_2], \ldots, [x_{i-1}, x_i], \ldots, [x_{n-1}, x_n]\}$$

of $[a, b]$, and suppose that P' contains an additional partition point c in the particular subinterval $[\alpha, \beta]$ of P.

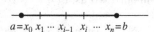

Note that c must be an interior point of $[\alpha, \beta]$. For we are assuming that all the partition points of partitions are distinct; and, if c were an end-point α or β, then P' would contain that point twice in its set of partition points.

Since $[\alpha, c] \subset [\alpha, \beta]$, it follows from Lemma 1 that

$$\inf_{x \in [\alpha, \beta]} f \leq \inf_{x \in [\alpha, c]} f \quad \text{and} \quad \inf_{x \in [\alpha, \beta]} f \leq \inf_{x \in [c, \beta]} f. \tag{2}$$

We use α and β here rather than x_{i-1} and x_i simply in order to avoid sub-subscripts.

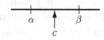

Now, the terms in the lower sums $L(f, P)$ and $L(f, P')$ are the same, except that the contribution to $L(f, P)$ associated with the interval $[\alpha, \beta]$ is

$$\inf_{x \in [\alpha, \beta]} f \times (\beta - \alpha),$$

whereas the contribution to $L(f, P')$ associated with the intervals $[\alpha, c]$ and $[c, \beta]$ is

$$\inf_{x \in [\alpha, c]} f \times (c - \alpha) + \inf_{x \in [c, \beta]} f \times (\beta - c).$$

It then follows, from the inequalities (2), that

$$\inf_{x \in [\alpha, c]} f \times (c - \alpha) + \inf_{x \in [c, \beta]} f \times (\beta - c) \geq \inf_{x \in [\alpha, \beta]} f \times (c - \alpha) + \inf_{x \in [\alpha, \beta]} f \times (\beta - c)$$

$$= \inf_{x \in [\alpha, \beta]} f \times \{(c - \alpha) + (\beta - c)\}$$

$$= \inf_{x \in [\alpha, \beta]} f \times (\beta - \alpha).$$

Since the lower sums $L(f, P)$ and $L(f, P')$ are the same, apart from these contributions to each, it follows that $L(f, P') \geq L(f, P)$, as required.

The proof that $U(f, P') \leq U(f, P)$ is similar, so we omit it. ∎

Lemma 2 shows that the addition of just one point to a partition increases the lower Riemann sum and decreases the upper Riemann sum. By applying this fact a finite number of times, we deduce the following general result.

Theorem 2 Let f be a bounded function on an interval $[a,b]$, and let P and P' be partitions of $[a,b]$, where P' is a refinement of P. then

$$L(f,P) \leq L(f,P') \quad \text{and} \quad U(f,P') \leq U(f,P).$$

Loosely speaking, refining a partition increases L and decreases U.

We now compare the Riemann sums for two different partitions, and discover that *all* the lower Riemann sums of a bounded function on an interval are less than or equal to *all* the upper Riemann sums. This is a significant improvement on the result of Theorem 1 that, for a given partition P of $[a,b]$, $L(f, P) \leq U(f, P)$.

Theorem 3 Let f be a bounded function on an interval $[a, b]$, and let P and P' be partitions of $[a, b]$. Then

$$L(f,P) \leq U(f,P').$$

Proof Let Q be the common partition of P and P'. Then, since Q is a refinement of both P and P', we have:

$L(f, P) \leq L(f, Q), \qquad$ by Theorem 2,

$L(f, Q) \leq U(f, Q), \qquad$ by Theorem 1,

$U(f, Q) \leq U(f, P'), \qquad$ by Theorem 2.

It follows from this chain of inequalities that $L(f, P) \leq U(f, P')$, as required. ∎

This result is exactly what makes our definitions of lower and upper Riemann sums useful – whatever the integral of f over an interval $[a, b]$ might be, if it even exists, we are certainly using lower Riemann sums, all of which provide underestimates, and upper Riemann sums, all of which provide overestimates.

Definitions   Let f be a bounded function on an interval $[a, b]$. Then we define:

- the **lower integral of** f on $[a, b]$ to be $\underline{\int_a^b} f = \sup_P L(f, P)$,

- the **upper integral of** f on $[a, b]$ to be $\overline{\int_a^b} f = \inf_P U(f, P)$,

where P denotes the set of all partitions of $[a, b]$.

Further, if the lower and upper integrals are equal, we define the **integral of** f on $[a, b]$, $\int_a^b f$, to be their common value; that is

$$\int_a^b f = \underline{\int_a^b} f = \overline{\int_a^b} f.$$

Sometimes written as $\underline{\int_a^b} f(x)dx$.

Sometimes written as $\overline{\int_a^b} f(x)dx$.

Often written as $\int_a^b f(x)dx$.

It is all too easy to be lulled into a sense of false security by a firmly stated definition! So we now prove the following result that assures us that the above definitions make sense.

Theorem 4   Let f be a bounded function on an interval $[a, b]$. Then:
(a) The lower integral $\underline{\int_a^b} f$ and the upper integral $\overline{\int_a^b} f$ both exist;
(b) $\underline{\int_a^b} f \leq \overline{\int_a^b} f$.

Proof

(a) Since f is bounded on $[a, b]$, there is some number M such that $|f(x)| \leq M$ on $[a, b]$; in particular,

$$f(x) \leq M, \quad \text{for } x \in [a, b].$$

Thus for any partition $P = \{[x_{i-1}, x_i]: 1 \leq i \leq n\}$ of $[a, b]$, we have

$$f(x) \leq M, \quad \text{for } x \in [x_{i-1}, x_i] \text{ and } 1 \leq i \leq n;$$

in particular, we have

$$m_i = \inf_{[x_{i-1}, x_i]} f(x) \leq M.$$

For, $\inf_{[x_{i-1}, x_i]} f(x) \leq f(x_i) \leq M$.

It follows that

$$L(f, P) = \sum_{i=1}^n m_i \delta x_i \leq \sum_{i=1}^n M \delta x_i$$

$$= M \sum_{i=1}^n \delta x_i = M(b - a).$$

Since all the lower sums $L(f, P)$ are bounded above by $M(b - a)$, it follows that the greatest lower bound of the lower sums, $\sup_P L(f, P)$, must exist. This greatest lower bound is precisely the lower integral $\underline{\int_a^b} f$.

Also, $\underline{\int_a^b} f \leq M(b - a)$.

The proof of the existence of the upper integral $\overline{\int_a^b} f$ is very similar, so we omit it here.

(b) Let P and P' be any two partitions of $[a, b]$. We know, from Theorem 3, that

$$L(f,P) \leq U(f,P').$$

If we fix P' for the moment, then we know that $U(f, P')$ serves as an upper bound for all the lower Riemann sums $L(f, P)$, whichever partition P may be. It follows, then, that the least upper bound of the $L(f, P)$, the lower integral $\underline{\int_a^b} f$, must satisfy the inequality

$$\underline{\int_a^b} f \leq U(f,P'). \tag{3}$$

It follows from the inequality (3) that $\underline{\int_a^b} f$ serves as a lower bound for all the upper Riemann sums $U(f, P')$, whichever partition P' may be. It follows, then, that the greatest lower bound of the $U(f, P')$, the upper integral $\overline{\int_a^b} f$, must satisfy the inequality

$$\underline{\int_a^b} f \leq \overline{\int_a^b} f. \qquad\qquad\blacksquare$$

Remark

It follows, from the definition of the integral (when it exists) as

$$\int_a^b f = \sup_P L(f,P) = \inf_P U(f,P),$$

that, for *any* partition P of $[a, b]$

$$L(f,P) \leq \int_a^b f \leq U(f,P).$$

Now, we have already seen that, if $f(x) = x$, $x \in [0, 1]$, and $P_n = \{[0,\frac{1}{n}], [\frac{1}{n},\frac{2}{n}], \ldots, [1-\frac{1}{n},1]\}$ is a standard partition of $[0, 1]$, then   Example 2, above.

$$\lim_{n\to\infty} L(f,P_n) = \frac{1}{2} \quad \text{and} \quad \lim_{n\to\infty} U(f,P_n) = \frac{1}{2}. \tag{4}$$

Since f is bounded on $[0, 1]$, we know that the lower integral $\underline{\int_0^1} f$ exists. But $L(f,P_n) \leq \underline{\int_0^1} f$, so that, by the Limit Inequality Rule for sequences, it follows from (4) that

$$\frac{1}{2} \leq \underline{\int_0^1} f. \tag{5}$$

Similarly, since f is bounded on $[0, 1]$, we know that the upper integral $\overline{\int_0^1} f$ exists. But $\overline{\int_0^1} f \leq U(f,P_n)$, so that, by the Limit Inequality Rule for sequences, it follows from (4) that

$$\overline{\int_0^1} f \leq \frac{1}{2}. \tag{6}$$

We then observe that

$$\int_0^{\overline{1}} f \leq \frac{1}{2} \leq \int_{\underline{0}}^1 f, \text{ in light of (5) and (6), and}$$

$$\int_{\underline{0}}^1 f \leq \int_0^{\overline{1}} f, \text{ by part (b) of Theorem 4.}$$

It follows that we must have $\int_{\underline{0}}^1 f = \int_0^{\overline{1}} f = \frac{1}{2}$, so that f is integrable on $[0, 1]$

and $\int_0^1 f = \frac{1}{2}$.   In other words, $\int_0^1 x\,dx = \frac{1}{2}$.

> **Problem 4**   Use the result of Problem 2 to prove that the function
> $f(x) = x^2$ is integrable on $[0, 1]$, and evaluate $\int_0^1 f$.

> **Problem 5**   Let f be the constant function $x \mapsto k$ on $[0, 1]$, and
> $P_n = \{[0, \frac{1}{n}], [\frac{1}{n}, \frac{2}{n}], \ldots, [1 - \frac{1}{n}, 1]\}$ a standard partition of $[0, 1]$. Cal-
> culate the Riemann sums $L(f, P_n)$ and $U(f, P_n)$. Hence prove that f is
> integrable on $[0, 1]$, and evaluate $\int_0^1 f$.

However, not all bounded functions defined on closed intervals are integrable!

Example 3   Prove that the function

$$f(x) = \begin{cases} 1, & 0 \leq x \leq 1, \quad x \text{ rational,} \\ 0, & 0 \leq x \leq 1, \quad x \text{ irrational,} \end{cases}$$

is not integrable on $[0, 1]$.

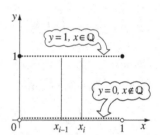

Solution   Let $P = \{[x_0, x_1], [x_1, x_2], \ldots, [x_{i-1}, x_i], \ldots, [x_{n-1}, x_n]\}$ be any
partition of $[0, 1]$.
    Then

$$L(f, P) = \sum_{i=1}^n m_i \delta x_i = \sum_{i=1}^n 0 \times \delta x_i$$
$$= 0,$$

and

$$U(f, P) = \sum_{i=1}^n M_i \delta x_i = \sum_{i=1}^n 1 \times \delta x_i$$
$$= \sum_{i=1}^n \delta x_i = 1.$$

For, each subinterval $[x_{i-1}, x_i]$
contains both rational and
irrational points.

Since all the lower Riemann sums are 0, their least upper bound $\int_{\underline{0}}^1 f$ is also
zero. Similarly, since all the upper Riemann sums are 1, their greatest lower

bound $\int_0^{\overline{1}} f$ is also 1.

    Since $\int_{\underline{0}}^1 f \neq \int_0^{\overline{1}} f$, it follows that f is not integrable on $[0, 1]$.   □

7.1.2   Criteria for integrability

It would be tedious to have to go through the ab initio discussion of integrability on each occasion that we wished to determine whether a given function
was integrable on a given closed interval. Therefore we now meet three criteria
that we often use to avoid that process.

We suggest that at a first
reading you omit ALL the
proofs in this sub-section but
read the rest of the text.

The first criterion is of particular interest in that its statement says nothing about the value of the integral itself: it mentions only the difference between the upper and the lower Riemann sums.

Theorem 5   Riemann's Criterion for integrability

Let f be a bounded function on an interval $[a, b]$. Then

$$f \text{ is integrable on } [a, b]$$

if and only if:

for each positive number ε, there is a partition P of $[a, b]$ for which

$$U(f, P) - L(f, P) < \varepsilon. \tag{7}$$

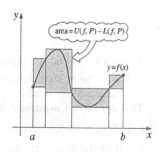

Proof   Suppose, first, that f is integrable on $[a, b]$. Then

$$\sup_{P} L(f, P) = \inf_{P} U(f, P), \text{ over all partitions } P \text{ of } [a, b];$$

denote by I the common value of these two quantities.

It follows that, for any given positive number ε, there are partitions Q and Q' of $[a, b]$ for which

$$L(f, Q) > I - \frac{1}{2}\varepsilon \quad \text{and} \quad U(f, Q') < I + \frac{1}{2}\varepsilon. \tag{8}$$

Now let P be the common refinement of Q and Q'. It follows from (8) and Theorem 2 that

$$L(f, P) \geq L(f, Q) > I - \frac{1}{2}\varepsilon,$$

$$U(f, P) \leq U(f, Q') < I + \frac{1}{2}\varepsilon.$$

Recall that $L(f, P) \leq U(f, P)$, for any partition P.

That is, that refining a partition increases L and decreases U.

Hence, we may deduce from subtracting these inequalities that

$$U(f, P) - L(f, P) < \left\{ I + \frac{1}{2}\varepsilon \right\} - \left\{ I - \frac{1}{2}\varepsilon \right\}$$

$$= \varepsilon.$$

This is precisely the assertion (7).

Suppose next that the statement (7) holds; that is, that for each positive number ε, there is some partition P of $[a, b]$ for which

$$U(f, P) - L(f, P) < \varepsilon. \tag{9}$$

Since $\overline{\int_a^b} f \leq U(f, P)$ and $\underline{\int_a^b} f \geq L(f, P)$, whatever partition P is, it follows from (9) that

$$\overline{\int_a^b} f - \underline{\int_a^b} f \leq U(f, P) - L(f, P)$$

$$< \varepsilon. \tag{10}$$

By (9).

Since the left-hand side of (10) is some non-negative number independent of ε, it follows that $\overline{\int_a^b} f - \underline{\int_a^b} f = 0$. In other words, f is integrable on $[a, b]$, as required. ∎

Problem 6 Use Riemann's Criterion to determine the integrability of the following functions on [0, 1]:

(a) $f(x) = \begin{cases} -2, & 0 \le x < 1, \\ 3, & x = 1; \end{cases}$

(b) $f(x) = \begin{cases} 1, & 0 \le x \le 1, \quad x \text{ rational}, \\ 0, & 0 \le x \le 1, \quad x \text{ irrational}. \end{cases}$

Our next criterion for integrability is phrased in terms of a *sequence* of partitions and its statement involves a value for the integral.

Theorem 6   Common Limit Criterion

Let f be a bounded function on an interval $[a, b]$.

(a) If f is integrable on $[a, b]$, then there is a sequence $\{P_n\}$ of partitions of $[a, b]$ such that

$$\lim_{n\to\infty} L(f, P_n) = \int_a^b f \quad \text{and} \quad \lim_{n\to\infty} U(f, P_n) = \int_a^b f. \quad (11)$$

(b) If there is a sequence $\{P_n\}$ of partitions of $[a, b]$ such that

$$\lim_{n\to\infty} L(f, P_n) \quad \text{and} \quad \lim_{n\to\infty} U(f, P_n) \text{ both exist and are equal,} \quad (12)$$

then f is integrable on $[a, b]$ and the common value of these two limits is $\int_a^b f$.

Proof

(a) Suppose first that f is integrable on $[a, b]$, and let I denote $\int_a^b f$.

Then, corresponding to each integer $n \ge 1$, there exist some partitions of $[a, b]$, Q_n and Q_n' say, for which

$$L(f, Q_n) > I - \frac{1}{n} \quad \text{and} \quad U(f, Q_n') < I + \frac{1}{n}. \quad (13)$$

Now let P_n be the common refinement of Q_n and Q_n'. It follows that

$$I - \frac{1}{n} < L(f, Q_n) \quad \text{(by (13))}$$
$$\le L(f, P_n) \quad \text{(since } P_n \text{ is a refinement of } Q_n)$$
$$\le U(f, P_n)$$
$$\le U(f, Q_n') \quad \text{(since } P_n \text{ is a refinement of } Q_n')$$
$$< I + \frac{1}{n}. \quad \text{(by (13))}$$

Since the sequences $\{I - \frac{1}{n}\}$ and $\{I + \frac{1}{n}\}$ both converge to I as $n \to \infty$, it follows from the Squeeze Rule for sequences and the above inequalities that

$$L(f, P_n) \to I \quad \text{and} \quad U(f, P_n) \to I \quad \text{as } n \to \infty.$$

This is precisely the statement (11).

(b) Suppose next that the statement (12) holds. Let I denote the common value of the two limits.

It follows from (12) that, for any given positive number ε, there are partitions Q and Q' of $[a, b]$ for which

$$L(f, Q) > I - \frac{1}{2}\varepsilon \quad \text{and} \quad U(f, Q') < I + \frac{1}{2}\varepsilon.$$

So, if we let P be the common refinement of Q and Q', it follows that

$$L(f, P) \geq L(f, Q) > I - \frac{1}{2}\varepsilon,$$

$$U(f, P) \leq U(f, Q') < I + \frac{1}{2}\varepsilon.$$

Hence, we may deduce from subtracting these inequalities that

$$U(f, P) - L(f, P) < \left\{I + \frac{1}{2}\varepsilon\right\} - \left\{I - \frac{1}{2}\varepsilon\right\}$$

$$= \varepsilon.$$

It follows from Theorem 5 that the function f is therefore integrable on $[a, b]$.

Finally, we use the sequence $\{P_n\}$ of partitions mentioned in the statement of part (b). For them, if we let $n \to \infty$ in the inequality

> We have just proved that f is integrable on $[a, b]$. Here we determine the value of its integral over $[a, b]$.

$$L(f, P_n) \leq \int_a^b f \leq U(f, P_n),$$

and use the assumption (12), it follows that $\lim_{n \to \infty} L(f, P_n) = \int_a^b f$ and $\lim_{n \to \infty} U(f, P_n) = \int_a^b f$. This concludes the proof. ∎

Now the statement of Theorem 6 can be loosely paraphrased as saying that a function f is integrable on $[a, b]$ if and only if there is a sequence of partitions $\{P_n\}$ such that the corresponding lower and upper Riemann sums tend to a common value. This gives us no information on what sort of sequences we should examine to verify that f is integrable. Our next result says that we only need to examine any sequence of partitions that we please *whose mesh tends to zero*. For instance, a function f is integrable on $[a, b]$ if and only if the lower and upper Riemann sums for the sequence of standard partitions $\{P_n\}$ of $[a, b]$ tend to a common value.

> This is a wonderful simplication!

Theorem 7   Null Partitions Criterion

Let f be a bounded function on an interval $[a, b]$, and $\{P_n\}$ any sequence of partitions of $[a, b]$ with $\|P_n\| \to 0$.

(a) If f is integrable on $[a, b]$, then

$$\lim_{n \to \infty} L(f, P_n) = \int_a^b f \quad \text{and} \quad \lim_{n \to \infty} U(f, P_n) = \int_a^b f.$$

(b) If there is a number I such that $\lim_{n \to \infty} L(f, P_n)$ and $\lim_{n \to \infty} U(f, P_n)$ both exist and equal I, then f is integrable on $[a, b]$ and $I = \int_a^b f$.

> Often the partitions P_n will be standard partitions of $[a, b]$.

Part (b) is simply a special case of part (b) of Theorem 6, for it is true even without the additional assumption that $\|P_n\| \to 0$.

Part (a) must be less straight-forward, for, if we drop the assumption that $\|P_n\| \to 0$, then it may not be the case that $\lim_{n \to \infty} L(f, P_n) = \int_a^b f$ or $\lim_{n \to \infty} U(f, P_n) = \int_a^b f$. To see this, consider the function $f(x) = x$, $x \in [0, 1]$.

We have already seen that $\int_0^1 f = \frac{1}{2}$. However, if we choose for each partition P_n simply the trivial partition $P_n = \{[0,1]\}$, then $L(f, P_n) = 0$ and $U(f, P_n) = 1$ for each n.

For $\inf_{[0,1]} f = 0$ and $\sup_{[0,1]} f = 1$.

We have not used a condition such as $\|P_n\| \to 0$ before, so we need to establish the following preliminary result before we can tackle the proof of part (a) of Theorem 7.

Lemma 3   Let f be a bounded function on an interval $[a, b]$, and let δ be any positive number. Let P be a partition of $[a, b]$ with $\|P\| < \delta$, and P' a partition of $[a, b]$ with the same partition points as P together with at most N additional partition points. Then

$$U(f, P') \geq U(f, P) - N(M - m)\delta,$$

where $m = \inf_{x \in [a,b]} f(x)$ and $M = \sup_{x \in [a,b]} f(x)$.

There is an analogous result for lower Riemann sums, but we do not need it.

Remark

We already know, since P' is a refinement of P, that $U(f, P') \leq U(f, P)$. Lemma 3 gives us some lower bound to *how much smaller* $U(f, P')$ can be than $U(f, P)$.

Theorem 2, Sub-section 7.1.1.

Proof   Let c be the first partition point of P' that is not a partition point of P. Then, if $P = \{[x_0, x_1], [x_1, x_2], \ldots, [x_{n-1}, x_n]\}$, there is some integer i for which $x_{i-1} < c < x_i$. Denote by Q the partition of $[a, b]$ whose partition points are those of P together with the additional point c.

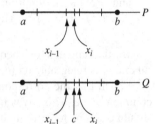

Now the upper Riemann sums $U(f, P)$ and $U(f, Q)$ are the same, apart from the contributions

$$\sup_{x \in [x_{i-1}, x_i]} f \times (x_i - x_{i-1}) \text{ to } U(f, P)$$

and

$$\sup_{x \in [x_{i-1}, c]} f \times (c - x_{i-1}) + \sup_{x \in [c, x_i]} f \times (x_i - c) \text{ to } U(f, Q).$$

Now

$$\sup_{x \in [x_{i-1}, x_i]} f \leq M, \text{ by Lemma 1,}$$

and

$$\sup_{x \in [x_{i-1}, c]} f \geq m \text{ and } \sup_{x \in [c, x_i]} f \geq m,$$

since m is a lower bound for f on all of $[a, b]$.
It follows that

$$\begin{aligned}
U(f, Q) - U(f, P) &= \sup_{x \in [x_{i-1}, c]} f \times (c - x_{i-1}) + \sup_{x \in [c, x_i]} f \times (x_i - c) \\
&\quad - \sup_{x \in [x_{i-1}, x_i]} f \times (x_i - x_{i-1}) \\
&\geq m(c - x_{i-1}) + m(x_i - c) - M(x_i - x_{i-1}) \\
&= m(x_i - x_{i-1}) - M(x_i - x_{i-1}) \\
&= -(M - m)(x_i - x_{i-1}) \\
&\geq -(M - m)\delta,
\end{aligned}$$

Lemma 1 (in Sub-section 7.1.1) said that, for intervals I and J where $I \subseteq J$

$$\inf_{x \in J} f \leq \inf_{x \in I} f \quad \text{and}$$

$$\sup_{x \in I} f \leq \sup_{x \in J} f.$$

For $0 < x_i - x_{i-1} \leq \delta$.

so that

$$U(f,Q) \geq U(f,P) - (M-m)\delta.$$

We can interpret this inequality as follows: the insertion of one additional point into the partition P (to obtain a new partition, Q) does not decrease the upper Riemann sum by more than $(M-m)\delta$.

We can repeat this insertion process up to a further $N-1$ times to end up with the final partition P'; the effect of this is that the upper Riemann sum of P does not decrease by more than $N(M-m)\delta$. In other words

$$U(f,P') \geq U(f,P) - N(M-m)\delta. \qquad \blacksquare$$

We now use this result to prove part (a) of Theorem 7.

Theorem 7, part (a)   Let f be a bounded function on an interval $[a, b]$, and $\{P_n\}$ any sequence of partitions of $[a, b]$ with $\|P_n\| \to 0$. If f is integrable on $[a, b]$, then

$$\lim_{n \to \infty} L(f,P_n) = \int_a^b f \quad \text{and} \quad \lim_{n \to \infty} U(f,P_n) = \int_a^b f.$$

Proof   We are considering a bounded function f integrable on an interval $[a,b]$, and $\{P_n\}$ any sequence of partitions of $[a, b]$ with $\|P_n\| \to 0$. We will use Lemma 3 to prove that

$$\lim_{n \to \infty} U(f,P_n) = \int_a^b f.$$

(The proof that $\lim_{n \to \infty} L(f,P_n) = \int_a^b f$ is similar, so we omit it.)

First, note that we can assume that f is non-constant on $[a, b]$, since the result is clearly immediately true in that situation.

Denote by I the value of the integral $\int_a^b f$, and let $m = \inf f(x)$ and $M = \sup f(x)$.

For each $n \geq 1$, we can choose a partition Q_n of $[a, b]$ for which

$$U(f,Q_n) < I + \frac{1}{n}.$$

Since f is integrable on $[a, b]$.

Let N_n denote the number of partition points in Q_n, and let

$$\delta = \frac{1}{N_n n(M-m)}.$$

Such a number δ exist since $M \neq m$, for we are assuming that f is non-constant.

Next, choose any partition P_n of $[a, b]$ for which $\|P_n\| < \delta$.

We then apply Lemma 3, with P_n in place of P, and with P_n' as the common refinement of P_n and Q_n in place of P'. It follows that

Notice that P_n' has at most N_n extra points as compared with P_n.

$$U(f,P_n') \geq U(f,P_n) - N_n(M-m)\delta$$

By Lemma 3.

$$= U(f,P_n) - \frac{1}{n}.$$

From the definition of δ.

We can now obtain our crucial estimate for $U(f, P_n)$ from this inequality, as follows

$$U(f, P_n) \le U(f, P_n') + \frac{1}{n}$$

This is the last inequality, rewritten.

$$\le U(f, Q_n) + \frac{1}{n}$$

For P_n' is a refinement of Q_n.

$$< I + \frac{2}{n}.$$

By our choice of Q_n.

We can therefore deduce that, for each $n \ge 1$

$$I \le U(f, P_n) < I + \frac{2}{n}.$$

It follows, by the Squeeze Rule for sequences, that $U(f, P_n) \to I$ as $n \to \infty$. ∎

We shall use these various criteria in the next section.

7.2   Properties of integrals

In Section 7.1 you met the definition of integrability, and saw that the constant function and the identity function were integrable on any interval $[a, b]$. But we want to know that many more functions than that are integrable! In this section we extend our family of integrable functions to a very large family indeed – much wider than, for example, just the continuous functions.

7.2.1   Infimum and Supremum of functions (revisited)

You met these earlier, in Subsection 1.4.2 and again in Section 7.1.

In order to capitalise on the definition of integrability via lower and upper Riemann sums, we need some further properties of the infimum and supremum of a function. So, first, we remind you of their definitions.

Definitions   Let f be a function defined on an interval $I \subseteq \mathbb{R}$. Then:
- A real number m is the **greatest lower bound**, or **infimum**, of f on I if:
  1. m is a lower bound of $f(I)$;
  2. if $m' > m$, then m' is not a lower bound of $f(I)$.
- A real number M is the **least upper bound**, or **supremum**, of f on I if:
  1. M is an upper bound of $f(I)$;
  2. if $M' < M$, then M' is not an upper bound of $f(I)$.

These are the definitions for the least upper bound and the greatest lower bound of a function on an interval I; there is a similar definition of these bounds on a general set S in $\mathbb{R}$.

Remark   Notice that, if m is a lower bound of f on I, then $\inf_I f \ge m$; and, if M is an upper bound of f on I, then $\sup_I f \le M$.

We will also sometimes need a strategy for verifying that a particular number is the infimum or supremum of a function on a particular interval.

> **Strategies**   Let f be a function defined on an interval $I \subseteq \mathbb{R}$. Then:
> - To show that m is the **greatest lower bound**, or **infimum**, of f on I, check that:
>   1. $f(x) \geq m$,   for *all* $x \in I$;
>   2. if $m' > m$,   then there is *some* $x \in I$ such that $f(x) < m'$.
> - To show that M is the **least upper bound**, or **supremum**, of f on I, check that:
>   1. $f(x) \leq M$,   for *all* $x \in I$;
>   2. if $M' < M$,   then there is *some* $x \in I$ such that $f(x) > M'$.

When using these strategies, we shall often use the numbers $m + \varepsilon$ and $M - \varepsilon$ in place of m' and M', to fit in with our increased use of ε as a positive number that can be as small as we please.

Problem 1   Let f be a bounded function on an interval I. Prove that, for any constant k

$$\inf_{x \in I}\{k + f(x)\} = k + \inf_{x \in I}\{f(x)\} \quad \text{and}$$

$$\sup_{x \in I}\{k + f(x)\} = k + \sup_{x \in I}\{f(x)\}.$$

In fact you have already met one property of $\inf f$ and $\sup f$ in your study of integration.

Lemma 1, Sub-section 7.1.1.

> **Lemma 1**   For any bounded function f defined on intervals I and J where $I \subseteq J$, we have
> $$\inf_{x \in J} f \leq \inf_{x \in I} f \quad \text{and} \quad \sup_{x \in I} f \leq \sup_{x \in J} f.$$

Loosely speaking, the larger interval gives the function more space to get smaller and more space to get larger.

A key tool in our work will be the following innocuous-looking result.

> **Lemma 2**   Let f be a bounded function on an interval I. If, for some number K, $f(x) - f(y) \leq K$ for all x and y in I, then $\sup_I f - \inf_I f \leq K$.

Note that K must be *independent* of the choice of x and y in I.

Proof   Since $f(x) - f(y) \leq K$ for all x and y in I, we have

$$f(x) \leq K + f(y), \quad \text{for all } x \text{ and } y \text{ in } I. \tag{1}$$

It follows from (1) that, for any choice whatsoever of y in I, then $K + f(y)$ serves as an upper bound for the set of all possible values of $f(x)$ for x in I. It follows that

$$\sup\{f(x) : x \in I\} \leq K + f(y),$$

which we may rewrite in the form

$$\sup\{f(x) : x \in I\} - K \leq f(y) \tag{2}$$

For $\sup\{f(x) : x \in I\}$ is simply the *least* upper bound. (See also the previous margin note.)

In a similar way, it follows from (2) that $\sup\{f(x) : x \in I\} - K$ serves as a lower bound for the set of all possible values of $f(y)$ for y in I. It follows that

$$\sup\{f(x) : x \in I\} - K \leq \inf\{f(y) : y \in I\}. \tag{3}$$

For $\inf\{f(y) : y \in I\}$ is simply the *greatest* lower bound.

We may rearrange the inequality (3) in the form $\sup\{f(x) : x \in I\} - \inf\{f(y) : y \in I\} \leq K$. This is precisely the required result, since the letters x and y in this last inequality are simply 'dummy variables'.   ∎

Remarks

1. If f is bounded on I and $f(x) - f(y) < K$ for all x and y in I, then it may not be true that $\sup_I f - \inf_I f < K$. For example, if f is the identity function on $I = (0, 1)$, then

$f(x) - f(y) < 1$, for all x and y in $(0,1)$, but $\sup\limits_{[0,1]} f - \inf\limits_{[0,1]} f = 1 - 0 = 1$.

2. The conclusion of Lemma 2 also holds if we know that, for some number K, $|f(x) - f(y)| \leq K$, for all x and y in I, since

$$f(x) - f(y) \leq |f(x) - f(y)|.$$

In some situations, the following result related to Lemma 2 is also useful.

Lemma 3   Let f be a bounded function on an interval I. Then
$$\sup_{x,y \in I} \{f(x) - f(y)\} = \sup_I f - \inf_I f.$$

Proof   We use the strategy for supremum mentioned earlier.

Let x and y be any numbers in I. Then, in particular

$$f(x) \leq \sup_I f, \quad \text{for all } x \in I;$$

and, since $f(y) \geq \inf_I f$, for all $y \in I$

$$-f(y) \leq -\inf_I f, \quad \text{for all } y \in I.$$

If we add these inequalities, we obtain

$$f(x) - f(y) \leq \sup_I f - \inf_I f, \quad \text{for all } x, y \in I,$$

so that

$$\sup_{x,y \in I} \{f(x) - f(y)\} \leq \sup_I f - \inf_I f.$$

To prove the desired result, we now need to prove that   By the strategy for supremum.

for each positive number ε, there are X and Y in I for which

$$f(X) - f(Y) > \left(\sup_I f - \inf_I f\right) - \varepsilon. \tag{4}$$

Now, by the definition of infimum and supremum on I, we know that, since $\frac{1}{2}\varepsilon > 0$, there exist X and Y in I such that

$$f(X) > \sup_I f - \frac{1}{2}\varepsilon \quad \text{and} \quad f(Y) < \inf_I f + \frac{1}{2}\varepsilon.$$

It follows from these two inequalities that

$$f(X) - f(Y) > \left(\sup_I f - \frac{1}{2}\varepsilon\right) - \left(\inf_I f + \frac{1}{2}\varepsilon\right)$$
$$= \sup_I f - \inf_I f - \varepsilon.$$

This completes the proof.   ■

Problem 2   Let $f(x) = x^2$ on the interval $I = (-2, 3]$. Determine $\inf\limits_I f$, $\sup\limits_I f$, $\inf\limits_{x,y \in I} \{f(x) - f(y)\}$ and $\sup\limits_{x,y \in I} \{f(x) - f(y)\}$.

Theorem 1   Combination Rules

Let f and g be bounded functions on an interval I. Then:

Sum Rule $\inf_I(f+g) \geq \inf_I f + \inf_I g$ and $\sup_I(f+g) \leq \sup_I f + \sup_I g$;

Multiple Rule $\inf_I(\lambda f) = \lambda\left(\inf_I f\right)$ and $\sup_I(\lambda f) = \lambda\left(\sup_I f\right)$, for $\lambda > 0$;

Negative Rule $\inf_I(-f) = -\left(\sup_I f\right)$ and $\sup_I(-f) = -\left(\inf_I f\right)$.

Proof   We will prove only the first part of each Rule, for the proofs of the second parts are similar.

Sum Rule

For any x and y in I

$$f(x) \geq \inf_I f \quad \text{and} \quad g(x) \geq \inf_I g;$$

so that, by adding these two inequalities, we obtain

$$f(x) + g(x) \geq \inf_I f + \inf_I g.$$

Since $\inf_I f + \inf_I g$ is a lower bound for $f(x) + g(x)$ on I, it follows that

$$\inf_I(f+g) \geq \inf_I f + \inf_I g.$$

Multiple Rule

For any x in I, $f(x) \geq \inf_I f$ so that

$$\lambda f(x) \geq \lambda \inf_I f, \quad \text{for all } x \in I. \qquad \text{Since } \lambda > 0.$$

Thus

$$\inf_{x \in I}(\lambda f(x)) \geq \lambda \inf_I f.$$

To prove that $\inf_{x \in I}(\lambda f(x)) = \lambda\left(\inf_I f\right)$, we now need to prove that:

for each positive number ε, there is some X in I for which

$$\lambda f(X) < \lambda\left(\inf_I f\right) + \varepsilon. \qquad (5)$$

Now, since $\varepsilon > 0$ and $\lambda > 0$, we have $\frac{\varepsilon}{\lambda} > 0$. It follows, from the definition of infimum of f on I, that there exists an X in I for which

$$f(X) < \inf_I f + \frac{\varepsilon}{\lambda}.$$

Multiplying both sides by the positive number λ, we obtain the desired result (5).

Negative Rule

For any x in I, $f(x) \leq \sup_I f$ so that

$$-f(x) \geq -\sup_I f, \quad \text{for all } x \in I.$$

Thus

$$\inf_{x \in I}(-f(x)) \geq -\sup_I f.$$

To prove that $\inf_{x\in I}(-f(x)) = -\sup_I f$, we now need to prove that:

for each positive number ε, there is some X in I for which

$$-f(X) < -\sup_I f + \varepsilon. \tag{6}$$

Now, since $\varepsilon > 0$ it follows, from the definition of supremum of f on I, that there exists an X in I for which

$$f(X) > \sup_I f - \varepsilon$$

so that

$$-f(X) < -\sup_I f + \varepsilon.$$

This is precisely the inequality (6). ∎

Problem 3 Prove that if f is a bounded function on an interval I, then:

(a) $\sup_I(\lambda f) = \lambda \sup_I f$, for $\lambda > 0$; (b) $\sup_I(-f) = -\inf_I f$.

We will now use these results on greatest lower bounds and least upper bounds to address the main topic of this section: integration!

7.2.2 Monotonic and continuous functions

We now prove that bounded functions on an interval $[a, b]$ are integrable if they are either monotonic on $[a, b]$ or continuous on $[a, b]$. This provides us with very many integrable functions!

Neither of these two classes of functions includes the other. For example, the function

$$f(x) = x - x^2, \quad x \in [0, 1],$$

is continuous on $[0,1]$ but is not monotonic on $[0, 1]$; while the function

$$f(x) = \begin{cases} x, & 0 \le x < 1, \\ 2, & x = 1, \end{cases} \quad x \in [0, 1],$$

is monotonic on $[0,1]$ but is not continuous on $[0, 1]$.

You should sketch the graphs of these two functions to appreciate these statements.

> **Theorem 2** Let f be a bounded function on an interval $[a, b]$. If f is monotonic on $[a, b]$, then it is integrable on $[a, b]$.

Proof We shall assume that f is increasing on $[a, b]$; if it is decreasing, the proof is similar. If f is constant on $[a, b]$, it is certainly integrable on $[a, b]$. So, we shall assume, in addition, that f is also non-constant on $[a, b]$; it follows, in particular, that $f(a) \neq f(b)$.

We will prove that:

for each positive number ε, there is a partition P of $[a, b]$ for which

$$U(f, P) - L(f, P) < \varepsilon.$$

It then follows from the Riemann Criterion for integrability that f is integrable on $[a, b]$.

Theorem 5, Sub-section 7.1.2.

For any given $\varepsilon > 0$, let P be any partition of $[a, b]$ with mesh $\|P\|$ such that $\|P\| < \frac{\varepsilon}{f(b)-f(a)}$. Then, with the usual notation for Riemann sums

$$U(f,P) - L(f,P) = \sum_{i=1}^{n} (M_i - m_i)\delta x_i = \sum_{i=1}^{n} (f(x_i) - f(x_{i-1}))\delta x_i \qquad \text{Since } f \text{ is increasing on } [a,b].$$

$$\leq \|P\| \times \sum_{i=1}^{n} (f(x_i) - f(x_{i-1}))$$

$$= \|P\| \times (f(b) - f(a))$$

$$< \varepsilon.$$

This completes the proof.   ■

Notice that the brevity of the above proof hides the fact that within the proof we are using quite a lot of work needed to prove the Riemann Criterion!

> **Theorem 3**   Let f be a bounded function on an interval $[a, b]$. If f is continuous on $[a, b]$, then it is integrable on $[a, b]$.

Proof   We will prove that:

for each positive number ε, there is a partition P of $[a, b]$ for which

$$U(f,P) - L(f,P) < \varepsilon.$$

It then follows from the Riemann Criterion for integrability that f is integrable on $[a, b]$.

Our key new tool here is the fact that, since f is continuous on an interval $[a, b]$, which is a closed interval, it is therefore *uniformly* continuous on $[a, b]$. It follows that, for any given $\varepsilon > 0$, since $\frac{\varepsilon}{2(b-a)} > 0$ there is a positive number δ for which

$$|f(x) - f(y)| < \frac{\varepsilon}{2(b-a)}, \quad \text{for all } x \text{ and } y \text{ in } [a,b] \text{ satisfying } |x - y| < \delta. \quad (7)$$

You met uniform continuity in Section 5.5; this assertion is Theorem 2 there.

Here the choice of δ depends ONLY on ε, the same choice whatever x and y may be. We choose $\frac{\varepsilon}{2(b-a)}$ rather than ε, for convenience later on.

Now, let $P = \{[x_{i-1}, x_i]\}_{i=1}^{n}$ be any partition of $[a, b]$ with mesh $\|P\|$ such that $\|P\| < \delta$. Then, for each i we have, for all x and y in $[x_{i-1}, x_i]$

$$|x - y| \leq x_i - x_{i-1}$$

$$\leq \|P\| < \delta.$$

It follows, from (7), that

$$|f(x) - f(y)| < \frac{\varepsilon}{2(b-a)}, \quad \text{for all } x \text{ and } y \text{ in } [x_{i-1}, x_i],$$

so that, by Remark 2 following Lemma 2

$$\sup_{x \in [x_{i-1}, x_i]} f(x) - \inf_{x \in [x_{i-1}, x_i]} f(x) \leq \frac{\varepsilon}{2(b-a)}.$$

In Sub-section 7.2.1. Recall that the conclusion of Lemma 2 is a *weak* inequality, not a strong inequality.

Then, with the usual notation for Riemann sums

$$U(f,P) - L(f,P) = \sum_{i=1}^{n} (M_i - m_i)\delta x_i$$

$$\leq \frac{\varepsilon}{2(b-a)} \times \sum_{i=1}^{n} \delta x_i$$

$$= \frac{\varepsilon}{2(b-a)} \times (b-a)$$

$$= \frac{1}{2}\varepsilon < \varepsilon.$$

This completes the proof.   ■

7.2.3 Rules for integration

Our first task is to prove the Combination Rules for integrable functions.

You may omit the details
of all the proofs in this
sub-section at a first reading.

Theorem 4   Combination Rules

Let f and g be integrable on $[a, b]$. Then so are:

Sum Rule $f + g$,   and $\int_a^b (f + g) = \int_a^b f + \int_a^b g$;

Multiple Rule λf,   and $\int_a^b (\lambda f) = \lambda \int_a^b f$;

Product Rule fg;

Modulus Rule $|f|$.

Proof   We use the usual notation for Riemann sums and integrals.

Sum Rule

Let $\{P_n\}$ be any sequence of partitions of $[a, b]$, where $||P_n|| \to 0$ as $n \to \infty$. Since f and g are integrable on $[a, b]$, it follows, from the Null Partitions Criterion for integrability, that

$P_n = \{[x_{i-1}, x_i]\}_{i=1}^n$.

Theorem 7, Sub-section 7.1.2.

$$L(f, P_n) \quad \text{and} \quad U(f, P_n) \to \int_a^b f \quad \text{as } n \to \infty,$$

and

$$L(g, P_n) \quad \text{and} \quad U(g, P_n) \to \int_a^b g \quad \text{as } n \to \infty.$$

Then

$$L(f, P_n) + L(g, P_n) = \sum_{i=1}^n \inf_{[x_{i-1}, x_i]} f \times \delta x_i + \sum_{i=1}^n \inf_{[x_{i-1}, x_i]} g \times \delta x_i$$

By the definition of lower Riemann sum.

$$= \sum_{i=1}^n \left\{ \inf_{[x_{i-1}, x_i]} f + \inf_{[x_{i-1}, x_i]} g \right\} \times \delta x_i$$

$$\leq \sum_{i=1}^n \inf_{[x_{i-1}, x_i]} (f + g) \times \delta x_i \qquad (= L(f + g, P_n))$$

By the Combination Rules for inf and sup, Theorem 1, Sub-section 7.2.1.

$$\leq \sum_{i=1}^n \sup_{[x_{i-1}, x_i]} (f + g) \times \delta x_i \qquad (= U(f + g, P_n))$$

$$\leq \sum_{i=1}^n \left\{ \sup_{[x_{i-1}, x_i]} f + \sup_{[x_{i-1}, x_i]} g \right\} \times \delta x_i$$

By the Combination Rules for inf and sup, again.

$$= U(f, P_n) + U(g, P_n).$$

We now let $n \to \infty$ in this last set of inequalities. Then, since

$$L(f, P_n) + L(g, P_n) \to \int_a^b f + \int_a^b g \quad \text{and} \quad U(f, P_n) + U(g, P_n) \to \int_a^b f + \int_a^b g,$$

it follows, from the Limit Inequality Rule for sequences, that

$$L(f + g, P_n) \to \int_a^b f + \int_a^b g \quad \text{and} \quad U(f + g, P_n) \to \int_a^b f + \int_a^b g, \quad \text{as } n \to \infty.$$

Hence, by part (b) of the Null Partitions Criterion for integrability, we deduce that $f + g$ is integrable on $[a, b]$ and $\int_a^b (f + g) = \int_a^b f + \int_a^b g$.

Theorem 7, Sub-section 7.1.2.

Multiple Rule

Let $\{P_n\}$ be any sequence of partitions of $[a, b]$, where $\|P_n\| \to 0$ as $n \to \infty$. $P_n = \{[x_{i-1}, x_i]\}_{i=1}^n$.
Since f is integrable on $[a, b]$, it follows from the Null Partitions Criterion for
integrability that   Theorem 7, Sub-section 7.1.2.

$$L(f, P_n) \quad \text{and} \quad U(f, P_n) \to \int_a^b f \quad \text{as } n \to \infty.$$

Then, if $\lambda > 0$

$$L(\lambda f, P_n) = \sum_{i=1}^n \inf_{[x_{i-1}, x_i]} (\lambda f) \times \delta x_i$$

$$= \lambda \sum_{i=1}^n \inf_{[x_{i-1}, x_i]} f \times \delta x_i \qquad\qquad \text{By the Combination Rules for}$$
                                                                                                 inf and sup, Theorem 1,
                                                                                                 Sub-section 7.2.1.
$$= \lambda L(f, P_n) \to \lambda \int_a^b f \quad \text{as } n \to \infty;$$

while, if $\lambda < 0$

$$L(\lambda f, P_n) = \sum_{i=1}^n \inf_{[x_{i-1}, x_i]} (\lambda f) \times \delta x_i$$

$$= \lambda \sum_{i=1}^n \sup_{[x_{i-1}, x_i]} f \times \delta x_i \qquad\qquad \text{By the Combination Rules for}$$
                                                                                                 inf and sup.
$$= \lambda U(f, P_n) \to \lambda \int_a^b f \quad \text{as } n \to \infty.$$

It follows that, for all real λ, $L(\lambda f, P_n) \to \lambda \int_a^b f$ as $n \to \infty$.   For, if $\lambda = 0$, the result is
A similar argument shows that, for all real λ, $U(\lambda f, P_n) \to \lambda \int_a^b f$ as $n \to \infty$.   trivial.
Since the limits of the two sequences of Riemann sums are equal, it follows
from part (b) of the Null Partitions Criterion for integrability, that λf is   Theorem 7, Sub-section 7.1.2.
integrable on $[a, b]$ and

$$\int_a^b (\lambda f) = \lambda \int_a^b f.$$

Product Rule

Since f and g are bounded on $[a, b]$, there is some number M, say, such that

$$|f(x)| \le M \quad \text{and} \quad |g(x)| \le M, \quad \text{for all } x \in [a, b].$$

Now let $\{P_n\}$ be any sequence of partitions of $[a, b]$, where $\|P_n\| \to 0$ as $P_n = \{[x_{i-1}, x_i]\}_{i=1}^n$.
$n \to \infty$. It follows that, if $x, y \in [x_{i-1}, x_i]$ for some i, then

$$|f(x)g(x) - f(y)g(y)| = |\{f(x)g(x) - f(x)g(y)\} + \{f(x)g(y) - f(y)g(y)\}|$$
$$= |f(x)\{g(x) - g(y)\} + g(y)\{f(x) - f(y)\}|$$
$$\le M|g(x) - g(y)| + M|f(x) - f(y)|$$
$$\le M\left\{ \sup_{[x_{i-1}, x_i]} g - \inf_{[x_{i-1}, x_i]} g \right\} + M\left\{ \sup_{[x_{i-1}, x_i]} f - \inf_{[x_{i-1}, x_i]} f \right\},$$

so that, by Lemma 2 in Sub-section 7.2.1 and Remark 2 following that Lemma

$$\sup_{[x_{i-1},x_i]} (fg) - \inf_{[x_{i-1},x_i]} (fg) \leq M\left\{ \sup_{[x_{i-1},x_i]} g - \inf_{[x_{i-1},x_i]} g \right\} + M\left\{ \sup_{[x_{i-1},x_i]} f - \inf_{[x_{i-1},x_i]} f \right\}. \qquad (8)$$

In terms of the Riemann sums for fg, it follows that

$$U(fg,P_n) - L(fg,P_n) = \sum_{i=1}^{n} \left(\sup_{[x_{i-1},x_i]} (fg) - \inf_{[x_{i-1},x_i]} (fg) \right) \delta x_i$$

By definition of U and L.

$$\leq M \sum_{i=1}^{n} \left\{ \sup_{[x_{i-1},x_i]} g - \inf_{[x_{i-1},x_i]} g \right\} \delta x_i + M \sum_{i=1}^{n} \left\{ \sup_{[x_{i-1},x_i]} f - \inf_{[x_{i-1},x_i]} f \right\} \delta x_i$$

By (8).

$$= M\{U(g,P_n) - L(g,P_n)\} + M\{U(f,P_n) - L(f,P_n)\}$$

$$\to 0 \quad \text{as } n \to \infty.$$

It then follows, from part (b) of the Null Partitions Criterion for integrability, that fg is integrable on $[a, b]$.

By part (a) of the Null Partitions Criterion for integrability (Theorem 7, Sub-section 7.1.2), since f and g are integrable on $[a, b]$.

Modulus Rule

Let $\{P_n\}$ be any sequence of partitions of $[a, b]$, where $\|P_n\| \to 0$ as $n \to \infty$.
  Then, if $x, y \in [x_{i-1}, x_i]$ for some i, we have

$$\big| |f(x)| - |f(y)| \big| \leq |f(x) - f(y)|,$$

so that

$$\big| |f(x)| - |f(y)| \big| \leq \sup_{[x_{i-1},x_i]} f - \inf_{[x_{i-1},x_i]} f$$

and therefore

$$\sup_{[x_{i-1},x_i]} |f| - \inf_{[x_{i-1},x_i]} |f| \leq \sup_{[x_{i-1},x_i]} f - \inf_{[x_{i-1},x_i]} f. \qquad (9)$$

$P_n = \{[x_{i-1}, x_i]\}_{i=1}^{n}$.

This follows from the 'reverse form' of the Triangle Inequality, that you met in Sub-section 1.3.1.

By Remark 2 following Lemma 2, Sub-section 7.2.1.

In terms of the Riemann sums for $|f|$, it follows that

$$U(|f|,P_n) - L(|f|,P_n) = \sum_{i=1}^{n} \left(\sup_{[x_{i-1},x_i]} |f| - \inf_{[x_{i-1},x_i]} |f| \right) \delta x_i$$

By definition of U and L.

$$\leq \sum_{i=1}^{n} \left(\sup_{[x_{i-1},x_i]} f - \inf_{[x_{i-1},x_i]} f \right) \delta x_i$$

By (9).

$$= U(f,P_n) - L(f,P_n)$$

$$\to 0 \quad \text{as } n \to \infty.$$

It then follows, from part (b) of the Null Partitions Criterion for integrability, that $|f|$ is integrable on $[a, b]$. ■

By part (a) of the Null Partitions Criterion for integrability (Theorem 7, Sub-section 7.1.2), since f is integrable on $[a, b]$.

Our next result is often overlooked as obvious – which it is not!

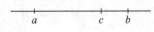

Theorem 5   The Sub-interval Theorem

(a) Let f be integrable on $[a, b]$, and let $a < c < b$. Then f is integrable on both $[a, c]$ and $[c, b]$, and $\int_a^b f = \int_a^c f + \int_c^b f$.

(b) Let f be integrable on $[a, c]$ and $[c, b]$, where $a < c < b$. Then f is integrable on $[a, b]$.

Proof

(a) Let $\{P_n\}$, where $P_n = \{[x_{i-1}, x_i]\}_{i=1}^n$, be any sequence of partitions of $[a, b]$ such that c is a partition point of each P_n, for $n \geq 2$, and such that $||P_n|| \to 0$ as $n \to \infty$.

For $n \geq 2$, let Q_n consist of those subintervals in P_n that lie in $[a, c]$, and Q_n' consist of those subintervals in P_n that lie in $[c, b]$. Thus $\{Q_n\}$ is a partition of $[a, c]$ with $||Q_n|| \to 0$ as $n \to \infty$, and $\{Q_n'\}$ is a partition of $[c, b]$ with $||Q_n'|| \to 0$ as $n \to \infty$.

Now, with the usual notation for Riemann sums, we have

$$U(f, Q_n) - L(f, Q_n) = \text{the sum of those terms in } \sum_{i=1}^n (M_i - m_i)\delta x_i$$
$$\text{that correspond to subintervals } [x_{i-1}, x_i]$$
$$\text{that lie in } [a, c]$$
$$\leq \sum_{i=1}^n (M_i - m_i)\delta x_i$$
$$= U(f, P_n) - L(f, P_n) \to 0 \quad \text{as } n \to \infty.$$

It follows from the Null Partitions Criterion that f is integrable on $[a, c]$.

A similar argument proves that f is integrable on $[c, b]$.

In particular, it follows that

$$L(f, Q_n) \to \int_a^c f \quad \text{and} \quad L(f, Q_n') \to \int_c^b f \quad \text{as } n \to \infty.$$

So, if we let $n \to \infty$ in the identity

$$L(f, P_n) = L(f, Q_n) + L(f, Q_n'),$$

we deduce that

$$\int_a^b f = \int_a^c f + \int_c^b f.$$

This identity follows from the definitions of P_n, Q_n and Q_n'.

By the definition of integral.

(b) In the opposite direction, suppose that f is integrable on $[a, c]$ and $[c, b]$.

Let $\{P_n\}$, where $P_n = \{[x_{i-1}, x_i]\}_{i=1}^n$, be any sequence of partitions of $[a, b]$ such that c is a partition point of each P_n, for $n \geq 2$, and $||P_n|| \to 0$ as $n \to \infty$.

For $n \geq 2$, let Q_n consist of those subintervals in P_n that lie in $[a, c]$, and Q_n' consist of those subintervals in P_n that lie in $[c, b]$. Thus $\{Q_n\}$ is a partition of $[a, c]$ with $||Q_n|| \to 0$ as $n \to \infty$, and $\{Q_n'\}$ is a partition of $[c, b]$ with $||Q_n'|| \to 0$ as $n \to \infty$.

Then, with the usual notation for Riemann sums, we have

$$U(f, P_n) - L(f, P_n) = \text{the sum of those terms in } \sum_{i=1}^n (M_i - m_i)\delta x_i \text{ that}$$
$$\text{correspond to subintervals } [x_{i-1}, x_i] \text{ that lie in } [a, c]$$
$$+ \text{ the sum of those terms in } \sum_{i=1}^n (M_i - m_i)\delta x_i \text{ that}$$
$$\text{correspond to subintervals } [x_{i-1}, x_i] \text{ that lie in } [c, b]$$
$$= \{U(f, Q_n) - L(f, Q_n)\} + \{U(f, Q_n') - L(f, Q_n')\}$$
$$\to 0 \quad \text{as } n \to \infty.$$

It follows, from the Null Partitions Criterion, that f is integrable on $[a, b]$. ∎

In light of Theorem 5 we now introduce some notational conventions.

Conventions   Let f be integrable on $[a, b]$, where $a < b$. Then we make the following definitions

$$\int_a^a f = 0 \quad \text{and} \quad \int_b^a f = -\int_a^b f.$$

These conventions apply where the upper and lower limits of the integral are the same, or are in reverse order on the real line.

With these conventions, we can assert that if f is integrable on any interval that contains the points a, b and c, then $\int_a^b f = \int_a^c f + \int_c^b f$.

To prove this assertion, we need to look at all the possible orderings of the three points on $\mathbb{R}$, which is rather tedious; so we omit the details!

7.3   Fundamental Theorem of Calculus

7.3.1   Fundamental Theorem of Calculus

It would be very tedious if, each time that we wished to evaluate an integral, we had to calculate upper and lower sums and find their greatest lower bound and least upper bound, respectively. Fortunately, this is usually unnecessary, as there is a short-cut using the idea of a *primitive*.

Definition   Let f be a function defined on an interval I. Then the function F is a **primitive** of f on I if F is differentiable on I and

$$F' = f.$$

Sometimes we denote a primitive of f by $\int f$, and we denote $F(x)$ by $\int f(x)dx$.

For example, if

$$f(x) = \frac{1}{4 - x^2}, \quad x \in [-1, 1],$$

then

$$F(x) = \frac{1}{4}\log_e\left(\frac{2 + x}{2 - x}\right), \quad x \in [-1, 1],$$

is a primitive of f on $[-1, 1]$, since

$$F'(x) = \frac{1}{4} \times \frac{2 - x}{2 + x} \times \frac{(2 - x) + (2 + x)}{(2 - x)^2}$$

$$= \frac{1}{4} \times \frac{4}{(2 + x)(2 - x)} = \frac{1}{4 - x^2}.$$

Problem 1

(a) Prove that the function $f(x) = \frac{1}{\sqrt{x^2 - 4}}$, $x \in (2, \infty)$, has a primitive $F(x) = \log_e\left(x + \sqrt{x^2 - 4}\right)$.

(b) Prove that the function $f(x) = \sqrt{4 - x^2}$, $x \in (-2, 2)$, has a primitive $F(x) = \frac{1}{2}x\sqrt{4 - x^2} + 2\sin^{-1}\left(\frac{x}{2}\right)$.

The connection between primitives of a function f and the integral of f on an interval I is given by the following result, which is one of the most important theorems in Analysis.

Theorem 1   Fundamental Theorem of Calculus
Let f be integrable on $[a, b]$, and F a primitive of f on $[a, b]$. Then

$$\int_a^b f = F(b) - F(a).$$

Often $F(b) - F(a)$ is written as

$$[F(x)]_a^b \text{ or } F(x)|_a^b.$$

Proof Since f is integrable on $[a, b]$, there is a sequence

$$P_n = \{[x_0, x_1], [x_1, x_2], \ldots, [x_{n-1}, x_n]\}$$

of partitions of $[a, b]$, with $\|P_n\| \to 0$ as $n \to \infty$, such that

$$\lim_{n\to\infty} L(f, P_n) = \int_a^b f \quad \text{and} \quad \lim_{n\to\infty} U(f, P_n) = \int_a^b f. \qquad (1)$$

We use the standard notation for lower and upper sums and partitions.

Now, the function F satisfies the conditions of the Mean Value Theorem on each subinterval $[x_{i-1}, x_i]$, for $i = 1, 2, \ldots, n$. It follows that there exists some point $c_i \in (x_{i-1}, x_i)$ such that

$$F(x_i) - F(x_{i-1}) = F'(c_i)(x_i - x_{i-1})$$
$$= f(c_i)\delta x_i. \qquad (2)$$

For $F' = f$ and $\delta x_i = x_i - x_{i-1}$.

Next, since $m_i \le f(c_i) \le M_i$, for $i = 1, 2 \ldots, n$, it follows that

$$\sum_{i=1}^n m_i \delta x_i \le \sum_{i=1}^n f(c_i)\delta x_i \le \sum_{i=1}^n M_i \delta x_i. \qquad (3)$$

Recall that

$$m_i = \inf\{f(x) : x \in [x_{i-1}, x_i]\},$$
$$M_i = \sup\{f(x) : x \in [x_{i-1}, x_i]\}.$$

We may now apply (2) to the general term in the sum $\sum_{i=1}^n f(c_i)\delta x_i$; thus it follows from (3) that

$$L(f, P_n) \le \sum_{i=1}^n \{F(x_i) - F(x_{i-1})\} \le U(f, P_n).$$

Since the middle term is a 'telescoping' sum, we deduce that

$$L(f, P_n) \le F(b) - F(a) \le U(f, P_n). \qquad (4)$$

Now let $n \to \infty$ in (4), using the facts in (1); this thus gives

$$\int_a^b f \le F(b) - F(a) \le \int_a^b f.$$

By the Limit Inequality Rule for sequences.

In other words, $F(b) - F(a) = \int_a^b f$, as required. ∎

Example 1 Evaluate $\int_0^1 2^x dx$.

Solution The function $f(x) = 2^x$ is continuous on $[0, 1]$ and so is integrable on $[0, 1]$; from the Table of Standard Primitives, it has a primitive $F(x) = \frac{2^x}{\log_e 2}$ on $[0, 1]$.

The Table of Standard Primitives appears in Appendix 2.

It follows from the Fundamental Theorem of Calculus that

$$\int_0^1 2^x dx = \left[\frac{2^x}{\log_e 2}\right]_0^1 = \frac{1}{\log_e 2}. \qquad \square$$

Problem 2 Using the Table of Standard Primitives, evaluate the following integrals:

(a) $\int_0^4 (x^2 + 9)^{\frac{1}{2}} dx$;   (b) $\int_1^e \log_e x \, dx$.

7.3.2 Finding primitives

The Fundamental Theorem of Calculus asserts that, if f is integrable on $[a, b]$, and F a primitive of f on $[a, b]$, then $\int_a^b f = F(b) - F(a)$. This does not make it clear whether a primitive of f is unique.

In fact, for any given function f, a primitive of f is NOT unique. For example, the functions

$$x \mapsto \sin^{-1} x \quad \text{and} \quad x \mapsto -\cos^{-1} x, \quad x \in (-1, 1),$$

are both primitives of the function $x \mapsto \frac{1}{\sqrt{1-x^2}}$; so that, for instance

$$\int_0^{\frac{1}{\sqrt 2}} \frac{1}{\sqrt{1 - x^2}} \, dx = \left[\sin^{-1} x\right]_0^{\frac{1}{\sqrt 2}} = \left[-\cos^{-1} x\right]_0^{\frac{1}{\sqrt 2}}$$
$$= \frac{1}{4}\pi.$$

However, two primitives of a given function f *are* related; they can only differ by a constant function.

For example

$$\sin^{-1} x - \left(-\cos^{-1} x\right) = \frac{1}{2}\pi.$$

Theorem 2   Uniqueness Theorem for Primitives

Let F_1 and F_2 be primitives of a function f on an interval I. Then there exists some constant c such that

$$F_2(x) = F_1(x) + c, \quad \text{for } x \in I.$$

Proof   Since F_1 and F_2 are primitives of f on I

$$F_1'(x) = f(x) \quad \text{and} \quad F_2'(x) = f(x), \quad \text{for } x \in I,$$

so that

$$F_2'(x) = F_1'(x), \quad \text{for } x \in I.$$

It follows that there exists some constant c such that

$$F_2(x) = F_1(x) + c, \quad \text{for } x \in I. \qquad \blacksquare$$

By Corollary 2, Subsection 6.4.2.

Our stock of primitives can be considerably extended by use of the following Combination Rules. For convenience, we include here an extra rule which applies to a composite function $f \circ g$ for which the function g is just an x-scaling.

Theorem 3   Combination Rules

Let F and G be primitives of f and g, respectively, on an interval I, and $\lambda \in \mathbb{R}$. Then, on I:

Sum Rule	$f + g$	has primitive $F + G$;
Multiple Rule	λf	has primitive λF;
Scaling Rule	$x \mapsto f(\lambda x)$	has primitive $x \mapsto \frac{1}{\lambda} F(\lambda x)$.

These rules are easily proved using the corresponding rules for derivatives.

For example, it follows from the Table of Standard Primitives and the Combination Rules that the function $x \mapsto 3 \sinh(2x) + \frac{1}{x}$, $x \in (0, \infty)$, has a primitive $x \mapsto \frac{3}{2} \cosh(2x) + \log_e x$.

In applications of Theorem 3 we do not usually bother to mention the theorem explicitly.

Problem 3   Using the Table of Standard Primitives and the Combination Rules, find a primitive of each of the following functions:

(a) $f(x) = 4 \log_e x - \frac{2}{4+x^2}$, $\quad x \in (0, \infty)$;

(b) $f(x) = 2 \tan(3x) + e^{2x} \sin x$, $\quad x \in \left(-\frac{1}{6}\pi, \frac{1}{6}\pi\right)$.

7.3.3 Techniques of integration

We are now in a position to use the Fundamental Theorem of Calculus to give rigorous proofs of some standard techniques for integration.

Theorem 4   Integration by Parts

If f and g are differentiable on an interval $[a, b]$, and f' and g' are continuous on $[a, b]$, then

$$\int_a^b fg' = [fg]_a^b - \int_a^b f'g.$$

Proof   We may reformulate the Product Rule for derivatives in the form

$$(fg)' = f'g + fg';$$

in other words, that fg is a primitive on $[a,b]$ of $f'g + fg'$.

It follows from the Fundamental Theorem of Calculus that

$$\int_a^b (f'g + fg') = [fg]_a^b,$$

so that $\int_a^b fg' = [fg]_a^b - \int_a^b f'g$, as required.   ∎

Theorem 1, Sub-section 6.2.1.

Recall the notation that

$$[fg]_a^b = f(b)g(b) - f(a)g(a).$$

Strategy to evaluate an integral using integration by parts

1. Write the original function in the form fg', where f is a function that you can differentiate and g' a function that you can integrate.
2. Use the formula $\int_a^b fg' = [fg]_a^b - \int_a^b f'g$.

There is a similar strategy for finding a primitive by integration by parts.

Example 2   Evaluate the integral $\int_0^1 \tan^{-1} x\,dx$.

Solution   Here we use a common trick: we consider $g'(x)$ to be the factor 1, and use integration by parts. Thus

$$\int_0^1 \tan^{-1} x\,dx = \int_0^1 1 \times \tan^{-1} x\,dx$$

$$= \left[x \tan^{-1} x\right]_0^1 - \int_0^1 \frac{x\,dx}{1 + x^2}$$

$$= \tan^{-1} 1 - \left[\frac{1}{2}\log_e\left(1 + x^2\right)\right]_0^1$$

$$= \frac{1}{4}\pi - \frac{1}{2}\log_e 2 + \frac{1}{2}\log_e 1 = \frac{1}{4}\pi - \frac{1}{2}\log_e 2. \quad \square$$

Here

$$f(x) = \tan^{-1} x, g'(x) = 1,$$

so that

$$f'(x) = \frac{1}{1 + x^2}, g(x) = x.$$

Problem 4

(a) Find a primitive of the function $f(x) = x^{\frac{1}{3}}\log_e x$, $x \in (0, \infty)$.

(b) Evaluate the integral $\int_0^{\frac{\pi}{2}} x^2 \cos x\,dx$.

   Hint: Use integration by parts *twice*.

You will meet further instances of integration by parts in the next section.

Next we look at the method of integration by substitution, and discover that one needs to be just a little careful in making substitutions.

Theorem 5   Integration by Substitution

If f is continuous on $[a, b]$, g differentiable on $[\alpha, \beta]$, g' continuous on $[\alpha, \beta]$, and $g([\alpha, \beta]) \subseteq [a, b]$, then

$$\int_{g(\alpha)}^{g(\beta)} f(x)dx = \int_{\alpha}^{\beta} f(g(t))g'(t)dt.$$

If, in addition, $g(\alpha) = a$, $g(\beta) = b$ and g possesses an inverse function g^{-1} on $[a, b]$, then

$$\int_{a}^{b} f(x)dx = \int_{g^{-1}(a)}^{g^{-1}(b)} f(g(t))g'(t)dt.$$

Note that, in this situation, for g to possess an inverse function, g must be strictly increasing.

There is an analogous result in the situation that g is strictly decreasing.

Proof   Let F be a primitive of f on $[a, b]$, and define the function h by

$$h(t) = F(g(t)), \quad t \in [\alpha, \beta]. \tag{5}$$

By the Composition Rule for derivatives, applied to the composite $h(t) = F(g(t))$, we have that

$$h'(t) = F'(g(t))g'(t). \tag{6}$$

Theorem 2, Sub-section 6.2.2.

By the Fundamental Theorem of Calculus applied to h on $[\alpha, \beta]$, we have

$$\int_{\alpha}^{\beta} h'(t)dt = h(\beta) - h(\alpha). \tag{7}$$

If we now substitute for h and h' from (5) and (6) into the two sides of equation (7), we obtain

$$\int_{\alpha}^{\beta} F'(g(t))g'(t)dt = F(g(\beta)) - F(g(\alpha)). \tag{8}$$

Since F is a primitive of f on $[a, b]$, and so a primitive of f on $g([\alpha, \beta])$, we can rewrite the left-hand side of equation (8) as $\int_{\alpha}^{\beta} f(g(t))g'(t)dt$.

Recall that, by hypothesis $g([\alpha, \beta]) \subseteq [a, b]$.

Further, since F is a primitive of f on $[a, b]$, and so a primitive of f on $g([\alpha, \beta])$, we can apply the Fundamental Theorem of Calculus to the function F on $g([\alpha, \beta])$ to express the right-hand side of equation (8) as

$$F(g(\beta)) - F(g(\alpha)) = \int_{g(\alpha)}^{g(\beta)} F'(x)dx$$

$$= \int_{g(\alpha)}^{g(\beta)} f(x)dx.$$

So, if we make these substitutions on both sides of equation (8), we deduce that

$$\int_{\alpha}^{\beta} f(g(t))g'(t)dt = \int_{g(\alpha)}^{g(\beta)} f(x)dx. \tag{9}$$

This is the first part of the theorem.

Suppose next that g possesses an inverse function g^{-1} on $[\alpha, \beta]$, and

$$a = g(\alpha) \quad \text{and} \quad b = g(\beta);$$

then

$$\alpha = g^{-1}(a) \quad \text{and} \quad \beta = g^{-1}(b).$$

Making these substitutions into equation (9), we obtain the second part of the theorem.   ∎

> **Strategy to evaluate an integral $\int_\alpha^\beta f(g(t))g'(t)dt$ using integration by substitution:**
>
> 1. Choose a function $x = g(t)$ such that $\frac{dx}{dt} = g'(t)$, and express dt in terms of x and dx.
> 2. Make the necessary substitutions to give an integral in terms of x and dx.
> 3. Calculate this integral.

There is a similar strategy for finding a primitive by integration by substitution.

When *evaluating an integral*, remember to check that g is one-one; this is not necessary when *finding a primitive*.

Example 3 Evaluate $\int_0^{\frac{3}{2}} \frac{t+1}{(2t+1)^{\frac{1}{2}}} dt$,

Solution Let $x = g(t) = 2t + 1$, $t \in \mathbb{R}$. The function g is one–one on $\mathbb{R}$.

Then $\frac{dx}{dt} = 2$, so that $dx = 2dt$. Making the various substitutions into the integral, we get

Here

when $t = 0$, $x = 1$;

when $t = \frac{3}{2}$, $x = 4$.

Also, $t = \frac{1}{2}(x - 1)$.

$$\int_0^{\frac{3}{2}} \frac{t+1}{(2t+1)^{\frac{1}{2}}} dt = \int_1^4 \frac{\frac{1}{2}(x-1)+1}{x^{\frac{1}{2}}} \frac{1}{2} dx$$

$$= \int_1^4 \frac{\frac{1}{2}(x+1)}{x^{\frac{1}{2}}} \frac{1}{2} dx$$

$$= \frac{1}{4} \int_1^4 \left(x^{\frac{1}{2}} + x^{-\frac{1}{2}} \right) dx$$

$$= \frac{1}{4} \left[\frac{2}{3} x^{\frac{3}{2}} + 2x^{\frac{1}{2}} \right]_1^4$$

$$= \frac{1}{4} \left(\frac{16}{3} + 4 \right) - \frac{1}{4} \left(\frac{2}{3} + 2 \right) = \frac{5}{3}. \qquad \square$$

Problem 5

(a) Find a primitive of the function $f(x) = \sin(2 \sin 3x) \cos 3x$, $x \in \mathbb{R}$.

(b) Evaluate the integral $\int_0^1 \frac{e^x}{(1+e^x)^2} dx$.

Remark

If you use the second assertion of Theorem 5, you must verify that the function g does have an inverse; you may end up with a contradiction if you simply calculate thoughtlessly.

For example, take $\alpha = -1$, $\beta = 2$, $g(t) = t^2$, and $f(x) = x^{\frac{1}{2}}$; then $a = g(-1) = 1$ and $b = g(2) = 4$. Then

$$\int_a^b f(x)dx = \int_1^4 x^{\frac{1}{2}} dx = \left[\frac{2}{3} x^{\frac{3}{2}} \right]_1^4 = \left(\frac{16}{3} \right) - \left(\frac{2}{3} \right) = \frac{14}{3},$$

but

For, if we make the substitution $x = t^2$, then

$t = -1 \Rightarrow x = 1$,

$t = 2 \Rightarrow x = 4$.

$$\int_{-1}^2 f(g(t))g'(t)dt = \int_{-1}^2 (t^2)^{\frac{1}{2}} 2tdt = \int_{-1}^2 2t^2 dt$$

$$= \left[\frac{2}{3} t^3 \right]_{-1}^2 = \left(\frac{16}{3} \right) - \left(-\frac{2}{3} \right) = 6.$$

This contradiction arises since the function g does NOT have an inverse on $[1, 4]$ that maps $[1, 4]$ to $[-1, 2]$.

Our second substitution technique applies when we let $t = g(x)$, where the function g has an inverse, so that $x = g^{-1}(t)$, and we express x in terms of t, and dx in terms of t and dt.

Strategy to evaluate an integral $\int_a^b f(x)dx$ using integration by substitution

1. Choose a function $t = g(x)$, where g has an inverse, so that $x = g^{-1}(t)$; express dx in terms of t and dt.
2. Make the necessary substitutions to give an integral in terms of t and dt.
3. Calculate this integral.

There is a similar strategy for finding a primitive by integration by substitution.

When *finding a primitive* it is not necessary to check that g is one-one.

Example 4   Evaluate $\int_0^{\log_e 5} \frac{e^{2x}}{(e^x-1)^{\frac{1}{2}}}dx$.

Solution   Let $t = g(x) = (e^x - 1)^{\frac{1}{2}}$, $x \in [0, \infty)$. The function g is one–one on $[0, \infty)$.

Then $t^2 = e^x - 1$, so that $e^x = t^2 + 1$ and $x = \log_e(t^2 + 1)$. It follows that

$$\frac{dx}{dt} = \frac{2t}{t^2 + 1}, \quad \text{so that } dx = \frac{2t}{t^2 + 1}dt.$$

Hence

$$\int_0^{\log_e 5} \frac{e^{2x}}{(e^x-1)^{\frac{1}{2}}}dx = \int_0^2 \frac{(t^2+1)^2}{t} \times \frac{2t}{t^2+1}dt$$

$$= \int_0^2 2(t^2 + 1)dt$$

$$= \left[\frac{2}{3}t^3 + 2t\right]_0^2$$

$$= \frac{16}{3} + 4 = \frac{28}{3}. \qquad \square$$

Here

when $x = 0$, $t = 0$;
when $x = \log_e 5$, $t = 2$.

Problem 6

(a) Find a primitive of the function $f(x) = \frac{1}{3(x-1)^{\frac{3}{2}} + x(x-1)^{\frac{1}{2}}}$, $x \in (1, \infty)$.

(b) Evaluate the integral $\int_0^{\log_e 3} e^x \sqrt{1 + e^x}dx$.

7.4   Inequalities for integrals and their applications

Often it is not possible to evaluate an integral explicitly, and a numerical estimate for its value is sufficient for our purposes. This can occur both in applications of mathematics and in the proofs of theorems that involve integration.

One of the goals of Numerical Analysis is to obtain good estimates for integrals.

7.4.1   The key inequalities

Our principal tool connecting inequalities and integrals is the following.

Theorem 1   Fundamental Inequality for Integrals

If f is integrable on $[a, b]$, and $f(x) \geq 0$ on $[a, b]$, then $\int_a^b f(x)dx \geq 0$.

Proof   Since f is integrable on $[a, b]$, the sequence $\{P_n\}$ of standard partitions of $[a, b]$ has the property that

$$\lim_{n \to \infty} L(f, P_n) = \int_a^b f.$$

Now, for each value of n, we have

By the Null Partitions Criterion, Theorem 7, Sub-section 7.1.2.

$$L(f, P_n) = \sum_{i=1}^{n} m_i \delta x_i, \quad \text{where } m_i = \inf\{f(x) : x \in [x_{i-1}, x_i]\}.$$

We use the usual notation for partitions and lower and upper Riemann sums.

But, since $f(x) \geq 0$ on $[a, b]$, it follows that $f(x) \geq 0$ on each subinterval $[x_{i-1}, x_i]$, so that 0 is a lower bound for f on $[x_{i-1}, x_i]$, for each i. Since m_i is the greatest lower bound for f on $[x_{i-1}, x_i]$, we must therefore have $m_i \geq 0$, for each i.

Since each term $m_i \delta x_i$ in the sum for $L(f, P_n)$ is non-negative, it follows that

$$L(f, P_n) = \sum_{i=1}^{n} m_i \delta x_i \geq 0. \tag{1}$$

Letting $n \to \infty$ in (1) and using the Limit Inequality Rule for sequences, we deduce that

Theorem 3, Sub-section 2.3.3.

$$\int_a^b f = \lim_{n \to \infty} L(f, P_n) \geq 0.\qquad\blacksquare$$

We can now use the Fundamental Inequality for Integrals to prove the most commonly used inequalities for integrals.

Theorem 2   Inequality Rule for Integrals

Let f and g be integrable on $[a, b]$. Then:

(a) If $f(x) \leq g(x)$ on $[a, b]$, then $\int_a^b f(x)dx \leq \int_a^b g(x)dx$.

(b) If $m \leq f(x) \leq M$ on $[a, b]$, then $m(b-a) \leq \int_a^b f(x)dx \leq M(b-a)$.

Recall that earlier we motivated our discussion of integrals in terms of areas. The following diagrams illustrate the results of Theorem 1 in the special case that the functions are positive so that we can interpret the integrals as areas between the curves and the x-axis.

Section 7.1.1.

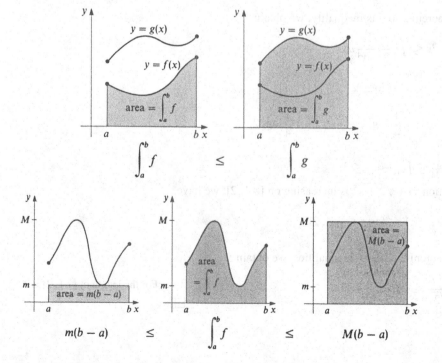

Proof of Theorem 2

(a) Since f and g are integrable on $[a, b]$, so is the function $g - f$.

By the Combination Rules for integrals, Theorem 4 of Section 7.2.3.

Since $f(x) \le g(x)$ on $[a, b]$, it follows that

$$g(x) - f(x) \ge 0, \quad x \in [a, b];$$

so, by Theorem 1, we deduce that

$$\int_a^b (g - f) \ge 0,$$

from which it follows at once that $\int_a^b f \le \int_a^b g$.

(b) First, the constant function $x \mapsto M, x \in [a, b]$, is continuous on $[a, b]$ and so is integrable on $[a, b]$.

We can therefore apply part (a) of the Theorem to the inequality $f(x) \le M$ on $[a, b]$ to obtain

$$\int_a^b f \le \int_a^b M dx = M(b - a).$$

Next, the constant function $x \mapsto m, x \in [a, b]$, is continuous on $[a, b]$ and so is integrable on $[a, b]$.

We can therefore apply part (a) of the Theorem to the inequality $f(x) \ge m$ on $[a, b]$ to obtain

$$\int_a^b f \ge \int_a^b m dx = m(b - a). \qquad \blacksquare$$

Example 1 Prove that $\int_0^1 \frac{x^3}{2 - \sin^4 x} dx \le \frac{1}{4} \log_e 2$.

Solution Since, by the Sine Inequality, $|\sin x| \le |x|$, for $x \in \mathbb{R}$, it follows that $\sin^4 x \le x^4$, for $x \in \mathbb{R}$. Hence

Theorem 2, Sub-section 4.1.3.

$$\frac{x^3}{2 - \sin^4 x} \le \frac{x^3}{2 - x^4}, \quad \text{for } x \in [0, 1].$$

If we apply part (a) of Theorem 2 to this inequality, we obtain

$$\int_0^1 \frac{x^3}{2 - \sin^4 x} dx \le \int_0^1 \frac{x^3}{2 - x^4} dx$$

$$= \left[-\frac{1}{4} \log_e (2 - x^4) \right]_0^1$$

$$= -\frac{1}{4} (\log_e 1 - \log_e 2) = \frac{1}{4} \log_e 2. \qquad \square$$

Example 2 Prove that $\frac{3}{\sqrt{34}} \le \int_{-1}^2 \frac{dx}{\sqrt{2 + x^5}} \le 3$.

Solution Since the function $x \mapsto \sqrt{2 + x^5}$ is increasing on $[-1, 2]$, we have

$$\frac{1}{\sqrt{34}} \le \frac{1}{\sqrt{2 + x^5}} \le 1, \quad \text{for } x \in [-1, 2].$$

If we apply part (b) of Theorem 2 to these inequalities, we obtain

$$\frac{3}{\sqrt{34}} \le \int_{-1}^2 \frac{dx}{\sqrt{2 + x^5}} \le 3. \qquad \square$$

For the length of $[-1, 2]$ is 3.

Problem 1 Use the Inequality Rule for integrals to prove that $\int_1^3 x\sin\left(\frac{1}{x^{10}}\right)dx \le 4$.

Problem 2 Use the Inequality Rule for integrals to prove that $\frac{1}{2} \le \int_0^{\frac{1}{2}} e^{x^2}dx \le \frac{2}{3}$.

Hint: Use the fact that $1 + x \le e^x \le \frac{1}{1-x}$, for $x \in [0, 1)$.

Our final result is of great value in many applications. We saw earlier that, if the function f is integrable on $[a, b]$, so is the function $|f|$. We can then use Theorem 2 to obtain a connection between the values of the integrals of f and $|f|$.

> Theorem 4, The Exponential Inequalities, Sub-section 4.1.3.
> By the Modulus Rule, Theorem 4, Sub-section 7.2.3.

Theorem 3 Triangle Inequality

If f is integrable on $[a, b]$, then
$$\left|\int_a^b f\right| \le \int_a^b |f|.$$
Furthermore, if $|f(x)| \le M$ on $[a, b]$, then
$$\left|\int_a^b f\right| \le M(b - a).$$

> The name arises because of the similarity between this inequality and the Triangle Inequality for numbers
> $$\left|\sum_{i=1}^n a_i\right| \le \sum_{i=1}^n |a_i|.$$

Proof From the definition of the modulus function, we have
$$-|f(x)| \le f(x) \le |f(x)|.$$
Thus, since $|f|$ is integrable on $[a, b]$, it follows from part (a) of Theorem 2 that
$$-\int_a^b |f| \le \int_a^b f \le \int_a^b |f|.$$
We can then rewrite this pair of inequalities in the desired form
$$\left|\int_a^b f\right| \le \int_a^b |f|.$$

Next, we assume that $|f(x)| \le M$ on $[a,b]$. Then, from the definition of the modulus function, it follows that
$$-M \le f(x) \le M, \quad \text{for } x \in [a, b].$$
It follows from part (b) of Theorem 2 that
$$-M(b - a) \le \int_a^b f \le M(b - a).$$
We can then rewrite this pair of inequalities in the desired form
$$\left|\int_a^b f\right| \le M(b - a). \qquad \blacksquare$$

Example 3 Prove that $\left|\int_0^{\frac{\pi}{2}} \frac{x - \frac{\pi}{2}}{2 + \cos x}dx\right| \le \frac{\pi^2}{16}$.

Solution By the Triangle Inequality for Integrals
$$\left|\int_0^{\frac{\pi}{2}} \frac{x - \frac{\pi}{2}}{2 + \cos x}dx\right| \le \int_0^{\frac{\pi}{2}} \frac{\left|x - \frac{\pi}{2}\right|}{|2 + \cos x|}dx$$
$$= \int_0^{\frac{\pi}{2}} \frac{\frac{\pi}{2} - x}{2 + \cos x}dx.$$

> So we must now examine this last integral.

But $2 + \cos x \geq 2$, for $x \in \left[0, \frac{\pi}{2}\right]$, so that

$$\frac{1}{2 + \cos x} \leq \frac{1}{2}, \quad \text{for } x \in \left[0, \frac{\pi}{2}\right].$$

By applying this inequality to the integrand in the last integral, we obtain from the Inequality Rule for Integrals that

$$\left| \int_0^{\frac{\pi}{2}} \frac{x - \frac{\pi}{2}}{2 + \cos x} dx \right| \leq \int_0^{\frac{\pi}{2}} \frac{1}{2} \left(\frac{\pi}{2} - x \right) dx$$

$$= \frac{1}{2} \left[\frac{\pi}{2} x - \frac{1}{2} x^2 \right]_0^{\frac{\pi}{2}}$$

$$= \frac{1}{2} \left(\frac{\pi^2}{4} - \frac{\pi^2}{8} \right) = \frac{\pi^2}{16}. \qquad \square$$

Problem 3 Prove the following inequalities:

(a) $\left| \int_1^4 \frac{\sin\left(\frac{1}{x}\right)}{2 + \cos\left(\frac{1}{x}\right)} dx \right| \leq 3;$ (b) $\left| \int_0^{\frac{\pi}{4}} \frac{\tan x}{3 - \sin(x^2)} dx \right| \leq \frac{1}{4} \log_e 2.$

We end this sub-section with a nice application of the Triangle Inequality, which is closely related to the Fundamental Theorem of Calculus.

Theorem 1, Sub-section 7.3.1.

Theorem 4 Let f be continuous on $[a, b]$, and F be defined on $[a, b]$ by the formula

$$F(x) = \int_a^x f(t)dt.$$

Then F is differentiable on $[a, b]$, and its derivative is

$$F'(x) = f(x).$$

Proof We will prove that F is differentiable at c and $F'(c) = f(c)$ in the case that $a < c < b$. (The cases $c = a$ and $c = b$ are similar, with the appropriate one-sided derivatives being used.)

We must prove that:

     for each positive number ε, there is a positive number δ such that

$$\left| \frac{F(x) - F(c)}{x - c} - f(c) \right| < \varepsilon, \quad \text{for all } x \text{ satisfying } 0 < |x - c| < \delta.$$

You may omit this proof at a first reading.

We will assume that δ is chosen small enough that $(c - \delta, c + \delta) \subset [a, b]$.

Now

$$\frac{F(x) - F(c)}{x - c} = \frac{1}{x - c} \left\{ \int_a^x f(t)dt - \int_a^c f(t)dt \right\}$$

$$= \frac{1}{x - c} \int_c^x f(t)dt,$$

and clearly

$$f(c) = \frac{1}{x - c} \int_c^x f(c)dt.$$

It follows that

$$\frac{F(x) - F(c)}{x - c} - f(c) = \frac{1}{x - c} \int_c^x \{f(t) - f(c)\}dt$$

so that

$$\left| \frac{F(x) - F(c)}{x - c} - f(c) \right| = \frac{1}{|x - c|} \left| \int_c^x \{f(t) - f(c)\}dt \right|.$$

We now suppose that $x > c$; it then follows from the above equation, using the Triangle Inequality for integrals, that

$$\left| \frac{F(x) - F(c)}{x - c} - f(c) \right| = \frac{1}{x - c} \left| \int_c^x \{f(t) - f(c)\}dt \right|$$

$$\leq \frac{1}{x - c} \int_c^x |f(t) - f(c)|dt \qquad \text{(since } x > c\text{).}$$

But, since f is continuous at c, we know that there must exist some positive number δ such that

$$|f(t) - f(c)| < \frac{1}{2}\varepsilon, \quad \text{for all } x \text{ satisfying } |x - c| < \delta.$$

It follows that, for x satisfying $c < x < c + \delta$, we have

$$\left| \frac{F(x) - F(c)}{x - c} - f(c) \right| \leq \frac{1}{x - c} \int_c^x \frac{1}{2}\varepsilon dt$$

$$= \frac{1}{2}\varepsilon$$

$$< \varepsilon.$$

A similar argument applies in the case that $x < c$; so that, for all x satisfying $0 < |x - c| < \delta$, we have   We omit the details.

$$\left| \frac{F(x) - F(c)}{x - c} - f(c) \right| < \varepsilon.$$

This completes the proof.   ■

7.4.2   Wallis's Formula

In this sub-section we use the method of Reduction of Order together with various inequalities between integrals to establish a remarkable formula for the number π.

Reduction of Order method

Quite often we need to evaluate an integral I_n that involves a non-negative integer n. A common approach to such integrals is to relate the value of I_n to the value of I_{n-1} or I_{n-2} by a *reduction formula*, using integration by parts.   This is just another name for a *recurrence formula*.

Example 4   Let $I_n = \int_0^{\frac{\pi}{2}} \sin^n x dx$, $n = 0, 1, 2, \ldots$.

(a) Evaluate I_0 and I_1.

(b) Prove that $I_n = \frac{n-1}{n} I_{n-2}$, for $n \geq 2$.

(c) Deduce from part (b) that, for $n \geq 1$

$$I_{2n} = \frac{1.3. \ldots .(2n-3)(2n-1)}{2.4. \ldots .(2n-2)(2n)} \times \frac{\pi}{2} \quad \text{and}$$

$$I_{2n+1} = \frac{2.4. \ldots .(2n-2)(2n)}{3.5. \ldots .(2n-1)(2n+1)}.$$

Solution

(a) We can evaluate the first two integrals easily

$$I_0 = \int_0^{\frac{\pi}{2}} 1 dx = \frac{\pi}{2}, \quad \text{and}$$

$$I_1 = \int_0^{\frac{\pi}{2}} \sin x dx = [-\cos x]_0^{\frac{\pi}{2}} = 1.$$

(b) We first write I_n in the form $I_n = \int_0^{\frac{\pi}{2}} \sin x \sin^{n-1} x dx$. Using integration by parts, we find that, for $n \geq 2$

We will integrate $\sin x$ and differentiate $\sin^{n-1} x$.

$$I_n = \left[(-\cos x)\sin^{n-1}x\right]_0^{\frac{\pi}{2}} - \int_0^{\frac{\pi}{2}} (-\cos x)(n-1)\sin^{n-2}x \cos x dx$$

$$= 0 + (n-1)\int_0^{\frac{\pi}{2}} \cos^2 x \sin^{n-2} x dx$$

$$= (n-1)\int_0^{\frac{\pi}{2}} \left(1 - \sin^2 x\right)\sin^{n-2} x dx$$

$$= (n-1)\{I_{n-2} - I_n\}.$$

We can then rewrite this result in the form $nI_n = (n-1)I_{n-2}$, so that

$$I_n = \frac{n-1}{n} I_{n-2}.$$

(c) If we replace n in the formula for I_n in part (b) by $2n, 2n-2, 2n-4, \ldots,$ in turn, we obtain

$$I_{2n} = \frac{2n-1}{2n} I_{2n-2}, \quad I_{2n-2} = \frac{2n-3}{2n-2} I_{2n-4}, \quad I_{2n-4} = \frac{2n-5}{2n-4} I_{2n-6}, \ldots$$

Hence

$$I_{2n} = \frac{2n-1}{2n} I_{2n-2}$$

$$= \frac{(2n-3)(2n-1)}{(2n-2)(2n)} I_{2n-4}$$

$$= \frac{(2n-5)(2n-3)(2n-1)}{(2n-4)(2n-2)(2n)} I_{2n-6},$$

and so on. Continuing this process, we obtain

$$I_{2n} = \frac{1.3.\ldots.(2n-3)(2n-1)}{2.4.\ldots.(2n-2)(2n)} I_0$$

$$= \frac{1.3.\ldots.(2n-3)(2n-1)}{2.4.\ldots.(2n-2)(2n)} \times \frac{\pi}{2}.$$

For $I_0 = \frac{\pi}{2}$.

Similarly

$$I_{2n+1} = \frac{2n}{2n+1} I_{2n-1}$$

$$= \frac{(2n-2)(2n)}{(2n-1)(2n+1)} I_{2n-3}$$

$$\vdots$$

$$= \frac{2.4.\ldots.(2n-2)(2n)}{3.5.\ldots.(2n-1)(2n+1)} I_1$$

$$= \frac{2.4.\ldots.(2n-2)(2n)}{3.5.\ldots.(2n-1)(2n+1)}.$$

For $I_1 = 1$.

$\square$

Problem 4 Let $I_n = \int_0^1 e^x x^n dx$, $n = 0, 1, 2, \ldots$.

(a) Evaluate I_0.

(b) Prove that $I_n = e - nI_{n-1}$, for $n \geq 1$.

(c) Deduce the values of I_1, I_2, I_3 and I_4.

Wallis's Formula

We now use the various results that we have just proved for the integral $I_n = \int_0^{\frac{\pi}{2}} \sin^n x dx$ to verify some surprising results.

Theorem 5 Wallis's Formula

(a) $\displaystyle\lim_{n \to \infty} \left(\frac{2}{1} \cdot \frac{2}{3} \cdot \frac{4}{3} \cdot \frac{4}{5} \cdot \frac{6}{5} \cdot \frac{6}{7} \cdot \ldots \cdot \frac{2n}{2n-1} \cdot \frac{2n}{2n+1} \right) = \frac{\pi}{2}.$

(b) $\displaystyle\lim_{n \to \infty} \frac{(n!)^2 2^{2n}}{(2n)! \sqrt{n}} = \sqrt{\pi}.$

Each of the two limits is called Wallis's Formula.

In the next problem, we ask you to establish a number of relationships between the terms of the two sequences in the statement of Theorem 5.

Problem 5 Let $a_n = \frac{2}{1} \cdot \frac{2}{3} \cdot \frac{4}{3} \cdot \frac{4}{5} \cdot \frac{6}{5} \cdot \frac{6}{7} \cdot \ldots \cdot \frac{2n}{2n-1} \cdot \frac{2n}{2n+1}$ and $b_n = \frac{(n!)^2 2^{2n}}{(2n)! \sqrt{n}}$, $n \geq 1$.

(a) Evaluate a_n and b_n, for $n = 1, 2$ and 3.

(b) Verify that $b_n^2 = \frac{2n+1}{n} a_n$, for $n = 1, 2$ and 3.

(c) Prove that $b_n^2 = \frac{2n+1}{n} a_n$, for all $n \geq 1$.

You may omit tackling this problem and the following proof at a first reading.

Proof of Theorem 5   Let $I_n = \int_0^{\frac{\pi}{2}} \sin^n x\,dx$, for all $n \geq 0$, and let the sequences $\{a_n\}$ and $\{b_n\}$ be as given in Problem 5.

(a) Using the formulas for I_{2n} and I_{2n+1} in part (c) of Example 4 above, we obtain

The numerator of I_{2n+1} is the same as the denominator of I_{2n}.

$$\frac{I_{2n}}{I_{2n+1}} = \frac{1.3.3.5.5.\ldots.(2n-1)(2n-1)(2n+1)}{2.2.4.4.6.\ldots.(2n-2)(2n)(2n)} \times \frac{\pi}{2},$$

which we arrange in the form

$$a_n = \frac{I_{2n+1}}{I_{2n}} \times \frac{\pi}{2}.$$

Notice that, in order to complete the proof of part (a), it is sufficient to show that

$$\frac{I_{2n+1}}{I_{2n}} \to 1 \quad \text{as } n \to \infty. \tag{2}$$

We do this as follows.

Since $0 \leq \sin x \leq 1$ for $x \in \left[0, \frac{\pi}{2}\right]$, we have

$$\sin^{2n} x \geq \sin^{2n+1} x \geq \sin^{2n+2} x, \quad \text{for } x \in \left[0, \frac{\pi}{2}\right].$$

It follows, from the Inequality Rule for integrals, that

$$I_{2n} \geq I_{2n+1} \geq I_{2n+2}.$$

Thus

$$1 \geq \frac{I_{2n+1}}{I_{2n}} \geq \frac{I_{2n+2}}{I_{2n}}$$

Note that
$$I_{2n+2} = \frac{2n+1}{2n+2} I_{2n},$$
by the reduction formula for I_n.

$$= \frac{2n+1}{2n+2}. \tag{3}$$

By letting $n \to \infty$ in (3) and using the Squeeze Rule for sequences, we deduce that the limit (2) holds, as desired.

(b) We know, from part (c) of Problem 5, that

$$b_n^2 = \frac{2n+1}{n} a_n, \quad \text{for all } n \geq 1.$$

We also know, from the proof of part (a) of this theorem, that

$$a_n \to \frac{\pi}{2} \quad \text{as } n \to \infty.$$

It follows, by applying the Product Rule for sequences to the above formula for b_n^2, that

$$b_n^2 \to 2 \times \frac{\pi}{2} = \pi \quad \text{as } n \to \infty.$$

Hence, by the continuity of the square root function, we conclude that $b_n \to \sqrt{\pi}$ as $n \to \infty$. ∎

Unfortunately the sequences in Theorem 5 converge rather slowly as n increases, so they are of little value in determining π or $\sqrt{\pi}$ to any reasonable degree of approximation. In the next chapter, we shall meet practical ways of estimating π.

Section 8.5.

7.4.3 Maclaurin Integral Test

We now introduce a method for determining the convergence or divergence of certain series of the form $\sum_{n=1}^{\infty} f(n)$, where the terms are positive and decrease to zero. In particular, we show that the series $\sum_{n=1}^{\infty} \frac{1}{n^p}$ converges if $p > 1$ and diverges if $0 < p \le 1$.

You saw in Sub-section 3.2.1 that this series converges if $p \ge 2$.

We use the fact that each term in such a series can be regarded as a contribution to a lower Riemann sum or upper Riemann sum for a suitable integral.

Theorem 6   Maclaurin Integral Test

Let f be positive and decreasing on $[1, \infty)$, and let $f(x) \to 0$ as $x \to \infty$. Then:

(a) $\sum_{n=1}^{\infty} f(n)$ converges if the sequence $\{\int_1^n f : n \in \mathbb{N}\}$ is bounded above;

(b) $\sum_{n=1}^{\infty} f(n)$ diverges if the sequence $\{\int_1^n f : n \in \mathbb{N}\}$ tends to ∞ as $n \to \infty$.

In these results, the number 1 may be replaced by any convenient positive integer.

Before proving the theorem, we illustrate the underlying ideas.

Example 5   Let P_{n-1} be the standard partition of $[1, n]$ with $n - 1$ subintervals

$$\{[1,2], [2,3], \ldots, [i, i+1], \ldots, [n-1, n]\}.$$

(a) Determine the lower and upper Riemann sums, $L(f, P_{n-1})$ and $U(f, P_{n-1})$, for the function $f(x) = \frac{1}{x^2}, x \in [1, \infty)$.

(b) Deduce that the series $\sum_{n=1}^{\infty} \frac{1}{n^2}$ converges, and that $1 \le \sum_{n=1}^{\infty} \frac{1}{n^2} \le 2$.

In fact, the sum of this series is $\frac{\pi^2}{6}$, but to prove this goes outside the scope of this book.

Solution

(a) Since f is decreasing on $[1, \infty)$, it follows that, for $i = 1, 2, \ldots, n - 1$,

$$m_i = f(i + 1) = \frac{1}{(i + 1)^2},$$

$$M_i = f(i) = \frac{1}{i^2}.$$

Also, each subinterval in the partition has length 1.

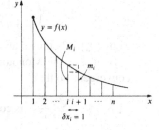

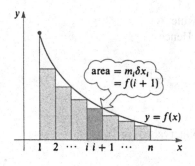

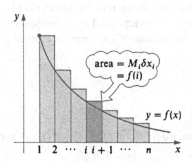

Hence the lower and upper Riemann sums for f are

$$L(f, P_{n-1}) = \sum_{i=1}^{n-1} m_i \times 1 = \frac{1}{2^2} + \frac{1}{3^2} + \cdots + \frac{1}{n^2},$$

$$U(f, P_{n-1}) = \sum_{i=1}^{n-1} M_i \times 1 = \frac{1}{1^2} + \frac{1}{2^2} + \cdots + \frac{1}{(n-1)^2}.$$

(b) Let s_n denote the nth partial sum of the series $\sum_{n=1}^{\infty} \frac{1}{n^2}$; thus

$$s_n = \frac{1}{1^2} + \frac{1}{2^2} + \cdots + \frac{1}{(n-1)^2} + \frac{1}{n^2}.$$

It follows that

$$L(f, P_{n-1}) = s_n - 1 \quad \text{and} \quad U(f, P_{n-1}) = s_n - \frac{1}{n^2}.$$

Since f is monotonic on $[1, n]$, it is integrable on $[1, n]$; hence we have

$$L(f, P_{n-1}) \le \int_1^n f \le U(f, P_{n-1}),$$

so that

$$s_n - 1 \le \int_1^n \frac{dx}{x^2} \le s_n - \frac{1}{n^2}. \tag{4}$$

Now, the sequence $\{s_n\}$ is increasing, since the series has positive terms. Also, from (4), we have

$$s_n \le \int_1^n \frac{dx}{x^2} + 1$$

$$= \left[-\frac{1}{x} \right]_1^n + 1$$

$$= \left(1 - \frac{1}{n} \right) + 1 = 2 - \frac{1}{n} \le 2. \tag{5}$$

Thus $\{s_n\}$ is bounded above.

Hence, by the Monotone Convergence Theorem for sequences, $\{s_n\}$ is convergent, so that

$$\sum_{n=1}^{\infty} \frac{1}{n^2} \quad \text{is convergent.}$$

Since $s_1 = 1$ and $\{s_n\}$ is increasing, the sum of the series is at least 1.

Finally, we deduce from (5), using the Limit Inequality Rule for sequences, that $\lim_{n \to \infty} s_n \le 2$; so the sum of the series is at most 2. Hence

$$1 < \sum_{n=1}^{\infty} \frac{1}{n^2} < 2. \qquad \square$$

Example 6 Let P_{n-1} be the standard partition of $[1, n]$ with $n - 1$ subintervals
$$\{[1, 2], [2, 3], \ldots, [i, i+1], \ldots, [n-1, n]\}.$$

(a) Determine the lower and upper Riemann sums, $L(f, P_{n-1})$ and $U(f, P_{n-1})$, for the function $f(x) = \frac{1}{x}, \quad x \in [1, \infty)$.

(b) Deduce that the series $\sum_{n=1}^{\infty} \frac{1}{n}$ diverges.

Solution

(a) Since f is decreasing on $[1, \infty)$, it follows that, for $i = 1, 2, \ldots, n-1$,

$$m_i = f(i+1) = \frac{1}{i+1},$$

$$M_i = f(i) = \frac{1}{i}.$$

Also, each subinterval in the partition has length 1.

Hence the lower and upper Riemann sums for f are

$$L(f, P_{n-1}) = \sum_{i=1}^{n-1} m_i \times 1 = \frac{1}{2} + \frac{1}{3} + \cdots + \frac{1}{n},$$

$$U(f, P_{n-1}) = \sum_{i=1}^{n-1} M_i \times 1 = \frac{1}{1} + \frac{1}{2} + \cdots + \frac{1}{n-1}.$$

(b) Let s_n denote the nth partial sum of the series $\sum_{n=1}^{\infty} \frac{1}{n}$; thus

$$s_n = \frac{1}{1} + \frac{1}{2} + \cdots + \frac{1}{(n-1)} + \frac{1}{n}.$$

It follows that

$$L(f, P_{n-1}) = s_n - 1 \quad \text{and} \quad U(f, P_{n-1}) = s_n - \frac{1}{n}.$$

Since f is monotonic on $[1, n]$, it is integrable on $[1, n]$; hence we have

$$L(f, P_{n-1}) \leq \int_1^n f \leq U(f, P_{n-1}),$$

so that

$$s_n - 1 \leq \int_1^n \frac{dx}{x} \leq s_n - \frac{1}{n},$$

or

$$s_n - 1 \leq \log_e n \leq s_n - \frac{1}{n}. \tag{6}$$

In particular, $s_n \geq \log_e n$. Then, since $\log_e n \to \infty$ as $n \to \infty$, it follows, from the Squeeze Rule for sequences that tend to infinity, that $s_n \to \infty$ as $n \to \infty$.

Hence, the series $\sum_{n=1}^{\infty} \frac{1}{n}$ diverges. $\qquad\square$

Remarks

We can in fact get more information from the above argument than just the result of the example.

1. If we define the sequence $\{\gamma_n\}$ by the expression

$$\gamma_n = 1 + \frac{1}{2} + \frac{1}{3} + \cdots + \frac{1}{n} - \log_e n$$

then

$$\gamma_{n+1} - \gamma_n = \frac{1}{n+1} - \log_e \frac{n+1}{n}$$

$$= \frac{1}{n+1} - \int_n^{n+1} \frac{dx}{x}$$

$$= \int_n^{n+1} \left(\frac{1}{n+1} - \frac{1}{x}\right) dx \le 0,$$

so that $\{\gamma_n\}$ is decreasing. Also, it follows from (6), using the fact that $\gamma_n = s_n - \log_e n$, that

By definition of the sequence $\{\gamma_n\}$.

For $\log_e n \le s_n - \frac{1}{n}$.

$$\gamma_n \ge \frac{1}{n};$$

thus the sequence $\{\gamma_n\}$ is bounded below by 0.

Hence the sequence $\{\gamma_n\}$ tends to a limit, γ say, as $n \to \infty$. The value of γ clearly lies between 1 (for $\gamma_1 = 1$) and 0.

This number is called *Euler's constant*, and occurs often in Analysis. In fact,

$$\gamma = 0.57721\ 56649\ 01532\ 86060\ 65120\ 90082\ 40243\ldots.$$

We give here just the first 35 decimal places.

2. Although the sequence $s_n = 1 + \frac{1}{2} + \cdots + \frac{1}{n-1} + \frac{1}{n}$ tends to infinity, we can make more precise statements than that, arising from (6). We can rewrite the inequalities in (6) in the following form

$$\frac{1}{n} + \log_e n \le s_n \le 1 + \log_e n;$$

if we then divide throughout by $\log_e n$ and let $n \to \infty$, we obtain, by the Limit Inequality Rule for sequences, that

$$\frac{s_n}{\log_e n} \to 1 \quad \text{as } n \to \infty.$$

We can write this last limit in an equivalent form as '$s_n \sim \log_e n$ as $n \to \infty$' – in other words, 'the behaviour of s_n and $\log_e n$ are essentially the same for large n'.

We shall address this $\sim$ notation more carefully in the next section.

Problem 6   Use the Maclaurin Integral Test to determine the behaviour of the series $\sum\limits_{n=1}^{\infty} \frac{1}{n^p}$, for $p > 0$, $p \ne 1$.

Problem 7   Show that $\int \frac{dx}{x(\log_e x)^2} = -\frac{1}{\log_e x}$, for $x > 1$, and hence prove that the series $\sum\limits_{n=2}^{\infty} \frac{1}{n(\log_e n)^2}$ converges.

Problem 8   Show that $\int \frac{dx}{x \log_e x} = \log_e(\log_e x)$, for $x > 1$, and hence prove that the series $\sum\limits_{n=2}^{\infty} \frac{1}{n \log_e n}$ diverges.

Sequences revisited

We can apply many ideas similar to those in Example 5 to obtain useful information about certain types of convergent sequences.

Example 7 Prove that $\frac{1}{n+1} + \frac{1}{n+2} + \cdots + \frac{1}{2n} \to \log_e 2$ as $n \to \infty$.

Solution Let $f(x) = \frac{1}{1+x}$, $x \in [0, 1]$; and let $P_n = \left\{ \left[0, \frac{1}{n}\right], \left[\frac{1}{n}, \frac{2}{n}\right], \ldots, \left[\frac{n-1}{n}, 1\right] \right\}$ be the standard partition of $[0, 1]$ into n sub-intervals of equal length $\frac{1}{n}$. Since f is decreasing on $[0, 1]$, it follows that on the ith sub-interval $\left[\frac{i-1}{n}, \frac{i}{n}\right]$ we have

$$m_i = f\left(\frac{i}{n}\right) = \frac{1}{1 + \frac{i}{n}}.$$

Recall that
$$m_i = \inf\left\{ f(x) : x \in \left[\frac{i-1}{n}, \frac{i}{n}\right] \right\}.$$

Thus the lower Riemann sum for f on P_n is

$$L(f, P_n) = \sum_{i=1}^{n} \frac{1}{1 + \frac{i}{n}} \times \frac{1}{n}$$
$$= \sum_{i=1}^{n} \frac{1}{n+i} = \frac{1}{n+1} + \frac{1}{n+2} + \cdots + \frac{1}{2n}.$$

Since f is decreasing on $[0, 1]$, it is integrable on $[0, 1]$. Hence, as $n \to \infty$, we have

$$\frac{1}{n+1} + \frac{1}{n+2} + \cdots + \frac{1}{2n} = L(f, P_n) \to \int_0^1 f = \int_0^1 \frac{dx}{1+x}$$
$$= [\log_e(1 + x)]_0^1 = \log_e 2. \quad \square$$

Example 7 illustrates a general technique that is often useful.

Strategy If f is positive and decreasing on $[0, 1]$, then: $\frac{1}{n} \sum_{i=1}^{n} f\left(\frac{i}{n}\right) \to \int_0^1 f$ as $n \to \infty$.

The trick in applying this strategy is to make a good choice of f.

Problem 9 Prove that $n\left(\frac{1}{n^2+1^2} + \frac{1}{n^2+2^2} + \cdots + \frac{1}{n^2+n^2}\right) \to \frac{\pi}{4}$ as $n \to \infty$.

You have already met the corresponding ideas for divergent sequences in Example 6 and Remark 2 after that example, so we state without comment the general strategy illustrated there.

Strategy If f is positive and decreasing on $[1, \infty)$, $f(x) \to 0$ as $x \to \infty$, and $\int_1^n f \to \infty$ as $n \to \infty$, then $\sum_{i=1}^{n} f(i) \sim \int_1^n f$ as $n \to \infty$.

Recall that this means that
$$\frac{\sum_{i=1}^{n} f(i)}{\int_1^n f} \to 1 \quad \text{as } n \to \infty.$$

Remark

In fact under the hypotheses of this strategy $\sum_{i=1}^{n} f(i) \sim \left(\int_1^n f\right) + c$, for any fixed number c; in particular situations we may choose c in any convenient way.

The reason for this will be explained in Section 7.5.1.

Problem 10 Prove that $1 + \frac{1}{\sqrt{2}} + \frac{1}{\sqrt{3}} + \cdots + \frac{1}{\sqrt{n}} \sim 2\sqrt{n}$ as $n \to \infty$.

Here you will use both the strategy and the subsequent remark.

Proof of the Maclaurin Integral Test

Theorem 6   Maclaurin Integral Test

Let f be positive and decreasing on $[1, \infty)$, and let $f(x) \to 0$ as $x \to \infty$. Then:

(a) $\displaystyle\sum_{n=1}^{\infty} f(n)$ converges if the sequence $\{\int_1^n f : n \in \mathbb{N}\}$ is bounded above;

(b) $\displaystyle\sum_{n=1}^{\infty} f(n)$ diverges if the sequence $\{\int_1^n f : n \in \mathbb{N}\}$ tends to ∞ as $n \to \infty$.

Proof   Let I_n denote the integral $\int_1^n f$, let $s_n = f(1) + f(2) + \cdots + f(n)$ denote the nth partial sum of the series, and let P_{n-1} be the standard partition of $[1, n]$ with $n - 1$ subintervals

$$\{[1, 2], [2, 3], \ldots, [i, i+1], \ldots, [n-1, n]\}.$$

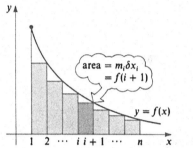 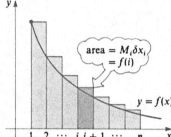

Since f is decreasing on $[1, \infty)$, it follows that, for $i = 1, 2, \ldots, n-1$,

$$m_i = f(i+1) \quad \text{and} \quad M_i = f(i).$$

Also, each subinterval in the partition has length 1.

Hence the lower and upper Riemann sums for f are

$$L(f, P_{n-1}) = \sum_{i=1}^{n-1} m_i \times 1 = f(2) + f(3) + \cdots + f(n)$$
$$= s_n - f(1),$$

$$U(f, P_{n-1}) = \sum_{i=1}^{n-1} M_i \times 1 = f(1) + f(2) + \cdots + f(n-1)$$
$$= s_n - f(n).$$

Since f is monotonic on $[1, n]$, it is integrable on $[1, n]$; hence we have

$$L(f, P_{n-1}) \le I_n \le U(f, P_{n-1}),$$

so that

$$s_n - f(1) \le I_n \le s_n - f(n). \tag{7}$$

Case 1: $\{I_n\}$ is bounded above

We are now assuming that, for some M, $I_n \le M$, for all n. It follows from (7) that

$$s_n \le f(1) + M, \quad \text{for all } n.$$

Thus the increasing sequence $\{s_n\}$ is bounded above, and so, by the Monotone Convergence Theorem, it is convergent.

Hence the series $\sum_{n=1}^{\infty} f(n)$ is convergent.

Case 2: $\{I_n\}$ is not bounded above

The sequence $\{I_n\}$ is increasing, since $I_{n+1} - I_n = \int_n^{n+1} f \geq 0$. Since we are now assuming that $\{I_n\}$ is not bounded above, it follows that $I_n \to \infty$ as $n \to \infty$.

Now, from (7), $s_n \geq I_n$; so, by the Squeeze Rule for sequences which tend to infinity,

$$s_n \to \infty \quad \text{as } n \to \infty.$$

Hence the series $\sum_{n=1}^{\infty} f(n)$ is divergent. ∎

7.5 Stirling's Formula for *n!*

For small values of n, we can evaluate $n!$ directly by multiplication or by using a scientific calculator.

Problem 1 Complete the following table of values of $n!$

n	$n!$	n	$n!$	n	$n!$
1	1	6	720	20	*
2	2	7	5040	30	*
3	6	8	40 320	40	*
4	24	9	362 880	50	*
5	120	10	3 628 800	60	*

As n increases, $n!$ grows very quickly; for instance:

- around 14! seconds have elapsed since the birth of Christ;
- around 18! seconds have elapsed since the formation of the Earth.

$14! \simeq 8.7 \times 10^{10}$

$18! \simeq 6.4 \times 10^{15}$

A calculator soon becomes useless for evaluating $n!$; thus the author's current scientific calculator (2005) gives that $69! \simeq 1.7 \times 10^{98}$, but an error message when asked the value of 70!.

Stirling's Formula gives us a way of estimating $n!$ for large values of n. In order to state the formula, though, we must first introduce some notation that enables us to compare the behaviour of positive functions of n for large values of n.

Often an estimate is sufficient for our purposes.

7.5.1 The tilda notation

For many numbers that arise in the normal way, we often use the symbol '$\simeq$' to denote 'is approximately equal to'. For example, $\sqrt{2} = 1.41421356237309504880\ldots$ (where we have given only the first 20 decimal places); for many purposes it is sufficient to use estimates such as $\sqrt{2} \simeq 1.414$ or even $\sqrt{2} \simeq 1.4$, since the error involved is small. But what do we mean by 'the error involved is small'? Sometimes we mean that the *error*, the difference between the exact value and the approximation, is a small number; sometimes we mean that the *percentage error*, namely

$$\text{percentage error} = \frac{\text{actual error}}{\text{exact value}} \times 100,$$

is a small number.

'For example, the error is less than 1'.

For example, 'the percentage error is less than 1%'.

To handle the behaviour of positive functions of n for large values of n, we introduce a very precise mathematical notation.

> **Notation**   For positive functions f and g with domain $\mathbb{N}$, we write
> $$f(n) \sim g(n) \quad \text{as } n \to \infty,$$
> to mean that $\frac{f(n)}{g(n)} \to 1$ as $n \to \infty$.

The notation is ONLY used for positive functions.

We say 'f tilda g' or 'f twiddles g'.

You can think of this as saying that the percentage error involved by replacing $f(n)$ by $g(n)$ is small for large n.

For example, $n^2 + 1000n + 10 \sim n^2$ as $n \to \infty$, since $\frac{n^2+1000n+10}{n^2} \to 1$ as $n \to \infty$; and $\sin\left(\frac{1}{n}\right) \sim \frac{1}{n}$ as $n \to \infty$, since $\frac{\sin\left(\frac{1}{n}\right)}{\frac{1}{n}} \to 1$ as $n \to \infty$.

For, $\lim_{x\to 0} \frac{\sin x}{x} = 1$.

For, $\frac{f(n)}{g(n)} \to 1 \Leftrightarrow \frac{g(n)}{f(n)} \to 1$.

Notice that, if $f(n) \sim g(n)$ as $n \to \infty$, then $g(n) \sim f(n)$ as $n \to \infty$.

Also, if $f(n) \sim g(n)$ as $n \to \infty$, $g(n) \to \infty$ and c is a given number, then $f(n) + c \sim g(n)$ as $n \to \infty$, since, as $n \to \infty$

$$\frac{f(n)+c}{g(n)} = \frac{f(n)}{g(n)} + \frac{c}{g(n)} \to 1 + 0 = 1.$$

Notice, however, that the statement $f(n) \sim g(n)$ as $n \to \infty$ does NOT mean that $f(n) - g(n) \to 0$ or even that $f(n) - g(n)$ is bounded. For example, $n^2 + 1000n + 10 \sim n^2$ as $n \to \infty$, yet

$$\left(n^2 + 1000n + 10\right) - \left(n^2\right) = 1000n + 10 \to \infty \quad \text{as } n \to \infty.$$

In situations like this example, '$\sim$' compares the sizes of the dominant terms in f and g.

> **Problem 2**   Find two pairs from the following functions such that $f_i(n) \sim f_j(n)$ as $n \to \infty$
> $$f_1(n) = \sin(n^2), \quad f_2(n) = \sin\left(\frac{1}{n^2}\right), \quad f_3(n) = 1 - \cos\left(\frac{1}{n}\right),$$
> $$f_4(n) = \frac{2}{n^2}, \qquad f_5(n) = \frac{1}{n^2}, \qquad f_6(n) = 1 - \frac{1}{n}, \qquad f_7(n) = \frac{1}{2n^2}.$$

Remark

If ℓ is finite and positive, then the statements

$$\text{'}f(n) \to \ell \text{ as } n \to \infty\text{'} \quad \text{and} \quad \text{'}f(n) \sim \ell \text{ as } n \to \infty\text{'}$$

are equivalent, since each is equivalent to the statement

$$\text{'}\frac{f(n)}{\ell} \to 1 \quad \text{as } n \to \infty\text{'}.$$

We can also, in this situation, legitimately say that $f(n) \sim \ell$ for large n.

For example, we may write

$$\left(1 + \frac{1}{n}\right)^n \sim e \quad \text{as } n \to \infty,$$

since

$$\left(1 + \frac{1}{n}\right)^n \to e \quad \text{as } n \to \infty.$$

The Combination Rules for handling tilda are direct consequences of the corresponding Combination Rules for sequences.

Note that we only use $\sim$ for positive funtions.

Combination Rules

Let $f_1(n) \sim g_1(n)$ and $f_2(n) \sim g_2(n)$ as $n \to \infty$. Then:

Sum Rule $f_1(n) + f_2(n) \sim g_1(n) + g_2(n)$;

Multiple Rule $\lambda f_1(n) \sim \lambda g_1(n)$, for any number $\lambda > 0$;

Product Rule $f_1(n) \times f_2(n) \sim g_1(n) \times g_2(n)$;

Reciprocal Rule $\frac{1}{f_1(n)} \sim \frac{1}{g_1(n)}$.

For example, if

$$f_1(n) = n^2 + n, \ g_1(n) = n^2,$$
$$f_2(n) = n^3 + n, \ g_2(n) = n^3,$$

then

$$(n^2 + n) + (n^3 + n) \sim n^2 + n^3;$$
$$5(n^2 + n) \sim 5n^2 \ (\lambda = 5);$$
$$(n^2 + n) \times (n^3 + n) \sim n^2 \times n^3 = n^5;$$
$$\frac{1}{n^2 + n} \sim \frac{1}{n^2}.$$

Example 1   Prove that, if $f(n) \sim g(n)$ as $n \to \infty$, then $(f(n))^{\frac{1}{n}} \sim (g(n))^{\frac{1}{n}}$ as $n \to \infty$.

Solution   Since $f(n) \sim g(n)$ as $n \to \infty$, we have

$$\frac{f(n)}{g(n)} \to 1 \quad \text{as } n \to \infty,$$

so that

$$\log_e \frac{f(n)}{g(n)} \to 0 \quad \text{as } n \to \infty,$$

and therefore

$$\frac{1}{n} \log_e \frac{f(n)}{g(n)} \to 0 \quad \text{as } n \to \infty.$$

Here we use the fact that the function $\log_e$ is continuous at 1.

But

$$\frac{1}{n} \log_e \frac{f(n)}{g(n)} = \log_e \left(\frac{f(n)}{g(n)} \right)^{\frac{1}{n}}$$
$$= \log_e \frac{(f(n))^{\frac{1}{n}}}{(g(n))^{\frac{1}{n}}},$$

so that

$$\log_e \frac{(f(n))^{\frac{1}{n}}}{(g(n))^{\frac{1}{n}}} \to 0 \quad \text{as } n \to \infty.$$

We now use the fact that the exponential function is continuous at 0; it follows from the previous limit that

$$\frac{(f(n))^{\frac{1}{n}}}{(g(n))^{\frac{1}{n}}} \to 1 \quad \text{as } n \to \infty,$$

which is precisely the statement that $(f(n))^{\frac{1}{n}} \sim (g(n))^{\frac{1}{n}}$ as $n \to \infty$.   □

Problem 3   Give specific examples of functions f and g to show that $(f(n))^{\frac{1}{n}} \sim (g(n))^{\frac{1}{n}}$ as $n \to \infty$ does not imply that $f(n) \sim g(n)$ as $n \to \infty$.
Hint: Try $f(n) = n^2$ and $g(n) = n$.

7.5.2   Stirling's Formula

Stirling's Formula was discovered in the eighteenth century, in an analysis of a gambling problem!

Theorem 1   Stirling's Formula

$$n! \sim \sqrt{2\pi n}\left(\frac{n}{e}\right)^n \quad \text{as } n \to \infty.$$

Problem 4   Use your calculator to evaluate $\sqrt{2\pi n}\left(\frac{n}{e}\right)^n$ for $n=5$. For this value of n does the expression approximate $n!$ to within 1%?

For small values of n, Stirling's Formula gives reasonable approximations to $n!$, and the percentage error quickly decreases as n increases:

n	$n!$	Stirling's approximation	Error
10	$3\,628\,800$	$3\,598\,696$	0.83% (1 in 120)
20	2.433×10^{18}	2.423×10^{18}	0.42% (1 in 240)
52	8.066×10^{67}	8.053×10^{67}	0.16% (1 in 620)
100	9.333×10^{157}	9.325×10^{157}	0.09% (1 in 1170)

We illustrate the use of Stirling's Formula as follows.

If 200 coins are tossed, then the probability of there being *exactly* 100 heads and 100 tails is

This fact is easily proved using Probability Theory.

$$\binom{200}{100} \times \left(\frac{1}{2}\right)^{200} = \frac{200!}{(100!)^2 \times 2^{200}}.$$

It follows from Stirling's Formula that this probability is

$$\frac{200!}{(100!)^2 \times 2^{200}} \simeq \frac{\sqrt{400\pi} \times \left(\frac{200}{e}\right)^{200}}{\left(\sqrt{200\pi} \times \left(\frac{100}{e}\right)^{100}\right)^2 \times 2^{200}}$$

Here we substitute $n=200$ and $n=100$ into Stirling's Formula.

$$= \frac{1}{10\sqrt{\pi}}$$

$$= \frac{1}{17.724\ldots}.$$

In other words, the probability of there being exactly 100 heads and 100 tails is about 1 in 18 – rather higher than you might expect.

Problem 5   Prove that $\lim\limits_{n\to\infty} \left(\frac{n^n}{n!}\right)^{\frac{1}{n}} = e$.

Hint: Use the result of Example 1.

Problem 6   Use Stirling's Formula to estimate the following numbers to two significant figures:

(a) $\binom{300}{150} \times \left(\frac{1}{2}\right)^{300}$;   (b) $\dfrac{300!}{(100!)^3} \times \left(\frac{1}{3}\right)^{300}$.

Problem 7   Use Stirling's Formula to determine a number λ such that

$$\binom{4n}{2n} \Big/ \binom{2n}{n} \sim \lambda 2^{2n} \quad \text{as } n \to \infty.$$

7.5.3   Proof of Stirling's Formula

You may omit this proof at a first reading.

> **Theorem 1   Stirling's Formula**
>
> $$n! \sim \sqrt{2\pi n}\left(\frac{n}{e}\right)^n \quad \text{as } n \to \infty.$$

Proof   We divide up the proof into a number of steps, for clarity.

Step 1: Setting things up

We consider the function

$$f(x) = \log_e x, \quad x \in [1, n],$$

and the standard partition P_{n-1} of the interval $[1, n]$ with $n - 1$ subintervals

$$\{[1, 2], [2, 3], \ldots, [i, i+1], \ldots, [n-1, n]\}.$$

We consider also the sequence of numbers $\{c_n\}_{n=2}^{\infty}$, where c_n is the total area between the concave curve $y = \log_e x$, for $x \in [1, n]$, and the polygonal line with vertices $(1, 0)$, $(2, \log_e 2)$, $(3, \log_e 3)$, ..., $(n, \log_e n)$, as illustrated below. This consists of $n - 1$ small slivers.

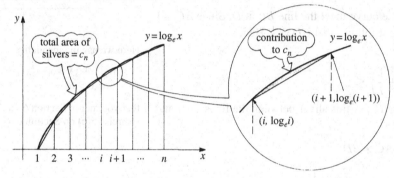

Step 2: Calculating areas

The area between the curve $y = \log_e x$ and the x-axis, for $x \in [1, n]$, is

$$\int_1^n \log_e x\, dx = [x \log_e x - x]_1^n$$
$$= n \log_e n - (n - 1). \tag{1}$$

Next, the area between the polygonal line and the x-axis is

$$\tfrac{1}{2}\{L(f, P_{n-1}) + U(f, P_{n-1})\}; \tag{2}$$

and, since f is increasing, we have

$$L(f, P_{n-1}) = \log_e 1 + \log_e 2 + \cdots + \log_e(n-1)$$
$$= \log_e(n-1)!$$
$$= \log_e n! - \log_e n \tag{3}$$

and

$$U(f, P_{n-1}) = \log_e 2 + \cdots + \log_e n$$
$$= \log_e n!. \tag{4}$$

Substituting from (3) and (4) into (2), we find that the area between the polygonal line and the x-axis is

$$\log_e n! - \tfrac{1}{2}\log_e n. \tag{5}$$

It follows from (1) and (5) that

$$c_n = n\log_e n - (n-1) - \log_e n! + \tfrac{1}{2}\log_e n$$

$$= \log_e \frac{n^{n+\frac{1}{2}}}{e^{n-1}n!}. \tag{6}$$

Step 3: Behaviour of the sequence $\{c_n\}$

It is obvious from the definition of c_n that the sequence $\{c_n\}$ is positive and increasing. In order to apply the Monotone Convergence Theorem to the sequence, we must prove next that $\{c_n\}$ is bounded above.

So, let i be an integer with $1 \le i \le n-1$, and let

$$A = (i, \log_e i), B = (i+1, \log_e(i+1)), \quad\text{and}\quad C = (i+1, \log_e i).$$

Then, as illustrated in the diagram in the margin

$$BC = \log_e(i+1) - \log_e i$$

$$= \log_e\left(1 + \frac{1}{i}\right).$$

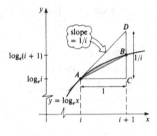

Next, let the tangent at A to the curve meet the line BC at D. Since $AC = 1$ and the slope of the line AD is $\frac{1}{i}$, it follows that $CD = \frac{1}{i}$. Hence

$$BD = CD - BC$$

Here we use the inequality

$$\log_e(1+x) > x - \tfrac{1}{2}x^2,$$
$$\text{for } x > 0,$$

$$= \frac{1}{i} - \log_e\left(1 + \frac{1}{i}\right) \le \frac{1}{2i^2}.$$

Hence the contribution to the area c_n of this sliver between the lines $x = i$ and $x = i+1$ is at most

that you met in Section 6.7, Exercise 5(c) on Section 6.4.

$$\text{area of } \triangle ABD = \frac{1}{2}AC \times BD$$

$$\le \frac{1}{2} \times 1 \times \frac{1}{2i^2} = \frac{1}{4i^2}.$$

If we now sum such areas over $i = 1, 2, \ldots, n-1$, we obtain

$$c_n \le \frac{1}{4}\sum_{i=1}^{n-1}\frac{1}{i^2}$$

$$\le \frac{1}{4}\sum_{i=1}^{\infty}\frac{1}{i^2}.$$

Since this infinite series converges, it follows that the sequence $\{c_n\}$ is bounded above.

The actual value of the bound does not matter here.

It follows from the Monotone Convergence Theorem that the sequence $\{c_n\}$ is convergent.

Step 4: Properties of the sequence $\{a_n\}$, where $a_n = e^{c_n}$

Since $\{c_n\}$ is convergent and the exponential function is continuous, it follows that, if we set $a_n = e^{c_n}$, for $n = 2, 3, \ldots$, then the sequence $\{a_n\}$ is also convergent. Thus, using the expression (6) for c_n, we have

We now introduce $\{a_n\}$ so that we can use Wallis's Formula in the next Step.

$$a_n = \frac{n^{n+\frac{1}{2}}}{e^{n-1}n!} \to L, \quad \text{as } n \to \infty, \tag{7}$$

for some non-zero number L.

$L \neq 0$ since the exponential does not take the value 0.

It follows from the formula for a_n in (7) that

$$\frac{a_n^2}{a_{2n}} = \frac{n^{2n+1}}{e^{2n-2}(n!)^2} \times \frac{e^{2n-1} \times (2n)!}{(2n)^{2n+\frac{1}{2}}}$$

$$= \frac{(2n)! n^{\frac{1}{2}} e}{(n!)^2 2^{2n+\frac{1}{2}}}, \qquad (8)$$

after some cancellation.

Step 5: Proving Stirling's Formula

We can rewrite (8) in the form

$$\frac{a_n^2}{a_{2n}} = \frac{(2n)! n^{\frac{1}{2}}}{(n!)^2 2^{2n}} \times \frac{e}{\sqrt{2}}. \qquad (9)$$

But, by Wallis's Formula, the first quotient on the right-hand side of (9) tends to $\frac{1}{\sqrt{\pi}}$ as $n \to \infty$. It follows that, if we let $n \to \infty$ on both sides of (9), we obtain

$$\frac{L^2}{L} = \frac{1}{\sqrt{\pi}} \times \frac{e}{\sqrt{2}},$$

so that

$$L = \frac{e}{\sqrt{2\pi}}.$$

You met Wallis's Formula in Theorem 5 of Sub-section 7.4.2:

$$\lim_{n \to \infty} \frac{(n!)^2 2^{2n}}{(2n)! \sqrt{n}} = \sqrt{\pi}.$$

Knowing the value of L, we can then rewrite (7) in the form

$$\frac{n^{n+\frac{1}{2}}}{e^{n-1} n!} \to \frac{e}{\sqrt{2\pi}};$$

by the Reciprocal Rule for sequences, it follows that

$$\frac{e^{n-1} n!}{n^n \sqrt{n}} \to \frac{\sqrt{2\pi}}{e}.$$

· By the definition of tilda, this limit is exactly equivalent to the relation

$$n! \sim \sqrt{2\pi n} \left(\frac{n}{e}\right)^n \quad \text{as } n \to \infty,$$

that we set out to prove. ∎

Remark

It is quite remarkable how many of the techniques that we have met so far in the book are needed in order to prove this apparently straight-forward result.

7.6 Exercises

Section 7.1

1. Sketch the graph of the function

$$f(x) = \begin{cases} 1 - |x|, & -1 < x < 1, \\ 1, & x = \pm 1. \end{cases}$$

Determine the minimum, maximum, infimum and supremum of f on $[-1, 1]$.

2. Let f be the function
$$f(x) = \begin{cases} |x|, & -1 < x < 1, \\ \frac{1}{2}, & x = \pm 1. \end{cases}$$
Evaluate $L(f, P)$ and $U(f, P)$ for each of the following partitions P of $[-1, 1]$:
(a) $P = \left\{ \left[-1, -\frac{1}{2}\right], \left[-\frac{1}{2}, 0\right], \left[0, \frac{1}{2}\right], \left[\frac{1}{2}, 1\right] \right\}$;
(b) $P = \left\{ \left[-1, -\frac{1}{4}\right], \left[-\frac{1}{4}, \frac{1}{3}\right], \left[\frac{1}{3}, 1\right] \right\}$.

3. Let f be the function
$$f(x) = \begin{cases} 1 - x, & 0 \leq x < 1, \\ 2, & x = 1. \end{cases}$$

(a) Using the standard partition P_n of $[0, 1]$ with n equal subintervals, evaluate $L(f, P_n)$ and $U(f, P_n)$.

(b) Deduce that f is integrable on $[0, 1]$, and evaluate $\int_0^1 f$.

4. Let f be the function $f(x) = x^3$, $x \in [0, 1]$.

(a) Using the standard partition P_n of $[0, 1]$ with n equal subintervals, evaluate $L(f, P_n)$ and $U(f, P_n)$.

(b) Deduce that f is integrable on $[0, 1]$, and evaluate $\int_0^1 f$.

5. Let f be the function $f(x) = \sin x$, $x \in \left[0, \frac{\pi}{2}\right]$.

(a) Using the standard partition P_n of $\left[0, \frac{\pi}{2}\right]$ with n equal subintervals, evaluate $L(f, P_n)$ and $U(f, P_n)$.

(b) Deduce that f is integrable on $\left[0, \frac{\pi}{2}\right]$, and evaluate $\int_0^{\frac{\pi}{2}} f$.

Hint: Use the formula

> This formula can be proved by Mathematical Induction.

$$\sin A + \sin(A + B) + \cdots + \sin(A + (n-1)B)$$
$$= \frac{\sin\left(\frac{1}{2}nB\right)}{\sin\left(\frac{1}{2}B\right)} \sin\left(A + \frac{n-1}{2}B\right), \quad B \neq 0.$$

6. Prove that the function
$$f(x) = \begin{cases} 1 + x, & 0 \leq x \leq 1, & x \text{ rational}, \\ 1 - x, & 0 \leq x \leq 1, & x \text{ irrational}, \end{cases}$$
is not integrable on $[0, 1]$.

7. Prove that Riemann's function

> You met this function in Sub-section 5.4.2.

$$f(x) = \begin{cases} \frac{1}{q}, & \text{if } x \text{ is a rational number } \frac{p}{q}, \text{ in lowest terms, with } q > 0, \\ 0, & \text{if } x \text{ is irrational}. \end{cases}$$
is integrable on $[0, 1]$.

Hint: Choose an $n \in \mathbb{N}$ with $\frac{1}{n} < \frac{1}{2}\varepsilon$, then a partition P of $[0,1]$ such that the total length of its subintervals containing a point x for which $f(x) > \frac{1}{n}$ is $< \frac{1}{2}\varepsilon$.

Section 7.2

1. For bounded functions f and g on an interval I, prove that
$$\sup_I (f + g) \leq \sup_I f + \sup_I g.$$

For the interval $I = [-1, 2]$, write down functions f and g for which strict inequality holds in this result.

2. Write down functions f and g and an interval I for which it is not true that

$$\sup_I (fg) = \sup_I f \times \sup_I g.$$

3. Write down functions f and g on $[0, 1]$ that are not integrable on $[0, 1]$ but such that $f + g$ is integrable on $[0, 1]$.

4. (a) Write down functions f and g on $[0, 1]$ such that f is integrable, g is not integrable, and fg is integrable.

   (b) Write down a function f on $[0, 1]$ such that $|f|$ is integrable but f is not integrable on $[0, 1]$.

5. Prove that, if the functions f and g are integrable on $[a, b]$, then so is the function $\max\{f, g\}$.

   Hint: Use the formula $\max\{a, b\} = \frac{1}{2}\{a + b + |a - b|\}$, for $a, b \in \mathbb{R}$.

6. Prove that, if f and g are integrable on $[a, b]$, then

$$\left(\int_a^b fg \right)^2 \le \left(\int_a^b f^2 \right) \left(\int_a^b g^2 \right).$$

This is known as the Cauchy–Schwarz Inequality for integrals.

   Hint: Integrate the inequality $(\lambda f + g)^2 \ge 0$, then choose a suitable value of λ.

The solution is similar to the proof of the Cauchy-Schwarz Inequality, Theorem 2, Sub-section 1.3.3.

Section 7.3

1. Write down a primitive of each of the following functions:

   (a) $f(x) = \sqrt{x^2 - 9}, \quad x \in (3, \infty)$;

   (b) $f(x) = \sin(2x + 3) - 4\cos(3x - 2), \quad x \in \mathbb{R}$;

   (c) $f(x) = e^{2x} \sin(3x), \quad x \in \mathbb{R}$.

2. Using the result of part (c) of Exercise 1, write down a primitive F of the function $f(x) = e^{2x} \sin(3x), x \in \mathbb{R}$, for which $F(\pi) = 0$.

3. Show that the following functions are all primitives of the function $f(x) = \operatorname{sech} x, x \in \left(-\frac{\pi}{2}, \frac{\pi}{2}\right)$:

   (a) $F_1(x) = \tan^{-1}(\sinh x)$;   (b) $F_2(x) = 2 \tan^{-1}(e^x)$;

   (c) $F_3(x) = \sin^{-1}(\tanh x)$;   (d) $F_4(x) = 2 \tan^{-1}\left(\tanh\left(\frac{1}{2}x\right)\right)$.

4. Let $I_n = \int_1^e x(\log_e x)^n dx$, for $n \ge 0$.

   (a) Prove that $I_n = \frac{1}{2}e^2 - \frac{1}{2}nI_{n-1}$, for $n \ge 1$.

   (b) Evaluate I_0, I_1, I_2 and I_3.

5. Evaluate each of the following integrals, using the suggested substitution where given:

   (a) $\int_0^{\frac{\pi}{2}} \tan(\sin x) \cos x\, dx$;   (b) $\int_0^1 \frac{(\tan^{-1} x)^2}{1+x^2} dx$ (try $u = \tan^{-1} x$);

   (c) $\int_0^1 (x^5 + 2x^2) \sqrt{x^6 + 4x^3 + 4}\, dx$;

   (d) $\int_0^{\frac{\pi}{2}} \frac{dx}{2 + \cos x}$ $\left(\text{try } u = \tan\left(\frac{1}{2}x\right) \text{ and use the identity } \cos x = \frac{1 - \tan^2\left(\frac{1}{2}x\right)}{1 + \tan^2\left(\frac{1}{2}x\right)}\right)$;

   (e) $\int_1^e 8x^7 \log_e x\, dx$;   (f) $\int_e^{e^2} \frac{\log_e(\log_e x)}{x} dx$;

   (g) $\int_0^{\frac{\pi}{2}} \frac{\sin(2x)}{1 + 3\cos^2 x} dx$;   (h) $\int_1^4 \frac{dx}{(1+x)\sqrt{x}}$ (try $u = \sqrt{x}$).

Section 7.4

1. Prove the following inequalities:
   (a) $\int_0^1 x^3 \sqrt{2(1+x^{99})}\,dx \le \frac{1}{2}$;   (b) $\int_0^1 \frac{x^4}{\sqrt{1+3x^{97}}}\,dx \ge \frac{1}{10}$.

2. Prove that $\frac{1}{2} \le \int_0^1 \frac{1+x^{30}}{2-x^{99}}\,dx \le 2$.

3. Prove that $\left| \int_0^2 \frac{x^2(x-3)\sin(99x)}{1+x^{20}}\,dx \right| \le 4$.

4. Show that $\int \frac{dx}{x(\log_e x)^{\frac{3}{2}}} = -2(\log_e x)^{-\frac{1}{2}}$, and deduce that $\sum_{n=2}^{\infty} \frac{1}{n(\log_e n)^{\frac{3}{2}}}$ is convergent.

5. Show that $\int \frac{dx}{x(\log_e x)^{\frac{1}{2}}} = 2(\log_e x)^{\frac{1}{2}}$, and deduce that $\sum_{n=2}^{\infty} \frac{1}{n(\log_e n)^{\frac{1}{2}}}$ is divergent.

6. Prove that $\lim_{n\to\infty} n\left(\frac{1}{(n+1)^2} + \frac{1}{(n+2)^2} + \cdots + \frac{1}{(2n)^2} \right) = \frac{1}{2}$.

7. Prove that $\lim_{n\to\infty} \left(1 + \frac{1}{3} + \frac{1}{5} + \cdots + \frac{1}{2n-1} - \frac{1}{2}\log_e n \right) = \log_e 2 + \frac{1}{2}\gamma$, where γ is Euler's constant.

8. Prove that $\sum_{k=2}^{n} \frac{\log_e k}{k} \sim \frac{1}{2}(\log_e n)^2$, as $n \to \infty$.

9. Prove that, if f is integrable on $[a,b]$, where $a < b$, and $f(x) > 0$, for all $x \in [a,b]$, then $\int_a^b f > 0$.

   You have already seen that if $f(x) \ge 0$, for all $x \in [a,b]$, then $\int_a^b f \ge 0$.

   Hint: Use the following steps:
   (a) Assume that the result is false, so that in fact $\int_a^b f = 0$;
   (b) Verify that, if $[c,d] \subseteq [a,b]$, then $\int_c^d f = 0$;
   (c) Prove that there is a partition P of $[a,b]$ with $U(f,P) \le b-a$, and deduce that there is some subinterval $[a_1,b_1]$ of $[a,b]$ with $a_1 < b_1$ and $\sup_{[a_1,b_1]} f \le 1$;
   (d) Using the fact that $\int_{a_1}^{b_1} f = 0$, prove that there is a sequence $\{[a_n,b_n]\}_{n=1}^{\infty}$ of intervals with $[a_{n+1},b_{n+1}] \subset [a_n,b_n]$ and $\sup_{[a_n,b_n]} f \le \frac{1}{n}$;
   (e) Prove that there is some point c in $[a,b]$ such that $a_n \to c$ as $n \to \infty$, and that $c \in [a_n,b_n]$, for all $n \ge 1$;
   (f) Verify that $f(c) = 0$.

Section 7.5

1. Prove that $\lim_{n\to\infty} \left(\frac{(3n)!}{n^{3n}} \right)^{\frac{1}{n}} = \frac{27}{e^3}$.

2. Use Stirling's Formula to estimate each of the following numbers to two significant figures:
   (a) $\binom{400}{200} \times \left(\frac{1}{2} \right)^{400}$;   (b) $\frac{400!\sqrt{(800\pi)}}{(100!)^4 4^{400}}$.

3. Use Stirling's Formula to determine the number λ such that
$$\frac{(8n)!}{((2n)!)^4} \sim \lambda \frac{2^{16n}}{n^{\frac{3}{2}}} \quad \text{as } n \to \infty.$$

8 Power series

The evaluation of given functions at given points in their domains is of great importance. If we are dealing with a polynomial function, the calculation of the function's values presents no problem: it is simply a matter of arithmetic. For example, if

$$f(x) = 1 + \frac{1}{2}x - \frac{1}{2}x^2 - \frac{1}{6}x^3 + \frac{1}{4}x^4,$$

then

$$f(1) = 1 + \frac{1}{2} - \frac{1}{2} - \frac{1}{6} + \frac{1}{4} = \frac{13}{12}.$$

On the other hand, the sine function is rather different: there is no way of calculating its values precisely merely by the use of arithmetic.

It is important to be able to estimate values of functions that cannot be evaluated exactly, and to know how close the estimates are to the actual values of the function.

In this chapter we are primarily concerned with a procedure for calculating approximate values of functions, like the sine function, whose values cannot be calculated easily at all points in their domains. We see how, in principle, we can use a certain sequence of polynomials to calculate the values of the sine function, for example, to any desired degree of accuracy; and how we can represent $\sin x$, for any x, as the sum of a *series*. We shall see, for instance, that the polynomial $p(x) = x - \frac{1}{6}x^3$ approximates $f(x) = \sin x$ to within 10^{-5} for all x in the interval $[0, 0.1]$, and that, in general

Section 8.2, Example 1, and Theorem 3.

$$\sin x = x - \frac{x^3}{3!} + \frac{x^5}{5!} - \frac{x^7}{7!} + \cdots, \quad \text{for } x \in \mathbb{R}.$$

In Section 8.1 we define the Taylor polynomial $T_n(x)$ and discuss some particular examples of functions for which the Taylor polynomials appear to provide useful approximations.

In Section 8.2 we investigate how closely Taylor polynomials approximate a given function in the neighbourhood of a point a in its domain, and we establish a criterion for when we can say that

$$f(x) = \lim_{n \to \infty} T_n(x) = \sum_{n=0}^{\infty} a_n(x-a)^n.$$

In this case, we say that f is the *sum function* of the *power series* $\sum_{n=0}^{\infty} a_n(x-a)^n$.

In Sections 8.3 and 8.4 we look at the behaviour of power series in their own right; that is, we consider functions which are *defined* by power series. In particular, in Section 8.3 we see that a power series $\sum_{n=0}^{\infty} a_n(x-a)^n$ behaves in one of three ways: it converges only for $x = a$, or it converges for all x, or it

313

converges in an interval $(a - R, a + R)$, where R is a positive number called the *radius of convergence*.

In Section 8.4 we discuss various rules for power series, including the Sum, Multiple and Product Rules. We also find that it is valid to differentiate or integrate a given power series term-by-term, and that this does not affect the radius of convergence.

Finally, in Section 8.5 we look briefly at various methods for estimating the number π, and prove that π is irrational.

Some of the proofs in this chapter are not particularly illuminating, and you may wish to leave them till a second reading of the chapter. You will need a calculator handy while you work through Sections 8.1 and 8.2.

8.1   Taylor polynomials

8.1.1   What are Taylor polynomials?

Let f be a *continuous* function defined on an open interval I containing the point a. It follows from the definition of continuity that

$$\lim_{x \to a} f(x) = f(a).$$

Thus we can write

$$f(x) \simeq f(a), \quad \text{for } x \text{ near } a.$$

In geometrical terms, this means that we can approximate the graph $y = f(x)$ near a by the horizontal line $y = f(a)$ through the point $(a, f(a))$. For most continuous functions, this does not give a very good approximation.

However, if the function is *differentiable* on I, then we can obtain a better approximation by using the *tangent line* through $(a, f(a))$ instead of the horizontal line. The tangent to the graph at $(a, f(a))$ has equation

$$\frac{y - f(a)}{x - a} = f'(a)$$

or

$$y = f(a) + f'(a)(x - a);$$

so, for x near a, we can write

$$f(x) \simeq f(a) + f'(a)(x - a).$$

This approximation is called the **tangent approximation** to f at a.

Notice that the function f and the approximating polynomial

$$f(a) + f'(a)(x - a)$$

have the same value at a and the same first derivative at a.

Example 1   Determine the tangent approximation to the function $f(x) = e^x$ at 0.

Solution   Here

$$f(x) = e^x, \quad f(0) = 1,$$

$$f'(x) = e^x, \quad f'(0) = 1.$$

That is, I is a *neighbourhood* of a.

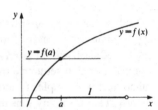

We can think of the tangent at $(a, f(a))$ as the *line of best approximation* to the graph near a.

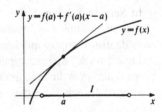

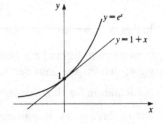

Hence the tangent approximation to f at 0 is

$$e^x \simeq f(0) + f'(0)(x - 0) = 1 + x.$$ □

Problem 1   Determine the tangent approximation to each of the following functions f at the given point a:

(a) $f(x) = e^x$, $a = 2$;   (b) $f(x) = \cos x$, $a = 0$.

So far we have seen two approximations to $f(x)$ for x near a:

$$f(x) \simeq f(a) \qquad\qquad \text{(a constant function)};$$
$$f(x) \simeq f(a) + f'(a)(x - a) \quad \text{(a linear function)}.$$

If the function f is *twice differentiable* on a neighbourhood I of a, then we can find an even better approximation to $f(x)$ by considering the expression

$$\frac{f(x) - \{f(a) + f'(a)(x - a)\}}{(x - a)^2}.$$

Let

$$F(x) = f(x) - \{f(a) + f'(a)(x - a)\}, \quad \text{for } x \in I,$$
$$G(x) = (x - a)^2, \qquad\qquad\qquad\quad \text{for } x \in I.$$

Then F and G are differentiable on I, and

$$F'(x) = f'(x) - f'(a),$$
$$G'(x) = 2(x - a).$$

Also, $F(a) = G(a) = 0$.

Hence, by l'Hôpital's Rule, it follows that   Theorem 2, Sub-section 6.5.2.

$$\lim_{x \to a} \frac{F(x)}{G(x)} \quad \text{exists and equals} \quad \lim_{x \to a} \frac{F'(x)}{G'(x)},$$

provided that this latter limit exists.

Now

$$\lim_{x \to a} \frac{F'(x)}{G'(x)} = \lim_{x \to a} \frac{f'(x) - f'(a)}{2(x - a)};$$

and, since the function f is twice differentiable at a, this limit exists and equals $\frac{1}{2} f''(a)$. Hence

$$\lim_{x \to a} \frac{f(x) - \{f(a) + f'(a)(x - a)\}}{(x - a)^2} \quad \text{exists and equals} \quad \frac{1}{2} f''(a).$$

We can reformulate this result as follows: for x near a

$$f(x) \simeq f(a) + f'(a)(x - a) + \frac{1}{2} f''(a)(x - a)^2 \text{ (a quadratic function)}.$$

Notice that the function f and the approximating polynomial $f(a) + f'(a)\,(x - a) + \frac{1}{2} f''(a)(x - a)^2$ have the same value at a and the same first and second derivatives at a.

This suggests that, if the function is *n-times differentiable* on *I*, then we may be able to find a better approximating polynomial of degree *n* whose value at *a* and whose first *n* derivatives at *a* are equal to those of *f*. This leads to the following definition.

Definition Let *f* be *n*-times differentiable on an open interval containing the point *a*. Then the **Taylor polynomial of degree *n* for *f* at *a*** is the polynomial

$$T_n(x) = f(a) + \frac{f'(a)}{1!}(x-a)$$
$$+ \frac{f''(a)}{2!}(x-a)^2 + \cdots + \frac{f^{(n)}(a)}{n!}(x-a)^n.$$

Strictly speaking, we should use more complicated notation to indicate that the Taylor polynomial $T_n(x)$ depends on n, a and f.

Notice that

$$T_n(a) = f(a), \; T_n'(a) = f'(a), \ldots, \; T_n^{(n)}(a) = f^{(n)}(a);$$

that is, *f* and T_n have the same value at *a* and have equal derivatives at *a* of all orders up to and including *n*.

Example 2 Determine the Taylor polynomials $T_1(x)$, $T_2(x)$ and $T_3(x)$ for the function $f(x) = \sin x$ at each of the following points:

(a) $a = 0$;   (b) $a = \frac{\pi}{2}$.

Solution Here

$$f(x) = \sin x, \qquad f(0) = 0, \qquad f\left(\frac{\pi}{2}\right) = 1;$$

$$f'(x) = \cos x, \qquad f'(0) = 1, \qquad f'\left(\frac{\pi}{2}\right) = 0;$$

$$f''(x) = -\sin x, \quad f''(0) = 0, \qquad f''\left(\frac{\pi}{2}\right) = -1;$$

$$f'''(x) = -\cos x, \quad f'''(0) = -1, \quad f'''\left(\frac{\pi}{2}\right) = 0.$$

Hence:

(a) $T_1(x) = x$, $T_2(x) = x$ and $T_3(x) = x - \frac{x^3}{3!}$;

(b) $T_1(x) = 1$, $T_2(x) = 1 - \frac{1}{2}\left(x - \frac{\pi}{2}\right)^2$ and $T_3(x) = 1 - \frac{1}{2}\left(x - \frac{\pi}{2}\right)^2$. □

We do not usually multiply out such brackets, since that would make the results less clear.

Problem 2 Determine the Taylor polynomials $T_1(x)$, $T_2(x)$ and $T_3(x)$ for each of the following functions *f* at the given point *a*:

(a) $f(x) = e^x$, $a = 2$;   (b) $f(x) = \cos x$, $a = 0$.

Problem 3 Determine the Taylor polynomial of degree 4 for each of the following functions *f* at the given point *a*:

(a) $f(x) = 7 - 6x + 5x^2 + x^3$, $a = 1$;   (b) $f(x) = \frac{1}{1-x}$, $a = 0$;

(c) $f(x) = \log_e(1+x)$, $a = 0$;   (d) $f(x) = \sin x$, $a = \frac{\pi}{4}$;

(e) $f(x) = 1 + \frac{1}{2}x - \frac{1}{2}x^2 - \frac{1}{6}x^3 + \frac{1}{4}x^4$, $a = 0$.

Problem 4 Determine the percentage error involved in using the Taylor polynomial of degree 3 for the function $f(x) = \tan x$ at 0 to evaluate $\tan 0.1$. (Use your calculator to evaluate $\tan 0.1$.)

8.1.2 Approximation by Taylor polynomials

We now look at some specific examples of functions to investigate the assertion that Taylor polynomials provide good approximations for a large class of functions.

The function $f(x) = 1 + \frac{1}{2}x - \frac{1}{2}x^2 - \frac{1}{6}x^3 + \frac{1}{4}x^4$

It follows from the result of Problem 3(e) above that the Taylor polynomials of degrees 1, 2, 3 and 4 for f at 0 are:

$$T_1(x) = 1 + \frac{1}{2}x, \qquad\qquad T_2(x) = 1 + \frac{1}{2}x - \frac{1}{2}x^2,$$

$$T_3(x) = 1 + \frac{1}{2}x - \frac{1}{2}x^2 - \frac{1}{6}x^3, \quad \text{and} \quad T_4(x) = 1 + \frac{1}{2}x - \frac{1}{2}x^2 - \frac{1}{6}x^3 + \frac{1}{4}x^4.$$

Since $f^{(n)}(0) = 0$ for $n \geq 5$, it follows that for $n \geq 5$ the Taylor polynomial of degree n for f at 0 is just the same as the Taylor polynomial of degree 4 for f at 0.

The graphs of these Taylor polynomials are as follows:

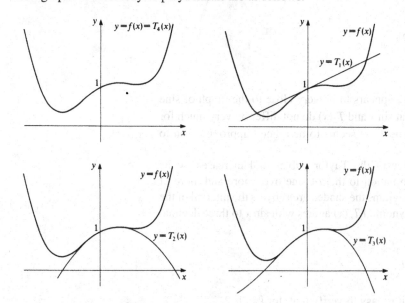

In this case, the polynomials $T_2(x)$ and $T_3(x)$ provide good approximations to $f(x)$ near 0, and $T_n(x) = f(x)$ for all $n \geq 4$.

In general, if f is a polynomial of degree N, then

$$T_n(x) = f(x) \quad \text{for all } n \geq N.$$

The function $f(x) = \sin x$

By calculating higher derivatives of the function $f(x) = \sin x$ at 0, we can show that the following are Taylor polynomials for f at 0

$$T_1(x) = T_2(x) = x, \qquad\qquad T_3(x) = T_4(x) = x - \frac{x^3}{3!},$$

$$T_5(x) = T_6(x) = x - \frac{x^3}{3!} + \frac{x^5}{5!}, \qquad T_7(x) = T_8(x) = x - \frac{x^3}{3!} + \frac{x^5}{5!} - \frac{x^7}{7!}.$$

The following graphs illustrate how the approximation to $f(x)$ given by $T_n(x)$ gets better as n increases.

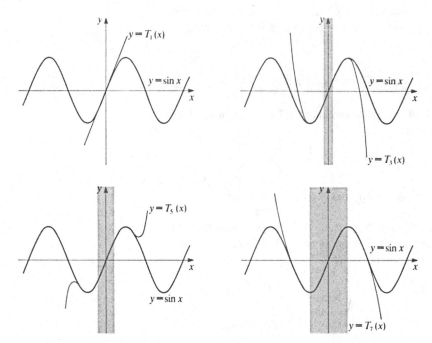

For example, the graph of T_5 appears to be very close to the graph of sine over the interval $\left(-\frac{\pi}{2}, \frac{\pi}{2}\right)$, so that $\sin x$ and $T_5(x)$ do not differ by very much for values of x in this interval. Thus $T_5(x)$ seems to be a good approximation to $\sin x$ in this interval.

It appears that, as the degree of the Taylor polynomial increases, so its graph becomes a good approximation to that of sine over more and more of $\mathbb{R}$. For instance, in the above diagrams the shaded area covers the interval of the x-axis on which the Taylor polynomial $T_n(x)$ agrees with $\sin x$ to three decimal places.

The function $f(x) = \frac{1}{1-x}$

By repeated differentiation it is easy to verify that, for $k = 1, 2, \ldots$

$$f^{(k)}(x) = \frac{k!}{(1-x)^{k+1}};$$

thus, in particular, $f^{(k)}(0) = k!$.

Hence the Taylor polynomial of degree n for f at 0 is

$$T_n(x) = \sum_{k=0}^{n} \frac{f^{(k)}(0)}{k!} x^k$$

$$= \sum_{k=0}^{n} x^k = 1 + x + x^2 + \cdots + x^n.$$

The following diagram shows the graphs of the Taylor polynomials for f at 0 of degrees 1, 2, 4 and 7.

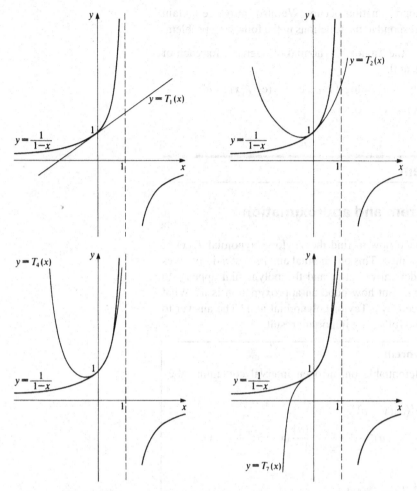

The graphs show that the nature of the approximation is very different from the previous examples. For sine, the interval over which the approximation is good seems to expand indefinitely as the degree of the polynomials increases; but for $f(x) = \frac{1}{1-x}$ the interval of good approximation always seems to be contained in the interval $(-1, 1)$.

Notice, however, that for this function f, the Taylor polynomials $T_n(x)$ are just the nth partial sums of the geometric series $\sum_{k=0}^{\infty} x^k$; this series converges with sum $\frac{1}{1-x}$ for $|x| < 1$, and diverges for $|x| \geq 1$. It follows that, if $|x| < 1$, then

For $|x| \geq 1$, the sequence $\{T_n(x)\}$ does *not* converge, and so *cannot* provide an approximation to $f(x)$.

$$T_n(x) \to f(x) \quad \text{as } n \to \infty;$$

so that, if $|x| < 1$, the polynomials $T_n(x)$ *do* provide better and better approximations to $f(x)$ as n increases.

Which functions can be approximated by Taylor polynomials?

In the next section we obtain a criterion for determining those functions f for which the Taylor polynomials provide useful approximating polynomials, and the intervals on which the approximation occurs. We also introduce certain basic power series which correspond to the functions in the following problem.

Problem 5 Determine the Taylor polynomial of degree n for each of the following functions at 0:

(a) $f(x) = \frac{1}{1-x}$;   (b) $f(x) = \log_e(1+x)$;   (c) $f(x) = e^x$;

(d) $f(x) = \sin x$;   (e) $f(x) = \cos x$.

8.2 Taylor's Theorem

8.2.1 Taylor's Theorem and approximation

In Section 8.1 we demonstrated how to find the Taylor polynomial $T_n(x)$ of degree n for a function f at a point a. This polynomial and its first n derivatives agree with f and its first n derivatives at a, and the polynomial appears to approximate f at points near a. But how good an approximation is it? What error is involved if we replace f by a Taylor polynomial at a? The answer to these questions is given by the following important result.

Theorem 1   Taylor's Theorem

Let f be $(n+1)$-times differentiable on an open interval containing the points a and x. Then

$$f(x) = f(a) + f'(a)(x-a)$$
$$+ \frac{f''(a)}{2!}(x-a)^2 + \cdots + \frac{f^{(n)}(a)}{n!}(x-a)^n + R_n(x),$$

where

$$R_n(x) = \frac{f^{(n+1)}(c)}{(n+1)!}(x-a)^{n+1},$$

and c is some point between a and x.

Remarks

1. When $n = 0$, Taylor's Theorem reduces to the assertion

$$f(x) = f(a) + f'(c)(x-a),$$

which we can rewrite (for $x \neq a$) in the form

$$\frac{f(x) - f(a)}{x - a} = f'(c), \quad \text{for some } c \text{ between } a \text{ and } x.$$

But this is just the Mean Value Theorem! It follows that Taylor's Theorem can be considered as a generalisation of the Mean Value Theorem.

2. The result of Theorem 1 can be expressed in the form

$$f(x) = T_n(x) + R_n(x),$$

where $R_n(x)$ is thought of as a 'remainder term' or 'error term' involved in approximating $f(x)$ by the estimate $T_n(x)$.

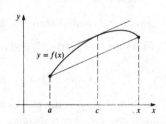

Strictly speaking, we should use more complicated notation to indicate that the remainder term $R_n(x)$ depends on n, a and f.

Proof For simplicity, we assume that $x > a$; the proof is similar if $x < a$. We use the auxiliary function

$$h(t) = f(t) - T_n(t) - A(t-a)^{n+1}, \quad t \in [a, x], \tag{1}$$

You may omit this proof at a first reading.

where T_n is the Taylor polynomial of degree n for f at a, and A is a constant chosen so that

$$h(a) = h(x). \tag{2}$$

This choice of A is made so that we can apply Rolle's Theorem to h on $[a, x]$.

Now

$$f(a) = T_n(a), f'(a) = T'_n(a), \ldots, \quad \text{and} \quad f^{(n)}(a) = T_n^{(n)}(a);$$

so that

$$h(a) = 0, \ h'(a) = 0, \ldots, \quad \text{and} \quad h^{(n)}(a) = 0.$$

The function h is continuous on the closed interval $[a, x]$ and differentiable on the open interval (a, x); also, $h(a) = h(x)$. Hence, by Rolle's Theorem, there exists some number c_1 between a and x for which

$$h'(c_1) = 0.$$

Next, we apply Rolle's Theorem to the function h' on the interval $[a, c_1]$. The function h' is continuous on $[a, c_1]$ and differentiable on (a, c_1); also, $h'(a) = h'(c_1) = 0$. Hence, by Rolle's Theorem, there exists some number c_2 between a and c_1 for which

$$h''(c_2) = 0.$$

In turn, we apply Rolle's Theorem to the functions

$$h'', h''', \ldots, h^{(n)}$$

on the intervals

$$[a, c_2], [a, c_3], \ldots, [a, c_n],$$

where $c_2 > c_3 > c_4 > \cdots > c_n > a$.

At the last stage, we find that there exists some point c between a and c_n for which

$$h^{(n+1)}(c) = 0. \tag{3}$$

By differentiating equation (1) $(n+1)$ times, we obtain

$$h^{(n+1)}(t) = f^{(n+1)}(t) - A(n+1)!. \tag{4}$$

From (3) and (4), we deduce that

$$0 = f^{(n+1)}(c) - A(n+1)!,$$

so that

$$A = \frac{f^{(n+1)}(c)}{(n+1)!}. \tag{5}$$

Finally, it follows from equations (1) and (2), with $h(a) = 0$, that

$$f(x) = T_n(x) + A(x-a)^{n+1}.$$

Hence, by equation (5)

$$f(x) = T_n(x) + \frac{f^{(n+1)}(c)}{(n+1)!}(x-a)^{n+1},$$

as required.

Problem 1   Obtain an expression for $R_1(x)$ when Taylor's Theorem is applied to the function $f(x) = \frac{1}{1-x}$ at $a = 0$. Calculate the value of c when $x = \frac{3}{4}$.

Problem 2   What can you say about $R_n(x)$ when f is a polynomial of degree at most n?

Problem 3   By applying Taylor's Theorem to the function $f(x) = \cos x$ at $a = 0$, prove that $\cos x = 1 - \frac{1}{2}x^2 + R_3(x)$, $x \in \mathbb{R}$, where $|R_3(x)| \leq \frac{1}{24}x^4$.

In most applications of Taylor's Theorem, we do not know the value of c explicitly. However, for many purposes this does not matter, since we can show that $|R_n(x)|$ is small by finding an estimate for $\left|f^{(n+1)}(c)\right|$ which is valid for *all* c between a and x, and then applying the following result.

Corollary 1   Remainder Estimate

Let f be $(n+1)$-times differentiable on an open interval containing the points a and x. If

$$\left|f^{(n+1)}(c)\right| \leq M,$$

for all c between a and x, then

$$f(x) = T_n(x) + R_n(x),$$

where

$$|R_n(x)| \leq \frac{M}{(n+1)!}|x - a|^{n+1}.$$

Strictly speaking, we should use more complicated notation to indicate that the upper bound M depends on n, a and f.

Proof   From Taylor's Theorem, we have

$$R_n(x) = \frac{f^{(n+1)}(c)}{(n+1)!}(x - a)^{n+1},$$

where c is some point between a and x. It follows that

$$|R_n(x)| = \frac{\left|f^{(n+1)}(c)\right|}{(n+1)!}|x - a|^{n+1}$$

$$\leq \frac{M}{(n+1)!}|x - a|^{n+1}. \qquad\blacksquare$$

Example 1   By applying the Remainder Estimate to the function $f(x) = \sin x$, with $a = 0$ and $n = 3$, calculate $\sin 0.1$ to four decimal places.

Solution   Here

$$f(x) = \sin x, \qquad f(0) = 0;$$

$$f'(x) = \cos x, \qquad f'(0) = 1;$$

$$f''(x) = -\sin x, \quad f''(0) = 0;$$

$$f'''(x) = -\cos x, \quad f'''(0) = -1.$$

Hence the Taylor polynomial of degree 3 for f at 0 is

$$T_3(x) = f(0) + f'(0)x + \frac{f''(0)}{2!}x^2 + \frac{f'''(0)}{3!}x^3$$

$$= x - \frac{1}{6}x^3.$$

Also, $f^{(4)}(x) = \sin x$, so that

$$\left| f^{(4)}(c) \right| = |\sin c| \leq 1, \quad \text{for } c \in \mathbb{R}.$$

Taking $M = 1$ in the Remainder Estimate, we deduce that

$$|R_3(0.1)| \leq \frac{1}{4!} \times (0.1)^4$$

$$= \frac{1}{24} \times 10^{-4}$$

$$< \frac{1}{2} \times 10^{-5}.$$

Now

$$\sin 0.1 \simeq T_3(0.1)$$

$$= 0.1 - \frac{1}{6} \times 10^{-3}$$

$$= 0.1 - 0.0001666\ldots$$

$$= 0.0998333\ldots.$$

It follows from the Remainder Estimate that $T_3(0.1)$ gives an estimate for $\sin 0.1$ to four decimal places. Hence $\sin 0.1 = 0.0998$ (to four decimal places). $\quad\square$

Problem 4 By applying the Remainder Estimate to the function $f(x) = \log_e(1+x)$, with $a = 0$ and $n = 2$, calculate $\log_e 1.02$ to four decimal places.

In many practical situations we do not know how many terms of the power series are needed in order to calculate the value of a given function to a prescribed number of decimal places. In such cases, we can use the Remainder Estimate to determine how many terms are needed.

Example 2 By applying the Remainder Estimate to the function $f(x) = e^x$, with $a = 0$, calculate the value of e to three decimal places.

Solution Since $f^{(k)}(x) = e^x$, for $k = 0, 1, \ldots$, we have

$$f^{(k)}(0) = 1, \quad \text{for } k = 0, 1, \ldots.$$

It follows that, for each n

$$T_n(x) = 1 + x + \frac{x^2}{2!} + \cdots + \frac{x^n}{n!}.$$

Also, $f^{(n+1)}(x) = e^x$, for all x, so that, for all $c \in (0, 1)$, we have

$$\left| f^{(n+1)}(c) \right| \leq e < 3.$$

We proved that $e < 3$ in Subsection 2.5.3.

It follows from the Remainder Estimate, with $x = 1$ and $M = 3$, that

$$|R_n(1)| \leq \frac{3}{(n+1)!} \times 1^{n+1}.$$

To calculate e to three decimal places, we choose n so that

$$\frac{3}{(n+1)!} < 10^{-4}, \quad \text{or} \quad 30{,}000 < (n+1)!.$$

Since $7! = 5{,}040$ and $8! = 40{,}320$, we may safely choose $n = 7$.

It follows that

$$e \simeq T_7(1)$$
$$= 1 + \frac{1}{1!} + \frac{1}{2!} + \frac{1}{3!} + \frac{1}{4!} + \frac{1}{5!} + \frac{1}{6!} + \frac{1}{7!}$$
$$= 1 + 1 + 0.5 + 0.1\overline{6} + 0.041\overline{6} + 0.008\overline{3} + 0.0013\overline{8}$$
$$+ 0.000198412\ldots$$
$$= 2.7182\ldots.$$

Hence, $e = 2.718$ (to three decimal places). $\qquad\square$

Problem 5 By applying the Remainder Estimate to the function $f(x) = \cos x$, with $a = 0$, calculate $\cos 0.2$ rounded to four decimal places.

Our next example illustrates how to obtain an approximation valid over an interval.

Example 3 Calculate the Taylor polynomial $T_3(x)$ for $f(x) = \frac{1}{x+2}$ at 1. Show that $T_3(x)$ approximates $f(x)$ with an error less than 5×10^{-3} on the interval $[1, 2]$.

Solution Here

$$f(x) = \frac{1}{x+2}, \quad f(1) = \frac{1}{3};$$
$$f'(x) = \frac{-1}{(x+2)^2}, \quad f'(1) = -\frac{1}{9};$$
$$f''(x) = \frac{2}{(x+2)^3}, \quad f''(1) = \frac{2}{27};$$
$$f'''(x) = \frac{-6}{(x+2)^4}, \quad f'''(1) = -\frac{2}{27}.$$

Hence the Taylor polynomial of degree 3 for f at 1 is

$$T_3(x) = \frac{1}{3} - \frac{1}{9}(x-1) + \frac{1}{27}(x-1)^2 - \frac{1}{81}(x-1)^3.$$

Also, $f^{(4)}(x) = \frac{24}{(x+2)^5}$, so that

$$\left|f^{(4)}(c)\right| \le \frac{24}{3^5}, \quad \text{for } c \in (1,2).$$

Taking $M = \frac{24}{3^5}$ in the Remainder Estimate, we deduce that

$$|R_3(x)| \le \frac{24}{3^5} \times \frac{(2-1)^4}{4!}$$
$$= \frac{1}{3^5} = 0.0041\ldots, \quad \text{for } x \in [1,2].$$

Since the remainder term is less than 0.005, it follows that $T_3(x)$ approximates $f(x)$ with an error less than 5×10^{-3} on $[1, 2]$. $\qquad\square$

Problem 6   Calculate the Taylor polynomial $T_4(x)$ for $f(x) = \cos x$ at π. Show that $T_4(x)$ approximates $f(x)$ with an error less than 3×10^{-3} on the interval $\left[\frac{3}{4}\pi, \frac{5}{4}\pi\right]$.

8.2.2   Taylor's Theorem and power series

From Taylor's Theorem, we know that, if a function f can be differentiated as often as we please on an open interval containing the points a and x, then, for any n

$$f(x) = T_n(x) + R_n(x)$$
$$= \sum_{k=0}^{n} \frac{f^{(k)}(a)}{k!}(x-a)^k + R_n(x),$$

where

$$R_n(x) = \frac{f^{(n+1)}(c)}{(n+1)!}(x-a)^{n+1},$$

for some c between a and x. It follows that, if $R_n(x) \to 0$ as $n \to \infty$, then we can express $f(x)$ as a power series in $(x-a)$.

Theorem 2   Let f have derivatives of all orders on an open interval containing the points a and x. If $R_n(x) \to 0$ as $n \to \infty$, then

$$f(x) = \sum_{n=0}^{\infty} \frac{f^{(n)}(a)}{n!}(x-a)^n.$$

We call f the **sum function** of the power series $\sum_{n=0}^{\infty} \frac{f^{(n)}(a)}{n!}(x-a)^n$, and we call this power series the **Taylor series for f at a**.

Warning   A dramatic example of what happens when the remainder term does not tend to zero is given by the function

$$f(x) = \begin{cases} e^{-\frac{1}{x^2}}, & x \neq 0, \\ 0, & x = 0. \end{cases}$$

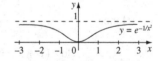

For this function, $f(0) = 0$, $f'(0) = 0$, $f''(0) = 0$, ... Thus the Taylor polynomial $T_n(x)$ is identically zero for each n, although the function f is not identically zero. In this case, $R_n(x) = f(x)$ for all n, and so the Taylor series for f at 0

We indicate a proof of this assertion in Section 8.6, Exercise 6 on Section 8.2.

$$\sum_{n=0}^{\infty} \frac{f^{(n)}(0)}{n!}x^n = 0 + 0x + 0x^2 + \cdots,$$

converges to $f(x)$ only at 0!

We can use Theorem 2 to obtain the following basic power series.

Theorem 3   Basic power series

(a) $\frac{1}{1-x} = 1 + x + x^2 + x^3 + \cdots \quad = \sum_{n=0}^{\infty} x^n, \qquad$ for $|x| < 1$;

(b) $\log_e(1+x) = x - \frac{x^2}{2} + \frac{x^3}{3} - \cdots = \sum_{n=1}^{\infty} (-1)^{n+1}\frac{x^n}{n}, \quad$ for $|x| < 1$;

(c) $e^x = 1 + x + \frac{x^2}{2!} + \frac{x^3}{3!} + \cdots = \sum_{n=0}^{\infty} \frac{x^n}{n!}$, for $x \in \mathbb{R}$;

(d) $\sin x = x - \frac{x^3}{3!} + \frac{x^5}{5!} - \cdots = \sum_{n=0}^{\infty} (-1)^n \frac{x^{2n+1}}{(2n+1)!}$, for $x \in \mathbb{R}$;

(e) $\cos x = 1 - \frac{x^2}{2!} + \frac{x^4}{4!} - \cdots = \sum_{n=0}^{\infty} (-1)^n \frac{x^{2n}}{(2n)!}$, for $x \in \mathbb{R}$.

Proof

(a) Here we use the fact that, for $x \neq 1$

$$\frac{1}{1-x} = 1 + x + x^2 + x^3 + \cdots + x^n + \frac{x^{n+1}}{1-x}.$$

This identity is easily proved by multiplying both sides by $1 - x$.

But the sequence $\left\{ \frac{x^{n+1}}{1-x} \right\}$ is null, if $|x| < 1$. Thus, by letting $n \to \infty$, we obtain

$$\frac{1}{1-x} = \sum_{n=0}^{\infty} x^n, \quad \text{for } |x| < 1.$$

(b) Similarly

$$\frac{1}{1+t} = 1 - t + t^2 - \cdots + (-1)^n t^n + \frac{(-1)^{n+1} t^{n+1}}{1+t}.$$

Integrating both sides from 0 to x, where $|x| < 1$, we obtain

$$\int_0^x \frac{dt}{1+t} = \int_0^x \left(1 - t + t^2 - \cdots + (-1)^n t^n + \frac{(-1)^{n+1} t^{n+1}}{1+t} \right) dt,$$

so that

$$[\log_e(1+t)]_0^x = \left[t - \frac{t^2}{2} + \frac{t^3}{3} - \cdots + (-1)^n \frac{t^{n+1}}{n+1} \right]_0^x$$
$$+ (-1)^{n+1} \int_0^x \frac{t^{n+1}}{1+t} dt.$$

Hence

$$\log_e(1+x) = x - \frac{x^2}{2} + \frac{x^3}{3} - \cdots + (-1)^n \frac{x^{n+1}}{n+1}$$
$$+ (-1)^{n+1} \int_0^x \frac{t^{n+1}}{1+t} dt.$$

When $0 \leq x < 1$, we have, by the Inequality Rule for integrals

Theorem 2, Sub-section 7.4.1.

$$\left| (-1)^{n+1} \int_0^x \frac{t^{n+1}}{1+t} dt \right| = \int_0^x \frac{t^{n+1}}{1+t} dt \leq \int_0^x t^{n+1} dt$$
$$= \left[\frac{t^{n+2}}{n+2} \right]_0^x$$
$$= \frac{x^{n+2}}{n+2} \leq \frac{1}{n+2} \to 0 \quad \text{as } n \to \infty.$$

When $-1 < x < 0$, we put $T = -t$ and $X = -x$, so that $0 < T < 1$. Then

$$\left| (-1)^{n+1} \int_0^x \frac{t^{n+1}}{1+t} dt \right| = \int_0^X \frac{T^{n+1}}{1-T} dT \le \frac{1}{1-X} \int_0^X T^{n+1} dT$$

For, $0 < X < 1$ and
$\frac{1}{1-T} \le \frac{1}{1-X}$ for $T \in [0, X]$.

$$= \frac{X^{n+2}}{(1-X)(n+2)} \le \frac{1}{(1-X)(n+2)} \to 0 \quad \text{as } n \to \infty.$$

Combining these two results, we deduce that

$$\log_e(1+x) = \sum_{n=1}^{\infty} (-1)^{n+1} \frac{x^n}{n}, \quad \text{for } -1 < x < 1.$$

(c) This was one of our definitions of the exponential function.

See Sub-section 3.4.3.

(d) Let $f(x) = \sin x$. Then, by Taylor's Theorem, we have $\sin x = T_n(x) + R_n(x)$, where

$$R_n(x) = \frac{f^{(n+1)}(c)}{(n+1)!} x^{n+1},$$

for some number c between 0 and x. We saw earlier that

Problem 5(d),
Sub-section 8.1.2.

$$f^{(n+1)}(c) = \pm \sin c \quad \text{or} \quad \pm \cos c,$$

so that, in particular, we can be sure that $\left| f^{(n+1)}(c) \right| \le 1$. It follows from the Remainder Estimate, with $M = 1$, that

$$|R_n(x)| \le \frac{1}{(n+1)!} |x|^{n+1} \to 0 \quad \text{as } n \to \infty;$$

so that, in particular, $R_n(x) \to 0$ as $n \to \infty$.

Hence, by letting $n \to \infty$ in the equation $\sin x = T_n(x) + R_n(x)$, we obtain

$$\sin x = x - \frac{x^3}{3!} + \frac{x^5}{5!} - \frac{x^7}{7!} + \cdots = \sum_{n=0}^{\infty} (-1)^n \frac{x^{2n+1}}{(2n+1)!}, \quad \text{for } x \in \mathbb{R}.$$

(e) The proof is similar to that of part (d), so we omit it. ∎

Remark

Probably the first definitions of $\sin x$ and $\cos x$ that you met were expressed in terms of a right-angled triangle, but those definitions only make sense for $x \in \left(0, \frac{\pi}{2}\right)$. We have now shown that $\sin x$ and $\cos x$ can be represented by the power series in parts (d) and (e) of Theorem 3, and we can use these power series to define $\sin x$ and $\cos x$ for all $x \in \mathbb{R}$.

Finally, we use Taylor's Theorem to prove an interesting limit which is a generalisation of two limits that you met earlier

Sub-section 2.5.3.

$$\lim_{n \to \infty} \left(1 + \frac{1}{n}\right)^n = e \quad \text{and} \quad \lim_{n \to \infty} \left(1 + \frac{x}{n}\right)^n = e^x, \quad \text{for any } x \in \mathbb{R}.$$

Example 4

(a) Calculate $T_1(x)$ and $R_1(x)$ for the function $f(x) = \log_e(1+x)$, $x \in (-1, 1)$, at 0.

(b) By replacing x in part (a) by $\frac{\alpha}{x}$, where $|x| > |\alpha|$, prove that, for any real numbers α and β

$$\lim_{x \to \infty} \left(1 + \frac{\alpha}{x}\right)^{\beta x} = e^{\alpha\beta}.$$

(c) Deduce from part (a) that

$$n \log_e \left(1 + \frac{1}{n}\right) \to 1 \quad \text{as } n \to \infty.$$

Solution

(a) For the function $f(x) = \log_e(1 + x)$, $x \in (-1, 1)$, we have

$$f(x) = \log_e(1 + x), \quad f(0) = 0;$$

$$f'(x) = \frac{1}{1+x}, \quad f'(0) = 1;$$

$$f''(x) = \frac{-1}{(1+x)^2}.$$

Hence

$$T_1(x) = f(0) + f'(0)x = x \quad \text{and} \quad R_1(x) = \frac{f''(c)}{2!}x^2 = \frac{-x^2}{2(1+c)^2},$$

for some number c between 0 and x.

(b) Replacing x by $\frac{\alpha}{x}$ in the equation

$$\log_e(1 + x) = x - \frac{x^2}{2(1+c)^2}, \quad \text{for } |x| < 1,$$

we obtain

$$\log_e\left(1 + \frac{\alpha}{x}\right) = \frac{\alpha}{x} - \frac{\alpha^2}{2(1+c)^2 x^2}, \quad \text{for } |x| > |\alpha|.$$

Hence

$$\beta x \log_e\left(1 + \frac{\alpha}{x}\right) = \alpha\beta - \frac{\alpha^2\beta}{2(1+c)^2 x}, \quad \text{for } |x| > |\alpha|.$$

Since $\frac{1}{x} \to 0$ as $x \to \infty$, it follows that

$$\lim_{x \to \infty} \beta x \log_e\left(1 + \frac{\alpha}{x}\right) = \alpha\beta,$$

so that

$$\lim_{x \to \infty} \log_e\left(1 + \frac{\alpha}{x}\right)^{\beta x} = \alpha\beta.$$

Since the exponential function is continuous on $\mathbb{R}$, and

$$\exp\left(\log_e\left(1 + \frac{\alpha}{x}\right)^{\beta x}\right) = \left(1 + \frac{\alpha}{x}\right)^{\beta x},$$

it follows, from the Composition Rule for limits, that

$$\lim_{x \to \infty} \left(1 + \frac{\alpha}{x}\right)^{\beta x} = e^{\alpha\beta}.$$

(c) For $x \in (-1, 1)$, we have, from part (a), that $\log_e(1 + x) = x - \frac{x^2}{2(1+c)^2}$, for some number c between 0 and x. Putting $x = \frac{1}{n}$, we obtain

$$\log_e\left(1 + \frac{1}{n}\right) = \frac{1}{n} - \frac{1}{2n^2(1 + c_n)^2},$$

for some number c_n between 0 and $\frac{1}{n}$.

Hence

$$n \log_e \left(1 + \frac{1}{n} \right) = 1 - \frac{1}{2n(1 + c_n)^2}$$

$$\to 1 \quad \text{as } n \to \infty,$$

since

$$0 \le \frac{1}{2n(1 + c_n)^2} \le \frac{1}{2n}$$

and $\left\{ \frac{1}{2n} \right\}$ is a null sequence. $\qquad \square$

8.3 Convergence of power series

8.3.1 The radius of convergence

In the previous section you saw that certain standard functions can be expressed as the sum functions of power series; for example

$$\frac{1}{1 - x} = \sum_{n=0}^{\infty} x^n \quad \text{and} \quad \log_e(1 + x) = \sum_{n=1}^{\infty} (-1)^{n+1} \frac{x^n}{n}, \quad \text{for } |x| < 1.$$

These power series are the Taylor series for the given functions at 0.

Conversely, power series can be used to *define* functions. Thus, for instance, we defined the exponential function $x \mapsto e^x$, $x \in \mathbb{R}$, to be the sum function for the power series $\sum_{n=0}^{\infty} \frac{x^n}{n!}$.

Section 3.4.

But there are many other functions which are defined as the sum functions of power series (as distinct from a power series obtained as the Taylor series for a given function). For example, the *Bessel function*

$$J_0(x) = \sum_{n=0}^{\infty} \frac{(-1)^n \left(\frac{x}{2} \right)^{2n}}{(n!)^2}, \quad \text{for } x \in \mathbb{R},$$

arises in analyses of the vibrations of a circular drum and of the radiation from certain types of radio antenna.

However, all such uses of power series depend on knowledge of those numbers x for which a power series $\sum_{n=0}^{\infty} a_n (x - a)^n$ converges. In each of the above examples, the series converges *on an interval*; indeed, *all* power series have this property.

> **Theorem 1 Radius of Convergence Theorem**
> For a given power series $\sum_{n=0}^{\infty} a_n (x - a)^n$, precisely one of the following possibilities occurs:
>
> (a) The series converges only when $x = a$;
>
> (b) The series converges for all x;

We give the proof of Theorem 1 in Subsection 8.3.3.

(c) There is a number $R > 0$, called the **radius of convergence** of the power series, such that the series converges if $|x - a| < R$ and diverges if $|x - a| > R$.

For example:

(a) $\sum_{n=0}^{\infty} n! x^n$ converges only when $x = 0$;

(b) $\sum_{n=0}^{\infty} \frac{x^n}{n!}$ converges for all x;

(c) $\sum_{n=0}^{\infty} x^n$ converges if $|x| < 1$ and diverges if $|x| > 1$ – its radius of convergence is 1.

For all non-zero x, the series $\sum_{n=0}^{\infty} n! x^n$ diverges, by the Non-null Test, as $\left\{\frac{1}{n! x^n}\right\} = \left\{\frac{\left(\frac{1}{x}\right)^n}{n!}\right\}$ is a basic null sequence.

Remarks

1. Sometimes we abuse our notation by writing $R = 0$, if a series converges only for $x = a$, or $R = \infty$, if a series converges for all x.

2. It is important to remember that Theorem 1, part (c), makes *no assertion* about the convergence or divergence of the power series at the end-points $a - R$, $a + R$ of the interval $(a - R, a + R)$.

For example, we shall see shortly that the three power series

$$\sum_{n=1}^{\infty} x^n, \quad \sum_{n=1}^{\infty} \frac{x^n}{n} \quad \text{and} \quad \sum_{n=1}^{\infty} \frac{x^n}{n^2}$$

Example 2.

all have radius of convergence 1. However, the intervals on which these series converge are, respectively, $(-1, 1)$, $[-1, 1)$ and $[-1, 1]$.

The **interval of convergence** of the power series is the interval $(a - R, a + R)$, together with any end-points of the interval at which the series converges.

The following diagram illustrates the various types of interval of convergence of $\sum_{n=0}^{\infty} a_n(x - a)^n$:

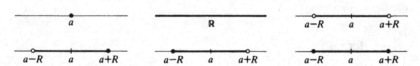

Theorem 1 is not concerned with the problem of evaluating the radius of convergence of a power series. However, a power series is simply a particular type of series, and so all the convergence tests for series can be applied to power series.

Sections 3.1–3.3.

In practice, we can tackle most commonly arising power series by using the following version of the Ratio Test.

Theorem 2   Ratio Test for Radius of Convergence

Suppose that $\sum_{n=0}^{\infty} a_n(x - a)^n$ is a given power series, and that

$$\left|\frac{a_{n+1}}{a_n}\right| \to L, \quad \text{as } n \to \infty.$$

(a) If L is ∞, the series converges only for $x = a$.

We give the proof of Theorem 2 in Sub-section 8.3.3.

$R = 0$.

(b) If $L = 0$, the series converges for all x. $R = \infty$.

(c) If $L > 0$, the series has radius of convergence $\frac{1}{L}$. $R = \frac{1}{L}$.

Example 1 Determine the radius of convergence of each of the following power series:

(a) $\displaystyle\sum_{n=1}^{\infty} \frac{n^n (x+1)^n}{n!}$;   (b) $\displaystyle\sum_{n=1}^{\infty} \frac{(x-2)^n}{n!}$.

Solution

(a) Here $a_n = \frac{n^n}{n!}$ for all n, so

$$\left| \frac{a_{n+1}}{a_n} \right| = \frac{(n+1)^{n+1}}{n^n} \times \frac{n!}{(n+1)!} = \left(1 + \frac{1}{n} \right)^n, \quad \text{for all } n.$$

Thus

$$\left| \frac{a_{n+1}}{a_n} \right| \to e \quad \text{as } n \to \infty.$$

Hence, by the Ratio Test, the radius of convergence is $\frac{1}{e}$.

(b) Here $a_n = \frac{1}{n!}$, for all n, so

$$\left| \frac{a_{n+1}}{a_n} \right| = \frac{n!}{(n+1)!} = \frac{1}{n+1}, \quad \text{for all } n.$$

Thus

$$\left| \frac{a_{n+1}}{a_n} \right| \to 0 \quad \text{as } n \to \infty.$$

Hence, by the Ratio Test, the power series $\displaystyle\sum_{n=1}^{\infty} \frac{(x-2)^n}{n!}$ converges for all x. $\square$

Problem 1 Determine the radius of convergence of each of the following power series:

(a) $\displaystyle\sum_{n=0}^{\infty} (2^n + 4^n) x^n$;   (b) $\displaystyle\sum_{n=1}^{\infty} \frac{(n!)^2}{(2n)!} x^n$;

(c) $\displaystyle\sum_{n=1}^{\infty} (n + 2^{-n})(x-1)^n$;   (d) $\displaystyle\sum_{n=1}^{\infty} \frac{x^n}{(n!)^{\frac{1}{n}}}$.

Hint for part (d): Use Stirling's Formula and the result of Example 1, Sub-section 7.5.1.

If we wish to find the *interval* of convergence (as opposed to the *radius* of convergence), then we may need to use some of the other tests for series in order to determine the behaviour at the end-points of the interval.

Example 2 Determine the interval of convergence of each of the following power series:

(a) $\displaystyle\sum_{n=1}^{\infty} x^n$;   (b) $\displaystyle\sum_{n=1}^{\infty} \frac{x^n}{n}$;   (c) $\displaystyle\sum_{n=1}^{\infty} \frac{x^n}{n^2}$.

Solution

(a) Here $a_n = 1$, for all n, so

$$\left| \frac{a_{n+1}}{a_n} \right| = 1, \quad \text{for all } n.$$

Hence, by the Ratio Test, the radius of convergence is 1; in other words

$$\sum_{n=1}^{\infty} x^n \quad \text{converges for } -1 < x < 1.$$

Next, we consider the behaviour of the power series at the end-points of this interval, namely -1 and 1. Since the sequences $\{1^n\}$ and $\{(-1)^n\}$ are both non-null, it follows that $\sum_{n=1}^{\infty} x^n$ diverges when $x = \pm 1$, by the Non-null Test.

Hence the interval of convergence of $\sum_{n=1}^{\infty} x^n$ is $(-1, 1)$.

(b) Here $a_n = \frac{1}{n}$, for all n, so

$$\left| \frac{a_{n+1}}{a_n} \right| = \frac{n}{n+1} = \frac{1}{1+\frac{1}{n}}, \quad \text{for all } n.$$

Thus

$$\left| \frac{a_{n+1}}{a_n} \right| \to 1 \quad \text{as } n \to \infty.$$

Hence, by the Ratio Test, the radius of convergence is 1; in other words

$$\sum_{n=1}^{\infty} \frac{x^n}{n} \quad \text{converges for } -1 < x < 1.$$

But we know that $\sum_{n=1}^{\infty} \frac{1}{n}$ diverges and $\sum_{n=1}^{\infty} \frac{(-1)^n}{n}$ converges. It follows that the interval of convergence of the power series $\sum_{n=1}^{\infty} \frac{x^n}{n}$ is $[-1, 1)$.

By Example 2 of Sub-section 3.2.1, and at the start of Sub-section 3.3.2, respectively.

(c) Here $a_n = \frac{1}{n^2}$, for all n, so

$$\left| \frac{a_{n+1}}{a_n} \right| = \frac{n^2}{(n+1)^2} = \frac{1}{\left(1+\frac{1}{n}\right)^2}, \quad \text{for all } n.$$

Thus

$$\left| \frac{a_{n+1}}{a_n} \right| \to 1 \quad \text{as } n \to \infty.$$

Hence, by the Ratio Test, the radius of convergence is 1; in other words

$$\sum_{n=1}^{\infty} \frac{x^n}{n^2} \quad \text{converges for } -1 < x < 1.$$

But we know that $\sum_{n=1}^{\infty} \frac{1}{n^2}$ converges; so, by the Absolute Convergence Test, the series $\sum_{n=1}^{\infty} \frac{(-1)^n}{n^2}$ also converges. It follows that the interval of convergence of the power series $\sum_{n=1}^{\infty} \frac{x^n}{n^2}$ is $[-1, 1]$. $\qquad \blacksquare$

Theorem 1, Sub-section 3.3.1.

Now, the Radius of Convergence Theorem tells us that, if a power series $\sum_{n=0}^{\infty} a_n (x-a)^n$ has radius of convergence R, then it converges for $|x - a| < R$. In fact, we can say more than this – namely, that, for all points x with $|x - a| < R$, the power series is actually *absolutely convergent*!

Theorem 3   Absolute Convergence Theorem

Let the power series $\sum_{n=0}^{\infty} a_n (x-a)^n$ have radius of convergence R. Then it is absolutely convergent for all x with $|x - a| < R$.

Proof Let x be a number such that $|x-a|<R$, and chose a number X such that $|x-a|<|X-a|<R$.

Now, the series $\sum_{n=0}^{\infty} a_n(X-a)^n$ is convergent, so that $a_n(X-a)^n \to 0$ as $n \to \infty$. In particular, there exists some number N such that $|a_n(X-a)^n|<1$, for all $n>N$. It follows that

$$|a_n(x-a)^n| = \left|\frac{x-a}{X-a}\right|^n \times |a_n(X-a)^n|$$
$$\leq \left|\frac{x-a}{X-a}\right|^n, \quad \text{for } n>N.$$

But the series $\sum_{n=0}^{\infty}\left|\frac{x-a}{X-a}\right|^n$ is convergent, so that the series $\sum_{n=0}^{\infty}|a_n(x-a)^n|$ is also convergent, by the Comparison Test for series.

It follows that the power series $\sum_{n=0}^{\infty} a_n(x-a)^n$ is absolutely convergent, as required. ∎

> Such a choice is always possible – for example, choose X to be the midpoint of x and the nearest end-point of the interval $(a-R, a+R)$.

> For $\sum_{n=0}^{\infty}\left|\frac{x-a}{X-a}\right|^n$ is a geometric series.

Problem 2 Determine the interval of convergence of each of the following power series:

(a) $\sum_{n=0}^{\infty} nx^n$; (b) $\sum_{n=1}^{\infty} \frac{1}{n3^n}x^n$.

Problem 3 Determine the radius of convergence of the power series

$$1+\alpha x+\frac{\alpha(\alpha-1)}{2!}x^2+\cdots = \sum_{n=0}^{\infty}\frac{\alpha(\alpha-1)\ldots(\alpha-n+1)}{n!}x^n,$$

where $\alpha \neq 0,1,2,\ldots$.

8.3.2 Abel's Limit Theorem

We can sometimes use the ideas of power series to sum interesting and commonly arising series such as

$$\sum_{n=1}^{\infty}\frac{(-1)^{n+1}}{n}=1-\frac{1}{2}+\frac{1}{3}-\frac{1}{4}+\cdots \quad \text{and} \quad \sum_{n=0}^{\infty}\frac{(-1)^n}{2n+1}=1-\frac{1}{3}+\frac{1}{5}-\frac{1}{7}+\cdots.$$

A major tool in this connection in the following result.

Theorem 4 Abel's Limit Theorem

Let $f(x)$ be the sum function of the power series $\sum_{n=0}^{\infty} a_nx^n$, which has radius of convergence 1; and let $\sum_{n=0}^{\infty} a_n$ be convergent. Then

$$\lim_{x\to 1^-} f(x) = \sum_{n=0}^{\infty} a_n.$$

> A similar result holds for power series of the form
> $$\sum_{n=0}^{\infty} a_n(x-a)^n, a \neq 0.$$
> Equivalently
> $$\lim_{x\to 1^-}\left(\sum_{n=0}^{\infty} a_nx^n\right)=\sum_{n=0}^{\infty} a_n.$$

Proof This proof is a wonderful illustration of the sheer power and magic of the $\varepsilon \to \delta$ method for proving results in Analysis! We use the standard approach for proving that a limit exists as $x \to 1^-$.

Let $s_n = a_0+a_1+\cdots+a_{n-1}$ be the nth partial sum of the series $\sum_{n=0}^{\infty} a_n$; and let s denote the sum $\sum_{n=0}^{\infty} a_n$.

> You may omit this proof at a first reading.

Now,

$$a_0 = s_1 \text{ and } a_n = s_{n+1} - s_n, \text{ for all } n \geq 1. \tag{1}$$

It follows that, for $|x| < 1$

$$(1-x)\sum_{n=0}^{\infty} s_{n+1}x^n = (1-x)\left(s_1 + s_2 x + s_3 x^2 + \cdots + s_{n+1}x^n + \cdots\right)$$

$$= s_1 + s_2 x + s_3 x^2 + s_4 x^3 \cdots + s_{n+1}x^n +$$
$$\qquad - s_1 x - s_2 x^2 - s_3 x^3 - \cdots - s_n x^n - s_{n+1}x^{n+1} - \cdots$$
$$= s_1 + (s_2 - s_1)x + (s_3 - s_2)x^2 + \cdots + (s_{n+1} - s_n)x^n + \cdots$$
$$= \sum_{n=0}^{\infty} a_n x^n = f(x).$$

Using (1).

We now study closely the identity

$$(1-x)\sum_{n=0}^{\infty} s_{n+1}x^n = f(x). \tag{2}$$

We want to prove that:

for each positive number ε, there is a positive number δ such that

$$|f(x) - s| < \varepsilon, \quad \text{for all } x \text{ satisfying } 1 - \delta < x < 1.$$

This is simply the definition of $\lim_{x \to 1^-} f(x) = s$.

Now, we use equation (2) in the following way to get a convenient expression for $f(x) - s$

$$f(x) - s = \sum_{n=0}^{\infty} a_n x^n - \sum_{n=0}^{\infty} a_n$$

$$= (1-x)\sum_{n=0}^{\infty} s_{n+1}x^n - s$$

$$= (1-x)\sum_{n=0}^{\infty} s_{n+1}x^n - (1-x)\sum_{n=0}^{\infty} s x^n$$

$$= (1-x)\sum_{n=0}^{\infty} (s_{n+1} - s)x^n. \tag{3}$$

Here we use the definition of s, equation (2), and the fact that the sum of the series $\sum_{n=0}^{\infty} x^n$ is $\frac{1}{1-x}$.

Next, choose a number N such that $|s_{n+1} - s| < \frac{1}{2}\varepsilon$ for all $n > N$. We can then apply the Triangle Inequality to equation (3), for $x \in (0,1)$, to see that

Such a choice of N is possible, since $s_{n+1} \to s$ as $n \to \infty$.

$$|f(x) - s| = (1-x)\left|\sum_{n=0}^{N}(s_{n+1} - s)x^n + \sum_{n=N+1}^{\infty}(s_{n+1} - s)x^n\right|$$

We split the infinite sum into two parts.

$$\leq (1-x)\left|\sum_{n=0}^{N}(s_{n+1} - s)x^n\right| + (1-x)\left|\sum_{n=N+1}^{\infty}(s_{n+1} - s)x^n\right|$$

Using the Triangle Inequality.

$$\leq (1-x)\sum_{n=0}^{N}|s_{n+1} - s|x^n + (1-x) \times \frac{1}{2}\varepsilon \times \sum_{n=N+1}^{\infty} x^n$$

For $|s_{n+1} - s| < \frac{1}{2}\varepsilon$, for all $n > N$; also, we use that $0 < x < 1$ so that $\sum_{n=0}^{\infty} x^n = \frac{1}{1-x}$.

$$\leq (1-x) \sum_{n=0}^{N} |s_{n+1} - s| x^n + \frac{1}{2}\varepsilon$$

$$\leq (1-x) \sum_{n=0}^{N} |s_{n+1} - s| + \frac{1}{2}\varepsilon \qquad (4) \qquad \text{Again we use that } 0 < x < 1.$$

Finally, since the linear function $x \mapsto (1-x) \sum_{n=0}^{N} |s_{n+1} - s|$ is continuous on $\mathbb{R}$, and takes the value 0 at 1, it follows that we can choose a positive number δ (with $\delta \in (0, 1)$) such that

We add in the extra requirement that $\delta < 1$ in order to ensure that we are only considering values of x in $(0, 1)$.

$$(1-x) \sum_{n=0}^{N} |s_{n+1} - s| < \frac{1}{2}\varepsilon, \quad \text{for all } x \text{ satisfying } 1 - \delta < x < 1.$$

If we substitute this upper bound $\frac{1}{2}\varepsilon$ for $(1-x) \sum_{n=0}^{N} |s_{n+1} - s|$ into the inequality (4), we find that we have proved that

for each positive number ε, there is a positive number δ such that
$$|f(x) - s| < \varepsilon, \quad \text{for all } x \text{ satisfying } 1 - \delta < x < 1.$$

This is what we set out to prove.   ∎

As an application of Abel's Limit Theorem, we use it to evaluate

$$\sum_{n=1}^{\infty} \frac{(-1)^{n+1}}{n} = 1 - \frac{1}{2} + \frac{1}{3} - \frac{1}{4} + \cdots.$$

Let $f(x) = \sum_{n=1}^{\infty} \frac{(-1)^{n+1}}{n} x^n$. We have seen already that this power series converges in $(-1, 1)$ to the sum $\log_e(1+x)$. Further, we saw earlier that the series $\sum_{n=1}^{\infty} \frac{(-1)^{n+1}}{n}$ is convergent, by the Alternating Test. It follows, from Abel's Limit Theorem, that

Part (b) of Theorem 3, Sub-section 8.2.2.

Sub-section 3.3.2.

$$\sum_{n=1}^{\infty} \frac{(-1)^{n+1}}{n} = \lim_{x \to 1^-} \sum_{n=1}^{\infty} \frac{(-1)^{n+1}}{n} x^n = \lim_{x \to 1^-} \log_e(1+x) = \log_e 2.$$

Remark

It is always necessary to check that the conditions of Abel's Theorem apply before using it. For example, a thoughtless application to the identity

$$1 - x + x^2 - x^3 + \cdots = \sum_{n=0}^{\infty} (-1)^n x^n = \frac{1}{1+x}, \quad \text{where } |x| < 1,$$

of taking the limit as $x \to 1^-$, would give the following absurd conclusion

$$1 - 1 + 1 - 1 + \cdots = \sum_{n=0}^{\infty} (-1)^n = \frac{1}{2}!$$

We did not check that the series $\sum_{n=0}^{\infty} (-1)^n x^n$ was convergent at 1!

Problem 4 Use Abel's Limit Theorem to evaluate

$$\sum_{n=0}^{\infty} \frac{(-1)^n}{2n+1} = 1 - \frac{1}{3} + \frac{1}{5} - \frac{1}{7} + \cdots.$$

Hint: Use the facts that $\tan^{-1} x = x - \frac{x^3}{3} + \frac{x^5}{5} - \frac{x^7}{7} + \cdots$ for $|x| < 1$, and that the latter series has radius of convergence 1.

We prove these facts in Sub-section 8.4.1.

8.3.3 Proofs

You may omit this sub-
section at a first reading.

We now supply the proofs that we omitted from Sub-section 8.3.1.

Theorem 1 Radius of Convergence Theorem

For a given power series $\sum\limits_{n=0}^{\infty} a_n(x-a)^n$, precisely one of the following possibilities occurs:

(a) The series converges only when $x = a$;

(b) The series converges for all x;

(c) There is a number $R > 0$ such that the series converges if $|x - a| < R$ and diverges if $|x - a| > R$.

Proof For simplicity, we shall assume that $a = 0$.

For the general proof, replace
x throughout by $(x - a)$.

Clearly the possibilities (a) and (b) are mutually exclusive; we shall therefore assume that for a given power series neither possibility occurs, and then prove that possibility (c) must occur.

First, let

$$S = \left\{ x : \sum_{n=0}^{\infty} a_n x^n \text{ converges} \right\}.$$

Since the possibility (b) has been excluded, we deduce that S must be bounded.

For, if $\sum\limits_{n=0}^{\infty} a_n X^n$ diverges, then, in view of the Absolute Convergence Theorem for power series, $\sum\limits_{n=0}^{\infty} a_n x^n$ cannot converge for any x with $|x| > |X|$, since otherwise $\sum\limits_{n=0}^{\infty} a_n X^n$ would then have to be convergent – which is not the case.

Theorem 3, Sub-section 8.3.1.

Also, S is non-empty, since the power series $\sum\limits_{n=0}^{\infty} a_n x^n$ converges at 0.

It then follows, from the Least Upper Bound Property of $\mathbb{R}$, that the set S must have a least upper bound, R say. We now prove that the series $\sum\limits_{n=0}^{\infty} a_n x^n$ is convergent if $|x| < R$ and divergent if $|x| > R$.

Here we identify a number R
that will be the desired radius
of convergence.

First, notice that $R > 0$. For, since possibility (a) has been excluded, there must be at least one non-zero value of x, x_1 say, such that $\sum\limits_{n=0}^{\infty} a_n x_1^n$ is convergent; hence we have $x_1 \in S$. It follows, from the Absolute Convergence Theorem, that $\sum\limits_{n=0}^{\infty} a_n x^n$ must converge for those x with $|x| < |x_1|$, so that $(-|x_1|, |x_1|)$ lies in S.

Now, choose any x for which $|x| < R$. Then by the definition of R as sup S, there exists some number x_2 with $|x| < x_2 < R$ such that $\sum\limits_{n=0}^{\infty} a_n x_2^n$ converges. It follows, from the Absolute Convergence Theorem, that $\sum\limits_{n=0}^{\infty} a_n x^n$ converges.

Finally, choose any x for which $|x| > R = \sup S$. It follows that $x \notin S$, so that $\sum\limits_{n=0}^{\infty} a_n x^n$ must be divergent. ∎

In view of the Absolute Convergence Theorem, we can slightly strengthen the conclusion of the Radius of Convergence Theorem, as follows.

Theorem 3, Sub-section 8.3.1.

> **Corollary** If the power series $\sum_{n=0}^{\infty} a_n(x-a)^n$ has radius of convergence $R > 0$, then it is absolutely convergent if $|x - a| < R$. If the series converges for all x, then it is absolutely convergent for all x.

This result is sometimes also called the **Absolute Convergence Theorem**.

Finally we prove the Ratio Test for Radius of Convergence of power series.

> **Theorem 2 Ratio Test for Radius of Convergence**
>
> Suppose that $\sum_{n=0}^{\infty} a_n(x-a)^n$ is a given power series, and that
> $$\left|\frac{a_{n+1}}{a_n}\right| \to L \quad \text{as } n \to \infty.$$
> (a) If L is ∞, the series converges only for $x = a$.
> (b) If $L = 0$, the series converges for all x.
> (c) If $L > 0$, the series has radius of convergence $\frac{1}{L}$.

Proof For simplicity, we shall assume that $a = 0$.

For the general proof, replace x throughout by $(x-a)$.

(a) Suppose that $\left|\frac{a_{n+1}}{a_n}\right| \to \infty$ as $n \to \infty$. Then, for any non-zero value of x, the sequence $\left|\frac{a_{n+1}x^{n+1}}{a_n x^n}\right| \to \infty$, so that the sequence $\{a_n x^n\}$ is unbounded. It follows, from the Non-null Test, that the series $\sum_{n=0}^{\infty} a_n x^n$ must be divergent.

(b) Suppose that $\left|\frac{a_{n+1}}{a_n}\right| \to 0$ as $n \to \infty$. Then, for any non-zero value of x
$$\left|\frac{a_{n+1}x^{n+1}}{a_n x^n}\right| = \left|\frac{a_{n+1}}{a_n}\right| \times |x|$$
$$\to 0 \times |x| = 0,$$
so that $\sum_{n=0}^{\infty} a_n x^n$ is absolutely convergent, and so is convergent.

(c) Suppose that $\left|\frac{a_{n+1}}{a_n}\right| \to L$ as $n \to \infty$, where $L > 0$.

First, suppose that $|x| > \frac{1}{L}$. Then
$$\left|\frac{a_{n+1}x^{n+1}}{a_n x^n}\right| = \left|\frac{a_{n+1}}{a_n}\right| \times |x|$$
$$\to L \times |x| > 1,$$
so that $\sum_{n=0}^{\infty} a_n x^n$ is not absolutely convergent. It follows, from the above Corollary to the Radius of Convergence Theorem, that the radius of convergence of $\sum_{n=0}^{\infty} a_n x^n$ must be less than or equal to $\frac{1}{L}$.

Next, suppose that $|x| < \frac{1}{L}$. Then
$$\left|\frac{a_{n+1}x^{n+1}}{a_n x^n}\right| = \left|\frac{a_{n+1}}{a_n}\right| \times |x|$$
$$\to L \times |x| < 1,$$

so that $\sum\limits_{n=0}^{\infty} a_n x^n$ is absolutely convergent, and so is convergent.

Hence the radius of convergence of $\sum\limits_{n=0}^{\infty} a_n x^n$ must be at least equal to $\frac{1}{L}$.

Combining these two facts, it follows that the desired radius of convergence is exactly $\frac{1}{L}$. ∎

8.4 Manipulating power series

It would be tedious to have to apply Taylor's Theorem every time that we wished to determine the Taylor series of a given function. While sometimes this really has to be done, in many commonly arising situations we can use standard rules for power series and the list of basic power series to avoid most of the effort.

Theorem 3, Sub-section 8.2.2.

We now set out to establish the rules for manipulating power series.

8.4.1 Rules for power series

Many of the rules for manipulating power series are similar to the corresponding rules for manipulating 'ordinary' series.

Theorem 1   Combination Rules

Let
$$f(x) = \sum_{n=0}^{\infty} a_n (x-a)^n, \quad \text{for } |x-a| < R, \quad \text{and}$$

$$g(x) = \sum_{n=0}^{\infty} b_n (x-a)^n, \quad \text{for } |x-a| < R'.$$

Then:

Sum Rule $(f+g)(x) = \sum\limits_{n=0}^{\infty} (a_n + b_n)(x-a)^n, \quad \text{for } |x-a| < r,$
where $r = \min\{R, R'\}$;

Multiple Rule $\lambda f(x) = \sum\limits_{n=0}^{\infty} \lambda a_n (x-a)^n, \quad \text{for } |x-a| < R, \text{ where } \lambda \in \mathbb{R}.$

Thus, if the power series for both f and g converge at a particular point x, then so does the series for f + g.

We do not supply a proof, as Theorem 1 is a simple restatement of the Sum and Multiple Rules for 'ordinary' series.

Theorem 2, Sub-section 3.1.4.

We can use the Combination Rules to find the Taylor series at 0 for the function $\cosh x$. We start with the power series for the exponential function

$$e^x = 1 + x + \frac{x^2}{2!} + \frac{x^3}{3!} + \cdots = \sum_{n=0}^{\infty} \frac{x^n}{n!}, \quad \text{for } x \in \mathbb{R}.$$

Theorem 3, Sub-section 8.2.2.

It follows that $e^{-x} = 1 - x + \frac{x^2}{2!} - \frac{x^3}{3!} + \cdots = \sum\limits_{n=0}^{\infty} (-1)^n \frac{x^n}{n!}$, for $x \in \mathbb{R}$. We may then use the Sum Rule to obtain

$$e^x + e^{-x} = 2 \times \left(1 + \frac{x^2}{2!} + \frac{x^4}{4!} + \cdots \right) = 2 \times \sum_{n=0}^{\infty} \frac{x^{2n}}{(2n)!}, \quad \text{for } x \in \mathbb{R},$$

For the odd-powered terms cancel.

so that, by using the Multiple Rule with $\lambda = \frac{1}{2}$, we obtain

$$\cosh x = \frac{1}{2}(e^x + e^{-x}) = 1 + \frac{x^2}{2!} + \frac{x^4}{4!} + \cdots = \sum_{n=0}^{\infty} \frac{x^{2n}}{(2n)!}, \quad \text{for } x \in \mathbb{R}.$$

Problem 1   Find the Taylor series at 0 for each of the following functions, and state its radius of convergence:

(a) $f(x) = \sinh x, \quad x \in \mathbb{R}$;

(b) $f(x) = \log_e(1-x) + 2(1-x)^{-1}, \quad |x| < 1$.

Problem 2   Find the Taylor series at 0 for each of the following functions, and state its radius of convergence:

(a) $f(x) = \sinh x + \sin x, \quad x \in \mathbb{R}$;   (b) $f(x) = \log_e\left(\frac{1+x}{1-x}\right), \quad |x| < 1$;

(c) $f(x) = \frac{1}{1+2x^2}, \quad x \in \mathbb{R}$.

Remark

In Theorem 1, the radius of convergence of the power series for $f + g$ *may* be larger than $r = \min\{R, R'\}$. For example, we can use the basic power series and the Combination Rules to verify that the Taylor series at 0 for the functions $f(x) = \frac{1}{1-x}$ and $g(x) = \frac{-1}{1-x} + \frac{1}{1-\frac{x}{2}}$ are

$$f(x) = 1 + x + x^2 + x^3 + x^4 + \cdots = \sum_{n=0}^{\infty} x^n \quad \text{and}$$

$$g(x) = -\frac{1}{2}x - \frac{3}{4}x^2 - \frac{7}{8}x^3 - \frac{15}{16}x^4 - \cdots = \sum_{n=0}^{\infty}\left(-1 + \frac{1}{2^n}\right)x^n,$$

each with radius of convergence 1. It follows, from Theorem 1, that the power series for the function $(f + g)(x) = \frac{1}{1-\frac{x}{2}}$ is

$$(f + g)(x) = 1 + \frac{1}{2}x + \frac{1}{4}x^2 + \frac{1}{8}x^3 + \frac{1}{16}x^4 + \cdots = \sum_{n=0}^{\infty}\frac{1}{2^n}x^n,$$

> We simply add the power series term-by-term.

and that this power series has radius of convergence at least 1. In fact, it has radius of convergence 2.

We can find the Taylor series at 0 for the function $f(x) = \frac{1+x}{1-x}$, for $|x| < 1$, by writing $\frac{1+x}{1-x}$ in the form $\frac{2-(1-x)}{1-x} = \frac{2}{1-x} - 1$. Since the Taylor series at 0 for the function $\frac{2}{1-x}$ is $2 + 2x + 2x^2 + 2x^3 + \cdots = \sum_{n=0}^{\infty} 2x^n$, with radius of convergence 1, it follows that the Taylor series for f at 0 is $1 + 2x + 2x^2 + 2x^3 + \cdots = 1 + \sum_{n=1}^{\infty} 2x^n$, with radius of convergence 1.

However we could obtain the same result by multiplying together the Taylor series for the functions $x \mapsto 1 + x$ and $x \mapsto \frac{1}{1-x}$, since

$$(1+x)(1 + x + x^2 + x^3 + \cdots) = 1 + x + x^2 + x^3 + \cdots + x(1 + x + x^2 + x^3 + \cdots)$$
$$= 1 + 2x + 2x^2 + 2x^3 + \cdots.$$

> We now 'collect together' the multiples of successive various powers of x.

In fact, we can justify such multiplication together of power series to obtain further power series.

Theorem 2   Product Rule
Let

$$f(x) = \sum_{n=0}^{\infty} a_n(x-a)^n, \quad \text{for } |x-a| < R, \quad \text{and}$$

$$g(x) = \sum_{n=0}^{\infty} b_n(x-a)^n, \quad \text{for } |x-a| < R'.$$

Then

$$(fg)(x) = \sum_{n=0}^{\infty} c_n(x-a)^n, \quad \text{for } |x-a| < r, \text{where } r = \min\{R, R'\},$$

and

$$c_0 = a_0 b_0, c_1 = a_0 b_1 + a_1 b_0 \quad \text{and}$$
$$c_n = a_0 b_n + a_1 b_{n-1} + \cdots + a_{n-1} b_1 + a_n b_0.$$

As in Theorem 1, the radius of convergence of the product series $\sum_{n=0}^{\infty} c_n(x-a)^n$ *may* be greater than r.

Theorem 2 is an immediate consequence of the Product Rule for series, applied to the two series $\sum_{n=0}^{\infty} a_n(x-a)^n$ and $\sum_{n=0}^{\infty} b_n(x-a)^n$, both of which are absolutely convergent for $|x-a| < r$ – by the Absolute Convergence Theorem.

Theorem 5, Sub-section 3.3.4.

Theorem 3, Sub-section 8.3.1.

Example 1   Determine the Taylor series at 0 for the function $f(x) = \frac{1+x}{(1-x)^2}$.

Solution   We use our knowledge of the power series at 0 for the functions $x \mapsto \frac{1}{1-x}$ and $x \mapsto \frac{1+x}{1-x}$, both of which have radius of convergence 1. Thus, by the Product Rule, we have

$$\frac{1+x}{(1-x)^2} = \frac{1}{1-x} \times \frac{1+x}{1-x}$$
$$= (1 + x + x^2 + x^3 + x^4 + \cdots) \times (1 + 2x + 2x^2 + 2x^3 + 2x^4 + \cdots)$$
$$= \sum_{n=0}^{\infty} c_n x^n,$$

where $c_0 = 1 \times 1 = 1$, $c_1 = 1 \times 2 + 1 \times 1 = 3$ and $c_n = 1 \times 2 + 1 \times 2 + \ldots + 1 \times 2 + 1 \times 1 = 2n + 1$.

Thus the required power series for f at 0 is $\frac{1+x}{(1-x)^2} = \sum_{n=0}^{\infty} (2n+1)x^n$. □

Also, its radius of convergence is at least 1. (In fact it is easy to check that its radius of convergence is exactly 1.)

Problem 3   Determine the Taylor series at 0 for:
(a) $f(x) = (1+x)\log_e(1+x)$, for $|x| < 1$;
(b) $f(x) = \frac{1+x}{(1-x)^3}$, for $|x| < 1$.

Next, we have seen already that the hyperbolic functions have the following Taylor series at 0

At the start of this sub-section.

$$\sinh x = x + \frac{x^3}{3!} + \frac{x^5}{5!} + \cdots \quad \text{and} \quad \cosh x = 1 + \frac{x^2}{2!} + \frac{x^4}{4!} + \cdots,$$

each with radius of convergence ∞. Notice that the derivative of the function $\sinh x$ is $\cosh x$, and that term-by-term differentiation of the series $x + \frac{x^3}{3!} + \frac{x^5}{5!} + \cdots$ gives the series $1 + \frac{x^2}{2!} + \frac{x^4}{4!} + \cdots$. It looks as though we can obtain the Taylor series for the derivative of a function f simply by differentiating the Taylor series for f itself.

Our next result states that we can differentiate or integrate the Taylor series of a function f term-by-term to obtain the Taylor series of the corresponding function f' or $\int f$, respectively.

Theorem 3   Differentiation and Integration Rules

Let $f(x) = \sum\limits_{n=0}^{\infty} a_n(x-a)^n$,   for $|x-a| < R$. Then:

Differentiation Rule $f'(x) = \sum\limits_{n=1}^{\infty} na_n(x-a)^{n-1}$,   for $|x-a| < R$;

Integration Rule $\int f(x)\,dx = \sum\limits_{n=0}^{\infty} a_n \frac{(x-a)^{n+1}}{n+1} + \text{constant}$,

for $|x-a| < R$.

All three series have the same radius of convergence.

To find the constant, put $x = a$.

They may, however, behave differently at the end-points of their intervals of convergence.

For example, consider the Taylor series at 0 for the function $\tan^{-1}$. We know that

$$\frac{1}{1+x^2} = 1 - x^2 + x^4 - x^6 + \cdots$$

This is a geometric series.

$$= \sum_{n=0}^{\infty} (-1)^n x^{2n}, \quad \text{with radius of convergence } 1,$$

and that

$$\frac{d}{dx} \tan^{-1} x = \frac{1}{1+x^2}, \quad \text{for } x \in \mathbb{R}.$$

It follows, from the Integration Rule, that

$$\tan^{-1} x = x - \frac{x^3}{3} + \frac{x^5}{5} - \frac{x^7}{7} + \cdots + c, \quad \text{for } |x| < 1, \text{and some constant } c.$$

By the Integration Rule, the final Taylor series must have the same radius of convergence as the original Taylor series, namely 1.

Substituting $x = 0$ into this equation, we find that $c = \tan^{-1} 0 = 0$. It follows that

$$\tan^{-1} x = x - \frac{x^3}{3} + \frac{x^5}{5} - \frac{x^7}{7} + \cdots, \quad \text{for } |x| < 1.$$

Problem 4   Find the Taylor series at 0 for the following functions:

(a) $f(x) = (1-x)^{-2}$, for $|x| < 1$;   (b) $f(x) = (1-x)^{-3}$, for $|x| < 1$;

(c) $f(x) = \tanh^{-1} x$, for $|x| < 1$.

Problem 5   Find the first three non-zero terms in the Taylor series at 0 for the function $f(x) = e^x(1-x)^{-2}$, $|x| < 1$. State its radius of convergence.

Problem 6   Let f be the function $f(x) = x + \frac{x^3}{1.3} + \frac{x^5}{1.3.5} + \cdots + \frac{x^{2n+1}}{1.3.\cdots.(2n+1)} + \cdots$, $x \in \mathbb{R}$.

Determine the Taylor series at 0 for each of the following functions:

(a) $f'(x)$;   (b) $f'(x) - xf(x)$.

Problem 7   Determine the Taylor series at 0 for the function $f(x) = e^{-x^2}$, $x \in \mathbb{R}$.

Deduce that $\int_0^1 e^{-x^2} dx = 1 - \frac{1}{3} + \frac{1}{10} - \frac{1}{42} + \cdots + \frac{(-1)^n}{(2n+1)\cdot n!} + \cdots$.

3000

We have now found a whole variety of techniques for identifying Taylor series:

- Taylor's Theorem;

- Combination, Product, Differentiation and Integration Rules.

But how do we *know* that different techniques will always give us the same expression as the Taylor series? The following result states that there is *only one* Taylor series for a function f at a given point a – any valid method will give the same coefficients.

Theorem 4   Uniqueness Theorem

If $\displaystyle\sum_{n=0}^{\infty} a_n(x-a)^n = \sum_{n=0}^{\infty} b_n(x-a)^n$, for $|x-a| < R$, then $a_n = b_n$.

Proof   Let $\displaystyle f(x) = \sum_{n=0}^{\infty} a_n(x - a)^n$ and $\displaystyle g(x) = \sum_{n=0}^{\infty} b_n(x - a)^n$, for $|x-a| < R$.

If we differentiate both equations n times, using the Differentiation Rule, and put $x = a$, we obtain

$$f^{(n)}(a) = n! \times a_n \quad \text{and} \quad g^{(n)}(a) = n! \times b_n;$$

Since $f(x) = g(x)$, for $|x - a| < R$, we must have $f^{(n)}(a) = g^{(n)}(a)$. It follows that $a_n = b_n$, for all $n \geq 0$.   ∎

8.4.2   General Binomial Theorem

You will have already met the Binomial Theorem, which states that, for each positive integer n

1*See, for example, Appendix 1.*

$$(1+x)^n = \sum_{k=0}^{n} \binom{n}{k} x^k, \text{ where } \binom{n}{k} = \frac{n(n-1)\ldots(n-k+1)}{k!} = \frac{n!}{k!(n-k)!}.$$

In the Binomial Theorem, the power n is a positive integer and the result is true for all $x \in \mathbb{R}$. In fact, a similar result holds for more general values of the power but with a restriction on the values of x for which it is valid.

Theorem 5   General Binomial Theorem

For any $\alpha \in \mathbb{R}$

$$(1+x)^\alpha = \sum_{n=0}^{\infty} \binom{\alpha}{n} x^n, \text{ where } \binom{\alpha}{n} = \frac{\alpha(\alpha-1)\ldots(\alpha-n+1)}{n!} \text{ and } |x| < 1.$$

By convention, $\binom{\alpha}{0} = 1$.

For example

$$(1 + 2x)^{-6} = \sum_{n=0}^{\infty} \binom{-6}{n}(2x)^n, \quad \text{where}$$

$$\binom{-6}{n} = \frac{(-6)(-7)\ldots(-6-n+1)}{n!} \quad \text{and} \quad |x| < \frac{1}{2},$$

so that

$$(1+2x)^{-6} = 1 - 12x + 84x^2 - \cdots, \quad \text{for } |x| < \frac{1}{2}.$$

Another important type of application occurs when the power α is not an integer. For example

$$(1+x)^{-\frac{1}{3}} = \sum_{n=0}^{\infty} \binom{-\frac{1}{3}}{n} x^n, \quad \text{where } \binom{-\frac{1}{3}}{n} = \frac{\left(-\frac{1}{3}\right)\left(-\frac{4}{3}\right)\left(-\frac{7}{3}\right)\cdots\left(-\frac{1}{3}-n+1\right)}{n!},$$

so that

$$(1+x)^{-\frac{1}{3}} = 1 - \frac{1}{3}x + \frac{2}{9}x^2 - \frac{14}{81}x^3 + \cdots, \quad \text{for } |x| < 1.$$

Problem 8   Use the General Binomial Theorem to determine the first four non-zero terms in the Taylor series at 0 for the function $f(x) = (1+6x)^{\frac{1}{4}}$, $|x| < \frac{1}{6}$. State the radius of convergence of the Taylor series.

Problem 9

(a) Determine the Taylor series at 0 for the function $f(x) = (1-x)^{-\frac{1}{2}}$, $|x| < 1$.

(b) Hence determine the Taylor series at 0 for the function $f(x) = \sin^{-1} x$, $|x| < 1$.

In our proof of the General Binomial Theorem, we use the following lemma.

> You may omit the remainder of this sub-section at a first reading.

> **Lemma**   For any $\alpha \in \mathbb{R}$ and any $n \in \mathbb{N}$, $n\binom{\alpha}{n} + (n+1)\binom{\alpha}{n+1} = \alpha\binom{\alpha}{n}$.

Proof   Since $\binom{\alpha}{k} = \frac{\alpha(\alpha-1)\cdot\ldots\cdot(\alpha-k+1)}{k!}$, for any $k \in \mathbb{N}$, we may rewrite the left-hand side of the desired identity in the form

$$n\binom{\alpha}{n} + (n+1)\binom{\alpha}{n+1} = n\binom{\alpha}{n} + (n+1)\frac{\alpha(\alpha-1)\ldots(\alpha-n)}{(n+1)!}$$

Here we cancel the term $(n+1)$.

$$= n\binom{\alpha}{n} + \frac{\alpha(\alpha-1)\ldots(\alpha-n)}{n!}$$

We now bring the term $(\alpha - n)$ in front of the fraction.

$$= n\binom{\alpha}{n} + (\alpha-n)\frac{\alpha(\alpha-1)\ldots(\alpha-n+1)}{n!}$$

The last fraction is just $\binom{\alpha}{n}$.

$$= n\binom{\alpha}{n} + (\alpha-n)\binom{\alpha}{n}$$

$$= \alpha\binom{\alpha}{n}. \qquad \blacksquare$$

> **Theorem 5   General Binomial Theorem**
>
> For any $\alpha \in \mathbb{R}$
>
> $$(1+x)^\alpha = \sum_{n=0}^{\infty} \binom{\alpha}{n} x^n, \quad \text{where } \binom{\alpha}{n} = \frac{\alpha(\alpha-1)\ldots(\alpha-n+1)}{n!} \text{ and } |x| < 1.$$

Proof Let

$$f(x) = \sum_{n=0}^{\infty} \binom{\alpha}{n} x^n \quad \text{and} \quad g(x) = f(x)(1+x)^{-\alpha}, \quad \text{for } |x| < 1.$$

We want to prove that $g(x) = 1$, for all x with $|x| < 1$.

Using the rules for differentiation, we may differentiate the expression for g to obtain

$$g'(x) = f'(x)(1+x)^{-\alpha} - \alpha f(x)(1+x)^{-\alpha-1}$$
$$= [(1+x)f'(x) - \alpha f(x)](1+x)^{-\alpha-1}.$$

Now

$$(1+x)f'(x) = (1+x)\sum_{n=1}^{\infty} n\binom{\alpha}{n} x^{n-1}$$

$$= \sum_{n=1}^{\infty} n\binom{\alpha}{n} x^{n-1} + \sum_{n=1}^{\infty} n\binom{\alpha}{n} x^n$$

$$= \sum_{n=0}^{\infty} (n+1)\binom{\alpha}{n+1} x^n + \sum_{n=0}^{\infty} n\binom{\alpha}{n} x^n$$

$$= \sum_{n=0}^{\infty} \left[(n+1)\binom{\alpha}{n+1} + n\binom{\alpha}{n}\right] x^n$$

$$= \alpha \sum_{n=0}^{\infty} \binom{\alpha}{n} x^n$$

$$= \alpha f(x).$$

Here we differentiate the power series term-by-term. Then we split the initial bracket.

We now replace n by $n+1$ in the first sum, and check what the limits of summation are in both sums.

Now we combine the two sums into one.

We can apply the result of the Lemma to this square bracket.

By the definition of $f(x)$.

It follows from the earlier expression for $g'(x)$ that

$$g'(x) = [(1+x)f'(x) - \alpha f(x)](1+x)^{-\alpha-1}$$
$$= 0 \times (1+x)^{-\alpha-1} = 0.$$

So, $g(x)$ is a constant. Hence

$$g(x) = g(0)$$
$$= f(0) = 1,$$

as required. ∎

8.4.3 Proofs

Differentiation Rule The series

$$f(x) = \sum_{n=0}^{\infty} a_n(x-a)^n \quad \text{and} \quad \sum_{n=1}^{\infty} na_n(x-a)^{n-1}$$

have the same radius of convergence, R say. Also, f is differentiable on $(a-R, a+R)$, and

$$f'(x) = \sum_{n=1}^{\infty} na_n(x-a)^{n-1}.$$

You may omit this Sub-section at a first reading.

This is one part of Theorem 3, Sub-section 8.4.1.

Proof For simplicity, we assume that $a = 0$.

Let the series $\sum\limits_{n=0}^{\infty} a_n x^n$ and $\sum\limits_{n=1}^{\infty} n a_n x^{n-1}$ have radii of convergence R and R', respectively.

We start by proving that $R = R'$.

We first show that, if $|x| < R$, then the power series $\sum\limits_{n=1}^{\infty} n a_n x^{n-1}$ is convergent. (By the Corollary to the Radius of Convergence Theorem, this shows that $R' \geq R$.)

Sub-section 8.3.3.

To prove this, choose a real number c with $|x| < c < R$. Then the series $\sum\limits_{n=0}^{\infty} a_n c^n$ is convergent, and so the sequence $\{a_n c^n\}$ is a null sequence. Thus there is a number K such that

For example, $c = \frac{1}{2}(|x| + R)$.

$$|a_n c^n| \leq K, \quad \text{for } n = 0, 1, 2, \ldots. \tag{1}$$

Then

$$\left| n a_n x^{n-1} \right| = |a_n c^n| \times \frac{n}{c} \times \left| \frac{x}{c} \right|^{n-1}$$
$$\leq \frac{K}{c} \times n \times \left| \frac{x}{c} \right|^{n-1}, \tag{2}$$

by (1). Since $\left| \frac{x}{c} \right| < 1$, the series $\sum\limits_{n=1}^{\infty} n \left| \frac{x}{c} \right|^{n-1}$ converges. Hence, by the Comparison Test for series, the series $\sum\limits_{n=1}^{\infty} n a_n x^{n-1}$ is absolutely convergent, and so is convergent. This proves that $R' \geq R$.

Theorem 1, Sub-section 3.2.1.

Next, suppose that $R' > R$. Let c be any number such that $R < c < R'$. Then $\sum\limits_{n=1}^{\infty} n a_n c^{n-1}$ is absolutely convergent, by the Absolute Convergence Theorem. But

For example, $c = \frac{1}{2}(R + R')$.

$$|a_n c^n| = \left| n a_n c^{n-1} \right| \times \frac{c}{n}$$
$$\leq c \times \left| n a_n c^{n-1} \right|.$$

Hence, by the Comparison Test for series, the series $\sum\limits_{n=1}^{\infty} a_n c^n$ is absolutely convergent, and so is convergent. This contradicts the definition of R, and thus shows that $R' \not> R$.

It follows that $R = R'$.

Next, we show that f is differentiable on $(-R, R)$, and that f' is of the stated form.

So, choose any point $c \in (-R, R)$. Then choose a positive number $r < R$ such that $c \in (-r, r)$. Then, for all non-zero h such that $|h| < r - |c|$

For example, $r = \frac{1}{2}(|c| + R)$.

$$\frac{f(c+h) - f(c)}{h} - \sum_{n=1}^{\infty} n a_n c^{n-1} = \frac{1}{h} \sum_{n=1}^{\infty} a_n \left((c+h)^n - c^n - h n c^{n-1} \right). \tag{3}$$

Now, we apply Taylor's Theorem to the function $x \mapsto x^n$ on the interval with end-points c and $c + h$. We obtain

$$(c+h)^n = c^n + n h c^{n-1} + \frac{1}{2} n(n-1) h^2 c_n^{n-2},$$

where c_n is some number between c and $c + h$ (and, in particular, with $|c_n| \leq r$).

Now

$$\left| (c + h)^n - c^n - nhc^{n-1} \right| \leq \frac{1}{2}n(n-1)h^2 r^{n-2}. \tag{4}$$

It follows from (3) and (4) and the Triangle Inequality that

$$\left| \frac{f(c+h) - f(c)}{h} - \sum_{n=1}^{\infty} na_n c^{n-1} \right| \leq \frac{1}{2}|h| \sum_{n=2}^{\infty} n(n-1)|a_n| r^{n-2}. \tag{5}$$

Since the series $\sum_{n=1}^{\infty} na_n x^{n-1}$ has radius of convergence R, it follows that so does the series $\sum_{n=2}^{\infty} n(n-1)a_n x^{n-2}$; we conclude that the series $\sum_{n=2}^{\infty} n(n-1) a_n r^{n-2}$ is (absolutely) convergent.

For we have shown that term-by-term differentiation does not alter the radius of convergence of a power series.

Hence, it follows from (5) and the Limit Inequality Rule that

$$\lim_{h \to 0} \frac{f(c+h) - f(c)}{h} = \sum_{n=1}^{\infty} na_n c^{n-1}. \qquad \blacksquare$$

Integration Rule   The series

$$f(x) = \sum_{n=0}^{\infty} a_n(x-a)^n \quad \text{and} \quad F(x) = \sum_{n=0}^{\infty} \frac{a_n}{n+1}(x-a)^{n+1}$$

have the same radius of convergence, R say. Also, f is integrable on $(a - R, a + R)$, and

$$\int f(x)dx = F(x).$$

This is the other part of Theorem 3, Sub-section 8.4.1.

That is, F is a primitive of f on $(a - R, a + R)$

Proof   The two power series have the same radius of convergence, by the Differentiation Rule for power series, applied to F.

It also follows, from the Differentiation Rule for power series, that F is differentiable and $F' = f$, so that F is a primitive of f. $\blacksquare$

8.5   Numerical estimates for π

One of the most interesting problems in the history of mathematics throughout the last two thousand years has been the determination to any pre-assigned degree of accuracy of naturally arising irrational numbers such as $\sqrt{2}$, e and π. Here we look at various way of estimating π, and prove that π is irrational.

8.5.1   Calculating π

It is easy to check that π is slightly larger than 3 by wrapping a string round a circle of radius r and measuring the length of the string corresponding to one circumference; for this length $2\pi r$ is just slightly larger than $6r$.

Over hundreds of years mathematicians devised more sophisticated methods for estimating the value of π. Some of their results are as follows:

the Babylonians	c. 2000 BC	$\pi = 3\frac{1}{8} = 3.125$
the Egyptians	c. 2000 BC	$\pi = 3\frac{13}{81} \simeq 3.160$
the Old Testament	c. 550 BC	$\pi = 3$
Archimedes	c. 300–200 BC	between $3\frac{1}{7}$ and $3\frac{10}{71}$, $\pi \simeq 3.141$
the Chinese	c. 400–500 AD	$\pi \simeq 3.1415926$
the Hindus	c. 500–600 AD	$\pi \simeq \sqrt{10} \simeq 3.16$

1 Kings vii, 23;
2 Chronicles iv, 2.

With the development of the Calculus in the seventeenth century, new formulas for estimating π were discovered, including:

Wallis's Formula $\frac{\pi}{2} = \frac{2}{1} \cdot \frac{2}{3} \cdot \frac{4}{3} \cdot \frac{4}{5} \cdot \frac{6}{5} \cdot \frac{6}{7} \cdot \ldots \cdot \frac{2n}{2n-1} \cdot \frac{2n}{2n+1} \cdot \ldots$

Theorem 5, Sub-section 7.4.2.

Leibniz's Series $\tan^{-1} x = x - \frac{x^3}{3} + \frac{x^5}{5} - \frac{x^7}{7} + \cdots$

Just before Problem 4, Sub-section 8.4.1.

Gregory's Series $\frac{\pi}{4} = 1 - \frac{1}{3} + \frac{1}{5} - \frac{1}{7} + \cdots$

Problem 4, Sub-section 8.3.2.

However, neither Wallis's Formula nor Gregory's Series is very useful for calculating π beyond a few decimal places, as they converge far too slowly. However, we can use the Taylor series for $\tan^{-1}$ effectively for calculating π by choosing values of x closer to 0; in fact, the smaller the value of x, the faster the series converges, and so the fewer the number of terms we need to calculate π to a given degree of accuracy.

For example, if we choose $x = \frac{1}{\sqrt{3}}$ in Leibniz's Series, we obtain

$$\frac{\pi}{6} = \tan^{-1}\left(\frac{1}{\sqrt{3}}\right) = \frac{1}{\sqrt{3}} - \frac{1}{3}\left(\frac{1}{\sqrt{3}}\right)^3 + \frac{1}{5}\left(\frac{1}{\sqrt{3}}\right)^5 - \frac{1}{7}\left(\frac{1}{\sqrt{3}}\right)^7 + \cdots$$

$$= \frac{1}{\sqrt{3}}\left(1 - \frac{1}{9} + \frac{1}{45} - \frac{1}{189} + \cdots\right).$$

This is known as *Sharp's Formula*.

This formula can be used to calculate π to several decimal places without much effort; we gain about one extra place for every two extra terms.

To obtain series that are more effective for calculating π, we can use the *Addition Formula* for $\tan^{-1}$

$$\tan^{-1} x + \tan^{-1} y = \tan^{-1}\left(\frac{x+y}{1-xy}\right),$$

Problem 3(a) on Section 4.3, in Section 4.5.

provided that $\tan^{-1} x + \tan^{-1} y$ lies in the interval $\left(-\frac{\pi}{2}, \frac{\pi}{2}\right)$.

Problem 1   Use the Addition Formula for $\tan^{-1}$ to prove that

$$\tan^{-1}\left(\frac{1}{3}\right) + \tan^{-1}\left(\frac{1}{4}\right) + \tan^{-1}\left(\frac{2}{9}\right) = \frac{\pi}{4}.$$

Similar applications of the Addition Formula give further expressions for $\frac{\pi}{4}$, such as

$$6\tan^{-1}\left(\frac{1}{8}\right) + 2\tan^{-1}\left(\frac{1}{57}\right) + \tan^{-1}\left(\frac{1}{239}\right) = \frac{\pi}{4};$$

$$4\tan^{-1}\left(\frac{1}{5}\right) - \tan^{-1}\left(\frac{1}{239}\right) = \frac{\pi}{4}.$$

This is known as *Machin's Formula*.

Such formulas have been used to calculate π to great accuracy: 1 million decimal places were achieved in 1974. Recently, improved methods of numerical analysis have been used to calculate π correct to several million decimal places.

In 1768, Lambert showed that π is *irrational*; that is, π is not the solution of a linear equation $a_0 + a_1 x = 0$, where the coefficients a_0 and a_1 are integers. Then, in 1882, Lindemann showed that π is *transcendental*; that is, π is not a solution of any polynomial equation $a_0 + a_1 x + a_2 x^2 + \cdots + a_n x^n = 0$, where all the coefficients are integers.

We end with a couple of mnemonics that can be used to recall the first few digits for π; the word lengths give the successive digits:

For interest only, we list the first 1000 decimal places of π in Appendix 3.

This solved the ancient Greek problem of finding a ruler and compass method to construct a square with an area equal to that of a given circle.

May	I	have	a	large	container	of	coffee?
3.	1	4	1	5	9	2	6

and

How	I	need	a	drink,	alcoholic	of	course,
3.	1	4	1	5	9	2	6

after	all	those	formulas	involving	tangent	functions!
5	3	5	8	9	7	9

8.5.2 Proof that π is irrational

You may omit this at a first reading.

We now prove that π is irrational. Our proof is rather intricate; and, surprisingly, it uses many of the properties of integrals that you met in Chapter 7! It depends on the properties of a suitably chosen integral that we examine in Lemmas 1–3.

> **Lemma 1** Let $I_n = \int_{-1}^{1} (1-x^2)^n \cos\left(\frac{1}{2}\pi x\right) dx$, $n = 0, 1, 2, \ldots$. Then:
> (a) $I_n > 0$; (b) $I_n \leq 2$.

Proof

(a) The function $f(x) = (1-x^2)^n \cos\left(\frac{1}{2}\pi x\right)$, for $x \in [-1, 1]$, is non-negative, for $n = 0, 1, 2, \ldots$. It follows, from the Inequality Rule for integrals, that $I_n \geq 0$, for each n.

To prove that $I_n > 0$ we need to examine the integral in more detail.

By the Inequality Rule for integrals, we have

$$\int_{-1}^{1} f(x)dx = \left(\int_{-1}^{-\frac{1}{2}} + \int_{-\frac{1}{2}}^{\frac{1}{2}} + \int_{\frac{1}{2}}^{1}\right) f(x)dx$$

$$\geq \int_{-\frac{1}{2}}^{\frac{1}{2}} f(x)dx.$$

Now, if $x \in \left[-\frac{1}{2}, \frac{1}{2}\right]$, then $(1-x^2)^n \cos\left(\frac{1}{2}\pi x\right) \geq \left(\frac{3}{4}\right)^n \cos\left(\frac{1}{4}\pi\right)$, so that

$$\int_{-1}^{1} f(x)dx \geq \int_{-\frac{1}{2}}^{\frac{1}{2}} \left(\frac{3}{4}\right)^n \cos\left(\frac{1}{4}\pi\right) dx$$

$$= \left(\frac{3}{4}\right)^n \cos\left(\frac{1}{4}\pi\right) > 0,$$

as required.

For $\int_{-1}^{-\frac{1}{2}} f(x)dx \geq 0$ and $\int_{\frac{1}{2}}^{1} f(x)dx \geq 0$.

By the Inequality Rule for integrals.

For the length of the interval of integration is 1.

(b) For $x \in [-1, 1]$, $(1 - x^2)^n \cos(\frac{1}{2}\pi x) \leq 1 \times 1 = 1$, so that

$$\int_{-1}^{1} f(x)dx \leq \int_{-1}^{1} 1dx = 2.$$

■ By the Inequality Rule for integrals.

Problem 2   Prove that $\pi I_0 = 4$ and $\pi^3 I_1 = 32$.

Next, we obtain a reduction formula for I_n, using integration by parts.

Lemma 2   The integral I_n satisfies the following reduction formula
$$\pi^2 I_n = 8n(2n - 1)I_{n-1} - 16n(n - 1)I_{n-2}.$$

Proof   Using integration by parts twice, we obtain

$$I_n = \left[(1 - x^2)^n \frac{2}{\pi}\sin\left(\frac{1}{2}\pi x\right)\right]_{-1}^{1} - \frac{2}{\pi}\int_{-1}^{1} n(-2x)(1 - x^2)^{n-1}\sin\left(\frac{1}{2}\pi x\right)dx$$

The value of the first term is zero.

$$= \frac{4n}{\pi}\int_{-1}^{1} x(1 - x^2)^{n-1}\sin\left(\frac{1}{2}\pi x\right)dx$$

$$= \frac{4n}{\pi}\left[x(1 - x^2)^{n-1}\left(-\frac{2}{\pi}\right)\cos\left(\frac{1}{2}\pi x\right)\right]_{-1}^{1}$$

The value of the first term is zero.

$$+ \frac{8n}{\pi^2}\int_{-1}^{1}\left\{(1 - x^2)^{n-1} - 2x^2(n - 1)(1 - x^2)^{n-2}\right\}\cos\left(\frac{1}{2}\pi x\right)dx$$

$$= \frac{8n}{\pi^2}I_{n-1} - \frac{16}{\pi^2}n(n - 1)\int_{-1}^{1} x^2(1 - x^2)^{n-2}\cos\left(\frac{1}{2}\pi x\right)dx$$

$$= \frac{8n}{\pi^2}I_{n-1} - \frac{16}{\pi^2}n(n - 1)\int_{-1}^{1}\left\{(1 - x^2)^{n-2} - (1 - x^2)^{n-1}\right\}\cos\left(\frac{1}{2}\pi x\right)dx$$

For
$$x^2(1 - x^2)^{n-2}$$
$$= \{1 - (1-x^2)\}(1 - x^2)^{n-2}$$
$$= (1 - x^2)^{n-2} - (1 - x^2)^{n-1}.$$

$$= \frac{8n}{\pi^2}I_{n-1} - \frac{16}{\pi^2}n(n - 1)\{I_{n-2} - I_{n-1}\}.$$

Multiplying both sides by π^2, we obtain

$$\pi^2 I_n = 8nI_{n-1} - 16n(n - 1)I_{n-2} + 16n(n - 1)I_{n-1}$$
$$= 8n(2n - 1)I_{n-1} - 16n(n - 1)I_{n-2}.$$ ■

We now use the result of Lemma 2 to prove the crucial tool in our proof that π is irrational.

Lemma 3   For $n = 0, 1, 2, \ldots$, there exist integers $a_0, a_1, a_2, \ldots, a_n$ such that

$$\frac{\pi^{2n+1}}{n!}I_n = a_0 + a_1\pi + a_2\pi^2 + \cdots + a_n\pi^n \left(= \sum_{k=0}^{n} a_k\pi^k\right).$$

Proof   For simplicity, let $J_n = \frac{\pi^{2n+1}}{n!}I_n$. It follows from Problem 2 that

$$J_0 = \pi I_0 = 4 \quad \text{and} \quad J_1 = \pi^3 I_1 = 32.$$ (1)

Next, we rewrite the reduction formula in Lemma 2 in terms of J_n as

$$J_n = \frac{\pi^{2n+1}}{n!} I_n$$

$$= \frac{8n(2n-1)}{n!}\pi^{2n-1}I_{n-1} - \frac{16n(n-1)}{n!}\pi^{2n-1}I_{n-2}$$

$$= 8(2n-1)J_{n-1} - 16\pi^2 J_{n-2}. \tag{2}$$

The desired result now follows by Mathematical Induction. We know that it holds for $n=0$ and for $n=1$, by the statements (1) above. Using the reduction formula (2) we can prove that the statement of the Lemma holds for all $n \geq 2$. ∎

We omit 'the gory details'!

Finally, we can use the fact that, for any integer p, the sequence $\left\{\frac{p^{2n+1}}{n!}\right\}$ is null, to prove that π is irrational.

The sequence $\left\{\frac{(p^2)^n}{n!}\right\}$ is a basic null sequence.

Theorem 1 π is irrational.

Proof Suppose, on the contrary, that π is rational; so that $\pi = \frac{p}{q}$, for $p, q \in \mathbb{N}$. Then, the conclusion of Lemma 3 can be written in the form

This is a *proof by contradiction*.

$$\frac{p^{2n+1}}{q^{2n+1}n!} I_n = \sum_{k=0}^{n} a_k \frac{p^k}{q^k},$$

where the coefficients a_k are integers.

If we now multiply both sides by the expression q^{2n+1}, we find that

$$\frac{p^{2n+1}}{n!} I_n = \sum_{k=0}^{n} a_k p^k q^{2n+1-k},$$

so that

$$\frac{p^{2n+1}}{n!} I_n \text{ is an integer.} \tag{3}$$

However, we know that the sequence $\left\{\frac{p^{2n+1}}{n!}\right\}$ is a null sequence; and we know, from Lemma 1, that $|I_n| \leq 2$. It follows, from the Squeeze Rule for sequences, that

$$\frac{p^{2n+1}}{n!} I_n \to 0 \quad \text{as } n \to \infty. \tag{4}$$

It follows from (4) that eventually $\frac{p^{2n+1}}{n!} I_n < 1$; and so, from (3), that in fact I_n must equal 0. This contradicts the result of Lemma 1, part (a). This contradiction proves that, in fact, π must be rational. ∎

8.6 Exercises

Section 8.1

1. Determine the tangent approximation to the function $f(x) = 2 - 3x + x^2 + e^x$ at the given point a:

   (a) $a = 0$;   (b) $a = 1$.

2. Determine the Taylor polynomial of degree 3 for each of the following functions f at the given point a:

(a) $f(x) = \log_e(1+x)$, $a = 2$;   (b) $f(x) = \sin x$, $a = \frac{\pi}{6}$;
(c) $f(x) = (1+x)^{-2}$, $a = \frac{1}{2}$;   (d) $f(x) = \tan x$, $a = \frac{\pi}{4}$.

3. Determine the Taylor polynomial of degree 4 for each of the following functions f at the given point a:

(a) $f(x) = \cosh x$, $a = 0$;   (b) $f(x) = x^5$, $a = 1$.

4. Determine the percentage error involved in using the Taylor polynomial of degree 3 for the function $f(x) = e^x$ at 0 to evaluate $e^{0.1}$.

Section 8.2

1. Obtain an expression for the remainder term $R_1(x)$ when Taylor's Theorem is applied to the function $f(x) = e^x$ at 0. Show that, when $x = 1$ and $n = 1$, then the value of c in the statement of Taylor's Theorem is approximately 0.36.

2. By applying Taylor's Theorem to the function $f(x) = \sin x$ with $a = \frac{\pi}{4}$, prove that

$$\sin x = \frac{1}{\sqrt{2}}\left(1 + \left(x - \frac{\pi}{4}\right) - \frac{1}{2}\left(x - \frac{\pi}{4}\right)^2\right) + R_2(x), \quad x \in \mathbb{R},$$

where $|R_2(x)| \le \frac{1}{6}\left|x - \frac{\pi}{4}\right|^3$.

3. By applying the Remainder Estimate to the function $f(x) = \sinh x$ with $a = 0$, calculate $\sinh 0.2$ to four decimal places.

4. Calculate the Taylor polynomial $T_3(x)$ for the function $f(x) = \frac{x}{x+3}$ at 2. Show that $T_3(x)$ approximates $f(x)$ to within 10^{-4} on the interval $\left[2, \frac{5}{2}\right]$.

5. (a) Determine the Taylor polynomial of degree n for the function $f(x) = \log_e x$ at 1.

(b) Write down the remainder term $R_n(x)$ in Taylor's Theorem for this function, and show that $R_n(x) \to 0$ as $n \to \infty$ if $x \in (1, 2)$.

(c) By using Theorem 2 in Section 8.2, determine a Taylor series for f at 1 which is valid when $x \in (1, 2)$.

6. Let $f(x) = \begin{cases} e^{-\frac{1}{x^2}}, & x \ne 0, \\ 0, & x = 0. \end{cases}$

(a) Prove that $f'(0) = 0$.

(b) Prove that, for $x \ne 0$, $f'(x)$ is of the form

(a polynomial of degree at most 3 in $\frac{1}{x}$) $\times e^{-\frac{1}{x^2}}$.

(c) Prove that, for $x \ne 0$, $f^{(n)}(x)$ is of the form

(a polynomial of degree at most $3n$ in $\frac{1}{x}$) $\times e^{-\frac{1}{x^2}}$.

(d) Prove that, for any positive integer n, $f^{(n)}(0) = 0$.

For this function, the Taylor polynomial $T_n(x)$ is identically zero for each n, although the function f is not identically zero. In this case, $R_n(x) = f(x)$, for all n; and so the Taylor series for f at 0, $\sum_{n=0}^{\infty} \frac{f^{(n)}(0)}{n!}x^n$, converges to $f(x)$ only at 0.

Section 8.3

1. Determine the radius of convergence of each of the following power series:

(a) $\sum_{n=1}^{\infty} \frac{(3n)!}{(n!)^2}(x+1)^n$;   (b) $\sum_{n=1}^{\infty} \frac{n^n}{n!}\left(x - \frac{1}{e}\right)^n$;

(c) $\sum_{n=1}^{\infty} (n+1)^{-n} x^n$; (d) $\frac{1}{2}x + \frac{1.3}{2.5}x^2 + \frac{1.3.5}{2.5.8}x^3 + \cdots$;

(e) $1 + \frac{a.b}{1.c}x + \frac{a(a+1).b(b+1)}{1.2.c(c+1)}x^2 + \cdots$, where $a, b, c > 0$.

This series is often called the hypergeometric series.

2. Determine the Taylor series for the function $f(x) = \log_e(2+x)$ at the given point a; in each case, indicate the general term and state the radius of convergence of the series:

(a) $a = 1$; (b) $a = -1$.

Hint: Use the Taylor series at 0 for the function $x \mapsto \log_e(1+x)$, with $t = 3x+1$ and $t = x-1$, respectively.

3. Determine the interval of convergence of each of the following power series:

(a) $\sum_{n=1}^{\infty} \frac{2^n}{n!} x^n$; (b) $\sum_{n=1}^{\infty} (-1)^n \frac{2^n}{n}(x-1)^n$.

4. Give an example (if one exists) of each of the following:

(a) a power series $\sum_{n=0}^{\infty} a_n x^n$ which diverges at both $x=1$ and $x=-2$;

(b) a power series $\sum_{n=0}^{\infty} a_n x^n$ which diverges at $x=1$ but converges at $x=-2$;

(c) a power series $\sum_{n=0}^{\infty} a_n x^n$ which converges at $x=1$ but diverges at $x=-2$.

5. Give an example (if one exists) of each of the following:

(a) a power series $\sum_{n=0}^{\infty} a_n x^n$ which converges only if $-3 < x < 3$;

(b) a power series $\sum_{n=0}^{\infty} a_n x^n$ which converges only if $-3 \le x < 3$;

(c) a power series $\sum_{n=0}^{\infty} a_n x^n$ which converges only if $-3 < x \le 3$;

(d) a power series $\sum_{n=0}^{\infty} a_n x^n$ which converges only if $-3 \le x \le 3$.

Section 8.4

1. Determine the Taylor series for each of the following functions at 0; in each case, indicate the general term and state the radius of convergence of the series:

(a) $f(x) = \log_e(1+x+x^2) = \log_e\left(\frac{1-x^3}{1-x}\right)$; (b) $f(x) = \frac{\cosh x}{1-x}$.

2. Show that, for $x > 1$
$$2\log_e x - \log_e(x+1) - \log_e(x-1) = \sum_{n=1}^{\infty} \frac{1}{nx^{2n}}.$$

3. For the function $f(x) = x + \frac{2}{3}x^3 + \frac{2.4}{3.5}x^5 + \frac{2.4.6}{3.5.7}x^7 + \cdots$, for $|x| < 1$, prove that
$$(1-x^2)f'(x) - xf(x) = 1.$$

In fact, $f(x) = \frac{\sin^{-1} x}{\sqrt{1-x^2}}$.

4. By using the Integration Rule for power series, prove that
$$\sinh^{-1} x = x - \frac{1}{2}\frac{x^3}{3} + \frac{1.3}{2.4}\frac{x^5}{5} - \frac{1.3.5}{2.4.6}\frac{x^7}{7} + \cdots, \quad \text{for } |x| < 1.$$

5. By using the Integration Rule for power series, find an infinite series with
    sum $\int_0^1 \sin(t^2)\,dt$.

6. Use the General Binomial Theorem, applied to the function
    $f(x) = (1+x)^{-1/2}$, to prove that

$$\sqrt{2} = 1 + \frac{1}{2^2} + \frac{1 \cdot 3}{2^4 \cdot 2!} + \frac{1 \cdot 3 \cdot 5}{2^6 \cdot 3!} + \cdots + \frac{1 \cdot 3 \cdot 5 \cdot \ldots \cdot (2n-1)}{2^{2n} \cdot n!} + \cdots.$$

Appendix 1: Sets, functions and proofs

Sets

A **set** is a collection of objects, called **elements**. We use the symbol $\in$ to mean 'is a member of' or 'belongs to'; thus '$x \in A$' means 'the element x is a member of the set A'. Similarly, we use the symbol $\notin$ to mean 'is not a member of' or 'does not belong to'; thus '$x \notin A$' means 'the element x is not a member of the set A'.

We often use curly brackets (or braces) to list the elements of a set in some way. Thus $\{a, b, c\}$ denotes the set whose three elements are a, b and c. Similarly, $\{x : x^2 = 2\}$ denotes the set whose elements x are such that $x^2 = 2$; we would read this symbol in words as 'the set of x such that $x^2 = 2$'. When we define a set in this latter way, the symbol x is a **dummy variable**; that is, if we replace that symbol x by any other symbol, such as y, the set is exactly the same set; thus, for instance, the sets $\{x : x^2 = 2\}$ and $\{y : y^2 = 2\}$ are identical.

> Thus the colon ':' means 'such that'.

Two sets are **equal** if they contain the same elements. We say that a set A is a **subset** of a set B, if all the elements of A are elements of B, and we denote this by writing '$A \subset B$'. We may wish to indicate specifically the possibility that the subset A is equal to B by writing '$A \subseteq B$', or that A is a **proper subset** of B (that is, A is a subset of B but $A \neq B$), by writing '$A \subsetneq B$'.

Notice that, to show that two sets A and B are equal, it is necessary to prove that A is a subset of B and that B is a subset of A. In symbols

$A = B$ is equivalent to the two properties both holding: $A \subseteq B$ and $B \subseteq A$.

The **empty set** is the set that contains no elements; it is denoted by the symbol '$\emptyset$'. It may seem strange to define such a thing; but, in practice, it is often a convenient set to use.

The **union** of two sets A and B consists of the set of elements that belong to at least one of A and B; in symbols

$$A \cup B = \{x : x \in A \text{ or } x \in B\}.$$

> Here we also allow x to belong to both A and B.

The **intersection** of two sets A and B consists of the set of elements that belong to both A and B; in symbols

$$A \cap B = \{x : x \in A \text{ and } x \in B\}.$$

We denote the set of elements that belong to A but not to B by $A - B$; that is

$$A - B = \{x : x \in A, x \notin B\}.$$

There are some special symbols used that are used to denote commonly arising sets of real numbers:

$\mathbb{N}$, the set of all **natural** numbers; thus $\mathbb{N} = \{1, 2, 3, \ldots\}$;
$\mathbb{Z}$, the set of all **integers**; thus $\mathbb{Z} = \{0, \pm 1, \pm 2, \pm 3, \ldots\}$;
$\mathbb{Q}$, the set of all **rational** numbers; that is, numbers of the form $\frac{p}{q}$, where $p, q \in \mathbb{Z}$ but $q \neq 0$;
$\mathbb{R}$, the set of all **real** numbers;
$\mathbb{R}^+$, the set of all **positive real** numbers.

On the **real line** $\mathbb{R}$, an **interval** I is a set of real numbers such that, if a and b both lie in I, then all numbers between a and b also lie in I.

There are nine types of intervals in $\mathbb{R}$:

$$[a,b] = \{x : x \in \mathbb{R}, a \leq x \leq b\};$$
$$(a,b) = \{x : x \in \mathbb{R}, a < x < b\};$$
$$[a,b) = \{x : x \in \mathbb{R}, a \leq x < b\};$$
$$(a,b] = \{x : x \in \mathbb{R}, a < x \leq b\};$$
$$[a,\infty) = \{x : x \in \mathbb{R}, x \geq a\};$$
$$(a,\infty) = \{x : x \in \mathbb{R}, x > a\};$$
$$(-\infty,b] = \{x : x \in \mathbb{R}, x \leq b\};$$
$$(-\infty,b) = \{x : x \in \mathbb{R}, x < b\};$$
$$\mathbb{R}.$$

The first four types of intervals are **bounded** intervals.

The final five types of intervals are **unbounded** intervals.

An interval is said to be **closed** if it contains both of its end-points, **open** if it contains neither of its end-points, and **half-closed** or **half-open** if it is neither closed nor open.

Functions

A **function** is a mapping of some element of $\mathbb{R}$ onto another element of $\mathbb{R}$. Thus a function f is defined by specifying:

- a set A, called the **domain** of f;
- a set B, called the **codomain** of f;
- a **rule** $x \mapsto f(x)$ that associates with each element x of A a *unique* element $f(x)$ of B.

Symbolically, we write this in the following way:

$$f : A \rightarrow B$$
$$x \mapsto f(x).$$

The element $f(x)$ is called the **image of x under f**, and the set $f(A) = \{y : y = f(x)$ for some $x \in A\}$ the **image of A under f**.

A particularly simple function is the **identity function** that maps each element of a set to itself. Thus, the **identity function on a set A**, i_A, is the function

$$i_A : A \rightarrow B$$
$$x \mapsto x.$$

Sometimes *every* point of the codomain is in the image. We say that a function $f : A \rightarrow B$ is an **onto** function if $f(A) = B$.

Sometimes every point of the image is the image of *precisely* one point of the domain. We say that a function $f : A \rightarrow B$ is a **one–one** function if:

whenever $f(x) = f(y)$ for elements x, y of A, then *necessarily* $x = y$.

Notice that this does NOT mean that f is a one–one mapping of A onto B.

A given function $f : A \rightarrow B$ may be onto, one–one, both onto and one–one, or neither onto nor one–one. A function $f : A \rightarrow B$ is called a **bijection** if it is both onto and one–one.

If a function $f : A \rightarrow B$ is one–one, then it has an **inverse** function $f^{-1} : f(B) \rightarrow A$ with the defining rule

$$f^{-1}(y) = x, \quad \text{where } y = f(x).$$

Notice that a function f only has an inverse function f^{-1} if it is bijective.

Sometimes we want to 'change the domain' of a function to a larger set or a smaller set.

If $f: A \to B$, and C is a subset of A, we say that the function $g : C \to B$ is the **restriction** of f to C if

$$g(x) = f(x), \quad \text{for all } x \in C.$$

Similarly, if $f: A \to B$, and A is a subset of C, we say that the function $g : C \to B$ is the **extension** of f to C if

$$g(x) = f(x), \quad \text{for all } x \in A.$$

Sometimes we use the term *extension* to denote a function $g : C \to D$ where $A \subset C$, $B \subset D$, and $g(x) = f(x)$, for all $x \in A$.

Finally, the **composite function** or **composite** $g \circ f$ of two functions

$$f : A \to B \quad \text{and} \quad g : C \to D,$$

where $f(A) \subset C$, is the function

$$g \circ f : A \to D$$
$$x \mapsto g(f(x)).$$

Proofs

Logical implications are so important in Mathematics that we use some special symbols, namely $\Rightarrow$, $\Leftarrow$ and $\Leftrightarrow$, to denote implications.

Thus, if we are discussing two statements P and Q, we write

$P \Rightarrow Q$ to mean: 'if P is true, then Q is true' or 'P implies Q';

$P \Leftarrow Q$ to mean: 'P is implied by Q', or 'if Q is true, then P is true' or 'Q implies P';

$P \Leftrightarrow Q$ to mean: 'P is true if and only if Q is true';
this is equivalent to the two separate statements

$$P \Rightarrow Q \quad \text{and} \quad Q \Rightarrow P.$$

In this case, P and Q are **equivalent** statement.

Thus, to prove that $P \Leftrightarrow Q$, we need to prove that $P \Rightarrow Q$ and $Q \Rightarrow P$.

The implications $P \Rightarrow Q$ and $Q \Rightarrow P$ are called **converse** implications.

In order to prove that an assertion '$P \Rightarrow Q$' is true, we need to verify that, in all situations where the statement P holds, then the statement Q also holds. To prove that an assertion '$P \Rightarrow Q$' is not true, all that we need do is to find one example of a situation where the statement P holds but the statement Q does not. Such a situation is called a **counter-example** to the assertion.

This is sometimes called *disproof by counter-example*.

Sometimes we prove assertions P by simply checking all possible cases; generally, though, we devise logical arguments that deal with all cases at the same time.

This approach is sometimes called *proof by exhaustion*.

Sometimes our logical arguments appear slightly convoluted to non-mathematicians; two examples of these are:

- *Proof by contradiction*: In order to show that a statement P holds, we start by assuming that P is false; we then look at the consequences of that assumption, and identify some specific consequence that is untrue or contradictory. It then follows that P *must* hold, after all.

- *Proof by contraposition*: In order to show that an implication $P \Rightarrow Q$ is true, it is sufficient to prove that

    if Q does not hold, then P does not hold.

Sometimes mathematical proofs are long and complicated, and involve the use of a variety of approaches to prove the desired result.

This fact is sometimes (humorously) called *proof by perspiration*.

Notice that not all approaches to proving a result are necessarily valid! Some examples of such approaches are:

- *Proof by picture*: Here you make a *false claim* in your argument because it *appears to be true* in the *particular* diagram that you have drawn.

- *Proof by example*: For instance, you prove that some property holds for $n = 1$ and $n = 2$, and then assert that it *therefore* holds for all positive integers.

- *Proof by omission*: For instance, you prove one or two special cases of a result, and claim that the general proof is 'similar'.
- *Proof by superiority*: Here you claim that the result is 'obvious', rather than sit down to construct a logical proof of it.
- *Proof by mumbo-jumbo*: Here you write down a jumble of formulas and relevant words, ending up with the claim that you have proved the result.

Principle of Mathematical Induction

This is a standard method of proving statements involving an integer n, generally a positive integer.

Principle of Mathematical Induction

Let $P(n)$ denote a statement involving a positive integer n. If the following two conditions are satisfied:

1. the statement $P(1)$ is true,

2. whenever the statement $P(k)$ is true for a positive integer k, then the statement $P(k+1)$ is also true,

then the statement $P(n)$ is true for all positive integers n.

Thus, in order to use the Principle, we have a two stage strategy:

1. check $P(1)$,

2. check that $P(k) \Rightarrow P(k+1)$.

Sometimes we need to use an equivalent version of the Principle.

(Second) Principle of Mathematical Induction

Let $P(n)$ denote a statement involving a positive integer n. If the following two conditions are satisfied:

1. the statement $P(1)$ is true,

2. whenever the statements $P(1)$, $P(2)$, ..., $P(k)$ are true for a positive integer k, then the statement $P(k+1)$ is also true,

then the statement $P(n)$ is true for all positive integers n.

We have stated the Principle when $P(n)$ applies to all integers $n \geq 1$; analogous results hold when $P(n)$ applies to all integers $n \geq N$, whatever integer N may be.

We end with some useful results that can be proved using the Principle of Mathematical Induction.

Sums

$$\sum_{k=1}^{n} 1 = n; \qquad\qquad \sum_{k=1}^{n} k = \frac{n(n+1)}{2};$$

$$\sum_{k=1}^{n} k^2 = \frac{n(n+1)(2n+1)}{6}; \quad \sum_{k=1}^{n} k^3 = \left(\frac{n(n+1)}{2}\right)^2;$$

$$\sum_{k=0}^{n} (2k+1) = (n+1)^2;$$

$$\sin A + \sin(A+B) + \cdots + \sin(A+(n-1)B)$$

$$= \frac{\sin\left(\frac{1}{2}nB\right)}{\sin\left(\frac{1}{2}B\right)} \sin\left(A + \frac{n-1}{2}B\right), \quad B \neq 0.$$

Arithmetic progression

$$a + (a+d) + (a+2d) + \cdots + (a+(n-1)d) = n\left(a + \tfrac{n-1}{2}d\right).$$

Geometric progression

$$a + ar + ar^2 + \cdots + ar^{n-1} = a\frac{1-r^n}{1-r}, \quad r \neq 1.$$

Binomial Theorem

$$(1+x)^n = \sum_{k=0}^{n}\binom{n}{k}x^k = 1 + nx + \frac{n(n-1)}{2!}x^2 + \cdots + x^n;$$

$$(a+b)^n = \sum_{k=0}^{n}\binom{n}{k}a^{n-k}b^k$$

$$= a^n + na^{n-1}b + \frac{n(n-1)}{2!}a^{n-2}b^2 + \cdots + b^n.$$

It follows that the sum of the geometric series $\sum_{k=0}^{\infty} ar^k$, for $|r| < 1$, is $\frac{a}{1-r}$.

Here, $\binom{n}{k} = \frac{n!}{k!(n-k)!}$ and $0! = 1$.

Appendix 2: Standard derivatives and primitives

$f(x)$	$f'(x)$	Domain
k	0	$\mathbb{R}$
x	1	$\mathbb{R}$
$x^n, n \in \mathbb{Z} - \{0\}$	nx^{n-1}	$\mathbb{R}$
$x^\alpha, \alpha \in \mathbb{R}$	$\alpha x^{\alpha-1}$	$(0, \infty)$
$a^x, a > 0$	$a^x \log_e a$	$\mathbb{R}$
x^x	$x^x (1 + \log_e x)$	$(0, \infty)$
$\sin x$	$\cos x$	$\mathbb{R}$
$\cos x$	$-\sin x$	$\mathbb{R}$
$\tan x$	$\sec^2 x$	$\mathbb{R} - \left\{ \left(n + \frac{1}{2}\right)\pi : n \in \mathbb{Z}\right\}$
$\operatorname{cosec} x$	$-\operatorname{cosec} x \cot x$	$\mathbb{R} - \{n\pi : n \in \mathbb{Z}\}$
$\sec x$	$\sec x \tan x$	$\mathbb{R} - \left\{ \left(n + \frac{1}{2}\right)\pi : n \in \mathbb{Z}\right\}$
$\cot x$	$-\operatorname{cosec}^2 x$	$\mathbb{R} - \{n\pi : n \in \mathbb{Z}\}$
$\sin^{-1} x$	$\frac{1}{\sqrt{1-x^2}}$	$(-1, 1)$
$\cos^{-1} x$	$\frac{-1}{\sqrt{1-x^2}}$	$(-1, 1)$
$\tan^{-1} x$	$\frac{1}{1+x^2}$	$\mathbb{R}$
e^x	e^x	$\mathbb{R}$
$\log_e x$	$\frac{1}{x}$	$(0, \infty)$
$\sinh x$	$\cosh x$	$\mathbb{R}$
$\cosh x$	$\sinh x$	$\mathbb{R}$
$\tanh x$	$\operatorname{sech}^2 x$	$\mathbb{R}$
$\sinh^{-1} x$	$\frac{1}{\sqrt{1+x^2}}$	$\mathbb{R}$
$\cosh^{-1} x$	$\frac{1}{\sqrt{x^2-1}}$	$(1, \infty)$
$\tanh^{-1} x$	$\frac{1}{1-x^2}$	$(-1, 1)$

$f(x)$	Primitive $F(x)$	Domain
$x^n, n \in \mathbb{Z} - \{-1\}$	$\frac{x^{n+1}}{n+1}$	$\mathbb{R}$
$x^\alpha, \alpha \in \mathbb{R} - \{-1\}$	$\frac{x^{\alpha+1}}{\alpha+1}$	$(0, \infty)$
$a^x, a > 0$	$\frac{a^x}{\log_e a}$	$\mathbb{R}$
$\sin x$	$-\cos x$	$\mathbb{R}$
$\cos x$	$\sin x$	$\mathbb{R}$
$\tan x$	$\log_e (\sec x)$	$\left(-\frac{1}{2}\pi, \frac{1}{2}\pi\right)$
e^x	e^x	$\mathbb{R}$
$\frac{1}{x}$	$\log_e x$	$(0, \infty)$
$\frac{1}{x}$	$\log_e |x|$	$(-\infty, 0)$
$\log_e x$	$x \log_e x - x$	$(0, \infty)$
$\sinh x$	$\cosh x$	$\mathbb{R}$
$\cosh x$	$\sinh x$	$\mathbb{R}$
$\tanh x$	$\log_e (\cosh x)$	$\mathbb{R}$
$\frac{1}{a^2-x^2}, a > 0$	$\frac{1}{2a} \log_e\left(\frac{a+x}{a-x}\right)$	$(-a, a)$
$\frac{1}{a^2+x^2}, a > 0$	$\frac{1}{a} \tan^{-1}\left(\frac{x}{a}\right)$	$\mathbb{R}$
$\frac{1}{\sqrt{a^2-x^2}}, a > 0$	$\begin{cases} \sin^{-1}\left(\frac{x}{a}\right) \\ -\cos^{-1}\left(\frac{x}{a}\right) \end{cases}$	$(-a, a)$ $(-a, a)$
$\frac{1}{\sqrt{x^2-a^2}}, a > 0$	$\begin{cases} \log_e\left(x + \sqrt{x^2 - a^2}\right) \\ \cosh^{-1}\left(\frac{x}{a}\right) \end{cases}$	(a, ∞) (a, ∞)
$\frac{1}{\sqrt{a^2+x^2}}, a > 0$	$\begin{cases} \log_e\left(x + \sqrt{a^2 + x^2}\right) \\ \sinh^{-1}\left(\frac{x}{a}\right) \end{cases}$	$\mathbb{R}$ $\mathbb{R}$
$\sqrt{a^2 - x^2}, a > 0$	$\frac{1}{2}x\sqrt{a^2 - x^2} + \frac{1}{2}a^2 \sin^{-1}\left(\frac{x}{a}\right)$	$(-a, a)$
$\sqrt{x^2 - a^2}, a > 0$	$\frac{1}{2}x\sqrt{x^2 - a^2} - \frac{1}{2}a^2 \log_e\left(x + \sqrt{x^2 - a^2}\right)$	(a, ∞)
$\sqrt{a^2 + x^2}, a > 0$	$\frac{1}{2}x\sqrt{a^2 + x^2} + \frac{1}{2}a^2 \log_e\left(x + \sqrt{a^2 + x^2}\right)$	$\mathbb{R}$
$e^{ax} \sin bx, a,b \neq 0$	$\frac{e^{ax}}{a^2+b^2}(a \sin bx - b \cos bx)$	$\mathbb{R}$
$e^{ax} \cos bx, a,b \neq 0$	$\frac{e^{ax}}{a^2+b^2}(a \cos bx + b \sin bx)$	$\mathbb{R}$

Appendix 3: The first 1000 decimal places of $\sqrt{2}$, e and π

More than a million decimal places of these numbers, and various similar commonly arising numbers, are easily available on the internet.

Numerical Analysts enjoy adding more digits to such decimals, using ever-more sophisticated computing techniques!

$\sqrt{2} = 1.$

4142135623	7309504880	1688724209	6980785696	7187537694	8073176679
7379907324	7846210703	8850387534	3276415727	3501384623	0912297024
9248360558	5073721264	4121497099	9358314132	2266592750	5592755799
9505011527	8206057147	0109559971	6059702745	3459686201	4728517418
6408891986	0955232923	0484308714	3214508397	6260362799	5251407989
6872533965	4633180882	9640620615	2583523950	5474575028	7759961729
8355752203	3753185701	1354374603	4084988471	6038689997	0699004815
0305440277	9031645424	7823068492	9369186215	8057846311	1596668713
0130156185	6898723723	5288509264	8612494977	1542183342	0428568606
0146824720	7714358548	7415565706	9677653720	2264854470	1585880162
0758474922	6572260020	8558446652	1458398893	9443709265	9180031138
8246468157	0826301005	9485870400	3186480342	1948972782	9064104507
2636881313	7398552561	1732204024	5091227700	2269411275	7362728049
5738108967	5040183698	6836845072	5799364729	0607629969	4138047565
4823728997	1803268024	7442062926	9124859052	1810044598	4215059112
0249441341	7285314781	0580360337	1077309182	8693147101	7111168391
6581726889	4197587165	8215212822	9518488472	...	

For most practical purposes

$\sqrt{2} \simeq 1.414.$

$e = 2.$

7182818284	5904523536	0287471352	6624977572	4709369995	9574966967
6277240766	3035354759	4571382178	5251664274	2746639193	2003059921
8174135966	2904357290	0334295260	5956307381	3232862794	3490763233
8298807531	9525101901	1573834187	9307021540	8914993488	4167509244
7614606680	8226480016	8477411853	7423454424	3710753907	7744992069
5517027618	3860626133	1384583000	7520449338	2656029760	6737113200
7093287091	2744374704	7230696977	2093101416	9283681902	5515108657
4637721112	5238978442	5056953696	7707854499	6996794686	4454905987
9316368892	3009879312	7736178215	4249992295	7635148220	8269895193
6680331825	2886939849	6465105820	9392398294	8879332036	2509443117
3012381970	6841614039	7019837679	3206832823	7646480429	5311802328
7825098194	5581530175	6717361332	0698112509	9618188159	3041690351
5988885193	4580727386	6738589422	8792284998	9208680582	5749279610
4841984443	6346324496	8487560233	6248270419	7862320900	2160990235
3043699418	4914631409	3431738143	6405462531	5209618369	0888707016
7683964243	7814059271	4563549061	3031072085	1038375051	0115747704
1718986106	8739696552	1267154688	9570350354	...	

For most practical purposes
$e \simeq 2.718.$

A useful fact sometimes is that $e^3 \simeq 20$.

361

$\pi = 3.$

1415926535	8979323846	2643383279	5028841971	6939937510	5820974944
5923078164	0628620899	8628034825	3421170679	8214808651	3282306647
0938446095	5058223172	5359408128	4811174502	8410270193	8521105559
6446229489	5493038196	4428810975	6659334461	2847564823	3786783165
2712019091	4564856692	3460348610	4543266482	1339360726	0249141273
7245870066	0631558817	4881520920	9628292540	9171536436	7892590360
0113305305	4882046652	1384146951	9415116094	3305727036	5759591953
0921861173	8193261179	3105118548	0744623799	6274956735	1885752724
8912279381	8301194912	9833673362	4406566430	8602139494	6395224737
1907021798	6094370277	0539217176	2931767523	8467481846	7669405132
0005681271	4526356082	7785771342	7577896091	7363717872	1468440901
2249534301	4654958537	1050792279	6892589235	4201995611	2129021960
8640344181	5981362977	4771309960	5187072113	4999999837	2978049951
0597317328	1609631859	5024459455	3469083026	4252230825	3344685035
2619311881	7101000313	7838752886	5875332083	8142061717	7669147303
5982534904	2875546873	1159562863	8823537875	9375195778	1857780532
1712268066	1300192787	6611195909	2164201989	...	

For most practical purposes $\pi \simeq 3.1416.$

A useful fact sometimes is that $\pi^2 \simeq 10.$

Appendix 4: Solutions to the problems

Chapter 1

Section 1.1

1. $-1 < -\frac{17}{20} < -\frac{45}{53} < 0 < \frac{45}{53} < \frac{17}{20} < 1.$

2. Let a and b be two distinct rational numbers, where $a < b$. Let $c = \frac{1}{2}(a+b)$; then c is rational, and

$$c - a = \frac{1}{2}(b-a) > 0, \quad \text{and}$$

$$b - c = \frac{1}{2}(b-a) > 0,$$

so that $a < c < b$.

3. $\frac{1}{7} = 0.142857142857\ldots; \frac{2}{13} = 0.153846153846\ldots.$

4. (a) Let $x = 0.\overline{231}$.

Multiplying both sides by 10^3, we obtain

$$1000x = 231.\overline{231} = 231 + x.$$

Hence

$$999x = 231 \Rightarrow x = \frac{231}{999} = \frac{77}{333}.$$

(b) Let $x = 0.\overline{81}$.

Multiplying both sides by 10^2, we obtain

$$100x = 81.\overline{81} = 81 + x.$$

Hence

$$99x = 81 \Rightarrow x = \frac{81}{99} = \frac{9}{11}.$$

Thus

$$2.2\overline{81} = 2 + \frac{2}{10} + \frac{9}{110} = \frac{251}{110}.$$

5. $\frac{17}{20} = 0.85$ and $\frac{45}{53} = 0.84\ldots$, so that $\frac{45}{53} < \frac{17}{20}$.

6. Suppose that there exists a rational number x such that $x^3 = 2$. Then we can write $x = \frac{p}{q}$. By cancelling, if necessary, we may assume that p and q have no common factor. The equation $x^3 = 2$ now becomes

$$p^3 = 2q^3.$$

Now, the cube of an odd number is odd, because

$$(2k+1)^3 = 8k^3 + 12k^2 + 6k + 1$$
$$= 2(4k^3 + 6k^2 + 3k) + 1,$$

and so p must be even. Hence we can write $p = 2r$, say. Our equation now becomes

$$(2r)^3 = 2q^3,$$

so that

$$q^3 = 4r^3.$$

Hence q^3 and so q is also even, so that p and q do have a common factor 2, which contradicts our earlier statement that p and q have no common factors.

Thus, no such number x exists.

7. We may take, for example, $x = 0.34$ and $y = 0.34001000100001\ldots$.

8. Let a and b have decimal representations

$$a = a_0 \cdot a_1 a_2 a_3 \ldots \quad \text{and} \quad b = b_0 \cdot b_1 b_2 b_3 \ldots,$$

Here a_0, b_0 are non-negative integers, and $a_1, b_1, a_2, b_2, \ldots$ are digits.

where we arrange that a does not end in recurring 9s, whereas b does not terminate (this latter can be arranged by replacing a terminating representation by an equivalent representation that ends in recurring 9s).

Since $a < b$, there must be some integer n such that

$$a_0 = b_0, \, a_1 = b_1, \, \ldots, \, a_{n-1} = b_{n-1}, \text{ but } a_n < b_n.$$

Then $x = a_0 \cdot a_1 a_2 a_3 \ldots a_{n-1} b_n$ is rational, and $a < x < b$ as required.

Next, let $c = \frac{1}{2}(x + b)$, so that $x < c < b$. By repeating the same procedure, this time to the interval between the numbers c and b, we can find a rational number y with $c < y < b$.

Note that x does not end in recurring 9s, from the way in which it was constructed.

We have thus constructed two distinct rational numbers between a and b, as required.

Section 1.2

1. **Rule 1** For any $a, b \in \mathbb{R}$, $\quad a \leq b \Leftrightarrow b - a \geq 0$.
   Rule 2 For any $a, b, c \in \mathbb{R}$, $\quad a \leq b \Leftrightarrow a + c \leq b + c$.
   Rule 3
   - For any $a, b \in \mathbb{R}$ and any $c > 0$, $\quad a \leq b \Leftrightarrow ac \leq bc$;
   - For any $a, b \in \mathbb{R}$ and any $c < 0$, $\quad a \leq b \Leftrightarrow ac \geq bc$.

Remark

Note that the following results are NOT true in general:

- For any $a, b \in \mathbb{R}$ and any $c \geq 0$, $\quad a \leq b \Leftrightarrow ac \leq bc$;
- For any $a, b \in \mathbb{R}$ and any $c \leq 0$, $\quad a \leq b \Leftrightarrow ac \geq bc$.

For, if $c = 0$, then we can make no assertion as to whether $a \leq b$ or $a \geq b$ from the information that $ac \leq bc$ or $ac \geq bc$. To see this, take in turn $a = 2$, $b = 3$ and $a = 2$, $b = 1$.

Rule 4 (Reciprocal Rule) For any positive $a, b \in \mathbb{R}$, $a \leq b \Leftrightarrow \frac{1}{a} \geq \frac{1}{b}$.
Rule 5 (Power Rule) For any non-negative $a, b \in \mathbb{R}$, and any $p > 0$, $a \leq b \Leftrightarrow a^p \leq b^p$.

2. (a) $x + 3 > 5$.
   (b) $2 - x < 0$.
   (c) $5x + 2 > 12$.
   (d) $\frac{-1}{(5x+2)} > \frac{-1}{12}$.

3. (a) Rearranging the inequality, we obtain

$$\frac{4x - x^2 - 7}{x^2 - 1} \geq 3 \Leftrightarrow \frac{4x - x^2 - 7}{x^2 - 1} - 3 \geq 0 \quad \Leftrightarrow \frac{4x - 4x^2 - 4}{x^2 - 1} \geq 0$$

$$\Leftrightarrow \frac{x^2 - x + 1}{x^2 - 1} \leq 0 \qquad \Leftrightarrow \frac{\left(x - \frac{1}{2}\right)^2 + \frac{3}{4}}{x^2 - 1} \leq 0.$$

Since $\left(x-\frac{1}{2}\right)^2+\frac{3}{4}>0$, for all x, the inequality holds if and only if $x^2-1<0$. Hence the solution set is

$$\left\{x:\frac{4x-x^2-7}{x^2-1}\geq 3\right\}=(-1,1).$$

(b) Rearranging the inequality, we obtain

$$2x^2\geq(x+1)^2\Leftrightarrow 2x^2\geq x^2+2x+1$$
$$\Leftrightarrow x^2-2x-1\geq 0$$
$$\Leftrightarrow(x-1)^2\geq 2.$$

Hence the solution set is

$$\left\{x:2x^2\geq(x+1)^2\right\}=\left\{x:x-1\leq-\sqrt{2}\right\}\cup\left\{x:x-1\geq\sqrt{2}\right\}$$
$$=\left(-\infty,1-\sqrt{2}\right]\cup\left[1+\sqrt{2},\infty\right).$$

4. We can obtain an equivalent inequality by squaring, provided that both sides are non-negative.

Now, $\sqrt{2x^2-2}$ is defined when $2x^2-2\geq 0$; that is, for $x^2\geq 1$. So, $\sqrt{2x^2-2}$ is defined and non-negative if x lies in $(-\infty,-1]\cup[1,\infty)$.

Hence, for $x\geq 1$

$$\sqrt{2x^2-2}>x\Leftrightarrow 2x^2-2>x^2$$
$$\Leftrightarrow x^2>2.$$

So the part of the solution set in $[1,\infty)$ is $\left(\sqrt{2},\infty\right)$.

Suppose, next, that $x\leq-1$. Then

$$\sqrt{2x^2-2}\geq 0>x,$$

so that the whole of $(-\infty,-1]$ lies in the solution set. Hence the complete solution set is

$$\left\{x:\sqrt{2x^2-2}>x\right\}=(-\infty,-1]\cup\left(\sqrt{2},\infty\right).$$

5. (a) $|2x^2-13|<5\Leftrightarrow -5<2x^2-13<5$ (by Rule 6)
$$\Leftrightarrow 8<2x^2<18$$
$$\Leftrightarrow 4<x^2<9$$
$$\Leftrightarrow 2<|x|<3.$$

Hence the solution set is
$$\left\{x:|2x^2-13|<5\right\}=(-3,-2)\cup(2,3).$$

(b) $|x-1|\leq 2|x+1|\Leftrightarrow(x-1)^2\leq 4(x+1)^2$
$$\Leftrightarrow x^2-2x+1\leq 4x^2+8x+4$$
$$\Leftrightarrow 0\leq 3x^2+10x+3$$
$$\Leftrightarrow 0\leq(3x+1)(x+3).$$

Hence the solution set is
$$\left\{x:|x-1|\leq 2|x+1|\right\}=(-\infty,-3]\cup\left[-\frac{1}{3},\infty\right).$$

Section 1.3

1. (a) Suppose that $|a| \leq \frac{1}{2}$.
   The Triangle Inequality gives

   $$|a + 1| \leq |a| + 1$$
   $$\leq \frac{1}{2} + 1 \quad (\text{since } |a| \leq \frac{1}{2})$$
   $$= \frac{3}{2}.$$

   Hence

   $$|a + 1| \leq \frac{3}{2}.$$

   (b) Suppose that $|b| < \frac{1}{2}$.
   The 'reverse form' of the Triangle Inequality gives

   $$\left|b^3 - 1\right| \geq \left||b^3| - 1\right|$$
   $$= \left||b|^3 - 1\right|$$
   $$\geq 1 - |b|^3.$$

   But

   $$|b| < \frac{1}{2} \Rightarrow |b|^3 < \frac{1}{8} \Rightarrow 1 - |b|^3 > \frac{7}{8};$$

   so, from the previous chain of inequalities, we deduce that

   $$\left|b^3 - 1\right| > \frac{7}{8}.$$

2. Rearranging the inequality, we obtain

   $$\frac{3n}{n^2 + 2} < 1 \Leftrightarrow 3n < n^2 + 2$$
   $$\Leftrightarrow 0 < n^2 - 3n + 2$$
   $$\Leftrightarrow 0 < (n - 1)(n - 2),$$

   and this final inequality certainly holds for $n > 2$.

3. The result is true for all those a and b for which $a + b \leq 0$.
   We now consider those a and b for which $a + b > 0$. Rearranging the given inequality, we obtain

   $$\frac{a + b}{\sqrt{2}} \leq \sqrt{a^2 + b^2} \Leftrightarrow \frac{(a + b)^2}{2} \leq a^2 + b^2$$
   $$\Leftrightarrow a^2 + 2ab + b^2 \leq 2a^2 + 2b^2$$
   $$\Leftrightarrow 0 \leq a^2 - 2ab + b^2$$
   $$\Leftrightarrow 0 \leq (a - b)^2;$$

   and this final inequality certainly holds.
   This completes the proof of the required inequality.

4. (a) Using the rules for rearranging inequalities, we obtain

   $$\frac{1}{2}\left(a + \frac{2}{a}\right) < a \Leftrightarrow \frac{1}{2} \times \frac{2}{a} < a - \frac{1}{2}a$$
   $$\Leftrightarrow \frac{1}{a} < \frac{1}{2}a$$
   $$\Leftrightarrow 2 < a^2 \quad (\text{since } a > 0).$$

Since the final inequality is true, the first inequality must also be true. Hence

$$\frac{1}{2}\left(a+\frac{2}{a}\right)<a.$$

(b) In Example 3 and the subsequent remark, we saw that, if $a\neq b$, then

$$ab<\left(\frac{a+b}{2}\right)^2. \qquad (*)$$

Now let $b=\frac{2}{a}$. Then $a\neq b$, since $a>\frac{2}{a}$ (as $a^2>2$); so, by $(*)$, it follows that

$$a\times\frac{2}{a}<\left(\frac{1}{2}\left(a+\frac{2}{a}\right)\right)^2,$$

which gives the required inequality.

Alternatively, use a direct argument after squaring the expression $\frac{1}{2}\left(a+\frac{2}{a}\right)$.

5. Since $c,d\geq0$, we can choose non-negative numbers a and b with $c=a^2$ and $d=b^2$. Substituting into the result $\sqrt{a^2+b^2}\leq a+b$, for $a,b\geq0$, of Example 4, we obtain

$$\sqrt{c+d}\leq\sqrt{c}+\sqrt{d}.$$

6. In Problem 5 we saw that

$$\sqrt{c+d}\leq\sqrt{c}+\sqrt{d}, \quad\text{for numbers } c,d\geq0. \qquad (*)$$

Following the good advice in the margin note before the Problem, we use this!
For $a,b,c\geq0$, we deduce from $(*)$ that

$$\begin{aligned}
\sqrt{a+b+c}&=\sqrt{a+(b+c)}\\
&\leq\sqrt{a}+\sqrt{b+c} &&\text{(by an application of }(*))\\
&\leq\sqrt{a}+\sqrt{b}+\sqrt{c} &&\text{(by a further application of }(*)).
\end{aligned}$$

7. If we substitute $x=\frac{1}{n}$ in the Binomial Theorem for $(1+x)^n$, when $n\geq2$ we get

$$\begin{aligned}
\left(1+\frac{1}{n}\right)^n&=1+n\left(\frac{1}{n}\right)+\frac{n(n-1)}{2!}\left(\frac{1}{n}\right)^2+\cdots+\left(\frac{1}{n}\right)^n\\
&\geq1+1+\frac{n-1}{2n}\\
&=2+\frac{1}{2}-\frac{1}{2n}=\frac{5}{2}-\frac{1}{2n}.
\end{aligned}$$

When $n=1$, the inequality also holds. This completes the proof.

8. Let $P(n)$ be the statement

$$P(n):4^n>n^4.$$

STEP 1 First we show that $P(5)$ is true: $4^5\geq5^4$.

Since $4^5=1024$ and $5^4=625$, $P(5)$ is certainly true.

STEP 2 We now assume that $P(k)$ holds for some $k\geq5$, and deduce that $P(k+1)$ is then true.

So, we are assuming that $4^k>k^4$. Multiplying this inequality by 4 we get   This assumption is just $P(k)$.

$$4^{k+1}>4k^4,$$

so it is therefore sufficient for our purposes to prove that $4k^4\geq(k+1)^4$.   Since $P(k+1)$ is:
Now $4^{k+1}>(k+1)^4$.

$$4k^4\geq(k+1)^4\Leftrightarrow4\geq\left(1+\frac{1}{k}\right)^4. \qquad (*)$$

As k increases, the expression $1+\frac{1}{k}$ decreases. Since $k\geq5$, $1+\frac{1}{k}\leq1+\frac{1}{5}=\frac{6}{5}$; it follows that

$$\left(1+\frac{1}{k}\right)^4 \le \left(\frac{6}{5}\right)^4$$

$$=\frac{1296}{625}=2.0736<4.$$

Thus the inequality $(*)$ certainly holds for $k \ge 5$, and so it follows that $4^{k+1} \ge (k+1)^4$ also holds for $k \ge 5$.

In other words: $P(k)$ true for some $k \ge 5 \Rightarrow P(k+1)$ true.

It follows, by the Principle of Mathematical Induction, that $4^n > n^4$, for $n \ge 5$.

9. Substituting $x=-\frac{1}{(2n)}$ into Bernouilli's Inequality $(1+x)^n \ge 1+nx$ (which we may do since $-\frac{1}{(2n)} \ge -1$ for all natural numbers n), we obtain

$$\left(1-\frac{1}{2n}\right)^n \ge 1-n \times \frac{1}{2n}$$

$$=\frac{1}{2}.$$

If we take the nth root of both sides of this final inequality (which is permissible, by the Power Rule), we find that

$$1-\frac{1}{2n} \ge \frac{1}{2^{\frac{1}{n}}},$$

so that

$$2^{\frac{1}{n}} \ge \frac{1}{1-\frac{1}{2n}} = \frac{2n}{2n-1}$$

$$=1+\frac{1}{2n-1}.$$

10. If we apply the Cauchy–Schwarz Inequality, with $\sqrt{a_k}$ in place of a_k and $\frac{1}{\sqrt{a_k}}$ in place of b_k, we have

$$(a_1+a_2+\cdots+a_n)\left(\frac{1}{a_1}+\frac{1}{a_2}+\cdots+\frac{1}{a_n}\right)$$

$$\ge \left(\sqrt{a_1} \times \frac{1}{\sqrt{a_1}} + \sqrt{a_2} \times \frac{1}{\sqrt{a_2}} + \cdots + \sqrt{a_n} \times \frac{1}{\sqrt{a_n}}\right)^2$$

$$=(1+1+\cdots+1)^2 \quad \text{(with } n \text{ terms in the bracket)}$$

$$=n^2.$$

11. Applying Theorem 3 to the $n+1$ positive numbers $1, 1+\frac{1}{n}, 1+\frac{1}{n}, \ldots, 1+\frac{1}{n}$, we obtain

$$\left(\left(1+\frac{1}{n}\right)^n\right)^{\frac{1}{n+1}} \le \frac{1}{n+1} \times \left(n+1+n \times \frac{1}{n}\right)$$

$$=\frac{n+2}{n+1}$$

$$=1+\frac{1}{n+1};$$

taking the $(n+1)$th power of this last inequality, by the Power Rule, we deduce that

$$\left(1+\frac{1}{n}\right)^n \le \left(1+\frac{1}{n+1}\right)^{n+1}.$$

Section 1.4

1. (a)

E_1 is bounded above. For example, $M = 1$ is an upper bound of E_1, since

$$x \le 1, \quad \text{for all } x \in E_1.$$

Also, $\max E_1 = 1$, since $1 \in E_1$.

(b)

E_2 is bounded above. For example, $M = 1$ is an upper bound of E_2, since

$$1 - \frac{1}{n} \le 1, \quad \text{for } n = 1, 2, \dots.$$

However, E_2 has no maximum element. If $x \in E_2$, then $x = 1 - \frac{1}{n}$ for some positive integer n; and so there is another element of E_2, such as $1 - \frac{1}{n+1}$, with

$$1 - \frac{1}{n} < 1 - \frac{1}{n+1} \quad \left(\text{since } \frac{1}{n} > \frac{1}{n+1} \right).$$

Hence x is not a maximum element of E_2.

(c)

The set E_3 is not bounded above, and so it cannot have a maximum element. For each number M, there is a positive integer n such that $n^2 > M$ (for instance, take $n > M$, which implies that $n^2 \ge n > M$).

Hence M cannot be an upper bound of E_3.

2. (a) The set $E_1 = (-\infty, 1]$ is not bounded below, and so it cannot have a minimum element. For each number m, there is a (negative) number x in E_1 such that $x < m$. Since $x \in E_1$, m cannot be a lower bound of E_1.

(b) E_2 is bounded below by 0, since

$$1 - \frac{1}{n} \ge 0, \quad \text{for } n = 1, 2 \dots.$$

Also, $0 \in E_2$, and so $\min E_2 = 0$.

(c) E_3 is bounded below by 1, since

$$n^2 \ge 1, \quad \text{for } n = 1, 2 \dots.$$

Also, $1 \in E_3$, and so $\min E_3 = 1$.

3. f is increasing on the interval $[-3, -2)$. Since $-3 \le x < -2$ so that $x^2 \in (4, 9]$, we have $f(x) = \frac{1}{x^2} \in \left[\frac{1}{9}, \frac{1}{4} \right)$. Hence f is bounded above and bounded below.

Next, since $f(-3) = \frac{1}{9}$ and $\frac{1}{9}$ is a lower bound for f on the interval $[-3, -2)$, it follows that f has a minimum value of $\frac{1}{9}$ on this interval.

Finally, $\frac{1}{4}$ is an upper bound for f on the interval $[-3, -2)$ but there is no point x in $[-3, -2)$ for which $f(x) = \frac{1}{4}$. So $\frac{1}{4}$ cannot be a maximum of f on the interval.

However, if y is any number in $\left[\frac{1}{9}, \frac{1}{4} \right)$, there is a number $x > -\frac{1}{\sqrt{y}}$ in $\left(-\frac{1}{\sqrt{y}}, -2 \right) \subset [-3, -2)$ such that $f(x) = \frac{1}{x^2} > y$, so that no number in $\left[\frac{1}{9}, \frac{1}{4} \right)$ will serve as a maximum of f on the interval. It follows that f has no maximum value on $[-3, -2)$.

4. (a) The set $E_1 = (-\infty, 1]$ has a maximum element 1, and so

$$\sup E_1 = \max E_1 = 1.$$

(b) We know that $E_2 = \{1 - \frac{1}{n} : n = 1, 2, \ldots\}$ is bounded above by 1, since

$$1 - \frac{1}{n} \leq 1, \quad \text{for } n = 1, 2 \ldots.$$

To show that $M = 1$ is the least upper bound of E_2, we prove that, if $M' < 1$, then there is an element $1 - \frac{1}{n}$ of E_2 such that

$$1 - \frac{1}{n} > M'.$$

However, since $M' < 1$, we have

$$1 - \frac{1}{n} > M' \Leftrightarrow 1 - M' > \frac{1}{n}$$

$$\Leftrightarrow n > \frac{1}{1 - M'} \quad (\text{since } 1 - M' > 0).$$

Choosing a positive integer n so large that $n > \frac{1}{1-M'}$, we obtain $1 - \frac{1}{n} > M'$, as required. Hence 1 is the least upper bound of E_2.

(c) The set $E_3 = \{n^2 : n = 1, 2, \ldots\}$ is not bounded above, and so it cannot have a least upper bound.

5. (a) The set $E_1 = (1, 5]$ is bounded below by 1, since

$$1 \leq x, \quad \text{for all } x \in E_1.$$

To show that $m = 1$ is the greatest lower bound of E_1, we prove that, if $m' > 1$, then there is an element x in E_1 which is less than m'. Since $m' > 1$, there is a number x of the form $x = 1.00 \ldots 01$ such that

$$1 < x < m',$$

and clearly $x \in E_1$.

Hence 1 is the greatest lower bound of E_1.

(b) The set $E_2 = \{\frac{1}{n^2} : n = 1, 2, \ldots\}$ is bounded below by 0, since

$$0 < \frac{1}{n^2}, \quad \text{for } n = 1, 2, \ldots.$$

To show that $m = 0$ is the greatest lower bound of E_2, we prove that, if $m' > 0$, then there is an element $\frac{1}{n^2}$ in E_2 such that $\frac{1}{n^2} < m'$. Since $m' > 0$, we have

$$\frac{1}{n^2} < m' \Leftrightarrow n^2 > \frac{1}{m'}$$

$$\Leftrightarrow n > \frac{1}{\sqrt{m'}}.$$

Choosing a positive integer n so large that $n > \frac{1}{\sqrt{m'}}$, we obtain $\frac{1}{n^2} < m'$, as required. Hence 0 is the greatest lower bound of E_2.

6. f is decreasing on the interval $[1, 4)$. Since $1 \leq x < 4$ so that $x^2 \in [1, 16)$, we have $f(x) = \frac{1}{x^2} \in \left(\frac{1}{16}, 1\right]$. Hence f is bounded above by 1 and bounded below by $\frac{1}{16}$.

Next, since $f(1) = 1$ and 1 is an upper bound for f on the interval $[1, 4)$, it follows that f has least upper bound 1 on $[1, 4)$.

Finally, $\frac{1}{16}$ is a lower bound for f on the interval $[1, 4)$ but there is no point x in $[1, 4)$ for which $f(x) = \frac{1}{16}$.

However, if y is any number in $\left(\frac{1}{16}, 1\right)$, there is a number $x < \frac{1}{\sqrt{y}}$ in $[1, 4)$ such that $f(x) = \frac{1}{x^2} < y$, so that no number in $\left(\frac{1}{16}, 1\right)$ will serve as a lower bound of f on $[1, 4)$. It follows that f has $\frac{1}{16}$ as its greatest lower bound on $[1, 4)$.

Chapter 2

Section 2.1

1. (a) (i) 4, 7, 10, 13, 16;

  (ii) $\frac{1}{3}, \frac{1}{9}, \frac{1}{27}, \frac{1}{81}, \frac{1}{243}$;

  (iii) $-1, 2, -3, 4, -5$.

  (b) (i) $a_1 = 1$, $a_2 = 2$, $a_3 = 6$, $a_4 = 24$, $a_5 = 120$;

  (ii) $a_1 = 2$, $a_2 = 2.25$, $a_3 = 2.37$, $a_4 = 2.44$, $a_5 = 2.49$.

2. (a)

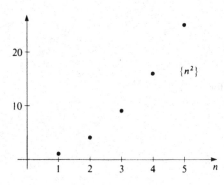

  (b)

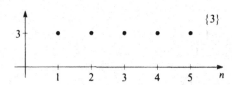

  (c)

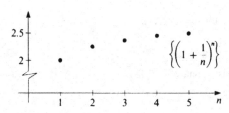

  (d)

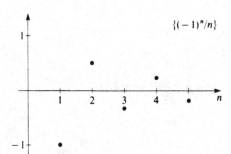

3. (a) $n!$ is monotonic, because

$$a_n = n! \quad \text{and} \quad a_{n+1} = (n+1)!,$$

so that

$$a_{n+1} - a_n = (n+1)! - n!$$
$$= n \times n! > 0, \quad \text{for } n = 1, 2, \ldots.$$

Thus $\{n!\}$ is increasing.

Alternatively, $a_n > 0$, for all n, and

$$\frac{a_{n+1}}{a_n} = \frac{(n+1)!}{n!} = n + 1 \geq 1, \quad \text{for } n = 1, 2, \ldots,$$

so that $\{n!\}$ is increasing.

(b) $\{2^{-n}\}$ is monotonic, because

$$a_n = 2^{-n} \quad \text{and} \quad a_{n+1} = 2^{-(n+1)},$$

so that

$$a_{n+1} - a_n = \frac{1}{2^{n+1}} - \frac{1}{2^n}$$
$$= \frac{1}{2^n}\left(\frac{1}{2} - 1\right) < 0, \quad \text{for } n = 1, 2, \ldots.$$

Thus $\{2^{-n}\}$ is decreasing.

Alternatively, $a_n > 0$, for all n, and

$$\frac{a_{n+1}}{a_n} = \frac{2^n}{2^{n+1}} = \frac{1}{2} < 1, \quad \text{for } n = 1, 2, \ldots,$$

so that $\{2^{-n}\}$ is decreasing.

(c) $\left\{n + \frac{1}{n}\right\}$ is monotonic, because

$$a_n = n + \frac{1}{n} \quad \text{and} \quad a_{n+1} = n + 1 + \frac{1}{n+1},$$

so that

$$a_{n+1} - a_n = \left(n + 1 + \frac{1}{n+1}\right) - \left(n + \frac{1}{n}\right)$$
$$= \frac{n(n+1) - 1}{n(n+1)} > 0, \quad \text{for } n = 1, 2, \ldots.$$

Thus $\left\{n + \frac{1}{n}\right\}$ is increasing.

4. (a) TRUE: $2^n > 1000$, for $n > 9$, since $\{2^n\}$ is increasing and $2^{10} = 1024$.

(b) FALSE: All the terms $a_1, a_3, a_5, \ldots$ are negative.

(c) TRUE: $\frac{1}{n} < 0.025$, for all $n > \frac{1}{0.025} = 40$.

(d) TRUE: $a_n > 0$ for all n, and $\frac{a_{n+1}}{a_n} = \frac{1}{4}\left(\frac{n+1}{n}\right)^4$.

Now

$$\frac{1}{4}\left(\frac{n+1}{n}\right)^4 \leq 1 \Leftrightarrow \left(\frac{n+1}{n}\right)^4 \leq 4 \quad \Leftrightarrow 1 + \frac{1}{n} \leq 4^{\frac{1}{4}}$$
$$\Leftrightarrow \frac{1}{n} \leq \sqrt{2} - 1 \quad \Leftrightarrow n \geq \frac{1}{\sqrt{2} - 1} \simeq 2.414.$$

So

$$\frac{a_{n+1}}{a_n} \leq 1, \quad \text{for } n \geq 3.$$

Hence

$$a_{n+1} \leq a_n, \quad \text{for } n \geq 3,$$

so that $\left\{\frac{n^4}{4^n}\right\}$ is eventually decreasing.

Section 2.2

1. (a) $\frac{1}{n} < \frac{1}{100} \Leftrightarrow n > 100$, by the Reciprocal Rule (for positive numbers). Hence we may take $X = 100$.

   Any value for X greater than 100 will also be valid.

   (b) $\frac{1}{n} < \frac{3}{1000} \Leftrightarrow n > \frac{1000}{3} = 333.3333\ldots$, by the Reciprocal Rule (for positive numbers). Hence we may take $X = 333$.

   Any value for X greater than 333 will also be valid.

2. (a)
$$\left| \frac{(-1)^n}{n^2} \right| < \frac{1}{100} \Leftrightarrow \frac{1}{n^2} < \frac{1}{100}$$
$$\Leftrightarrow n^2 > 100$$
$$\Leftrightarrow n > 10.$$

   Hence we may take $X = 10$.

   Any value for X greater than 10 will also be valid.

   (b)
$$\left| \frac{(-1)^n}{n^2} \right| < \frac{3}{1000} \Leftrightarrow \frac{1}{n^2} < \frac{3}{1000}$$
$$\Leftrightarrow n^2 > \frac{1000}{3}$$
$$\Leftrightarrow n > \sqrt{\frac{1000}{3}} \simeq 18.25.$$

   Hence we may take $X = 18$.

   Any value for X greater than 18 will also be valid.

3. (a) The sequence $\left\{ \frac{1}{2n-1} \right\}$ is a null sequence.

   To prove this, we want to show that:

   for each positive number ε, there is a number X such that
$$\left| \frac{1}{2n-1} \right| < \varepsilon, \quad \text{for all } n > X. \qquad (*)$$

   We know that
$$\left| \frac{1}{2n-1} \right| < \varepsilon \Leftrightarrow \frac{1}{2n-1} < \varepsilon \quad (\text{since } 2n - 1 > 0)$$
$$\Leftrightarrow 2n - 1 > \frac{1}{\varepsilon}$$
$$\Leftrightarrow n > \frac{1}{2}\left(1 + \frac{1}{\varepsilon}\right);$$

   it follows that $(*)$ holds if we take $X = \frac{1}{2}\left(1 + \frac{1}{\varepsilon}\right)$. Hence $\left\{ \frac{1}{2n-1} \right\}$ is null.

   (b) The sequence $\left\{ \frac{(-1)^n}{10} \right\}$ is not a null sequence.

   To prove this, we must find a positive value of ε such that the sequence does not eventually lie in the horizontal strip on the sequence diagram from $-\varepsilon$ up to ε. We can take $\varepsilon = \frac{1}{20}$, as the following sequence diagram shows:

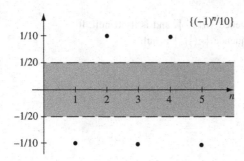

There is NO value of X such that the following statement holds

$$\left|\frac{(-1)^n}{10}\right| < \frac{1}{20}, \quad \text{for all } n > X.$$

(c) The sequence $\left\{\frac{(-1)^n}{n^4+1}\right\}$ is a null sequence.
To prove this, we want to show that:

for each positive number ε, there is a number X such that

$$\left|\frac{(-1)^n}{n^4+1}\right| < \varepsilon, \quad \text{for all } n > X. \qquad (*)$$

We know that

$$\left|\frac{(-1)^n}{n^4+1}\right| < \varepsilon \Leftrightarrow \frac{1}{n^4+1} < \varepsilon$$

$$\Leftrightarrow n^4 + 1 > \frac{1}{\varepsilon}$$

$$\Leftrightarrow n^4 > \frac{1}{\varepsilon} - 1.$$

Now, if $\varepsilon \geq 1$, then $\frac{1}{\varepsilon} - 1 \leq 0$, so that

$$n^4 > \frac{1}{\varepsilon} - 1, \quad \text{for } n = 1, 2, \ldots;$$

hence $(*)$ holds with $X = 1$.
On the other hand, if $0 < \varepsilon < 1$, then $\frac{1}{\varepsilon} - 1 > 0$, so that

$$n^4 > \frac{1}{\varepsilon} - 1 \Leftrightarrow n > \left(\frac{1}{\varepsilon} - 1\right)^{\frac{1}{4}};$$

hence $(*)$ holds with $X = \left(\frac{1}{\varepsilon} - 1\right)^{\frac{1}{4}}$.
Thus $(*)$ holds in either case. Hence $\left\{\frac{(-1)^n}{n^4+1}\right\}$ is null.

4. (a) We know that $\left\{\frac{1}{2n-1}\right\}$ is null, and so $\left\{\frac{1}{(2n-1)^3}\right\}$ is also null, by the Power Rule.

(b) The sequences $\left\{\frac{1}{n}\right\}$ and $\left\{\frac{1}{2n-1}\right\}$ are null, so $\left\{\frac{6}{\sqrt[3]{n}}\right\}$ and $\left\{\frac{5}{(2n-1)^7}\right\}$ are also null, by the Power Rule and Multiple Rule.

Hence the sequence $\left\{\frac{6}{\sqrt[3]{n}} + \frac{5}{(2n-1)^7}\right\}$ is null, by the Sum Rule.

(c) The sequences $\left\{\frac{1}{n}\right\}$ and $\left\{\frac{1}{2n-1}\right\}$ are null, so $\left\{\frac{1}{n^4}\right\}$ and $\left\{\frac{1}{(2n-1)^{\frac{1}{3}}}\right\}$ are also null, by the Power Rule.

Hence $\left\{\frac{1}{3n^4(2n-1)^{\frac{1}{3}}}\right\}$ is also null, by the Product Rule and Multiple Rule.

5. Here, using the Hint, we have

$$n\left(\frac{1}{2}\right)^n = n \times \frac{1}{2^n}$$

$$\leq n \times \frac{1}{n^2} = \frac{1}{n}.$$

Thus, the sequence $\left\{\frac{1}{n}\right\}$ dominates the sequence $\left\{n\left(\frac{1}{2}\right)^n\right\}$, and is itself null. It follows, from the Squeeze Rule, that the sequence $\left\{n\left(\frac{1}{2}\right)^n\right\}$ is null.

6. (a) We guess that $\left\{\frac{1}{n^2+n}\right\}$ is dominated by $\left\{\frac{1}{n}\right\}$.
To check this, we have to show that

$$\frac{1}{n^2+n} \leq \frac{1}{n}, \quad \text{for } n = 1, 2, \ldots;$$

this certainly holds, because

$$n^2 + n \geq n, \quad \text{for } n = 1, 2, \ldots.$$

Since $\left\{\frac{1}{n}\right\}$ is null, we deduce that $\left\{\frac{1}{n^2+n}\right\}$ is null, by the Squeeze Rule.

(b) We guess that $\left\{\frac{(-1)^n}{n!}\right\}$ is dominated by $\left\{\frac{1}{n}\right\}$.

To check this, we have to show that

$$\left|\frac{(-1)^n}{n!}\right| \leq \frac{1}{n}, \quad \text{for } n = 1, 2, \ldots;$$

this certainly holds, because

$$\left|\frac{(-1)^n}{n!}\right| = \frac{1}{n!}, \quad \text{for } n = 1, 2, \ldots,$$

and

$$n! \geq n, \quad \text{for } n = 1, 2, \ldots.$$

Since $\left\{\frac{1}{n}\right\}$ is null, we deduce that $\left\{\frac{(-1)^n}{n!}\right\}$ is null, by the Squeeze Rule.

(c) We guess that $\left\{\frac{\sin n^2}{n^2 + 2^n}\right\}$ is dominated by $\left\{\frac{1}{n^2}\right\}$.

To check this, we have to show that

$$\left|\frac{\sin n^2}{n^2 + 2^n}\right| \leq \frac{1}{n^2}, \quad \text{for } n = 1, 2, \ldots;$$

this certainly holds, because

$$\left|\sin n^2\right| \leq 1, \quad \text{for } n = 1, 2, \ldots,$$

and

$$n^2 + 2^n \geq n^2, \quad \text{for } n = 1, 2, \ldots.$$

Since $\left\{\frac{1}{n^2}\right\}$ is null, we deduce that $\left\{\frac{\sin n^2}{n^2 + 2^n}\right\}$ is null, by the Squeeze Rule.

Section 2.3

1. (a)

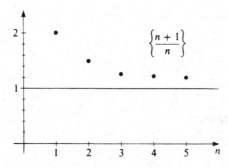

The sequence appears to converge to 1.

(b) Since $b_n = a_n - 1 = \frac{n+1}{n} - 1 = \frac{1}{n}$, it follows that $\{b_n\} = \left\{\frac{1}{n}\right\}$ is a null sequence.

2. (a)

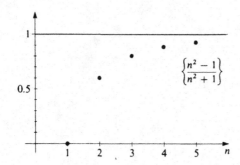

Here $a_n - 1 = \frac{n^2-1}{n^2+1} - 1 = \frac{-2}{n^2+1}$. We know that $\left\{\frac{-2}{n^2+1}\right\}$ is a null sequence, so it follows that $\{a_n - 1\}$ is also a null sequence.

Hence $\{a_n\}$ converges to 1.

(b)

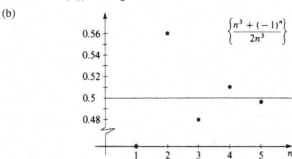

Here $a_n - \frac{1}{2} = \frac{n^3+(-1)^n}{2n^3} - \frac{1}{2} = \frac{(-1)^n}{2n^3}$. We know that $\left\{\frac{(-1)^n}{2n^3}\right\}$ is a null sequence, so it follows that $\left\{a_n - \frac{1}{2}\right\}$ is also a null sequence.

Hence $\{a_n\}$ converges to $\frac{1}{2}$.

3. (a) The dominant term is n^3, so we write a_n as

$$a_n = \frac{n^3 + 2n^2 + 3}{2n^3 + 1}$$

$$= \frac{1 + \frac{2}{n} + \frac{3}{n^3}}{2 + \frac{1}{n^3}}.$$

Since $\left\{\frac{1}{n}\right\}$ and $\left\{\frac{1}{n^3}\right\}$ are basic null sequences

$$\lim_{n\to\infty} a_n = \frac{1+0+0}{2+0} = \frac{1}{2},$$

by the Combination Rules.

(b) The dominant term is 3^n, so we write a_n as

$$a_n = \frac{n^2 + 2^n}{3^n + n^3}$$

$$= \frac{\frac{n^2}{3^n} + \left(\frac{2}{3}\right)^n}{1 + \frac{n^3}{3^n}}.$$

Since $\left\{\frac{n^2}{3^n}\right\}$, $\left\{\left(\frac{2}{3}\right)^n\right\}$ and $\left\{\frac{n^3}{3^n}\right\}$ are basic null sequences

$$\lim_{n\to\infty} a_n = \frac{0+0}{1+0} = 0,$$

by the Combination Rules.

(c) The dominant term is $n!$, so we write a_n as

$$a_n = \frac{n! + (-1)^n}{2^n + 3n!}$$

$$= \frac{1 + \frac{(-1)^n}{n!}}{\frac{2^n}{n!} + 3}.$$

Since $\left\{\frac{(-1)^n}{n!}\right\}$ and $\left\{\frac{2^n}{n!}\right\}$ are basic null sequences

$$\lim_{n\to\infty} a_n = \frac{1+0}{0+3} = \frac{1}{3},$$

by the Combination Rules.

4. (a) Now

$$n^{\frac{1}{n}} \leq 1 + \sqrt{\frac{2}{n-1}} \Leftrightarrow n \leq \left(1 + \sqrt{\frac{2}{n-1}}\right)^n.$$

Using the hint, with $x = \sqrt{\frac{2}{n-1}}$, we obtain

$$\left(1 + \sqrt{\frac{2}{n-1}}\right)^n \geq \frac{n(n-1)}{2!}\left(\sqrt{\frac{2}{n-1}}\right)^2$$
$$= \frac{n(n-1)}{2} \times \frac{2}{n-1} = n,$$

as required.

(b) Since $n \geq 1$, we have $n^{\frac{1}{n}} \geq 1$. Combining this inequality with that in part (a), we obtain

$$1 \leq n^{\frac{1}{n}} \leq 1 + \sqrt{\frac{2}{n-1}}, \quad \text{for } n \geq 2.$$

Now, $\left\{\sqrt{\frac{2}{n-1}}\right\}_{n=2}^{\infty}$ is a null sequence, by the Power Rule, so that

$$\lim_{n \to \infty} \left(1 + \sqrt{\frac{2}{n-1}}\right) = 1.$$

Hence, by the Squeeze Rule, $\lim_{n \to \infty}\left(n^{\frac{1}{n}}\right) = 1$.

Section 2.4

1. (a) $\{1 + (-1)^n\}$ is *bounded*, since the terms $1 + (-1)^n$ take only the values 0 and 2. Hence

$$|1 + (-1)^n| \leq 2, \quad \text{for } n = 1, 2, \ldots.$$

(b) $\{(-1)^n n\}$ is *unbounded*. Given any number K, there is a positive integer n such that

$$|(-1)^n n| > K;$$

for instance, $n = |K| + 1$.

(c) $\left\{\frac{2n+1}{n}\right\}$ is *bounded*, since $\frac{2n+1}{n} = 2 + \frac{1}{n}$ so that

$$\left|\frac{2n+1}{n}\right| = 2 + \frac{1}{n} \leq 3, \quad \text{for } n = 1, 2, \ldots.$$

(d) $\left\{\left(1 - \frac{1}{n}\right)^n\right\}$ is *bounded*, since

$$0 \leq 1 - \frac{1}{n} \leq 1, \quad \text{for } n = 1, 2, \ldots,$$

so that

$$\left|\left(1 - \frac{1}{n}\right)^n\right| = \left(1 - \frac{1}{n}\right)^n \leq 1, \quad \text{for } n = 1, 2, \ldots.$$

2. (a) $\{\sqrt{n}\}$ is *unbounded*, and hence *divergent*, by Corollary 1.

(b) $\left\{\frac{n^2+n}{n^2+1}\right\}$ is *convergent* (with limit 1), and hence *bounded*, by Theorem 1. In fact

$$\frac{n^2+n}{n^2+1} = \frac{1 + \frac{1}{n}}{1 + \frac{1}{n^2}} \leq 1 + \frac{1}{n} \leq 2, \quad \text{for } n = 1, 2, \ldots.$$

(c) $\{(-1)^n n^2\}$ is *unbounded*, and hence *divergent*, by Corollary 1.

(d) The terms of the sequence $\left\{n^{(-1)^n}\right\}$ are

$$1, 2, \frac{1}{3}, 4, \frac{1}{5}, 6, \ldots,$$

so the sequence is *unbounded*: given any number K, there is an even positive integer $2n$ such that $2n > K$. Hence the sequence is *divergent*, by Corollary 1.

3. (a) Each term of $\{\frac{2^n}{n}\}$ is positive, and $\{\frac{n}{2^n}\}$ is a basic null sequence. Hence $\frac{2^n}{n} \to \infty$, by the Reciprocal Rule.

(b) First, note that

$$2^n - n^{100} = 2^n\left(1 - \frac{n^{100}}{2^n}\right), \quad \text{for } n = 1, 2, \ldots,$$

and that $\{\frac{n^{100}}{2^n}\}$ is a basic null sequence. It follows that $\frac{n^{100}}{2^n}$ is eventually less than 1, so that $2^n - n^{100}$ is eventually positive.

Also

$$\lim_{n\to\infty} \frac{1}{2^n - n^{100}} = \lim_{n\to\infty} \frac{\frac{1}{2^n}}{1 - \frac{n^{100}}{2^n}} = \frac{0}{1-0} = 0,$$

by the Combination Rules.

Hence $2^n - n^{100} \to \infty$, by the Reciprocal Rule.

(c) We know that $\frac{2^n}{n} \to \infty$, by part (a), and that

$$\frac{2^n}{n} + 5n^{100} \geq \frac{2^n}{n}, \quad \text{for } n = 1, 2, \ldots.$$

Hence $\frac{2^n}{n} + 5n^{100} \to \infty$, by the Squeeze Rule.

Remark

You could have used the Reciprocal Rule or the Sum and Multiple Rules.

(d) Each term of the sequence $\{\frac{2^n+n^2}{n^{10}+n}\}$ is positive, and

$$\lim_{n\to\infty}\left(\frac{2^n+n^2}{n^{10}+n}\right)^{-1} = \lim_{n\to\infty} \frac{n^{10}+n}{2^n+n^2}$$

$$= \lim_{n\to\infty} \frac{\frac{n^{10}}{2^n} + \frac{n}{2^n}}{1 + \frac{n^2}{2^n}}$$

$$= \frac{0+0}{1+0} = 0,$$

by the Combination Rules.

Hence $\frac{2^n+n^2}{n^{10}+n} \to \infty$, by the Reciprocal Rule.

4. (a) (i) $a_2 = 4$, $a_4 = 16$, $a_6 = 36$, $a_8 = 64$, $a_{10} = 100$;

(ii) $a_3 = 9$, $a_7 = 49$, $a_{11} = 121$, $a_{15} = 225$, $a_{19} = 361$;

(iii) $a_1 = 1$, $a_4 = 16$, $a_9 = 81$, $a_{16} = 256$, $a_{25} = 625$.

(b) $a_1 = 1, a_3 = \frac{1}{3}, a_5 = \frac{1}{5}$;
$a_2 = 2, a_4 = 4, a_6 = 6$.

5. (a) If $a_n = (-1)^n + \frac{1}{n}$, for $n = 1, 2, \ldots$, then

$$a_{2k} = 1 + \frac{1}{2k} \quad \text{and} \quad a_{2k-1} = -1 + \frac{1}{2k-1},$$

so that

$$\lim_{k\to\infty} a_{2k} = 1, \quad \text{whereas} \quad \lim_{k\to\infty} a_{2k-1} = -1.$$

Hence $\{a_n\}$ is divergent, by the First Subsequence Rule.

(b) If $a_n = \frac{1}{3}n - [\frac{1}{3}n]$, for $n = 1, 2, \ldots$, then

$$a_{3k} = 0 \quad \text{and} \quad a_{3k+1} = \frac{1}{3}, \quad \text{for } k = 1, 2, \ldots,$$

so that

$$\lim_{k\to\infty} a_{3k} = 0, \quad \text{whereas} \quad \lim_{k\to\infty} a_{3k+1} = \frac{1}{3}.$$

Hence $\{a_n\}$ is divergent, by the First Subsequence Rule.

(c) If $a_n = n \sin(\frac{1}{2}n\pi)$, for $n = 1, 2, \ldots$, then
$$a_1 = 1, a_2 = 0, a_3 = -3, a_4 = 0, a_5 = 5, a_6 = 0, \ldots.$$

Now
$$a_{4k+1} = (4k + 1)\sin\left(2k\pi + \frac{1}{2}\pi\right)$$
$$= 4k + 1, \quad \text{for } k = 1, 2, \ldots,$$

so that $a_{4k+1} \to \infty$.

Hence $\{a_n\}$ is divergent, by the Second Subsequence Rule.

Section 2.5

1. Suppose that the sequence $\{a_n\}$ is decreasing and bounded below, so that it is necessarily convergent. We show that the limit of $\{a_n\}$ is the number
$$m = \inf\{a_n : n = 1, 2, \ldots\}.$$

To see this, let ℓ denote $\lim_{n \to \infty} a_n$. By the Limit Inequality Rule, since $a_n \geq m$ we know that $\ell \geq m$. Now assume that in fact $\ell > m$.

> We shall show that this assumption leads to a contradiction.

Since $\ell > m$, it follows that $\ell > \frac{1}{2}(\ell + m) > m$. It follows, from the definition of greatest lower bound, that there then exists some integer X such that $a_X < \frac{1}{2}(\ell + m)$; and so, since $\{a_n\}$ is decreasing, that
$$a_n < \frac{1}{2}(\ell + m), \quad \text{for all } n > X.$$

We deduce from the Limit Inequality Rule that $\ell \leq \frac{1}{2}(\ell + m)$. We may rearrange this inequality as $2\ell \leq \ell + m$, or $\ell \leq m$.

> By letting $n \to \infty$.

This contradicts our assumption that $\ell > m$. It follows that, in fact, $\ell = m$.

> For we cannot have $\ell < m$ and $\ell \geq m$!

2. We use the ideas in the discussion prior to the problem.

(a) Let $a_1 > 3$. Then, from equation (5) in the discussion before the statement of the problem
$$a_2 - a_1 = \frac{1}{4}(a_1 - 1)(a_1 - 3)$$
$$> 0,$$

so that $a_2 > a_1$. This suggests that, in general, it might be that $a_{n+1} > a_n$ in the case that $a_1 > 3$; we check this, using Mathematical Induction.

> We need a discussion of this type in order to identify the result that we wish to verify using Mathematical Induction.

Let $P(n)$ be the statement: If $a_1 > 3$, then $a_{n+1} > a_n$.

We have just seen that the statement $P(1)$ is true.

Now assume that the statements $P(1)$, $P(2)$, ..., $P(k)$ are true; that is, that $a_{k+1} > a_k > a_{k-1} > \ldots > a_2 > a_1 \,(>3)$ for some $k \geq 1$. It follows from equation (5) in the discussion before the statement of the problem that
$$a_{k+2} - a_{k+1} = \frac{1}{4}(a_{k+1} - 1)(a_{k+1} - 3) > 0,$$

so that $a_{k+2} > a_{k+1}$; in other words, the statement $P(k + 1)$ is true.

This proves that $P(n)$ holds for all $n \geq 1$.

Thus, if $a_1 > 3$, the sequence $\{a_n\}$ is increasing.

It follows, from the Monotone Convergence Theorem, that either $\{a_n\}$ converges to some (finite) limit ℓ or tends to infinity as $n \to \infty$.

But the only possible (finite) limits of the sequence are 1 and 3, so that since $a_1 > 3$ and the sequence is increasing, clearly $\{a_n\}$ cannot tend to a limit. It follows that $a_n \to \infty$ as $n \to \infty$.

(b) Let $0 \leq a_1 < 1$. Then, from equations (3) and (5) in the discussion before the statement of the problem,

$$a_2 - 1 = \frac{1}{4}(a_1^2 - 1) < 0 \text{ and } a_2 - a_1 = \frac{1}{4}(a_1 - 1)(a_1 - 3) > 0,$$

so that $a_2 < 1$ (and clearly $a_2 = \frac{1}{4}(a_1^2 + 3) > 0$) and $a_2 > a_1$. This suggests that, in general, it might be that $0 \le a_n < 1$ and $a_{n+1} > a_n$ for all $n \ge 1$. We now check the first of these possibilities, using Mathematical Induction.

Now, from the recursion formula $a_{n+1} = \frac{1}{4}(a_n^2 + 3)$ it is clear that $a_{n+1} \ge \frac{3}{4} > 0$ for all $n \ge 1$; hence in the case that $0 \le a_1 < 1$ we must have $a_n \ge 0$ for all $n \ge 1$.

Next, let $P(n)$ be the statement: If $0 \le a_1 < 1$, then $0 \le a_n < 1$ for all $n \ge 1$.
The statement $P(1)$ is certainly true.

Now assume that the statement $P(k)$ is true; that is, that $0 \le a_k < 1$ for some $k \ge 1$. It follows, from equation (3) in the discussion before the statement of the problem, namely $a_{k+1} - 1 = \frac{1}{4}(a_k^2 - 1)$, that $a_{k+1} < 1$; hence the statement $P(k+1)$ is true.

This proves that $P(n)$ holds for all $n \ge 1$.

A similar proof (which we omit) using Mathematical Induction shows that also $a_{n+1} > a_n$ for all $n \ge 1$. Thus, if $0 \le a_1 < 1$, the sequence $\{a_n\}$ is increasing.

It follows, from the Monotone Convergence Theorem, that either $\{a_n\}$ converges to some (finite) limit ℓ or tends to infinity as $n \to \infty$.

But the only possible (finite) limits of the sequence are 1 and 3, so that since $a_n < 1$ for all $n \ge 1$, clearly $\{a_n\}$ cannot tend to 3 or to infinity. It follows that $a_n \to 1$ as $n \to \infty$.

(c) Let $a_1 < 0$. Since $a_{n+1} = \frac{1}{4}(a_n^2 + 3)$, it is clear that the behaviour of the sequence as $n \to \infty$ depends only on the magnitude of a_1 and not on its sign.
It follows from this observation that:

$$\begin{aligned}
&\text{if } -1 < a_1 < 0, & &\text{then } a_n \to 1 & &\text{as } n \to \infty; \\
&\text{if } a_1 = -1, & &\text{then } a_n \to 1 & &\text{as } n \to \infty; \\
&\text{if } -3 < a_1 < -1, & &\text{then } a_n \to 1 & &\text{as } n \to \infty; \\
&\text{if } a_1 = -3, & &\text{then } a_n \to 3 & &\text{as } n \to \infty; \\
&\text{if } a_1 < -3, & &\text{then } a_n \to \infty & &\text{as } n \to \infty.
\end{aligned}$$

3. (a) The Binomial Theorem gives

$$\left(1 + \frac{x}{n}\right)^n = 1 + n\left(\frac{x}{n}\right) + \frac{n(n-1)}{2!}\left(\frac{x}{n}\right)^2 + \cdots + \left(\frac{x}{n}\right)^n$$

$$= 1 + x + \frac{1}{2!}\left(1 - \frac{1}{n}\right)x^2 + \cdots + \frac{x^n}{n^n}.$$

As n increases, the number of terms in this sum increases, and the new terms are all positive. Also, for each fixed $k \ge 1$ and any $n \ge k$, the $(k+1)$th term of this sum is

$$\frac{n(n-1)\ldots(n-k+1)}{k!}\left(\frac{x}{n}\right)^k = \frac{1}{k!}\left(1 - \frac{1}{n}\right)\left(1 - \frac{2}{n}\right)\cdots\left(1 - \frac{k-1}{n}\right)x^k.$$

Here the product on the right increases as n increases (since each of its factors is itself increasing). Hence the sequence $(1 + \frac{x}{n})^n$ is increasing if $x > 0$.

(b) By the Binomial Theorem

$$\left(1 + \frac{1}{n}\right)^k \ge 1 + k\left(\frac{1}{n}\right) = 1 + \frac{k}{n}, \quad \text{for } k = 1, 2, \ldots. \qquad (*)$$

To prove that $\left\{\left(1+\frac{x}{n}\right)^n\right\}$ is bounded above, we choose an integer $k \geq x$ and use the inequality $(*)$, as follows

$$\left(1+\frac{x}{n}\right)^n \leq \left(1+\frac{k}{n}\right)^n$$

$$\leq \left(\left(1+\frac{1}{n}\right)^k\right)^n = \left(\left(1+\frac{1}{n}\right)^n\right)^k$$

$$\leq e^k,$$

since the sequence $\left\{\left(1+\frac{1}{n}\right)^n\right\}$ is increasing with limit e. Hence $\left\{\left(1+\frac{x}{n}\right)^n\right\}$ is bounded above by e^k.

(c) Since $\left\{\left(1+\frac{x}{n}\right)^n\right\}$ is increasing and bounded above, it must be convergent, by the Monotone Convergence Theorem.

4. The first n terms of the sequence $\left\{\left(1+\frac{1}{n}\right)^n\right\}$ are

$$\left(\frac{2}{1}\right)^1, \left(\frac{3}{2}\right)^2, \left(\frac{4}{3}\right)^3, \ldots, \left(\frac{n+1}{n}\right)^n.$$

Taking the product of these terms, each of which is less than e, we obtain

$$\frac{2^1 3^2 4^3 \ldots (n+1)^n}{1^1 2^2 3^3 \ldots n^n} < e^n;$$

it follows, by cancellation, that

$$\frac{(n+1)^n}{n!} < e^n.$$

Hence

$$n! > \frac{(n+1)^n}{e^n} = \left(\frac{n+1}{e}\right)^n, \quad \text{for } n = 1, 2, \ldots.$$

5. We use the formulas:

area $= \frac{1}{2}ab \sin \theta$

area $= \frac{1}{2}ah$

(a) Area $= \frac{1}{2} \times 1 \times 1 \times \sin\left(\frac{1}{3}\pi\right) = \frac{1}{2} \times \frac{\sqrt{3}}{2} = \frac{\sqrt{3}}{4}$.

(b) Area $= \frac{1}{2} \times 1 \times 1 \times \sin\left(\frac{1}{6}\pi\right) = \frac{1}{2} \times \frac{1}{2} = \frac{1}{4}$.

(c) Area $= \frac{1}{2} \times 2\tan\left(\frac{1}{6}\pi\right) \times 1 = \frac{1}{2} \times 2 \times \frac{1}{\sqrt{3}} = \frac{1}{\sqrt{3}}$.

(d) Area $= \frac{1}{2} \times 2\tan\left(\frac{1}{12}\pi\right) \times 1 = \tan\left(\frac{1}{12}\pi\right)$.

To determine $\tan\left(\frac{1}{12}\pi\right)$, we use the formula

$$\tan\left(\frac{1}{6}\pi\right) = \frac{2\tan\left(\frac{1}{12}\pi\right)}{1 - \tan^2\left(\frac{1}{12}\pi\right)}.$$

Since $\tan\left(\frac{1}{6}\pi\right) = \frac{1}{\sqrt{3}}$, we obtain

$$\tan^2\left(\frac{1}{12}\pi\right) + 2\sqrt{3}\tan\left(\frac{1}{12}\pi\right) - 1 = 0,$$

so that

$$\tan\left(\frac{1}{12}\pi\right) = \frac{-2\sqrt{3} \pm \sqrt{\left(2\sqrt{3}\right)^2 + 4}}{2}$$

$$= -\sqrt{3} \pm 2.$$

Since $\tan\left(\frac{1}{12}\pi\right) > 0$, we must have $\tan\left(\frac{1}{12}\pi\right) = 2 - \sqrt{3}$; it follows that

$$\text{Area} = 2 - \sqrt{3}.$$

Chapter 3

Section 3.1

1. (a) Using the formula for summing a geometric progression, with $a = r = -\frac{1}{3}$, we obtain

$$s_n = \left(-\frac{1}{3}\right) + \left(-\frac{1}{3}\right)^2 + \left(-\frac{1}{3}\right)^3 + \cdots + \left(-\frac{1}{3}\right)^n$$

$$= -\frac{1}{3} \times \frac{1 - \left(-\frac{1}{3}\right)^n}{1 - \left(-\frac{1}{3}\right)}$$

$$= -\frac{1}{4}\left(1 - \left(-\frac{1}{3}\right)^n\right).$$

Since $\left\{\left(-\frac{1}{3}\right)^n\right\}$ is a basic null sequence

$$\lim_{n\to\infty} s_n = -\frac{1}{4},$$

and so

$$\sum_{n=1}^{\infty} \left(-\frac{1}{3}\right)^n \quad \text{is convergent, with sum} -\frac{1}{4}.$$

(b) Here

$$s_n = 1 + (-1)^1 + (-1)^2 + \cdots + (-1)^{n-1}$$

$$= 1 - 1 + 1 - 1 + \cdots + (-1)^{n-1},$$

so that

$$s_n = \begin{cases} 1, & n \text{ odd}, \\ 0, & n \text{ even}. \end{cases}$$

Hence $s_{2k+1} \to 1$ and $s_{2k} \to 0$ as $k \to \infty$, so that $\{s_n\}$ is divergent, by the First Subsequence Rule. Thus the series $\sum_{n=0}^{\infty} (-1)^n$ is divergent.

(c) Using the formula for summing a geometric progression with $a = 2$ and $r = \frac{1}{2}$, we obtain

$$s_n = 2 + 1 + \left(\frac{1}{2}\right) + \left(\frac{1}{2}\right)^2 + \cdots + \left(\frac{1}{2}\right)^{n-3}$$

$$= 2 \times \frac{1 - \left(\frac{1}{2}\right)^n}{1 - \left(\frac{1}{2}\right)}$$

$$= 4\left(1 - \left(\frac{1}{2}\right)^n\right).$$

Since $\left\{\left(\frac{1}{2}\right)^n\right\}$ is a basic null sequence, $\lim_{n\to\infty} s_n = 4$, and so

$$\sum_{n=-1}^{\infty} \left(\frac{1}{2}\right)^n \quad \text{is convergent, with sum } 4.$$

2. (a) We interpret $0.111\ldots$ as

$$\frac{1}{10^1} + \frac{1}{10^2} + \frac{1}{10^3} + \cdots.$$

This is a geometric series with $a = \frac{1}{10}$ and $r = \frac{1}{10}$. Since $\frac{1}{10} < 1$, the series is convergent with sum

$$\frac{a}{1-r} = \frac{\frac{1}{10}}{1 - \frac{1}{10}} = \frac{1}{9};$$

hence

$$0.111\ldots = \frac{1}{9}.$$

(b) We interpret $0.86363\ldots$ as

$$\frac{8}{10} + \frac{1}{10}\left(\frac{63}{100^1} + \frac{63}{100^2} + \cdots\right).$$

The series in the bracket is a geometric series with $a = \frac{63}{100}$ and $r = \frac{1}{100}$. Since $\frac{1}{100} < 1$, this series is convergent with sum

$$\frac{a}{1-r} = \frac{\frac{63}{100}}{1 - \frac{1}{100}} = \frac{63}{99} = \frac{7}{11};$$

hence

$$0.86363\ldots = \frac{8}{10} + \frac{7}{110} = \frac{19}{22}.$$

(c) We interpret $0.999\ldots$ as

$$\frac{9}{10^1} + \frac{9}{10^2} + \frac{9}{10^3} + \cdots.$$

This is a geometric series with $a = \frac{9}{10}$ and $r = \frac{1}{10}$. Since $\frac{1}{10} < 1$, this series is convergent with sum

$$\frac{a}{1-r} = \frac{\frac{9}{10}}{1 - \frac{1}{10}} = 1;$$

hence

$$0.999\ldots = 1.$$

3. $s_1 = \dfrac{1}{1 \times 2} = \dfrac{1}{2};$

$s_2 = \dfrac{1}{1 \times 2} + \dfrac{1}{2 \times 3} = \dfrac{1}{2} + \dfrac{1}{6} = \dfrac{2}{3};$

$s_3 = \dfrac{1}{1 \times 2} + \dfrac{1}{2 \times 3} + \dfrac{1}{3 \times 4} = \dfrac{2}{3} + \dfrac{1}{12} = \dfrac{3}{4};$

$s_4 = \dfrac{1}{1 \times 2} + \dfrac{1}{2 \times 3} + \dfrac{1}{3 \times 4} + \dfrac{1}{4 \times 5} = \dfrac{3}{4} + \dfrac{1}{20} = \dfrac{4}{5}.$

4. Since

$$\frac{1}{n(n+2)} = \frac{1}{2}\left(\frac{1}{n} - \frac{1}{n+2}\right), \quad \text{for } n = 1, 2, \ldots,$$

we have

$$\sum_{n=1}^{\infty} \frac{1}{n(n+2)} = \sum_{n=1}^{\infty} \frac{1}{2}\left(\frac{1}{n} - \frac{1}{n+2}\right),$$

and so

$$s_n = \frac{1}{1 \times 3} + \frac{1}{2 \times 4} + \frac{1}{3 \times 5} + \frac{1}{4 \times 6} + \cdots + \frac{1}{n(n+2)}$$

$$= \frac{1}{2}\left[\left(1 - \frac{1}{3}\right) + \left(\frac{1}{2} - \frac{1}{4}\right) + \left(\frac{1}{3} - \frac{1}{5}\right) + \left(\frac{1}{4} - \frac{1}{6}\right) + \cdots + \left(\frac{1}{n} - \frac{1}{n+2}\right)\right].$$

Most of the terms in alternate brackets cancel out, leaving

$$s_n = \frac{1}{2}\left(1 + \frac{1}{2} - \frac{1}{n+1} - \frac{1}{n+2}\right)$$
$$= \frac{3}{4} - \frac{1}{2(n+1)} - \frac{1}{2(n+2)}.$$

Since $\left\{\frac{1}{n+1}\right\}$ and $\left\{\frac{1}{n+2}\right\}$ are null sequences, $\lim_{n\to\infty} s_n = \frac{3}{4}$, and so

$$\sum_{n=1}^{\infty} \frac{1}{n(n+2)} \quad \text{is convergent, with sum } \frac{3}{4}.$$

5. The series $\sum_{n=1}^{\infty} \left(\frac{3}{4}\right)^n$ is a geometric series, with $a = r = \frac{3}{4}$. Hence, it is convergent, with sum $\frac{\frac{3}{4}}{1-\frac{3}{4}} = 3$.

The series $\sum_{n=1}^{\infty} \frac{1}{n(n+1)}$ is convergent, with sum 1 (cf. Sub-section 3.1.3). Hence, by the Combination Rules

$$\sum_{n=1}^{\infty} \left(\left(\frac{3}{4}\right)^n - \frac{2}{n(n+1)}\right) \quad \text{is convergent, with sum } 3 - (2 \times 1) = 1.$$

6. By the Combination Rules for sequences

$$\lim_{n\to\infty} \frac{n^2}{2n^2+1} = \lim_{n\to\infty} \frac{1}{2+\frac{1}{n^2}} = \frac{1}{2},$$

so that the sequence $\left\{\frac{n^2}{2n^2+1}\right\}$ is not null.

Hence, by the Non-null Test, $\sum_{n=1}^{\infty} \frac{n^2}{2n^2+1}$ is divergent.

Section 3.2

1. Let $s_n = 1 + \frac{1}{2^2} + \frac{1}{3^2} + \cdots + \frac{1}{n^2}$ and $t_n = 1 + \frac{1}{2} + \frac{1}{3} + \cdots + \frac{1}{n}$. The values of these partial sums are as listed below:

n	1	2	3	4	5	6	7	8
s_n	1	1.25	1.36	1.42	1.46	1.49	1.51	1.53
t_n	1	1.50	1.83	2.08	2.28	2.45	2.59	2.71

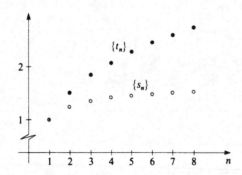

2. (a) We use the Comparison Test. Since

$$n^3 + n \geq n^3, \quad \text{for } n = 1, 2, \ldots,$$

we have

$$\frac{1}{n^3 + n} \leq \frac{1}{n^3}, \quad \text{for } n = 1, 2, \ldots.$$

Since $\sum_{n=1}^{\infty} \frac{1}{n^3}$ is convergent, we deduce, from the Comparison Test, that

$$\sum_{n=1}^{\infty} \frac{1}{n^3 + n} \quad \text{is convergent.}$$

(b) Let

$$a_n = \frac{1}{n + \sqrt{n}} \quad \text{and} \quad b_n = \frac{1}{n}, \quad \text{for } n = 1, 2, \ldots,$$

so that

$$\lim_{n \to \infty} \frac{a_n}{b_n} = \lim_{n \to \infty} \frac{n}{n + \sqrt{n}}$$

$$= \lim_{n \to \infty} \frac{1}{1 + \frac{1}{\sqrt{n}}}$$

$$= 1 \neq 0.$$

Since $\sum_{n=1}^{\infty} \frac{1}{n}$ is divergent, we deduce, from the Limit Comparison Test, that

$$\sum_{n=1}^{\infty} \frac{1}{n + \sqrt{n}} \quad \text{is divergent.}$$

(c) Let

$$a_n = \frac{n + 4}{2n^3 - n + 1} \quad \text{and} \quad b_n = \frac{1}{n^2}, \quad \text{for } n = 1, 2, \ldots,$$

so that

$$\lim_{n \to \infty} \frac{a_n}{b_n} = \lim_{n \to \infty} \frac{n^3 + 4n^2}{2n^3 - n + 1}$$

$$= \lim_{n \to \infty} \frac{1 + \frac{4}{n}}{2 - \frac{1}{n^2} + \frac{1}{n^3}}$$

$$= \frac{1}{2} \neq 0.$$

Since $\sum_{n=1}^{\infty} \frac{1}{n^2}$ is convergent, we deduce, from the Limit Comparison Test, that

$$\sum_{n=1}^{\infty} \frac{n + 4}{2n^3 - n + 1} \quad \text{is convergent.}$$

(d) We use the Comparison Test. Since

$$0 \leq \cos^2(2n) \leq 1, \quad \text{for } n = 1, 2, \ldots,$$

we have

$$0 \leq \frac{\cos^2(2n)}{n^3} \leq \frac{1}{n^3}, \quad \text{for } n = 1, 2, \ldots.$$

Since $\sum_{n=1}^{\infty} \frac{1}{n^3}$ is convergent, we deduce, from the Comparison Test, that

$$\sum_{n=1}^{\infty} \frac{\cos^2(2n)}{n^3} \quad \text{is convergent.}$$

3. (a) Let $a_n = \frac{n^3}{n!}$, for $n = 1, 2, \ldots$, so that

$$\frac{a_{n+1}}{a_n} = \left(\frac{(n+1)^3}{(n+1)!} \right) \times \left(\frac{n!}{n^3} \right)$$

$$= \frac{(n+1)^2}{n^3}$$

$$= \frac{n^2 + 2n + 1}{n^3} = \frac{1}{n} + \frac{2}{n^2} + \frac{1}{n^3}.$$

Hence, by the Combination Rules for sequences

$$\lim_{n\to\infty} \frac{a_{n+1}}{a_n} = 0;$$

it follows, from the Ratio Test, that $\sum_{n=1}^{\infty} \frac{n^3}{n!}$ is convergent.

(b) Let $a_n = \frac{n^2 2^n}{n!}$, for $n = 1, 2, \ldots$, so that

$$\frac{a_{n+1}}{a_n} = \left(\frac{(n+1)^2 2^{n+1}}{(n+1)!} \right) \times \left(\frac{n!}{n^2 2^n} \right)$$

$$= \frac{2(n+1)}{n^2}$$

$$= 2\left(\frac{1}{n} + \frac{1}{n^2} \right).$$

Hence, by the Combination Rules for sequences

$$\lim_{n\to\infty} \frac{a_{n+1}}{a_n} = 0;$$

it follows, from the Ratio Test, that $\sum_{n=1}^{\infty} \frac{n^2 2^n}{n!}$ is convergent.

(c) Let $a_n = \frac{(2n)!}{n^n}$, for $n = 1, 2, \ldots$, so that

$$\frac{a_{n+1}}{a_n} = \left(\frac{(2n+2)!}{(n+1)^{n+1}} \right) \times \left(\frac{n^n}{(2n)!} \right)$$

$$= \frac{(2n+2)(2n+1)n^n}{(n+1)^{n+1}}$$

$$= \frac{2(2n+1)n^n}{(n+1)^n}$$

$$= \frac{4n+2}{\left(1 + \frac{1}{n}\right)^n}.$$

We know that $\lim_{n\to\infty} \left(1 + \frac{1}{n}\right)^n = e$ and that $\left\{ \frac{1}{4n+2} \right\}$ is null, so that $\left\{ \frac{\left(1+\frac{1}{n}\right)^n}{4n+2} \right\}$ is null, by the Product Rule for sequences. It follows, from the Reciprocal Rule for sequences, that

$$\frac{a_{n+1}}{a_n} \to \infty \quad \text{as } n \to \infty.$$

It follows, from the Ratio Test, that $\sum_{n=1}^{\infty} \frac{(2n)!}{n^n}$ is divergent.

Remark

Notice that

$$\frac{(2n)!}{n^n} \geq \left(\frac{2n}{n} \right) \times \left(\frac{2n-1}{n} \right) \times \cdots \times \left(\frac{n+1}{n} \right)$$

$$\geq 1;$$

it follows, from the Non-null Test, that $\sum_{n=1}^{\infty} \frac{(2n)!}{n^n}$ is divergent.

4. (a) Let $a_n = \frac{1}{n \log_e n}$, $n = 2, \ldots$. Then a_n is positive; and, since $\{n\log_e n\}$ is an increasing sequence, $\{a_n\} = \left\{ \frac{1}{n \log_e n} \right\}$ is a decreasing sequence.

Next, let $b_n = 2^n a_{2^n}$; thus

$$b_n = 2^n \times \frac{1}{2^n \log_e(2^n)}$$

$$= \frac{1}{n \log_e 2} = \frac{1}{\log_e 2} \times \frac{1}{n}.$$

Since $\sum_{n=2}^{\infty} \frac{1}{n}$ is a basic divergent series, it follows by the Multiple Rule that $\sum_{n=2}^{\infty} b_n$ must be divergent. Hence, by the Condensation Test, the original series $\sum_{n=2}^{\infty} \frac{1}{n \log_e n}$ must also be divergent.

(b) Let $a_n = \frac{1}{n(\log_e n)^2}$, $n = 2, \ldots$. Then a_n is positive; and, since $\{n(\log_e n)^2\}$ is an increasing sequence, $\{a_n\} = \left\{\frac{1}{n(\log_e n)^2}\right\}$ is a decreasing sequence.

Next, let $b_n = 2^n a_{2^n}$; thus

$$b_n = 2^n \times \frac{1}{2^n (\log_e (2^n))^2}$$

$$= \frac{1}{(n \log_e 2)^2} = \frac{1}{(\log_e 2)^2} \times \frac{1}{n^2}.$$

Since $\sum_{n=2}^{\infty} \frac{1}{n^2}$ is a basic convergent series, it follows by the Multiple Rule that $\sum_{n=2}^{\infty} b_n$ must be convergent. Hence by the Condensation Test, the original series $\sum_{n=2}^{\infty} \frac{1}{n(\log_e n)^2}$ must also be convergent.

Section 3.3

1. (a) Let $a_n = \frac{(-1)^{n+1} n}{n^3 + 1}$, for $n = 1, 2, \ldots$, so that

$$|a_n| = \frac{n}{n^3 + 1}, \quad \text{for } n = 1, 2, \ldots.$$

Now

$$\frac{n}{n^3 + 1} \le \frac{n}{n^3} = \frac{1}{n^2}, \quad \text{for } n = 1, 2, \ldots,$$

and $\sum_{n=1}^{\infty} \frac{1}{n^2}$ is a basic convergent series. Hence, by the Comparison Test

$$\sum_{n=1}^{\infty} \frac{n}{n^3 + 1} \quad \text{is convergent.}$$

It follows, from the Absolute Convergence Test, that

$$\sum_{n=1}^{\infty} \frac{(-1)^{n+1} n}{n^3 + 1} \quad \text{is convergent.}$$

(b) Let $a_n = \frac{\cos n}{2^n}$, for $n = 1, 2, \ldots$; then

$$|a_n| \le \frac{1}{2^n}, \quad \text{for } n = 1, 2, \ldots,$$

since $|\cos n| \le 1$.

Now $\sum_{n=1}^{\infty} \frac{1}{2^n}$ is a basic convergent series (in fact, a convergent geometric series). Hence, by the Comparison Test

$$\sum_{n=1}^{\infty} \frac{\cos n}{2^n} \quad \text{is absolutely convergent.}$$

It follows, from the Absolute Convergence Test, that

$$\sum_{n=1}^{\infty} \frac{\cos n}{2^n} \quad \text{is convergent.}$$

2. The series

$$\frac{1}{2} + \frac{1}{4} - \frac{1}{8} + \frac{1}{16} + \frac{1}{32} - \frac{1}{64} + \cdots$$

is absolutely convergent, and hence convergent, by the Comparison Test, since the

series $\sum\limits_{n=1}^{\infty} \frac{1}{2^n}$ is a convergent geometric series.

By the infinite form of the Triangle Inequality, we have

$$\left| \frac{1}{2} + \frac{1}{4} - \frac{1}{8} + \frac{1}{16} + \frac{1}{32} - \frac{1}{64} + \cdots \right| \le \frac{1}{2} + \frac{1}{4} + \frac{1}{8} + \frac{1}{16} + \frac{1}{32} + \frac{1}{64} + \cdots$$

$$= 1.$$

It follows that the sum of the original series lies in the interval $[-1, 1]$.

3. (a) The sequence $\left\{ \frac{(-1)^{n+1}}{n^3} \right\}$ is of the form $\left\{ (-1)^{n+1} b_n \right\}$, where

$$b_n = \frac{1}{n^3}, \quad \text{for } n = 1, 2, \ldots.$$

Now:

1. $\frac{1}{n^3} \ge 0$, for $n = 1, 2, \ldots$;
2. $\left\{ \frac{1}{n^3} \right\}$ is a basic null sequence;
3. $\left\{ \frac{1}{n^3} \right\}$ is decreasing, because $\{ n^3 \}$ is increasing.

Hence, by the Alternating Test, $\sum\limits_{n=1}^{\infty} \frac{(-1)^{n+1}}{n^3}$ is convergent.

(b) The sequence $\left\{ (-1)^{n+1} \frac{n}{n+2} \right\}$ is is not a null sequence, since

$$\lim_{n \to \infty} \frac{n}{n+2} = 1,$$

and so the odd subsequence tends to 1.

Hence, by the Non-null Test

$$\sum_{n=1}^{\infty} (-1)^{n+1} \frac{n}{n+2} \quad \text{is divergent.}$$

(c) The sequence $\left\{ \frac{(-1)^{n+1}}{n^{\frac{1}{3}} + n^{\frac{1}{2}}} \right\}$ is of the form $\left\{ (-1)^{n+1} b_n \right\}$, where

$$b_n = \frac{1}{n^{\frac{1}{3}} + n^{\frac{1}{2}}}, \quad \text{for } n = 1, 2, \ldots.$$

Now:

1. $\frac{1}{n^{\frac{1}{3}} + n^{\frac{1}{2}}} \ge 0$, for $n = 1, 2, \ldots$;

2. $\left\{ \frac{1}{n^{\frac{1}{3}} + n^{\frac{1}{2}}} \right\}$ is a null sequence, by the Squeeze Rule, since

$$0 \le \frac{1}{n^{\frac{1}{3}} + n^{\frac{1}{2}}} \le \frac{1}{n^{\frac{1}{2}}},$$

where $\left\{ \frac{1}{n^{\frac{1}{2}}} \right\}$ is a basic null sequence;

3. $\left\{ \frac{1}{n^{\frac{1}{3}} + n^{\frac{1}{2}}} \right\}$ is decreasing, because $\left\{ n^{\frac{1}{3}} + n^{\frac{1}{2}} \right\}$ is increasing.

Hence, by the Alternating Test, $\sum\limits_{n=1}^{\infty} \frac{(-1)^{n+1}}{n^{\frac{1}{3}} + n^{\frac{1}{2}}}$ is convergent.

4. Let s_n and t_n denote the nth partial sums of the series

$$1 - \frac{1}{2} - \frac{1}{4} + \frac{1}{3} - \frac{1}{6} - \frac{1}{8} + \frac{1}{5} - \frac{1}{10} - \frac{1}{12} + \cdots \quad \text{and} \quad 1 - \frac{1}{2} + \frac{1}{3} - \frac{1}{4} + \frac{1}{5} - \frac{1}{6} + \cdots,$$

respectively. Denote by H_n the nth partial sum $\sum\limits_{k=1}^{n} \frac{1}{k} = 1 + \frac{1}{2} + \cdots + \frac{1}{n}$ of the harmonic series.

The terms of the series $\sum_{n=1}^{\infty} a_n$ come 'in a natural way' in threes, so it seems sensible to look at the partial sums s_{3n}

$$s_{3n} = 1 - \frac{1}{2} - \frac{1}{4} + \frac{1}{3} - \frac{1}{6} - \frac{1}{8} + \frac{1}{5} - \frac{1}{10} - \frac{1}{12} + \cdots + \frac{1}{2n-1} - \frac{1}{4n-2} - \frac{1}{4n}$$

$$= \left(1 - \frac{1}{2} - \frac{1}{4}\right) + \left(\frac{1}{3} - \frac{1}{6} - \frac{1}{8}\right) + \left(\frac{1}{5} - \frac{1}{10} - \frac{1}{12}\right) + \cdots + \left(\frac{1}{2n-1} - \frac{1}{4n-2} - \frac{1}{4n}\right)$$

$$= \left(1 + \frac{1}{3} + \frac{1}{5} + \cdots + \frac{1}{2n-1}\right) - \left(\frac{1}{2} + \frac{1}{4} + \frac{1}{6} + \frac{1}{8} + \cdots + \frac{1}{4n-2} + \frac{1}{4n}\right)$$

$$= \left(1 + \frac{1}{2} + \frac{1}{3} + \frac{1}{4} + \cdots + \frac{1}{2n-1} + \frac{1}{2n}\right) - \left(\frac{1}{2} + \frac{1}{4} + \cdots + \frac{1}{2n}\right)$$

$$\quad - \frac{1}{2}\left(1 + \frac{1}{2} + \frac{1}{3} + \cdots + \frac{1}{2n}\right)$$

$$= H_{2n} - \frac{1}{2}H_n - \frac{1}{2}H_{2n}$$

$$= \frac{1}{2}(H_{2n} - H_n).$$

But as we saw in Example 2

$$t_{2n} = H_{2n} - H_n,$$

so that

$$s_{3n} = \frac{1}{2}t_{2n}.$$

By assumption, $t_n \to \log_e 2$ and so $t_{2n} \to \log_e 2$ as $n \to \infty$. It follows, by the Multiple Rule for sequences, that

$$s_{3n} \to \frac{1}{2}\log_e 2 \quad \text{as } n \to \infty.$$

Finally

$$s_{3n-1} = s_{3n} + \frac{1}{4n} \to \frac{1}{2}\log_e 2 \quad \text{as } n \to \infty,$$

and

$$s_{3n-2} = s_{3n} + \frac{1}{4n-2} + \frac{1}{4n} \to \frac{1}{2}\log_e 2 \quad \text{as } n \to \infty.$$

It follows from these three results that $s_n \to \frac{1}{2}\log_e 2$ as $n \to \infty$.

5. We follow the pattern of the solution to Example 3, by finding a subsequence of partial sums of the series $\sum_{n=1}^{\infty} \frac{(-1)^{n+1}}{n} = 1 - \frac{1}{2} + \frac{1}{3} - \frac{1}{4} + \frac{1}{5} - \frac{1}{6} + \cdots$ that are greater than $2, 3, \ldots$ in turn and ensuring that the other partial sums are not much less than these values.

First, note that all terms of the series $\sum_{n=1}^{\infty} \frac{(-1)^{n+1}}{n}$ are at most 1 in modulus.

Next, since the 'positive' part of the series $\sum_{n=1}^{\infty} \frac{(-1)^{n+1}}{n}$ has partial sums that tend to ∞, we start to construct the desired rearranged series as follows. Take enough of the 'positive' terms $1, \frac{1}{3}, \frac{1}{5}, \ldots, \frac{1}{2N_1 - 1}$ so that the sum $1 + \frac{1}{3} + \frac{1}{5} + \cdots + \frac{1}{2N_1 - 1}$ is greater than 2, choosing N_1 so that it is the first integer such that this sum is greater than 2. Then these N_1 terms will form the first N_1 terms in our desired rearranged series.

Next, take one 'negative' term $-\frac{1}{2}$ and consider the expression

$$\left(1 + \frac{1}{3} + \frac{1}{5} + \cdots + \frac{1}{2N_1 - 1}\right) - \left(\frac{1}{2}\right).$$

It is certainly less than the partial sum $t_{N_1} = 1 + \frac{1}{3} + \frac{1}{5} + \cdots + \frac{1}{2N_1 - 1}$ of the rearranged series, for which $t_{N_1} > 2$. But, since all terms are at most 1 in magnitude, it follows that $t_{N_1 + 1} > 1$.

Now add in some more 'positive' terms $\frac{1}{2N_1+1}, \frac{1}{2N_1+3}, \ldots, \frac{1}{2N_2-1}$ so that the sum

$$\left(1+\frac{1}{3}+\frac{1}{5}+\cdots+\frac{1}{2N_1-1}\right)-\left(\frac{1}{2}\right)+\left(\frac{1}{2N_1+1}+\frac{1}{2N_1+3}+\cdots+\frac{1}{2N_2-1}\right)$$

is greater than 3, choosing N_2 so that it is the first available integer such that this sum is greater than 3. Then these N_2+1 terms will form the first N_2+1 terms in our desired rearranged series.

We then add in just one 'negative' term; this makes the next partial sum of the rearranged series satisfy $t_{N_2+2} > 2$. Then we add in sufficient positive terms to make the partial sum $t_{N_3+2} > 4$. And so on indefinitely.

In each set of two steps in this process we must use at least one of the 'positive' terms and one of the 'negative' terms, so that eventually all the 'positive' terms and all the 'negative' terms of the original series will be taken exactly once in the new series, which we denote by $\sum_{n=1}^{\infty} b_n$. So certainly the series $\sum_{n=1}^{\infty} b_n$ is a rearrangement of the original series $\sum_{n=1}^{\infty} \frac{(-1)^{n+1}}{n}$.

From our construction, we have ensured that:

for all $n > N_k$, for any k, the nth partial sums of $\sum_{n=1}^{\infty} b_n$ are $> k$.

It follows, then, that the nth partial sums of $\sum_{n=1}^{\infty} b_n$ tend to infinity as $n \to \infty$.

6. (a) The sequence $\{\frac{1}{2}n\}$ is not null; hence, by the Non-null Test, the series $\sum_{n=1}^{\infty} \frac{1}{2}n$ is divergent.

Thus $\sum_{n=1}^{\infty} \frac{1}{2}n$ is neither convergent nor absolutely convergent.

(b) We have

$$\frac{5n+2^n}{3^n} = 5n\left(\frac{1}{3}\right)^n+\left(\frac{2}{3}\right)^n, \quad \text{for } n = 1, 2, \ldots.$$

Now, $\sum_{n=1}^{\infty} n\left(\frac{1}{3}\right)^n$ and $\sum_{n=1}^{\infty} \left(\frac{2}{3}\right)^n$ are both basic convergent series; hence, by the Combination Rules for series

$$\sum_{n=1}^{\infty} \frac{5n+2^n}{3^n} \quad \text{is convergent (and absolutely convergent).} \qquad \text{Since all terms are non-negative.}$$

(c) The sequence $\{a_n\} = \left\{\frac{3}{2n^3-1}\right\}$ is null (so the Non-null Test is inappropriate), and contains only positive terms.

The Ratio Test fails, since

$$\lim_{n\to\infty} \frac{a_{n+1}}{a_n} = \lim_{n\to\infty} \frac{2n^3-1}{2(n+1)^3-1} = 1.$$

Instead, we use the Limit Comparison Test, with

$$b_n = \frac{1}{n^3}, \quad \text{for } n = 1, 2, \ldots.$$

Then

$$\lim_{n\to\infty} \frac{a_n}{b_n} = \lim_{n\to\infty} \frac{\frac{3}{2n^3-1}}{\frac{1}{n^3}}$$
$$= \lim_{n\to\infty} \frac{3n^3}{2n^3-1}$$
$$= \frac{3}{2} \neq 0.$$

Since $\sum_{n=1}^{\infty} \frac{1}{n^3}$ is a basic convergent series, we deduce, from the Limit Comparison Test, that

$$\sum_{n=1}^{\infty} \frac{3}{2n^3 - 1} \quad \text{is convergent.}$$

Since the terms in the series are non-negative, it follows that it is also absolutely convergent.

(d) The sequence $\left\{ \frac{(-1)^{n+1}}{n^{\frac{1}{3}}} \right\}$ is null, but $\sum_{n=1}^{\infty} \frac{1}{n^{\frac{1}{3}}}$ is a basic divergent series. Hence $\sum_{n=1}^{\infty} \frac{(-1)^{n+1}}{n^{\frac{1}{3}}}$ is not absolutely convergent.

However, $\left\{ \frac{(-1)^{n+1}}{n^{\frac{1}{3}}} \right\}$ is of the form $\left\{ (-1)^{n+1} b_n \right\}$, where

$$b_n = \frac{1}{n^{\frac{1}{3}}}, \quad \text{for } n = 1, 2, \ldots .$$

Now:

1. $\frac{1}{n^{\frac{1}{3}}} \geq 0, \quad$ for $n = 1, 2, \ldots$;

2. $\left\{ \frac{1}{n^{\frac{1}{3}}} \right\}$ is a null sequence;

3. $\left\{ \frac{1}{n^{\frac{1}{3}}} \right\}$ is decreasing, because $\left\{ n^{\frac{1}{3}} \right\}$ is increasing.

Hence, by the Alternating Test, $\sum_{n=1}^{\infty} \frac{(-1)^{n+1}}{n^{\frac{1}{3}}}$ is convergent.

(e) The sequence $a_n = \frac{(-1)^{n+1} n^2}{n^2 + 1}$, $n = 1, 2, \ldots$, is not null, since

$$\lim_{n \to \infty} \frac{n^2}{n^2 + 1} = \lim_{n \to \infty} \frac{1}{1 + \frac{1}{n^2}} = 1,$$

so that the odd subsequence tends to 1.

Hence, $\sum_{n=1}^{\infty} |a_n|$ and $\sum_{n=1}^{\infty} a_n$ are divergent, by the Non-null Test. It follows that

$$\sum_{n=1}^{\infty} \frac{(-1)^{n+1} n^2}{n^2 + 1} \quad \text{is neither convergent nor absolutely convergent.}$$

(f) Let $a_n = \frac{(-1)^{n+1} n}{n^3 + 5}$, $n = 1, 2, \ldots$; then

$$|a_n| = \frac{n}{n^3 + 5}, \quad n = 1, 2, \ldots,$$

so that

$$|a_n| \leq \frac{n}{n^3} = \frac{1}{n^2}, \quad \text{for } n = 1, 2, \ldots .$$

It follows, by the Comparison Test, that $\sum_{n=1}^{\infty} |a_n|$ is convergent; that is, that $\sum_{n=1}^{\infty} a_n$ is absolutely convergent. Since $\sum_{n=1}^{\infty} \frac{1}{n^2}$ is a basic convergent series.

Hence, by the Absolute Convergence Test, $\sum_{n=1}^{\infty} a_n$ is convergent.

(g) Since $\left\{ \frac{n^6}{2^n} \right\}$ is a basic null sequence of positive terms, it follows, from the Reciprocal Rule for sequences, that

$$\frac{2^n}{n^6} \to \infty \quad \text{as } n \to \infty.$$

Hence $\left\{ \frac{2^n}{n^6} \right\}$ is not a null sequence; it follows, from the Non-null Test, that

$$\sum_{n=1}^{\infty} \frac{2^n}{n^6} \quad \text{is divergent.}$$

Thus, $\sum_{n=1}^{\infty} \frac{2^n}{n^6}$ is neither convergent nor absolutely convergent.

(h) Let $a_n = \frac{(-1)^{n+1}n}{n^2+2}$, $n = 1, 2, \ldots$, so that

$$|a_n| = \frac{n}{n^2+2}, \quad \text{for } n = 1, 2, \ldots.$$

Now we try the Limit Comparison Test, with

$$b_n = \frac{1}{n}, \quad \text{for } n = 1, 2, \ldots.$$

We obtain

$$\lim_{n\to\infty} \frac{|a_n|}{b_n} = \lim_{n\to\infty} \frac{\frac{n}{n^2+2}}{\frac{1}{n}}$$

$$= \lim_{n\to\infty} \frac{n^2}{n^2+2} = 1 \neq 0.$$

Thus, $\sum_{n=1}^{\infty} |a_n|$ is divergent, since $\sum_{n=1}^{\infty} \frac{1}{n}$ is divergent; it follows that $\sum_{n=1}^{\infty} a_n$ is not absolutely convergent.

However, $\left\{\frac{(-1)^{n+1}n}{n^2+2}\right\}$ is of the form $\left\{(-1)^{n+1}b_n\right\}$, where

$$b_n = \frac{n}{n^2+2}, \quad \text{for } n = 1, 2, \ldots.$$

Now:

1. $\frac{n}{n^2+2} \geq 0$, for $n = 1, 2, \ldots$;

2. $\left\{\frac{n}{n^2+2}\right\}$ is a null sequence, since

$$\frac{n}{n^2+2} = \frac{\left(\frac{1}{n}\right)}{\left(1+\frac{2}{n^2}\right)} \to 0 \quad \text{as } n \to \infty;$$

3. $\left\{\frac{n}{n^2+2}\right\}$ is decreasing, since $\left\{\frac{n^2+2}{n}\right\}$ is increasing, because

$$\frac{(n+1)^2+2}{n+1} - \frac{n^2+2}{n} = \left(n+1+\frac{2}{n+1}\right) - \left(n+\frac{2}{n}\right)$$

$$= 1 - \frac{2}{n(n+1)} \geq 0.$$

Hence, by the Alternating Test, $\sum_{n=1}^{\infty} \frac{(-1)^{n+1}n}{n^2+2}$ is convergent.

(i) Let $a_n = \frac{1}{n(\log_e n)^{\frac{3}{4}}}$, $n = 2, \ldots$. Then a_n is positive; and, since $\left\{n(\log_e n)^{\frac{3}{4}}\right\}$ is an increasing sequence, $\{a_n\} = \left\{\frac{1}{n(\log_e n)^{\frac{3}{4}}}\right\}$ is a decreasing sequence.

Next, let $b_n = 2^n a_{2^n}$; thus

$$b_n = 2^n \times \frac{1}{2^n(\log_e 2^n)^{\frac{3}{4}}}$$

$$= \frac{1}{(n\log_e 2)^{\frac{3}{4}}}$$

$$= \frac{1}{(\log_e 2)^{\frac{3}{4}}} \times \frac{1}{n^{\frac{3}{4}}}.$$

Since $\sum_{n=2}^{\infty} \frac{1}{n^{\frac{3}{4}}}$ is a basic divergent series, it follows (for example, by the Ratio Test) that $\sum_{n=2}^{\infty} b_n$ must be divergent. Hence, by the Condensation Test, the original series $\sum_{n=2}^{\infty} \frac{1}{n(\log_e n)^{\frac{3}{4}}}$ must also be divergent. Hence the series is not convergent, and, since its terms are non-negative, is not absolutely convergent.

Section 3.4

1. Substituting $x = 2$ into the power series for e^x, we find that the seventh partial sum of the power series for e^2 gives

$$e^2 \simeq 1 + \frac{2}{1!} + \frac{2^2}{2!} + \frac{2^3}{3!} + \frac{2^4}{4!} + \frac{2^5}{5!} + \frac{2^6}{6!}$$
$$= 1 + 2 + \frac{4}{2} + \frac{8}{6} + \frac{16}{24} + \frac{32}{120} + \frac{64}{720}$$
$$= 1 + 2 + 2 + \frac{4}{3} + \frac{2}{3} + \frac{4}{15} + \frac{4}{45}$$
$$= 7 + \frac{16}{45} = 7.35\overline{5}.$$

Hence this method gives as an estimate for e^2 to 3 decimal places the number 7.355.

Remark

In fact, $e^2 = 7.389$, to 3 decimal places.

2. From equation (5) we know that

$$0 < e - s_n < \frac{1}{(n-1)!} \times \frac{1}{n-1},$$

so that we want n to satisfy the requirement that

$$\frac{1}{(n-1)!} \times \frac{1}{n-1} < 5 \times 10^{-11}, \text{ or } (n-1)! \times (n-1) > \frac{1}{5} \times 10^{11} = 2 \times 10^{10}.$$

But $13! \times 13 \simeq 8.095 \times 10^{10}$, so $n = 14$ is sufficient.

In other words, 14 terms will suffice.

Chapter 4

Section 4.1

1. (a) Let $\{x_n\}$ be any sequence in $\mathbb{R}$ that converges to 2. Then, by the Combination Rules for sequences, we have

$$f(x_n) = x_n^3 - 2x_n^2$$
$$\to 2^3 - 2 \times 2^2 = 8 - 8 = 0 \quad \text{as } n \to \infty.$$

But $f(0) = 0$, so that $f(x_n) \to f(0)$ as $n \to \infty$. Thus f is continuous at 2.

(b) Consideration of the graph of f near 1 suggests that f is not continuous at 1.

So, let $\{x_n\}$ be any sequence in $(0, 1)$ that tends to 1. Then $f(x_n) = 0$, for all n, so that   For example, $x_n = 1 - \frac{1}{2n}$.

$$f(x_n) \to 0 \quad \text{as } n \to \infty.$$

But $f(1) = 1$, so that $f(x_n) \nrightarrow f(1)$ as $n \to \infty$. Thus f is not continuous at 1.

Remark

If we had chosen $\{x_n\}$ to be any sequence in $(1, 2)$ that tends to 1, we would have found that $f(x_n) \to f(1)$ as $n \to \infty$. However this does NOT prove that f is continuous at 1, since the definition of continuity requires that $f(x_n) \to f(1)$ for ALL sequences that tend to 1.

2. (a) Let c be any point in $\mathbb{R}$.

Let $\{x_n\}$ be any sequence in $\mathbb{R}$ that converges to c. Then $f(x_n) = 1$, for all n, so that

$$f(x_n) \to 1 \quad \text{as } n \to \infty.$$

But $f(c) = 1$, so that $f(x_n) \to f(c)$ as $n \to \infty$. Thus f is continuous at c.

Since c is an arbitrary point of $\mathbb{R}$, it follows that f is continuous on $\mathbb{R}$.

(b) Let c be any point in $\mathbb{R}$.

Let $\{x_n\}$ be any sequence in $\mathbb{R}$ that converges to c. Then $f(x_n) = x_n$, for all n, so that

$$f(x_n) \to c \quad \text{as } n \to \infty.$$

But $f(c) = c$, so that $f(x_n) \to f(c)$ as $n \to \infty$. Thus f is continuous at c.

Since c is an arbitrary point of $\mathbb{R}$, it follows that f is continuous on $\mathbb{R}$.

3. The domain of f is the interval $I = \{x : x \geq 0\}$ if n is even and $\mathbb{R}$ if n is odd.

Thus, we have to show that for each c in I:

for each sequence $\{x_k\}$ in I such that $x_k \to c$, then $x_k^{\frac{1}{n}} \to c^{\frac{1}{n}}$.

First, let $c = 0$. We have already seen (in the Power Rule for null sequences and the subsequent Remark) that, for any null sequence $\{x_k\}$ in I, $\left\{x_k^{\frac{1}{n}}\right\}$ is also a null sequence. In other words, $f(x_k) \to f(0)$ as $k \to \infty$. Hence f is continuous at 0.

Next, let $c \neq 0$. We have to prove that if $\{x_k - c\}$ is a null sequence, then so is $\left\{x_k^{\frac{1}{n}} - c^{\frac{1}{n}}\right\}$.

Now, using the hint, we have

$$x_k^{\frac{1}{n}} - c^{\frac{1}{n}} = \frac{x_k - c}{x_k^{\frac{n-1}{n}} + x_k^{\frac{n-2}{n}} c^{\frac{1}{n}} + x_k^{\frac{n-3}{n}} c^{\frac{2}{n}} + \cdots + c^{\frac{n-1}{n}}}.$$

By the result of part (b) of Exercise 3 on Section 2.3, we obtain

$$x_k^{\frac{n-1}{n}} + x_k^{\frac{n-2}{n}} c^{\frac{1}{n}} + x_k^{\frac{n-3}{n}} c^{\frac{2}{n}} + \cdots + c^{\frac{n-1}{n}}$$
$$\to c^{\frac{n-1}{n}} + c^{\frac{n-2}{n}} c^{\frac{1}{n}} + c^{\frac{n-3}{n}} c^{\frac{2}{n}} + \cdots + c^{\frac{n-1}{n}} \quad \text{as } k \to \infty$$
$$= nc^{\frac{n-1}{n}};$$

since $c \neq 0$, we deduce that

$$x_k^{\frac{1}{n}} - c^{\frac{1}{n}} \to \frac{0}{nc^{\frac{n-1}{n}}} \quad \text{as } k \to \infty$$
$$= 0.$$

In other words, $\left\{x_k^{\frac{1}{n}} - c^{\frac{1}{n}}\right\}$ is null, as required. It follows that f is continuous at c.

4. (a) Let d be any point in $\mathbb{R}$.

Let $\{x_n\}$ be any sequence in $\mathbb{R}$ that converges to d. Then $f(x_n) = c$, for all n, so that

$$f(x_n) \to c \quad \text{as } n \to \infty.$$

But $f(d) = c$, so that $f(x_n) \to f(d)$ as $n \to \infty$. Thus f is continuous at d.

Since d is an arbitrary point of $\mathbb{R}$, it follows that f is continuous on $\mathbb{R}$.

(b) Let c be any point in $\mathbb{R}$.

Let $\{x_k\}$ be any sequence in $\mathbb{R}$ that converges to c. Then $f(x_k) = x_k^n$, for all k, so that, by the Product Rule for sequences

$$f(x_k) \to c^n \quad \text{as } k \to \infty.$$

But $f(c) = c^n$, so that $f(x_k) \to f(c)$ as $k \to \infty$. Thus f is continuous at c.

Since c is an arbitrary point of $\mathbb{R}$, it follows that f is continuous on $\mathbb{R}$.

(c) Let c be any point in $\mathbb{R}$.

Let $\{x_n\}$ be any sequence in $\mathbb{R}$ that converges to c. Then, by Theorem 4 of Sub-section 2.3.3, we have

$$f(x_n) = |x_n| \to |c| \quad \text{as } n \to \infty.$$

But $f(c) = |c|$, so that $f(x_n) \to f(c)$ as $n \to \infty$. Thus f is continuous at c.

Since c is an arbitrary point of $\mathbb{R}$, it follows that f is continuous on $\mathbb{R}$.

5. First, let $c = 0$. Consideration of the graph of f near 0 suggests that f is not continuous at 0. So, let $\{x_n\}$ be any sequence in $(0, 1)$ that tends to 0. Then $f(x_n) = 1$, for all n, so that

$$f(x_n) \to 1 \quad \text{as } n \to \infty.$$

But $f(0) = 0$, so that $f(x_n) \not\to f(0)$ as $n \to \infty$. Thus f is not continuous at 0.

Next, let $c > 0$. We will show that f is continuous at c.

So, let $\{x_n\}$ be any sequence that tends to c. Since $c > 0$, it follows that eventually $x_n > 0$; thus $x_n > 0$ for all $n > N$, for a suitable choice of N. Then $f(x_n) = 1$, for all $n > N$, so that

$$f(x_n) \to 1 \quad \text{as } n \to \infty.$$

But $f(c) = 1$, so that $f(x_n) \to f(c)$ as $n \to \infty$. Thus f is continuous at c.

Finally, let $c < 0$. We will show that f is continuous at c.

So, let $\{x_n\}$ be any sequence that tends to c. Since $c < 0$, it follows that eventually $x_n < 0$; thus $x_n < 0$ for all $n > N$, for a suitable choice of N. Then $f(x_n) = -1$, for all $n > N$, so that

$$f(x_n) \to -1 \quad \text{as } n \to \infty.$$

But $f(c) = -1$, so that $f(x_n) \to f(c)$ as $n \to \infty$. Thus f is continuous at c.

It follows that f is continuous on $\mathbb{R} - \{0\}$, and discontinuous at 0.

6. We have already seen, in Example 3 of Sub-section 4.1.1, that the function

$$g : x \mapsto x^{\frac{1}{2}}, \quad \text{for } x \geq 0,$$

is continuous on $[0, \infty)$. Also, it follows from Problem 4 that the function

$$h : x \mapsto x^3, \quad \text{for } x \in \mathbb{R},$$

is continuous on $\mathbb{R}$.

We may apply the Composition Rule to the functions g and h, since $g([0, \infty)) \subset \mathbb{R}$; it follows that

$$h \circ g : x \mapsto \left(x^{\frac{1}{2}} \right)^3 = x^{\frac{3}{2}}, \quad \text{for } x \geq 0,$$

is continuous on $[0, \infty)$.

7. By the Sum and Product Rules, and by using the fact that the constant function and the identity function are continuous on $\mathbb{R}$, we see that

$$g : x \mapsto x^2 + x + 1, \quad x \in \mathbb{R},$$

is continuous on $\mathbb{R}$. Since $g(\mathbb{R}) \subset (0, \infty)$ and the square root function h is continuous on $(0, \infty)$, it follows, by the Composition Rule, that the function

$$h \circ g : x \mapsto \sqrt{x^2 + x + 1}, \quad x \in \mathbb{R},$$

is continuous on $\mathbb{R}$.

> This also follows from the fact that all polynomials are continuous on $\mathbb{R}$.

Next, by the Sum, Multiple and Product Rules, and by using the fact that the constant function and the identity function are continuous on $\mathbb{R}$, we see that the functions

$$k : x \mapsto -5x \quad \text{and} \quad l : x \mapsto 1 + x^2, \quad x \in \mathbb{R},$$

are continuous on $\mathbb{R}$. Since $l(x) \neq 0$, it follows from the Product and Quotient Rules, that

$$m : x \mapsto \frac{-5x}{1 + x^2}, \quad x \in \mathbb{R},$$

is continuous on $\mathbb{R}$.

> This also follows from the fact that all polynomials are continuous on $\mathbb{R}$.

> This also follows from the fact that rational functions are continuous on their domains.

It then follows, by applying the Sum Rule to $h \circ g$ and m, that the function

$$f(x) = \sqrt{x^2 + x + 1} - \frac{5x}{1 + x^2}, \quad x \in \mathbb{R},$$

is continuous on $\mathbb{R}$.

8. (a) The graph of f suggests that we should find functions g and h that squeeze f near 0.

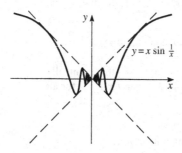

So, we define $g(x) = -|x|$, $x \in \mathbb{R}$, and $h(x) = |x|$, $x \in \mathbb{R}$. With these two chosen, we check the conditions of the Squeeze Rule.

Now we know that

$$-1 \le \sin\left(\frac{1}{x}\right) \le 1, \quad \text{for any } x \ne 0.$$

It follows that

$$-|x| \le x\sin\left(\frac{1}{x}\right) \le |x|, \quad \text{for any } x \ne 0,$$

so that

$$g(x)[=-|x|] \le f(x) \le [|x| =] \, h(x), \quad \text{for any } x \in \mathbb{R}.$$

So condition 1 of the Squeeze Rule is satisfied.

Next, the functions f, g and h all take the value 0 at the point 0. Thus condition 2 of the Squeeze Rule is satisfied.

Finally, the functions g and h are both continuous at 0.

It then follows, from the Squeeze Rule, that f is continuous at 0.

(b) Consideration of the graph of f near 0 suggests that f is not continuous at 0.

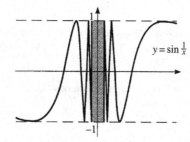

So, let $\{x_n\}$ be the sequence

$$\{x_n\} = \left\{\frac{1}{(2n+\frac{1}{2})\pi}\right\}.$$

Then $x_n \to 0$ as $n \to \infty$, and

$$f(x_n) = \sin\left(\left(2n+\frac{1}{2}\right)\pi\right)$$
$$= \sin\left(\frac{1}{2}\pi\right) = 1$$

for all n, so that

$$f(x_n) \to 1 \quad \text{as } n \to \infty.$$

But $f(0) = 0$, so that $f(x_n) \not\to f(0)$ as $n \to \infty$. Thus f is not continuous at 0.

9. Since the constant function and the identity function are continuous on $\mathbb{R}$, it follows from the Combination Rules that the function

$$g(x) = x^2 + 1, \quad x \in \mathbb{R},$$

is continuous on $\mathbb{R}$.

> This also follows from the fact that all polynomials are continuous on $\mathbb{R}$.

Next, since the sine function is continuous on $\mathbb{R}$, it follows by the Composition Rule, that

$$\sin \circ g(x) = \sin(x^2 + 1), \quad x \in \mathbb{R},$$

is continuous on $\mathbb{R}$.

It then follows from the Multiple and Sum Rules, that the function

$$f(x) = x^2 + 1 + 3\sin(x^2 + 1), \quad x \in \mathbb{R},$$

is continuous on $\mathbb{R}$.

10. The cosine function is continuous on $\mathbb{R}$, so that, by the Multiple Rule, the function

$$x \mapsto \frac{\pi}{2}\cos x, \quad x \in \mathbb{R},$$

is continuous on $\mathbb{R}$.

Then, since the sine function, also, is continuous on $\mathbb{R}$, we deduce, from the Composition Rule, that

$$f(x) = \sin\left(\frac{\pi}{2}\cos x\right), \quad x \in \mathbb{R},$$

is continuous on $\mathbb{R}$.

11. Since the identity function is continuous on $\mathbb{R}$, it follows, from the Combination Rules, that the function

$$g(x) = x^5 - 5x^2, \quad x \in \mathbb{R},$$

is continuous on $\mathbb{R}$.

> This also follows from the fact that all polynomials are continuous on $\mathbb{R}$.

Then, since the cosine function is continuous on $\mathbb{R}$, we deduce, from the Composition Rule, that

$$h(x) = \cos(x^5 - 5x^2), \quad x \in \mathbb{R},$$

is continuous on $\mathbb{R}$.

Next, since the identity function is continuous on $\mathbb{R}$, it follows, from the Combination Rules, that the function

$$k(x) = -x^2, \quad x \in \mathbb{R},$$

is continuous on $\mathbb{R}$.

> This also follows from the fact that all polynomials are continuous on $\mathbb{R}$.

Then, since the exponential function is continuous on $\mathbb{R}$, we deduce, from the Composition Rule, that

$$l(x) = e^{-x^2}, \quad x \in \mathbb{R},$$

is continuous on $\mathbb{R}$.

Finally, it follows, from the Combination Rules, that

$$f(x) = \cos(x^5 - 5x^2) + 7e^{-x^2}, \quad x \in \mathbb{R},$$

is continuous on $\mathbb{R}$.

Section 4.2

1. Consider the function

$$f(x) = \cos x - x, \quad x \in [0, 1].$$

We shall show that f has a zero c in $(0, 1)$.

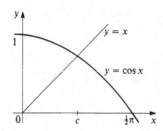

Certainly, f is continuous on $[0, 1]$, and

$$f(0) = \cos 0 - 0 = 1 > 0,$$
$$f(1) = \cos 1 - 1 < 0.$$

Thus, by the Intermediate Value Theorem, there is some number c in $(0, 1)$ such that $f(c) = 0$, and so such that

$$\cos c = c.$$

2. If $f(0) = 0$ or $f(1) = 1$, we can take $c = 0$ or $c = 1$, respectively.

Otherwise, we have $f(0) > 0$ and $f(1) < 1$, since the image of $[0, 1]$ under f lies in $[0, 1]$.

We now define the auxiliary function

$$g(x) = f(x) - x, \quad x \in [0, 1],$$

and show that g has a zero c in $(0, 1)$.

Certainly, g is continuous on $[0, 1]$, and

$$g(0) = f(0) - 0 > 0,$$
$$g(1) = f(1) - 1 < 0.$$

Thus, by the Intermediate Value Theorem, there is some number c in $(0, 1)$ such that $g(c) = 0$, and so such that

$$f(c) = c.$$

3. For the function $f(x) = x^5 + x - 1$, $x \in [0, 1]$, we have

$$f(0) = -1 < 0, \; f(1) = 1 + 1 - 1 = 1 > 0, \quad \text{and}$$

$$f\left(\frac{1}{2}\right) = \frac{1}{32} + \frac{1}{2} - 1 = -\frac{15}{32}.$$

It follows, by the Intermediate Value Theorem applied to $\left[\frac{1}{2}, 1\right]$, that f has a zero in $\left(\frac{1}{2}, 1\right)$.

Next, $f\left(\frac{3}{4}\right) = \frac{243}{1024} + \frac{3}{4} - 1 = -\frac{13}{1024} < 0$. Thus, since $f\left(\frac{3}{4}\right) < 0$ and $f(1) > 0$ it follows, by the Intermediate Value Theorem applied to $\left[\frac{3}{4}, 1\right]$, that f has a zero in $\left(\frac{3}{4}, 1\right)$.

Finally, $f\left(\frac{7}{8}\right) = \frac{16,807}{32,768} - \frac{1}{8} = \frac{12,711}{32,768} > 0$. Thus, since $f\left(\frac{3}{4}\right) < 0$ and $f\left(\frac{7}{8}\right) > 0$ it follows, by the Intermediate Value Theorem applied to $\left[\frac{3}{4}, \frac{7}{8}\right]$, that f has a zero in $\left(\frac{3}{4}, \frac{7}{8}\right)$. This interval $\left(\frac{3}{4}, \frac{7}{8}\right)$ is of length $\frac{1}{8}$, as required.

4. We compile the following table of values for the continuous function p:

x	-1	0	1	2
$p(x)$	-3	1	-1	3

Since $p(-1) < 0$ and $p(0) > 0$, it follows, from the Intermediate Value Theorem, that p has a zero in $(-1, 0)$.

Since $p(0) > 0$ and $p(1) < 0$, it follows, from the Intermediate Value Theorem, that p has a zero in $(0, 1)$.

Since $p(1) < 0$ and $p(2) > 0$, it follows, from the Intermediate Value Theorem, that p has a zero in $(1, 2)$.

5. For the polynomial $p(x) = x^5 + 3x^4 - x - 1$, $x \in \mathbb{R}$, that is continuous on $\mathbb{R}$, we have

$$M = 1 + \max\{|3|, |-1|, |-1|\} = 4,$$

so that, by the Zeros Localisation Theorem, all the zeros of p lie in $(-4, 4)$.

We now compile a table of values of $p(x)$, for $x = -4, -3, \ldots, 4$:

x	-4	-3	-2	-1	0	1	2	3	4
$p(x)$	-253	2	17	2	-1	2	77	482	$1,787$

By applying the Intermediate Value Theorem to p on each of the three intervals $[-4, -3]$, $[-1, 0]$ and $[0, 1]$, we deduce that p has a zero in each of the intervals

$$(-4, -3), (-1, 0) \quad \text{and} \quad (0, 1).$$

6. (a) First, we look separately at the two closed subintervals $[-1, 0]$ and $[0, 2]$, on each of which f is strictly monotonic. The function f is continuous on each of these subintervals, and so has a maximum and a minimum on each.

Since f is strictly decreasing on $[-1, 0]$, it follows that

$$\max\{f(x) : x \in [-1, 0]\} = f(-1) = 1,$$

and this is attained in $[-1, 0]$ only at -1; also

$$\min\{f(x) : x \in [-1, 0]\} = f(0) = 0,$$

and this is attained in $[-1, 0]$ only at 0.

Since f is strictly increasing on $[0, 2]$, it follows that

$$\max\{f(x) : x \in [0, 2]\} = f(2) = 4,$$

and this is attained in $[0, 2]$ only at 2; also

$$\min\{f(x) : x \in [0, 2]\} = f(0) = 0,$$

and this is attained in $[0, 2]$ only at 0.

Combining these results, we see that

$$\max\{f(x) : x \in [-1, 2]\} = 4,$$

and this is attained in $[-1, 2]$ only at 2; also

$$\min\{f(x) : x \in [-1, 2]\} = 0,$$

and this is attained in $[-1, 2]$ only at 0.

(b) First, we look separately at the three closed subintervals $\left[0, \frac{1}{2}\pi\right]$, $\left[\frac{1}{2}\pi, \frac{3}{2}\pi\right]$ and $\left[\frac{3}{2}\pi, 2\pi\right]$ on each of which f is strictly monotonic. The function f is continuous on each of these subintervals, and so has a maximum and a minimum on each.

Since f is strictly increasing on $\left[0, \frac{1}{2}\pi\right]$, it follows that

$$\max\{f(x) : x \in \left[0, \tfrac{1}{2}\pi\right]\} = f\left(\tfrac{1}{2}\pi\right) = 1,$$

and this is attained in $\left[0, \frac{1}{2}\pi\right]$ only at $\frac{1}{2}\pi$; also

$$\min\{f(x) : x \in \left[0, \tfrac{1}{2}\pi\right]\} = f(0) = 0,$$

and this is attained in $\left[0, \frac{1}{2}\pi\right]$ only at 0.

Since f is strictly decreasing on $\left[\frac{1}{2}\pi, \frac{3}{2}\pi\right]$, it follows that

$$\max\{f(x) : x \in \left[\tfrac{1}{2}\pi, \tfrac{3}{2}\pi\right]\} = f\left(\tfrac{1}{2}\pi\right) = 1,$$

and this is attained in $\left[\frac{1}{2}\pi, \frac{3}{2}\pi\right]$ only at $\frac{1}{2}\pi$; also

$$\min\{f(x) : x \in \left[\tfrac{1}{2}\pi, \tfrac{3}{2}\pi\right]\} = f\left(\tfrac{3}{2}\pi\right) = -1,$$

and this is attained in $\left[\frac{1}{2}\pi, \frac{3}{2}\pi\right]$ only at $\frac{3}{2}\pi$.

Next, since f is strictly increasing on $\left[\frac{3}{2}\pi, 2\pi\right]$, it follows that

$$\max\{f(x) : x \in \left[\tfrac{3}{2}\pi, 2\pi\right]\} = f(2\pi) = 0,$$

and this is attained in $\left[\frac{3}{2}\pi, 2\pi\right]$ only at 0; also

$$\min\{f(x) : x \in \left[\tfrac{3}{2}\pi, 2\pi\right]\} = f\left(\tfrac{3}{2}\pi\right) = -1,$$

and this is attained in $\left[\frac{3}{2}\pi, 2\pi\right]$ only at $\frac{3}{2}\pi$.
   Combining these results, we see that

$$\max\{f(x) : x \in [0, 2\pi]\} = 1,$$

and this is attained in $[0, 2\pi]$ only at $\frac{1}{2}\pi$; also

$$\min\{f(x) : x \in [0, 2\pi]\} = -1,$$

and this is attained in $[0, 2\pi]$ only at $\frac{3}{2}\pi$.

Section 4.3

1. (a) If $0 \le x_1 < x_2$, then $2x_1 < 2x_2$ and $x_1^4 < x_2^4$. Hence

$$x_1^4 + 2x_1 + 3 < x_2^4 + 2x_2 + 3,$$

so that f is strictly increasing on $\mathbb{R}$, and is thus one–one on $\mathbb{R}$.

(b) If $0 < x_1 < x_2$, then $x_1^2 < x_2^2$ and $\frac{1}{x_1} > \frac{1}{x_2}$, so that $-\frac{1}{x_1} < -\frac{1}{x_2}$. Hence

$$x_1^2 - \frac{1}{x_1} < x_2^2 - \frac{1}{x_2};$$

it follows that f is strictly increasing on $(0, \infty)$, and is thus one–one on $(0, \infty)$.

2. We perform the three steps of the strategy.

   1. We showed, in Problem 1(b) above, that f is strictly increasing on $(0, \infty)$.

   2. The function

   $$x \mapsto x^2 - \frac{1}{x} = \frac{x^3 - 1}{x}, \quad x \in \mathbb{R} - \{0\},$$

   is a rational function, and therefore is continuous on its domain $\mathbb{R} - \{0\}$. In particular, f is continuous on $(0, \infty)$.

   3. Choose the increasing sequence $\{n\}$, which tends to ∞, the right-hand end-point of $(0, \infty)$. Then

   $$f(n) = n^2 - \frac{1}{n} \to \infty \quad \text{as } n \to \infty,$$

   by the Reciprocal Rule for sequences. Thus the right-hand end-point of $J = f((0, \infty))$ is ∞.

   4. Choose the decreasing sequence $\{\frac{1}{n}\}$, which tends to 0, the left-hand end-point of $(0, \infty)$. Then

   $$f\left(\frac{1}{n}\right) = \frac{1}{n^2} - n \to -\infty \quad \text{as } n \to \infty,$$

   by the Reciprocal Rule for sequences. Thus the left-hand end-point of $J = f((0, \infty))$ is $-\infty$.
      Hence f has a continuous inverse $f^{-1} \colon \mathbb{R} \to (0, \infty)$, by the Inverse Function Rule.

3. (a) Since $\sin\left(\frac{1}{4}\pi\right) = \frac{1}{\sqrt{2}}$ and $\frac{1}{4}\pi$ lies in $\left[-\frac{1}{2}\pi, \frac{1}{2}\pi\right]$, we have $\sin^{-1}\left(\frac{1}{\sqrt{2}}\right) = \frac{1}{4}\pi$.

   Since $\cos\left(\frac{1}{3}\pi\right) = \frac{1}{2}$, we have $\cos\left(\frac{2}{3}\pi\right) = -\frac{1}{2}$, and $\frac{2}{3}\pi$ lies in $[0, \pi]$, so that $\cos^{-1}\left(-\frac{1}{2}\right) = \frac{2}{3}\pi$.

   Since $\tan\left(\frac{1}{3}\pi\right) = \sqrt{3}$ and $\frac{1}{3}\pi$ lies in $\left(-\frac{1}{2}\pi, \frac{1}{2}\pi\right)$, we have $\tan^{-1}\left(\sqrt{3}\right) = \frac{1}{3}\pi$.

(b) Following the Hint, we put $y = \sin^{-1} x$. Then
$$\cos(2\sin^{-1} x) = \cos(2y)$$
$$= 1 - 2\sin^2 y$$
$$= 1 - 2x^2,$$

since $x = \sin y$.

4. Following the Hint, we put $a = \log_e x$ and $b = \log_e y$. Then $x = e^a$ and $y = e^b$, so that
$$\log_e(xy) = \log_e\left(e^a e^b\right)$$
$$= \log_e\left(e^{a+b}\right)$$
$$= a + b$$
$$= \log_e x + \log_e y.$$

5. Let $y = \cosh^{-1} x$, where $x \geq 1$. Then
$$x = \cosh y = \tfrac{1}{2}(e^y + e^{-y}),$$

so that
$$e^{2y} - 2xe^y + 1 = 0.$$

This is a quadratic equation in e^y, and so
$$e^y = x \pm \sqrt{x^2 - 1}.$$

Both choices of $+$ or $-$ give a positive expression on the right, but recall that $y \geq 0$ and so $e^y \geq 1$. Since
$$\left(x + \sqrt{x^2 - 1}\right) \times \left(x - \sqrt{x^2 - 1}\right) = 1,$$

we choose the $+$ sign, because $x + \sqrt{x^2 - 1} \geq 1$, whereas $x - \sqrt{x^2 - 1} \leq 1$.

Hence
$$y = \cosh^{-1} x = \log_e\left(x + \sqrt{x^2 - 1}\right).$$

(The value $y = \log_e\left(x - \sqrt{x^2 - 1}\right)$ gives the negative solution of the equation $\cosh y = x$.)

Section 4.4

1. (a) For $x > 0$, $f(x) = x^\alpha = e^{\alpha \log_e x}$.
   Now, the functions
   $$x \mapsto \log_e x, \ x \in (0, \infty), \quad \text{and} \quad x \mapsto e^x, \ x \in \mathbb{R},$$
   are continuous; hence, by the Multiple Rule and the Composition Rule, f is continuous.

   (b) For $x > 0$, $f(x) = x^x = e^{x \log_e x}$.
   Now, the functions
   $$x \mapsto \log_e x, \quad x \in (0, \infty), \qquad x \mapsto x, \quad x \in \mathbb{R},$$
   $$\text{and } x \mapsto e^x, \quad x \in \mathbb{R},$$
   are continuous; hence, by the Product Rule and the Composition Rule, f is continuous.

2. Since $a^x = e^{x \log_e a}$ and $a^y = e^{y \log_e a}$, we have
$$a^x a^y = e^{x \log_e a} e^{y \log_e a}$$
$$= e^{x \log_e a + y \log_e a}$$
$$= e^{(x+y) \log_e a}$$
$$= a^{x+y}.$$

3. We use the inequality $e^x > 1 + x$, for $x > 0$. Applying this inequality with x replaced by $\left(\frac{x}{e}\right) - 1$, we obtain

$$e^{\left(\frac{x}{e}\right)-1} > 1 + \left(\frac{x}{e} - 1\right), \quad \text{for } x > e,$$
$$= \frac{x}{e} = xe^{-1}.$$

It follows that

$$e^{\frac{x}{e}} > x,$$

so that, by Rule 5 with $p = e$

$$e^x > x^e.$$

Chapter 5

Section 5.1

1. (a) The function

$$f(x) = \frac{x^2 + x}{x}$$

has domain $\mathbb{R} - \{0\}$, so that f is defined on any punctured neighbourhood of 0.

Next, notice that $f(x) = x + 1$, for $x \neq 0$. It follows that, if $\{x_n\}$ lies in $\mathbb{R} - \{0\}$ and $x_n \to 0$ as $n \to \infty$, then $f(x_n) \to 1$ as $n \to \infty$.

Hence

$$\lim_{x \to 0} f(x) = 1.$$

(b) First, notice that

$$f(x) = \frac{|x|}{x} = \begin{cases} 1, & x > 0, \\ -1, & x < 0. \end{cases}$$

The domain of f is $\mathbb{R} - \{0\}$, so that f is defined on any punctured neighbourhood of 0.

However, the two null sequences $\left\{\frac{1}{n}\right\}$ and $\left\{-\frac{1}{n}\right\}$ both have non-zero terms, but

$$\lim_{n \to \infty} f\left(\frac{1}{n}\right) = 1, \quad \text{whereas } \lim_{n \to \infty} f\left(-\frac{1}{n}\right) = -1.$$

Hence, $\lim_{x \to 0} f(x)$ does not exist.

2. (a) The function $f(x) = \sqrt{x}$ is defined on $[0, \infty)$, which contains the point 2; it is also continuous at 2.

Hence, by Theorem 2

$$\lim_{x \to 2} \sqrt{x} = \sqrt{2}.$$

(b) The function $f(x) = \sqrt{\sin x}$ is defined on $[0, \pi]$, which contains the point $\frac{\pi}{2}$; it is also continuous at $\frac{\pi}{2}$, by the Composition Rule for continuous functions.

Hence, by Theorem 2

$$\lim_{x \to \frac{\pi}{2}} \sqrt{\sin x} = \sqrt{\sin\left(\frac{\pi}{2}\right)} = 1.$$

(c) The function $f(x) = \frac{e^x}{1+x}$ is defined on $(-1, \infty)$, which contains the point 1; it is also continuous at 1, by the Composition Rule for continuous functions.

Hence, by Theorem 2

$$\lim_{x \to 1} \frac{e^x}{1+x} = \frac{1}{2}e.$$

3. (a) Since

$$\lim_{x\to 0}\frac{\sin x}{x}=1 \quad\text{and}\quad \lim_{x\to 0}\frac{1}{2+x}=\frac{1}{2},$$

we deduce from the Product Rule for limits that

$$\lim_{x\to 0}\frac{\sin x}{2x+x^2}=\frac{1}{2}.$$

(b) Let $f(x)=x^2$ and $g(x)=\frac{\sin x}{x}$; then

$$\lim_{x\to 0}x^2=0 \quad\text{and}\quad \lim_{x\to 0}\frac{\sin x}{x}=1.$$

Also, $f(x)=x^2\ne 0$ in $\mathbb{R}-\{0\}$; hence, by the Composition Rule

$$\lim_{x\to 0}g(f(x))=\lim_{x\to 0}\frac{\sin(x^2)}{x^2}=1.$$

(c) Let $f(x)=\frac{x}{\sin x}$ and $g(x)=\sqrt{x}$; then f is defined on the punctured neighbourhood $(-\pi,0)\cup(0,\pi)$ of 0, and, by the Quotient Rule for limits

$$\lim_{x\to 0}f(x)=\left(\lim_{x\to 0}\frac{\sin x}{x}\right)^{-1}=1.$$

Also, g is defined on $(0,\infty)$, and is continuous at 1; hence, by Theorem 2 and the Composition Rule

$$\lim_{x\to 0}g(f(x))=\lim_{x\to 0}\left(\frac{x}{\sin x}\right)^{\frac{1}{2}}=g(1)=1.$$

4. Here

$$(f\circ g)(x)=\begin{cases} f(0), & x=0, \\ f(1+x), & x\ne 0, \end{cases}$$

$$=\begin{cases} f(0), & x=0, \\ f(0), & x=-1, \\ f(1+x), & x\ne 0,\ -1 \quad\text{(so that } 1+x\ne 1,0) \end{cases}$$

$$=\begin{cases} 0, & x=0, \\ 0, & x=-1, \\ 2+(1+x), & x\ne 0,\ -1, \end{cases}$$

$$=\begin{cases} 0, & x=0,\ -1, \\ 3+x, & x\ne 0,\ -1. \end{cases}$$

In particular, $(f\circ g)(0)=0$.

Next, $\lim_{x\to 0}g(x)=1$, so that

$$f\left(\lim_{x\to 0}g(x)\right)=f(1)=-2.$$

Finally, since $f(g(x))=3+x$ on the punctured neighbourhood $(-1,0)\cup(0,1)$ of 0

$$\lim_{x\to 0}f(g(x))=\lim_{x\to 0}(3+x)=3.$$

5. (a) The inequalities

$$-x^2\le x^2\sin\left(\frac{1}{x}\right)\le x^2,\quad\text{for }x\ne 0,$$

show that condition 1 of the Squeeze Rule holds, with

$$g(x)=-x^2 \quad\text{and}\quad h(x)=x^2,\quad x\in\mathbb{R}.$$

Since
$$\lim_{x \to 0}(x^2) = 0 \quad \text{and} \quad \lim_{x \to 0}(-x^2) = 0,$$
it follows, from the Squeeze Rule, that
$$\lim_{x \to 0} x^2 \sin\left(\frac{1}{x}\right) = 0.$$

(b) The inequalities
$$-|x| \le x \cos\left(\frac{1}{x}\right) \le |x|, \quad \text{for } x \ne 0,$$
show that condition 1 of the Squeeze Rule holds, with
$$g(x) = -|x| \quad \text{and} \quad h(x) = |x|, \quad x \in \mathbb{R}.$$
Since
$$\lim_{x \to 0}(-|x|) = 0 \quad \text{and} \quad \lim_{x \to 0}(|x|) = 0,$$
it follows, from the Squeeze Rule, that
$$\lim_{x \to 0} x \cos\left(\frac{1}{x}\right) = 0.$$

6. Let
$$f(x) = \begin{cases} \sin\dfrac{1}{x}, & -1 \le x < 0, \\ 1 + x, & 0 \le x \le 0. \end{cases}$$
Then, for $n = 1, 2, \ldots$, we have
$$f\left(-\frac{1}{n\pi}\right) = \sin(-n\pi) = 0 \to 0 \quad \text{as } n \to \infty,$$
whereas
$$f\left(-\frac{1}{(2n + \frac{1}{2})\pi}\right) = \sin\left(-\left(2n + \tfrac{1}{2}\right)\pi\right) = -1 \to -1 \quad \text{as } n \to \infty.$$

It follows that $\lim_{x \to 0^-} f(x)$ does not exist.

However, for any sequence $\{x_n\}$ in $(0, 3)$ for which $x_n \to 0$, it follows, from the fact that the function $x \mapsto 1 + x$ is continuous on $[0, 3)$, that
$$\lim_{x \to 0^+} f(x) = f(0) = 1.$$

7. (a) Since $\lim_{x \to 0} \frac{\sin x}{x} = 1$, we have, by Theorem 7, that
$$\lim_{x \to 0^+} \frac{\sin x}{x} = 1.$$
Hence, by the Sum Rule
$$\lim_{x \to 0^+}\left(\frac{\sin x}{x} + \sqrt{x}\right) = \lim_{x \to 0^+} \frac{\sin x}{x} + \lim_{x \to 0^+} \sqrt{x}$$
$$= 1 + 0$$
$$= 1.$$

(b) Let $f(x) = \sqrt{x}, x \ge 0$, and $g(x) = \frac{\sin x}{x}, x \ne 0$. Then
$$\lim_{x \to 0^+} \sqrt{x} = 0 \quad \text{and} \quad \lim_{x \to 0} \frac{\sin x}{x} = 1.$$
Also, $f(x) = \sqrt{x} \ne 0$ in $(0, \infty)$. Hence, by the Composition Rule for limits
$$\lim_{x \to 0^+} g(f(x)) = \lim_{x \to 0^+} \frac{\sin(\sqrt{x})}{\sqrt{x}} = 1.$$

8. Since $1 + x \le e^x \le \frac{1}{1-x}$, for $|x| < 1$, we have

$$x \le e^x - 1 \le \frac{x}{1 - x}, \quad \text{for } |x| < 1, \qquad\qquad (*)$$

so that

$$1 \le \frac{e^x - 1}{x} \le \frac{1}{1 - x}, \quad \text{for } 0 < x < 1.$$

Then, since $\lim\limits_{x \to 0^+} 1 = 1$ and $\lim\limits_{x \to 0^+} \frac{1}{1-x} = 1$, it follows, from the Squeeze Rule, that

$$\lim_{x \to 0^+} \frac{e^x - 1}{x} = 1.$$

Next, it follows, from the inequalities $(*)$ above, that

$$1 \ge \frac{e^x - 1}{x} \ge \frac{1}{1 - x}, \quad \text{for } -1 < x < 0;$$

so that, by an argument similar to the one above, we have

$$\lim_{x \to 0^-} \frac{e^x - 1}{x} = 1.$$

Hence, by Theorem 7, we may combine these two one-sided limit results to obtain

$$\lim_{x \to 0} \frac{e^x - 1}{x} = 1.$$

Section 5.2

1. (a) Let $f(x) = |x|$. Then $f(x) > 0$ for $x \ne 0$, and, since f is continuous at 0

$$\lim_{x \to 0} |x| = 0.$$

Hence, by the Reciprocal Rule

$$\frac{1}{f(x)} = \frac{1}{|x|} \to \infty \quad \text{as } x \to 0.$$

(b) Let $f(x) = x^3 - 1$. Then $f(x) > 0$ for $x \in (1, \infty)$, and, since f is continuous at 1

$$\lim_{x \to 1^+} (x^3 - 1) = 0.$$

Hence, by the Reciprocal Rule

$$\frac{1}{f(x)} = \frac{1}{x^3 - 1} \to \infty \quad \text{as } x \to 1^+,$$

so that

$$-\frac{1}{f(x)} = \frac{1}{1 - x^3} \to -\infty \quad \text{as } x \to 1^+.$$

(c) Let $f(x) = \frac{x^3}{\sin x}$. Then $f(x) > 0$ for $x \in \left(-\frac{1}{2}\pi, 0\right) \cup \left(0, \frac{1}{2}\pi\right)$, and

$$\lim_{x \to 0} \frac{x^3}{\sin x} = \lim_{x \to 0} \left\{ \frac{x^2}{\frac{\sin x}{x}} \right\}$$

$$= \frac{\left\{ \lim\limits_{x \to 0} x^2 \right\}}{\left\{ \lim\limits_{x \to 0} \frac{\sin x}{x} \right\}}$$

$$= \frac{0}{1} = 0,$$

by the Quotient Rule for limits.

Hence, by the Reciprocal Rule

$$\frac{1}{f(x)} = \frac{\sin x}{x^3} \to \infty \quad \text{as } x \to 0.$$

2. (a) Let

$$f(x) = \frac{1}{x^2}, \quad x \in \mathbb{R} - \{0\}, \quad \text{and} \quad g(x) = \frac{1}{x^2}, \quad x \in \mathbb{R} - \{0\}.$$

Then $f(x) \to \infty$ as $x \to 0$, and $g(x) \to \infty$ as $x \to 0$; but

$$f(x) - g(x) = 0 \to 0 \quad \text{as } x \to 0.$$

(b) Let

$$f(x) = \frac{1}{x^2}, \quad x \in \mathbb{R} - \{0\}, \quad \text{and} \quad g(x) = x^2, \quad x \in \mathbb{R} - \{0\}.$$

Then $f(x) \to \infty$ as $x \to 0$, and $g(x) \to 0$ as $x \to 0$; but

$$f(x)g(x) = 1 \to 1 \quad \text{as } x \to 0.$$

3. (a) By the Sum Rule for limits as $x \to \infty$, we have

$$\lim_{x \to \infty} \frac{2x^3 + x}{x^3} = \lim_{x \to \infty} \left(2 + \frac{1}{x^2}\right)$$

$$= \lim_{x \to \infty} (2) + \lim_{x \to \infty} \left(\frac{1}{x^2}\right)$$

$$= 2 + 0 = 2.$$

(b) Since $-1 \le \sin x \le 1$ for $x \in \mathbb{R}$, we have

$$-\frac{1}{x} \le \frac{\sin x}{x} \le \frac{1}{x}, \quad \text{for } x \in (0, \infty).$$

Also

$$g(x) = -\frac{1}{x} \to 0 \quad \text{as } x \to \infty$$

and

$$h(x) = \frac{1}{x} \to 0 \quad \text{as } x \to \infty.$$

Hence, by the Squeeze Rule for limits as $x \to \infty$,

$$f(x) = \frac{\sin x}{x} \to 0 \quad \text{as } x \to \infty.$$

4. (a) Let $f(x) = \log_e x$ and $g(x) = \log_e x$, for $x \in (0, \infty)$. Then

$$f(x) \to \infty \quad \text{as } x \to \infty \quad \text{and} \quad g(x) \to \infty \quad \text{as } x \to \infty.$$

Hence, by the Composition Rule for limits

$$g(f(x)) = \log_e(\log_e x) \to \infty \quad \text{as } x \to \infty.$$

(b) Let $f(x) = \frac{1}{x}$ and $g(x) = \frac{e^x}{x}$, for $x \in \mathbb{R} - \{0\}$. Then

$$g(f(x)) = xe^{\frac{1}{x}}, \quad \text{for } x \in \mathbb{R} \quad \{0\}.$$

Now

$$f(x) \to -\infty \quad \text{as } x \to 0^-.$$

In order to use the Composition rule for limits, we need to determine the behaviour of $g(x)$ as $x \to -\infty$; that is, the behaviour of $g(-x)$ as $x \to \infty$.

Now, for $x > 0$

$$g(-x) = -\frac{e^{-x}}{x} = -\frac{1}{xe^x}$$

and

$$xe^x \to \infty \quad \text{as } x \to \infty.$$

It follows, by the Reciprocal Rule, that

$$g(-x) \to 0 \quad \text{as } x \to \infty.$$

Hence, by the Composition Rule for limits,

$$g(f(x)) = xe^{\frac{1}{x}} \to 0 \quad \text{as } x \to 0^-.$$

5. Let

$$f(x) = 1, \quad \text{for all } x \in \mathbb{R},$$

and

$$g(x) = \begin{cases} 2, & x = 1, \\ 3, & x \neq 1. \end{cases}$$

Then

$$f(x) \to 1 \quad \text{as } x \to \infty \quad (\text{so that } \ell = 1),$$

and

$$g(x) \to 3 \quad \text{as } x \to 1 \quad (\text{so that } m = 3);$$

however as $x \to \infty$, we have

$$g(f(x)) = g(1) = 2$$
$$\to 2 \neq 3(= m).$$

Section 5.3

1. (a) $|a_n - 0| < 0.1$

$\Leftrightarrow \left| \frac{(-1)^n}{n^2} \right| \quad < 0.1$

$\Leftrightarrow \frac{1}{n^2} \quad < 0.1$

$\Leftrightarrow n^2 \quad > 10.$

It follows that $|a_n - 0| < 0.1$ for all $n > X$, so long as $X \geq \sqrt{10} \simeq 3.16$.

(b) $|a_n - 0| < 0.01$

$\Leftrightarrow \left| \frac{(-1)^n}{n^2} \right| \quad < 0.01$

$\Leftrightarrow \frac{1}{n^2} \quad < 0.01$

$\Leftrightarrow n^2 \quad > 100$

$\Leftrightarrow n \quad > 10.$

It follows that $|a_n - 0| < 0.01$ for all $n > X$, so long as $X \geq 10$.

(c) $|a_n - 0| < \varepsilon$

$\Leftrightarrow \left| \frac{(-1)^n}{n^2} \right| \quad < \varepsilon$

$\Leftrightarrow \frac{1}{n^2} \quad < \varepsilon$

$\Leftrightarrow n^2 \quad > \frac{1}{\varepsilon}$

$\Leftrightarrow n \quad > \frac{1}{\sqrt{\varepsilon}}.$

It follows that $|a_n - 0| < \varepsilon$ for all $n > X$, so long as $X \geq \frac{1}{\sqrt{\varepsilon}}$.

2. (a) $|f(x) - 5| < 0.1$

     $\Leftrightarrow |2x - 2| \;\; < 0.1$

     $\Leftrightarrow |x - 1| \;\;\;\; < 0.05.$

It follows that $|f(x) - 5| < 0.1$ whenever $0 < |x - 1| < \delta$, so long as $\delta \leq 0.05$.

   (b) $|f(x) - 5| < 0.01$

     $\Leftrightarrow |2x - 2| \;\; < 0.01$

     $\Leftrightarrow |x - 1| \;\;\;\; < 0.005.$

It follows that $|f(x) - 5| < 0.01$ whenever $0 < |x - 1| < \delta$, so long as $\delta \leq 0.005$.

   (c) $|f(x) - 5| < \varepsilon$

     $\Leftrightarrow |2x - 2| \;\; < \varepsilon$

     $\Leftrightarrow |x - 1| \;\;\;\; < \dfrac{1}{2}\varepsilon.$

It follows that $|f(x) - 5| < \varepsilon$ whenever $0 < |x - 1| < \delta$, so long as $\delta \leq \frac{1}{2}\varepsilon$.

3. (a) The function $f(x) = 5x - 2$ is defined on every punctured neighbourhood of 3. We must prove that:

     for each positive number ε, there is a positive number δ such that
$$|(5x - 2) - 13| < \varepsilon, \quad \text{for all } x \text{ satisfying } 0 < |x - 3| < \delta;$$

that is

$$|x - 3| < \frac{1}{5}\varepsilon, \quad \text{for all } x \text{ satisfying } 0 < |x - 3| < \delta. \qquad (*)$$

Choose $\delta = \frac{1}{5}\varepsilon$; then the statement $(*)$ holds.
   Hence
$$\lim_{x \to 3}(5x - 2) = 13.$$

  (b) The function $f(x) = 1 - 7x^3$ is defined on every punctured neighbourhood of 0. We must prove that:

     for each positive number ε, there is a positive number δ such that
$$\left|(1 - 7x^3) - 1\right| < \varepsilon, \quad \text{for all } x \text{ satisfying } 0 < |x - 0| < \delta;$$

that is

$$\left|7x^3\right| < \varepsilon, \quad \text{for all } x \text{ satisfying } 0 < |x| < \delta. \qquad (*)$$

Choose $\delta = \sqrt[3]{\frac{1}{7}\varepsilon}$ (so that $7\delta^3 = \varepsilon$); then the statement $(*)$ holds.
Hence
$$\lim_{x \to 0}\left(1 - 7x^3\right) = 1.$$

  (c) The function $f(x) = x^2 \cos\left(\frac{1}{x^3}\right)$ is defined on every punctured neighbourhood of 0. We must prove that:

     for each positive number ε, there is a positive number δ such that
$$\left|x^2 \cos\left(\frac{1}{x^3}\right) - 0\right| < \varepsilon, \quad \text{for all } x \text{ satisfying } 0 < |x - 0| < \delta;$$

that is

$$\left|x^2 \cos\left(\frac{1}{x^3}\right)\right| < \varepsilon, \quad \text{for all } x \text{ satisfying } 0 < |x| < \delta. \qquad (*)$$

Choose $\delta = \sqrt{\varepsilon}$. Then, since $\left|\cos\left(\frac{1}{x^3}\right)\right| \leq 1$ for all non-zero x, it follows that, for $0 < |x| < \delta$

$$\left| x^2 \cos\left(\frac{1}{x^3}\right)\right| \leq x^2$$
$$< \delta^2 = \varepsilon,$$

so that the statement $(*)$ holds.

Hence

$$\lim_{x \to 0}\left(x^2 \cos\frac{1}{x^3}\right) = 0.$$

(d) We consider the cases that $x < 0$ and $x > 0$ separately; that is, the two one-sided limits.

Since $|x| = -x$ when $x < 0$, we have

$$\lim_{x \to 0^-}\frac{x + |x|}{x} = \lim_{x \to 0^-}\frac{0}{x} = 0;$$

also, since $|x| = x$ when $x > 0$, we have

$$\lim_{x \to 0^+}\frac{x + |x|}{x} = \lim_{x \to 0^+}\frac{2x}{x} = \lim_{x \to 0^+}(2) = 2.$$

Since the two one-sided limits are not equal, it follows that $\lim_{x \to 0}\frac{x + |x|}{x}$ does not exist.

4. The function $f(x) = x^3$, $x \in \mathbb{R}$, is defined on every punctured neighbourhood of 1. We must prove that:

for each positive number ε, there is a positive number δ such that
$$|x^3 - 1| < \varepsilon \text{ for all } x \text{ satisfying } 0 < |x - 1| < \delta;$$

that is

$$|x^2 + x + 1| \times |x - 1| < \varepsilon, \quad \text{for all } x \text{ satisfying } 0 < |x - 1| < \delta. \quad (*)$$

Choose $\delta = \min\{1, \tfrac{1}{7}\varepsilon\}$, so that:

(i) for $0 < |x - 1| < \delta$, we have $x \in (0, 2)$, and so
$$|x^2 + x + 1| = x^2 + x + 1$$
$$< 2^2 + 2 + 1 = 7;$$

(ii) $|x - 1| < \tfrac{1}{7}\varepsilon$.

It follows that

$$|x^2 + x + 1| \times |x - 1| < 7 \times \frac{1}{7}\varepsilon = \varepsilon, \quad \text{for all } x \text{ satisfying } 0 < |x - 1| < \delta;$$

that is, the statement $(*)$ holds.

Hence

$$\lim_{x \to 1}(x^3) = 1.$$

5. Since the limits for f and g exist, the functions f and g must be defined on some punctured neighbourhoods $(c - r_1, c) \cup (c, c + r_1)$ and $(c - r_2, c) \cup (c, c + r_2)$ of c, it follows that the function fg is certainly defined on the punctured neighbourhood $(c - r, c) \cup (c, c + r)$, where $r = \min\{r_1, r_2\}$.

We want to prove that:

for each positive number ε, there is a positive number δ such that

$$|(f(x)g(x)) - (\ell m)| < \varepsilon, \quad \text{for all } x \text{ satisfying } 0 < |x - c| < \delta. \quad (*)$$

In order to examine the first inequality in $(*)$, we write

$$f(x)g(x) - \ell m \quad \text{as} \quad (f(x) - \ell)(g(x) - m) + m(f(x) - \ell) + \ell(g(x) - m),$$

so that

$$|f(x)g(x) - \ell m| = |(f(x) - \ell)(g(x) - m) + m(f(x) - \ell) + \ell(g(x) - m)|$$
$$\leq |f(x) - \ell| \times |g(x) - m| + |m||f(x) - \ell| + |\ell||g(x) - m|. \quad (1)$$

We know that, since $\lim_{x \to c} f(x) = \ell$, there is a positive number δ_1 such that

$$|f(x) - \ell| < \varepsilon, \quad \text{for all } x \text{ satisfying } 0 < |x - c| < \delta_1; \quad (2)$$

and a positive number δ_2 such that

$$|f(x) - \ell| < 1, \quad \text{for all } x \text{ satisfying } 0 < |x - c| < \delta_2; \quad (3)$$

similarly, since $\lim_{x \to c} g(x) = m$, there is a positive number δ_3 such that

$$|g(x) - m| < \varepsilon, \quad \text{for all } x \text{ satisfying } 0 < |x - c| < \delta_3. \quad (4)$$

We now choose $\delta = \min\{\delta_1, \delta_2, \delta_3\}$. Then the statements (2), (3) and (4) hold for all x satisfying $0 < |x - c| < \delta$, so that, from (1), we deduce

$$|f(x)g(x) - \ell m| \leq |f(x) - \ell| \times |g(x) - m| + |m||f(x) - \ell| + |\ell||g(x) - m|$$
$$\leq 1 \times \varepsilon + |m|\varepsilon + |\ell|\varepsilon$$
$$= (1 + |m| + |\ell|)\varepsilon,$$

for all x satisfying $0 < |x - c| < \delta$.

This is equivalent to the statement $(*)$, in which the number ε has been replaced by $K\varepsilon$, where $K = 1 + |m| + |\ell|$. Since K is a constant, this is sufficient, by the $K\varepsilon$ Lemma.

It follows that $\lim_{x \to c} f(x)g(x) = \ell m$, as required.

6. Since $\lim_{x \to c} f(x) = \ell$ and $\ell \neq 0$, we may take $\varepsilon = \frac{1}{2}|\ell|$ in the definition of limit; it follows that there is a positive number δ such that

$$|f(x) - \ell| < \frac{1}{2}|\ell| \quad \text{for all } x \text{ satisfying } 0 < |x - c| < \delta;$$

in other words, for all x satisfying $0 < |x - c| < \delta$, we have

$$-\frac{1}{2}|\ell| < f(x) - \ell < \frac{1}{2}|\ell|,$$

which we may rewrite in the form

$$\ell - \frac{1}{2}|\ell| < f(x) < \ell + \frac{1}{2}|\ell|. \quad (*)$$

It follows, from $(*)$, that, if $\ell > 0$, then

$$f(x) > \ell - \frac{1}{2}|\ell|$$
$$= \ell - \frac{1}{2}\ell = \frac{1}{2}\ell > 0;$$

and, if $\ell < 0$, then

$$f(x) < \ell + \frac{1}{2}|\ell|$$
$$= \ell - \frac{1}{2}\ell = \frac{1}{2}\ell < 0.$$

In either case, we obtain that $f(x)$ has the same sign as ℓ on the punctured neighbourhood $\{x: 0 < |x - c| < \delta\}$ of c.

Section 5.4

1. The function $f(x) = x^3$, $x \in \mathbb{R}$, is defined on $\mathbb{R}$.
    We must prove that:

    for each positive number ε, there is a positive number δ such that
    $$|x^3 - 1| < \varepsilon, \quad \text{for all } x \text{ satisfying } |x - 1| < \delta;$$
    that is
    $$|x^2 + x + 1| \times |x - 1| < \varepsilon, \quad \text{for all } x \text{ satisfying } |x - 1| < \delta. \quad (*)$$

    Choose $\delta = \min\{1, \frac{1}{7}\varepsilon\}$, so that

    (i) for $|x - 1| < \delta$, we have $x \in (0, 2)$, and so
    $$|x^2 + x + 1| = x^2 + x + 1$$
    $$< 2^2 + 2 + 1 = 7;$$

    (ii) $|x - 1| < \frac{1}{7}\varepsilon.$

    It follows that
    $$|x^2 + x + 1| \times |x - 1| < 7 \times \frac{1}{7}\varepsilon = \varepsilon, \quad \text{for all } x \text{ satisfying } |x - 1| < \delta;$$

    that is, the statement $(*)$ holds.
    Hence f is continuous at 1.

2. The function $f(x) = \sqrt{x}$ is defined on $[0, \infty)$, and $f(4) = \sqrt{4} = 2$.
    We must prove that:

    for each positive number ε, there is a positive number δ such that
    $$|\sqrt{x} - 2| < \varepsilon, \quad \text{for all } x \text{ satisfying } |x - 4| < \delta. \quad (*)$$

    Now
    $$|\sqrt{x} - 2| = \left| \frac{x - 4}{\sqrt{x} + 2} \right|$$
    $$= \frac{|x - 4|}{\sqrt{x} + 2}$$
    $$\leq \frac{1}{2}|x - 4|, \quad \text{since } \sqrt{x} \geq 0.$$

    So, choose $\delta = \varepsilon$. It follows that, for all x satisfying $|x - 4| < \delta$, we have
    $$|\sqrt{x} - 2| \leq \frac{1}{2}|x - 4|$$
    $$< \frac{1}{2}\delta$$
    $$= \frac{1}{2}\varepsilon$$
    $$< \varepsilon.$$

    That is, the statement $(*)$ holds.
    Hence f is continuous at 4.

3. The graph of f suggests that, while f is 'well behaved' to the left of 2, we should examine the behaviour of f to the right of 2. If we choose ε to be any positive number less than $\frac{1}{4}$, say, then there will always be points x (with $x > 2$) as close as we please to 2 where $f(x)$ is not within a distance ε of $f(2) = 1$.

    So, take $\varepsilon = \frac{1}{4}$. Then, if f *were* continuous at 2, there would be some positive number δ such that

$$|f(x) - 1| < \frac{1}{4}, \quad \text{for all } x \text{ satisfying } |x - 2| < \delta. \qquad (*)$$

Now let $x_n = 2 + \frac{1}{n}$, for $n = 1, 2, \ldots$. Then

$$x_n \left(= 2 + \frac{1}{n} \right) \in (2, 2 + \delta), \quad \text{for all } n > X, \quad \text{where } X = \frac{1}{\delta},$$

but

$$|f(x_n) - 1| = \left| \left(2 + \frac{1}{n} - \frac{1}{2} \right) - 1 \right|$$

$$= \frac{1}{2} + \frac{1}{n}$$

$$\not< \frac{1}{4} = \varepsilon.$$

Thus, with this choice of ε, no value of δ exists such that requirement $(*)$ holds. This proves that f is discontinuous at 2.

4. Since f is continuous at an interior point c of I and $f(c) \neq 0$, it follows, by Theorem 2 of Sub-section 5.4.1, that there is a neighbourhood $N = (c - r, c + r)$ of c on which

$$|f(x)| > \frac{1}{2}|f(c)|. \qquad (1)$$

Since $f(c) \neq 0$, it follows from the inequality (1) that

$$|f(x)| > \frac{1}{2}|f(c)|$$

$$> 0, \quad \text{for all } x \text{ with } |x - c| < r.$$

In particular, $f(x) \neq 0$ on N, so that $\frac{1}{f}$ is defined on N.

To prove that $\frac{1}{f}$ is continuous at c, we must prove that,

for each positive number ε, there is a positive number δ such that

$$\left| \frac{1}{f(x)} - \frac{1}{f(c)} \right| < \varepsilon, \quad \text{for all } x \text{ satisfying } |x - c| < \delta. \qquad (*)$$

Now

$$\left| \frac{1}{f(x)} - \frac{1}{f(c)} \right| = \left| \frac{f(c) - f(x)}{f(x)f(c)} \right|$$

$$= \frac{|f(x) - f(c)|}{|f(x)| \times |f(c)|}. \qquad (2)$$

Since f is continuous at c, we know that, for the given value of ε in $(*)$, there is a positive number δ_1 such that

$$|f(x) - f(c)| < \varepsilon, \quad \text{for all } x \text{ satisfying } |x - c| < \delta_1. \qquad (3)$$

Now choose $\delta = \min\{r, \delta_1\}$, so that both (1) and (3) hold, for all x satisfying $|x - c| < \delta$. It then follows, from (1), (2) and (3), that

$$\left| \frac{1}{f(x)} - \frac{1}{f(c)} \right| = \frac{|f(x) - f(c)|}{|f(x)| \times |f(c)|}$$

$$< \frac{\varepsilon}{\frac{1}{2}|f(c)|^2}$$

$$= \frac{1}{\frac{1}{2}|f(c)|^2} \times \varepsilon.$$

This is equivalent to the statement $(*)$, in which the number ε has been replaced by $K\varepsilon$, where $K = \frac{1}{\frac{1}{2}|f(c)|^2}$. Since K is a constant, this is sufficient, by the $K\varepsilon$ Lemma.

It follows that $\frac{1}{f}$ is continuous at c, as required.

Section 5.5

1. Let ε be any given positive number. We want to find a positive number δ such that
$$|f(x) - f(c)| < \varepsilon, \quad \text{for all } x \text{ and } c \text{ in } [2,3] \text{ satisfying } |x - c| < \delta,$$
that is, that
$$|x^2 - c^2| < \varepsilon, \quad \text{for all } x \text{ and } c \text{ in } [2,3] \text{ satisfying } |x - c| < \delta. \quad (*)$$
Now, since $x^2 - c^2 = (x-c)(x+c)$, we have
$$|x^2 - c^2| = |x - c| \times |x + c|.$$
But, for x and c in $[2,3]$, we know that
$$4 \le x + c \le 6.$$
Hence, in order to prove the statement $(*)$, it is sufficient to prove that there exists a positive number δ such that
$$6|x - c| < \varepsilon, \quad \text{for all } x \text{ and } c \text{ in } [2,3] \text{ satisfying } |x - c| < \delta. \quad (**)$$
In view of $(**)$, take $\delta = \frac{1}{6}\varepsilon$ (note that this choice of δ depends only on ε, and not at all on x or c). With this choice of δ, it follows that if $x, c \in [2,3]$ and satisfy $|x - c| < \delta$, then
$$6|x - c| < 6\delta$$
$$= \varepsilon.$$

In other words, the statement $(**)$ holds, and hence the statement $(*)$ also holds. Hence the choice $\delta = \frac{1}{6}\varepsilon$ serves our purpose.

Chapter 6

Section 6.1

1. (a) Let c be any point of $\mathbb{R}$. Then, for any non-zero h, the difference quotient $Q(h)$ at c is
$$\begin{aligned} Q(h) &= \frac{f(c+h) - f(c)}{h} \\ &= \frac{(c+h)^n - c^n}{h} \\ &= \frac{1}{h}\left(nc^{n-1}h + \frac{1}{2}n(n-1)c^{n-2}h^2 + \cdots + nch^{n-1} + h^n\right) \end{aligned}$$
$$\text{(n terms in the bracket)}$$
$$\begin{aligned} &= nc^{n-1} + \frac{1}{2}n(n-1)c^{n-2}h + \cdots + nch^{n-2} + h^{n-1} \\ &\to nc^{n-1} \quad \text{as } h \to 0. \end{aligned}$$
It follows that f is differentiable at c, and that $f'(c) = nc^{n-1}$.

(b) Let $f(x) = k$, and let c be any point of $\mathbb{R}$. Then, for any non-zero h, the difference quotient $Q(h)$ at c is
$$\begin{aligned} Q(h) &= \frac{f(c+h) - f(c)}{h} \\ &= \frac{k - k}{h} \\ &= 0 \\ &\to 0 \quad \text{as } h \to 0. \end{aligned}$$
It follows that f is differentiable at c, and that $f'(c) = 0$.

2. Let c be any point of $\mathbb{R}$, with $c \neq 0$. Then, for any non-zero h, the difference quotient $Q(h)$ at c is

$$Q(h) = \frac{f(c+h) - f(c)}{h}$$
$$= \frac{1}{h}\left(\frac{1}{c+h} - \frac{1}{c}\right)$$
$$= \frac{c - (c+h)}{h(c+h)c}$$
$$= \frac{-1}{(c+h)c}$$
$$\rightarrow -\frac{1}{c^2} \quad \text{as } h \rightarrow 0.$$

It follows that f is differentiable at c, and that $f'(c) = -\frac{1}{c^2}$.

3. To prove that $f(x) = x^4$, $x \in \mathbb{R}$, is differentiable at 1, with derivative $f'(1) = 4$, we have to show that:

for each positive number ε, there is a positive number δ such that

$$\left|\frac{x^4 - 1}{x - 1} - 4\right| < \varepsilon, \quad \text{for all } x \text{ satisfying } 0 < |x - 1| < \delta;$$

that is

$$\left|x^3 + x^2 + x - 3\right| < \varepsilon, \quad \text{for all } x \text{ satisfying } 0 < |x - 1| < \delta.$$

We use the fact that

$$x^4 - 1 = (x^2 - 1)(x^2 + 1)$$
$$= (x - 1)(x + 1)(x^2 + 1)$$
$$= (x - 1)(x^3 + x^2 + x + 1).$$

Now, $x^3 + x^2 + x - 3 = (x - 1)(x^2 + 2x + 3)$; thus, for $0 < |x - 1| < 1$, we have $x \in (0, 2)$, so that

$$\left|x^2 + 2x + 3\right| < 2^2 + 2 \times 2 + 3 = 11.$$

Next, choose $\delta = \min\{1, \frac{1}{11}\varepsilon\}$. With this choice of δ, it follows that, for all x satisfying $0 < |x - 1| < \delta$, we have

$$\left|x^3 + x^2 + x - 3\right| = |x - 1| \times |x^2 + 2x + 3|$$
$$< \frac{1}{11}\varepsilon \times 11 = \varepsilon.$$

It follows that f is differentiable at 1, with derivative $f'(1) = 4$.

4. (a) Let $f(x) = |x|^{\frac{1}{2}}$. The graph of f near 0 suggests that f is not differentiable at 0. Then, for any non-zero h, the difference quotient $Q(h)$ at 0 is

$$Q(h) = \frac{f(h) - f(0)}{h}$$
$$= \frac{|h|^{\frac{1}{2}} - 0}{h}$$
$$= \frac{|h|^{\frac{1}{2}}}{h}.$$

For $h > 0$, it follows that $Q(h) = \frac{1}{h^{\frac{1}{2}}}$. So, in particular, if we take $h = \frac{1}{n^2}$ for $n \in \mathbb{N}$, we find that

$$Q\left(\frac{1}{n^2}\right) = n \rightarrow \infty \quad \text{as } n \rightarrow \infty.$$

It follows that f is not differentiable at 0.

(b) Let $f(x) = [x]$, the integer part of x. The graph of f near 1 suggests that f is not differentiable at 1.

Then, for any h with $-1 < h < 0$, the difference quotient $Q(h)$ at 1 is

$$Q(h) = \frac{f(1+h) - f(1)}{h}$$

$$= \frac{0 - 1}{h}$$

$$= -\frac{1}{h}.$$

It follows that

$$\lim_{h \to 0^-} Q(h) = \lim_{h \to 0^-} \left(-\frac{1}{h}\right)$$

$$= \infty.$$

Hence f is not differentiable at 1.

5.

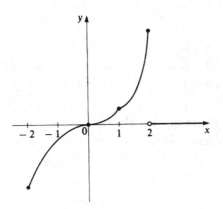

(a) For $2 > h > 0$, the difference quotient for f at -2 is

$$Q(h) = \frac{f(-2+h) - f(-2)}{h}$$

$$= \frac{-(-2+h)^2 + 4}{h}$$

$$= 4 - h$$

$$\to 4 \quad \text{as } h \to 0^+.$$

Hence f is differentiable on the right at -2, and $f_R'(-2) = 4$.
f is not differentiable at -2, since it is not defined to the left of 2.

(b) For $-1 < h < 1$, the difference quotient for f at 0 is

$$Q(h) = \frac{f(h) - f(0)}{h}$$

$$= \begin{cases} \frac{-h^2 - 0}{h}, & h < 0, \\ \frac{h^4 - 0}{h}, & h > 0, \end{cases}$$

$$= \begin{cases} -h, & h < 0, \\ h^3, & h > 0. \end{cases}$$

Thus

$$\lim_{h \to 0^-} Q(h) = 0 \quad \text{and} \quad \lim_{h \to 0^+} Q(h) = 0,$$

so that $f_L'(0)$ and $f_R'(0)$ both exist and equal 0.
It follows that f is differentiable at 0, and $f'(0) = 0$.

(c) The difference quotient for f at 1 is

$$Q(h) = \frac{f(1+h) - f(1)}{h}$$

$$= \begin{cases} \frac{(1+h)^4 - 1}{h}, & -1 < h < 0, \\ \frac{(1+h)^3 - 1}{h}, & 0 < h < 1, \end{cases}$$

$$= \begin{cases} 4 + 6h + 4h^2 + h^3, & -1 < h < 0, \\ 3 + 3h + h^2, & 0 < h < 1. \end{cases}$$

Thus

$$\lim_{h \to 0^-} Q(h) = 4 \quad \text{and} \quad \lim_{h \to 0^+} Q(h) = 3,$$

so that $f_L'(1) = 4$ and $f_R'(1) = 3$.

It follows that f is not differentiable at 1.

(d) The difference quotient for f at 2 is

$$Q(h) = \frac{f(2+h) - f(2)}{h}$$

$$= \begin{cases} \frac{(2+h)^3 - 8}{h}, & -1 < h < 0, \\ \frac{0-8}{h}, & 0 < h < 1, \end{cases}$$

$$= \begin{cases} 12 + 6h + h^2, & -1 < h < 0, \\ -\frac{8}{h}, & 0 < h < 1. \end{cases}$$

Since $\lim_{h \to 0^+} Q(h)$ does not exist, it follows that f is not differentiable at 2.
However f is differentiable on the left at 2, and $f_L'(2) = 12$.

6. For $n \in \mathbb{N}$

$$f\left(\frac{1}{(2n + \frac{1}{2})\pi}\right) = \sin\left(2n + \frac{1}{2}\right)\pi$$

$$= 1 \to 1 \quad \text{as } n \to \infty.$$

Hence, although $\frac{1}{(2n+\frac{1}{2})\pi} \to 0$ as $n \to \infty$

$$f\left(\frac{1}{(2n + \frac{1}{2})\pi}\right) \not\to 0$$

$$= f(0).$$

It follows that f is not continuous at 0; and so, by Corollary 1, f is not differentiable at 0.

7. For any non-zero h, the difference quotient $Q(h)$ at 0 is

$$Q(h) = \frac{f(h) - f(0)}{h}$$

$$= \frac{h^2 \sin\left(\frac{1}{h}\right)}{h}$$

$$= h \sin\left(\frac{1}{h}\right)$$

$$\to 0 \quad \text{as } h \to 0.$$

It follows that f is differentiable at 0, and that $f'(0) = 0$.

Now, for $x \neq 0$, $f'(x) = 2x \sin\frac{1}{x} - \cos\frac{1}{x}$. Hence, if $n \in \mathbb{N}$

$$f'\left(\frac{1}{2n\pi}\right) = \frac{1}{n\pi}\sin(2n\pi) - \cos(2n\pi)$$

$$= 0 - 1$$

$$\to -1, \quad \text{as } n \to \infty.$$

Since $f'(0) \neq -1$, it follows that f' is not continuous at 0.

8. Let c be any point in $\mathbb{R}$. At c, the difference quotient for f is

$$Q(h) = \frac{\cos(c+h) - \cos c}{h}$$

$$= \frac{\cos c \cos h - \sin c \sin h - \cos c}{h}$$

$$= \cos c \times \frac{\cos h - 1}{h} - \sin c \times \frac{\sin h}{h}$$

$$\rightarrow \cos c \times 0 - \sin c \times 1, \quad \text{as } h \rightarrow 0,$$

$$= -\sin c.$$

It follows that f is differentiable at c, and $f'(c) = -\sin c$. Since c is an arbitrary point of $\mathbb{R}$, it follows that f is differentiable on $\mathbb{R}$.

9. For $c > 0$, the different quotient for f at c is

$$Q(h) = \frac{(c+h)^3 - c^3}{h}$$

$$= \frac{1}{h}\left(3c^2 h + 3ch^2 + h^3\right)$$

$$= 3c^2 + 3ch + h^2 \rightarrow 3c^2 \quad \text{as } h \rightarrow 0,$$

so that f is differentiable at c and $f'(c) = 3c^2$.

> We assume that h is sufficiently small that $|h| < c$, so that $c + h > 0$; and hence that we are using the correct value for f at $c + h$.

Next, for $c < 0$, the different quotient for f at c is

$$Q(h) = \frac{(c+h)^2 - c^2}{h}$$

$$= \frac{1}{h}\left(2ch + h^2\right)$$

$$= 2c + h \rightarrow 2c \quad \text{as } h \rightarrow 0,$$

so that f is differentiable at c and $f'(c) = 2c$.

> We assume that h is sufficiently small that $|h| < -c$, so that $c + h < 0$; and hence that we are using the correct value for f at $c + h$.

Finally, the difference quotient for f at 0 is

$$Q(h) = \frac{f(h) - f(0)}{h}$$

$$= \begin{cases} \frac{h^2 - 0}{h}, & h < 0, \\ \frac{h^3 - 0}{h}, & h > 0, \end{cases}$$

$$= \begin{cases} h, & h < 0, \\ h^2, & h > 0, \end{cases}$$

$$\rightarrow 0, \quad \text{as} \quad h \rightarrow 0.$$

It follows that f is differentiable at 0, and that $f'(0) = 0$.

Therefore, the function f' is given by the following formula

$$f'(x) = \begin{cases} 2x, & x < 0, \\ 0, & x = 0, \\ 3x^2, & x > 0. \end{cases}$$

Similar arguments then show that f' is differentiable on the left at 0, with $f_L''(0) = 2$; and differentiable on the right at 0, with $f_R''(0) = 0$. Since $f_L''(0) \neq f_R''(0)$, f' is not differentiable at 0.

Section 6.2

1. (a) $f'(x) = 7x^6 - 8x^3 + 9x^2 - 5, \quad x \in \mathbb{R}$.

(b) $f'(x) = \dfrac{(x^3 - 1)2x - (x^2 + 1)3x^2}{(x^3 - 1)^2}$

$$= \frac{-x^4 - 3x^2 - 2x}{(x^3 - 1)^2}, \quad x \in \mathbb{R} - \{1\}.$$

(c) $f'(x) = 2\cos^2 x - 2\sin^2 x$

$\qquad = 2\cos 2x, \quad x \in \mathbb{R}.$

(d) $f'(x) = \dfrac{(3 + \sin x - 2\cos x)e^x - e^x(\cos x + 2\sin x)}{(3 + \sin x - 2\cos x)^2}$

$\qquad = \dfrac{e^x(3 - \sin x - 3\cos x)}{(3 + \sin x - 2\cos x)^2}, \quad x \in \mathbb{R}.$

2. $f'(x) = e^{2x} + 2xe^{2x} = e^{2x}(1 + 2x);$

$f''(x) = 2e^{2x}(1 + 2x) + 2e^{2x} = e^{2x}(4 + 4x);$

$f'''(x) = 2e^{2x}(4 + 4x) + 4e^{2x} = e^{2x}(12 + 8x).$

3. In each case, we use the Quotient Rule and the derivatives of sine and cosine.

(a) Here $f(x) = \frac{\sin x}{\cos x}$, so that

$$f'(x) = \frac{\cos x \cos x - \sin x(-\sin x)}{\cos^2 x}$$
$$= \frac{1}{\cos^2 x}$$
$$= \sec^2 x,$$

on the domain of f.

(b) Here $f(x) = \frac{1}{\sin x}$, so that

$$f'(x) = -\frac{\cos x}{\sin^2 x}$$
$$= -\operatorname{cosec} x \cot x,$$

on the domain of f.

(c) Here $f(x) = \frac{1}{\cos x}$, so that

$$f'(x) = \frac{\sin x}{\cos^2 x}$$
$$= \sec x \tan x,$$

on the domain of f.

(d) Here $f(x) = \frac{\cos x}{\sin x}$, so that

$$f'(x) = \frac{\sin x(-\sin x) - \cos x \cos x}{\sin^2 x}$$
$$= \frac{-1}{\sin^2 x}$$
$$= -\operatorname{cosec}^2 x,$$

on the domain of f.

4. (a) Here $f(x) = \frac{1}{2}(e^x - e^{-x})$, $x \in \mathbb{R}$, so that

$$f'(x) = \frac{1}{2}(e^x + e^{-x})$$
$$= \cosh x, \quad x \in \mathbb{R}.$$

(b) Here $f(x) = \frac{1}{2}(e^x + e^{-x})$, $x \in \mathbb{R}$, so that

$$f'(x) = \frac{1}{2}(e^x - e^{-x})$$
$$= \sinh x, \quad x \in \mathbb{R}.$$

(c) Here

$$f(x) = \frac{\sinh x}{\cosh x}, \quad x \in \mathbb{R},$$

so that

$$f'(x) = \frac{\cosh x \cosh x - \sinh x \sinh x}{\cosh^2 x}$$
$$= \frac{1}{\cosh^2 x}$$
$$= \operatorname{sech}^2 x, \quad x \in \mathbb{R}.$$

5. (a) Here

$$f(x) = \sinh(x^2), \quad x \in \mathbb{R},$$

so that

$$f'(x) = 2x \cosh(x^2), \quad x \in \mathbb{R}.$$

(b) Here

$$f(x) = \sin(\sinh(2x)), \quad x \in \mathbb{R},$$

so that

$$f'(x) = \cos(\sinh(2x))2\cosh(2x), \quad x \in \mathbb{R}.$$

(c) Here

$$f(x) = \sin\left(\frac{\cos 2x}{x^2}\right), \quad x \in (0, \infty),$$

so that

$$f'(x) = \cos\left(\frac{\cos 2x}{x^2}\right) \times \left(\frac{-x^2 2\sin 2x - 2x\cos 2x}{x^4}\right), \quad x \in (0, \infty).$$

6. (a) The function

$$f(x) = \cos x, \quad x \in (0, \pi),$$

is continuous and strictly decreasing on $(0, \pi)$, and

$$f((0, \pi)) = (-1, 1).$$

Also, f is differentiable on $(0, \pi)$, and its derivative $f'(x) = -\sin x$ is non-zero there.

So f satisfies the conditions of the Inverse Function Rule.

Hence $f^{-1} = \cos^{-1}$ is differentiable on $(-1, 1)$; and, if $y = f(x)$, then

$$(f^{-1})'(y) = \frac{1}{f'(x)} = -\frac{1}{\sin x}.$$

Since $\sin x > 0$ on $(0, \pi)$, and $\sin^2 x + \cos^2 x = 1$, it follows that

$$\sin x = \sqrt{1 - \cos^2 x} = \sqrt{1 - y^2}.$$

Hence

$$(f^{-1})'(y) = \frac{-1}{\sqrt{1 - y^2}}.$$

If we now replace the domain variable y by x, we obtain

$$(f^{-1})'(x) = \frac{-1}{\sqrt{1 - x^2}}, \quad x \in (-1, 1).$$

(b) The function

$$f(x) = \sinh x, \quad x \in \mathbb{R},$$

is continuous and strictly increasing on $\mathbb{R}$, and

$$f(\mathbb{R}) = \mathbb{R}.$$

Also, f is differentiable on $\mathbb{R}$, and its derivative $f'(x) = \cosh x$ is non-zero there. So f satisfies the conditions of the Inverse Function Rule.

Hence $f^{-1} = \sinh^{-1}$ is differentiable on $\mathbb{R}$; and, if $y = f(x)$, then

$$(f^{-1})'(y) = \frac{1}{f'(x)} = \frac{1}{\cosh x}.$$

Since $\cosh x > 0$ on $\mathbb{R}$, and $\cosh^2 x = 1 + \sinh^2 x$, it follows that

$$\cosh x = \sqrt{1 + \sinh^2 x} = \sqrt{1 + y^2}.$$

Hence

$$(f^{-1})'(y) = \frac{1}{\sqrt{1 + y^2}}.$$

If we now replace the domain variable y by x, we obtain

$$(f^{-1})'(x) = \frac{1}{\sqrt{1 + x^2}}, \quad x \in \mathbb{R}.$$

7. If $x_1 < x_2$, then $x_1^5 < x_2^5$; it follows that $f(x_1) < f(x_2)$, so that f is strictly increasing on $\mathbb{R}$.

Since f is a polynomial function, it is continuous and differentiable on $\mathbb{R}$. Also

$$f'(x) = 5x^4 + 1 \neq 0 \quad \text{on } \mathbb{R}.$$

Thus, f satisfies the conditions of the Inverse Function Theorem.

Now, $f(0) = -1, f(1) = 1$ and $f(-1) = -3$. Hence, by the Inverse Function Rule

$$(f^{-1})'(-1) = \frac{1}{f'(0)} = 1,$$

$$(f^{-1})'(1) = \frac{1}{f'(1)} = \frac{1}{6},$$

and

$$(f^{-1})'(-3) = \frac{1}{f'(-1)} = \frac{1}{6}.$$

8. By definition,

$$f(x) = x^x = \exp(x \log_e x).$$

The functions $x \mapsto x$ and $x \mapsto \log_e x$ are differentiable on $(0, \infty)$, and $\exp$ is differentiable on $\mathbb{R}$. It follows, by the Product Rule and the Composition Rule, that f is differentiable on $(0, \infty)$, and that

$$f'(x) = \exp(x \log_e x) \times \left(\log_e x + x \times \frac{1}{x} \right)$$
$$= x^x (\log_e x + 1).$$

Section 6.3

1. Since f is a polynomial function, f is continuous on $[-1, 2]$ and differentiable on $(-1, 2)$; also

$$f'(x) = x^3 - x^2 = x^2(x - 1).$$

Thus f' vanishes at 0 and 1.

First, we consider the behaviour of f near 0. For $x \in (-1, 1)$, for example

$$f(x) = \frac{1}{4}x^3 \left(x - \frac{4}{3} \right)$$

has the opposite sign to that of x, since $x - \frac{4}{3} < 0$; thus

$$f(x) > 0 \quad \text{for } x \in (-1, 0)$$

and

$$f(x) < 0 \quad \text{for } x \in (0, 1).$$

Since $f(0) = 0$, it follows that 0 is not a local extremum of f.

Next, we consider the behaviour of f near 1. Now

$$f(x) - f(1) = \left(\frac{1}{4}x^4 - \frac{1}{3}x^3\right) - \left(\frac{1}{4} - \frac{1}{3}\right)$$

$$= \frac{1}{4}(x^4 - 1) - \frac{1}{3}(x^3 - 1)$$

$$= \frac{1}{4}(x - 1)(x^3 + x^2 + x + 1) - \frac{1}{3}(x - 1)(x^2 + x + 1)$$

$$= \frac{1}{12}(x - 1)(3x^3 - x^2 - x - 1)$$

$$= \frac{1}{12}(x - 1)^2(3x^2 + 2x + 1).$$

It follows that, for $x \in (0, 2)$, for example

$$f(x) - f(1) \geq 0,$$

so that f has a local minimum at 1, with value $f(1) = -\frac{1}{12}$.

2. Since the sine and cosine functions are continuous and differentiable on $\mathbb{R}$, so also is f.

   Now

$$f'(x) = 2\sin x \cos x - \sin x = \sin x(2\cos x - 1);$$

   thus f' vanishes in $\left(0, \frac{1}{2}\pi\right)$ only when $\cos x = \frac{1}{2}$; that is, when $x = \frac{1}{3}\pi$.
   Since $f(0) = 1, f\left(\frac{1}{2}\pi\right) = 1$ and

$$f\left(\frac{1}{3}\pi\right) = \left(\frac{1}{2}\sqrt{3}\right)^2 + \frac{1}{2} = \frac{3}{4} + \frac{1}{2} = \frac{5}{4};$$

   it follows that, on $\left[0, \frac{1}{2}\pi\right]$:

   the minimum of f is 1, and occurs when $x = 0$ and $x = \frac{1}{2}\pi$;

   the maximum of f is $\frac{5}{4}$, and occurs when $x = \frac{1}{3}\pi$.

3. Since f is a polynomial function, f is continuous on $[1, 3]$ and differentiable on $(1, 3)$.
   Also, $f(1) = 2$ and $f(3) = 2$, so that $f(1) = f(3)$. Thus f satisfies the conditions of Rolle's Theorem on $[1, 3]$.

   Now

$$f'(x) = 4x^3 - 12x^2 + 6x = 2x(2x^2 - 6x + 3),$$

   so that $f'(x) = 0$ when $x = 0$ and $x = \frac{1}{2}(3 \pm \sqrt{3})$ (the roots of the quadratic equation $2x^2 - 6x + 3 = 0$). Thus the only value of x in $(1, 3)$ such that $f'(x) = 0$ is $x = \frac{1}{2}(3 + \sqrt{3})$, so we must take $c = \frac{1}{2}(3 + \sqrt{3}) \simeq 2.37$.

4. (a) NO: f is not defined at $\frac{1}{2}\pi$.

   (b) NO: f is not differentiable at 1, and $f(0) \neq f(2)$.

   (c) YES: All the conditions are satisfied.

   (d) NO: $f(0) \neq f\left(\frac{1}{2}\pi\right)$.

Section 6.4

1. (a) Since f is a polynomial function, f is continuous on $[-2, 2]$ and differentiable on $(-2, 2)$. Thus f satisfies the conditions of the Mean Value Theorem on $[-2, 2]$.

Now
$$f'(x) = 3x^2 + 2,$$
so that c satisfies the conclusion of the theorem when
$$3c^2 + 2 = \frac{f(2) - f(-2)}{2 - (-2)} = \frac{12 - (-12)}{2 - (-2)}$$
$$= 6;$$
that is, when $3c^2 = 4$ so that $c = \pm\sqrt{\frac{4}{3}} \simeq \pm 1.15$.
Thus there are two possible values of c.

(b) The function exp is continuous on $[0, 3]$ and differentiable on $(0, 3)$. Thus f satisfies the conditions of the Mean Value Theorem on $[0, 3]$.
Now, $f'(x) = e^x$, so that c satisfies the conclusion of the theorem when
$$e^c = \frac{f(3) - f(0)}{3 - 0} = \frac{1}{3}\left(e^3 - 1\right);$$
that is, when $c = \log_e\left(\frac{1}{3}(e^3 - 1)\right) \simeq 1.85$.

2. (a) Here $f'(x) = 4x^{\frac{1}{3}} - 4 = 4\left(x^{\frac{1}{3}} - 1\right)$, so that $f'(x) > 0$ on $(1, \infty)$. Hence, by the Increasing-Decreasing Theorem, f is strictly increasing on $[1, \infty)$.

(b) Here $f'(x) = 1 - \frac{1}{x}$, so that $f'(x) < 0$ on $(0, 1)$. Hence, by the Increasing-Decreasing Theorem, f is strictly decreasing on $(0, 1]$.

3. (a) Let
$$f(x) = \sin^{-1}x + \cos^{-1}x, \quad x \in [-1, 1].$$
Then f is continuous on $[-1, 1]$ and differentiable on $(-1, 1)$.
Now
$$f'(x) = \frac{1}{\sqrt{1 - x^2}} - \frac{1}{\sqrt{1 - x^2}} = 0, \quad \text{for } x \in (-1, 1).$$
It follows, from Corollary 1, that
$$\sin^{-1}x + \cos^{-1}x = c, \quad \text{for } x \in [-1, 1],$$
for some constant c.
Putting $x = 0$, we obtain $0 + \frac{1}{2}\pi = c$, so that $c = \frac{1}{2}\pi$. Hence
$$\sin^{-1}x + \cos^{-1}x = \frac{1}{2}\pi, \quad x \in [-1, 1].$$

(b) Let
$$f(x) = \tan^{-1}x + \tan^{-1}\frac{1}{x}, \quad x \in (0, \infty).$$
Then f is continuous and differentiable on $(0, \infty)$.
Now
$$f'(x) = \frac{1}{1 + x^2} - \frac{\frac{1}{x^2}}{1 + \left(\frac{1}{x}\right)^2} = 0, \quad \text{for } x \in (0, \infty).$$
It follows, from Corollary 1, that
$$\tan^{-1}x + \tan^{-1}\frac{1}{x} = c, \quad \text{for } x \in (0, \infty),$$
for some constant c.
Putting $x = 1$, we obtain $\frac{1}{4}\pi + \frac{1}{4}\pi = c$, so that $c = \frac{1}{2}\pi$. Hence
$$\tan^{-1}x + \tan^{-1}\frac{1}{x} = \frac{1}{2}\pi, \quad x \in (0, \infty).$$

4. We have
$$f'(x) = 3x^2 - 6x = 3x(x-2), \quad \text{for } x \in \mathbb{R}.$$

Thus $f'(x) = 0$ when $x = 0$ and $x = 2$, so that $c = 0, 2$.

Now
$$f''(x) = 6x - 6, \quad \text{for } x \in \mathbb{R},$$

so that $f''(0) = -6 < 0$ and $f''(2) = 6 > 0$. It follows, from the Second Derivative Test, that f has a local maximum that occurs at 0 and a local minimum that occurs at 2.

The value of the local maximum is $f(0) = 1$, and the value of the local minimum is $f(2) = -3$.

5. (a) Let
$$f(x) = x - \sin x, \quad x \in \left[0, \frac{1}{2}\pi\right].$$

Then f is continuous on $\left[0, \frac{1}{2}\pi\right]$ and differentiable on $\left(0, \frac{1}{2}\pi\right)$.

Now
$$f'(x) = 1 - \cos x > 0, \quad \text{for } x \in \left(0, \frac{1}{2}\pi\right),$$

and $f(0) = 0$.

It follows that $f(x) \geq 0$ for $x \in \left[0, \frac{1}{2}\pi\right]$, so that
$$x \geq \sin x, \quad \text{for } x \in \left[0, \frac{1}{2}\pi\right].$$

(b) Let
$$f(x) = \frac{2}{3}x + \frac{1}{3} - x^{\frac{2}{3}}, \quad x \in [0, 1].$$

Then f is continuous on $[0, 1]$ and differentiable on $(0, 1)$.

Now
$$\begin{aligned} f'(x) &= \frac{2}{3} - \frac{2}{3}x^{-\frac{1}{3}} \\ &= \frac{2}{3}\left(1 - x^{-\frac{1}{3}}\right) < 0, \quad \text{for } x \in (0, 1), \end{aligned}$$

and $f(1) = \frac{2}{3} + \frac{1}{3} - 1 = 0$.

It follows that $f(x) \geq 0$ for $x \in [0, 1]$, so that
$$\frac{2}{3}x + \frac{1}{3} \geq x^{\frac{2}{3}}, \quad \text{for } x \in [0, 1].$$

Section 6.5

1. Here
$$f(x) = x^3 + x^2 \sin x \quad \text{and} \quad g(x) = x \cos x - \sin x \text{ on } [0, \pi].$$

Since polynomial functions and the sine and cosine functions are continuous and differentiable on $\mathbb{R}$, it follows, by the Combination Rules, that f and g are continuous on $[0, \pi]$ and differentiable on $(0, \pi)$. It follows that Cauchy's Mean Value Theorem applies to the functions f and g on $[0, \pi]$.

Now, $g(0) = 0 \times 1 - 0 = 0$ and $g(\pi) = \pi\cos\pi - \sin\pi = -\pi$, so that $g(\pi) \neq g(0)$.

Also
$$\begin{aligned} f'(x) &= 3x^2 + 2x \sin x + x^2 \cos x \\ &= x^2(3 + \cos x) + 2x \sin x \end{aligned}$$

and
$$g'(x) = \cos x - x \sin x - \cos x$$
$$= -x \sin x$$

on $(0, \pi)$. In particular, $g'(x) \neq 0$ on $(0, \pi)$.

It therefore follows from Cauchy's Mean Value Theorem that there exists at least one point c in $(0, \pi)$ for which
$$\frac{c^2(3 + \cos c) + 2c \sin c}{-c \sin c} = \frac{(\pi^3) - (0)}{(-\pi) - (0)},$$

so that
$$\frac{c(3 + \cos c) + 2 \sin c}{- \sin c} = -\pi^2.$$

By cross-multiplying, we obtain
$$c(3 + \cos c) + 2 \sin c = \pi^2 \sin c,$$

so that
$$3c = (\pi^2 - 2) \sin c - c \cos c.$$

Hence the equation $3x = (\pi^2 - 2)\sin x - x\cos x$ has at least one root in $(0, \pi)$.

2. (a) Let
$$f(x) = \sinh 2x \quad \text{and} \quad g(x) = \sin 3x, \quad \text{for } x \in \mathbb{R}.$$

Then f and g are differentiable on $\mathbb{R}$, and
$$f(0) = g(0) = 0;$$
so that f and g satisfy the conditions of l'Hôpital's Rule at 0.

Now
$$f'(x) = 2\cosh 2x \quad \text{and} \quad g'(x) = 3\cos 3x,$$
so that
$$\frac{f'(x)}{g'(x)} = \frac{2\cosh 2x}{3\cos 3x}.$$

It follows, from l'Hôpital's Rule, that the desired limit $\lim_{x \to 0} \frac{f(x)}{g(x)}$ exists and equals
$$\lim_{x \to 0} \frac{f'(x)}{g'(x)} = \lim_{x \to 0} \frac{2\cosh 2x}{3\cos 3x},$$

provided that this last limit exists.

But, by the Combination Rules for continuous functions, the function
$$x \mapsto \frac{2\cosh 2x}{3\cos 3x}$$

is continuous at 0, so that
$$\lim_{x \to 0} \frac{f'(x)}{g'(x)} = \frac{f'(0)}{g'(0)}$$
$$= \frac{2}{3}.$$

It follows, from l'Hôpital's Rule, that the original limit exists, and that its value is $\frac{2}{3}$.

(b) Let
$$f(x) = (1 + x)^{\frac{1}{3}} - (1 - x)^{\frac{1}{3}}$$
and
$$g(x) = (1 + 2x)^{\frac{2}{3}} - (1 - 2x)^{\frac{2}{3}},$$
for $x \in \left(-\frac{1}{2}, \frac{1}{2}\right)$.

Then f and g are differentiable on $\left(-\frac{1}{2},\frac{1}{2}\right)$, and
$$f(0)=g(0)=0;$$
so that f and g satisfy the conditions of l'Hôpital's Rule at 0.
Now
$$f'(x)=\frac{1}{5}(1+x)^{-\frac{4}{5}}+\frac{1}{5}(1-x)^{-\frac{4}{5}}$$
and
$$g'(x)=\frac{4}{5}(1+2x)^{-\frac{3}{5}}+\frac{4}{5}(1-2x)^{-\frac{3}{5}},$$
so that
$$\frac{f'(x)}{g'(x)}=\frac{\frac{1}{5}(1+x)^{-\frac{4}{5}}+\frac{1}{5}(1-x)^{-\frac{4}{5}}}{\frac{4}{5}(1+2x)^{-\frac{3}{5}}+\frac{4}{5}(1-2x)^{-\frac{3}{5}}}.$$

It follows, from l'Hôpital's Rule, that the desired limit $\lim_{x\to 0}\frac{f(x)}{g(x)}$ exists and equals
$$\lim_{x\to 0}\frac{f'(x)}{g'(x)}=\lim_{x\to 0}\frac{\frac{1}{5}(1+x)^{-\frac{4}{5}}+\frac{1}{5}(1-x)^{-\frac{4}{5}}}{\frac{4}{5}(1+2x)^{-\frac{3}{5}}+\frac{4}{5}(1-2x)^{-\frac{3}{5}}},$$
provided that this last limit exists.

But, by the Combination Rules and the Power Rule for continuous functions, the functions f' and g' are continuous at 0; hence, by the Quotient Rule
$$\lim_{x\to 0}\frac{f'(x)}{g'(x)}=\frac{f'(0)}{g'(0)}$$
$$=\frac{\frac{1}{5}+\frac{1}{5}}{\frac{4}{5}+\frac{4}{5}}$$
$$=\frac{1}{4}.$$

It follows, from l'Hôpital's Rule, that the original limit exists, and equals $\frac{1}{4}$.

(c) Let
$$f(x)=\sin\left(x^2+\sin\left(x^2\right)\right)\quad\text{and}\quad g(x)=1-\cos 4x,\quad\text{for }x\in\mathbb{R}.$$

Then f and g are differentiable on $\mathbb{R}$, and
$$f(0)=g(0)=0;$$
so that f and g satisfy the conditions of l'Hôpital's Rule at 0.
Now
$$f'(x)=\cos\left(x^2+\sin\left(x^2\right)\right)\times\left(2x+2x\cos\left(x^2\right)\right)$$
and
$$g'(x)=4\sin 4x,$$
so that
$$\frac{f'(x)}{g'(x)}=\frac{\cos(x^2+\sin(x^2))\times(2x+2x\cos(x^2))}{4\sin 4x}.$$

It follows, from l'Hôpital's Rule, that the limit $\lim_{x\to 0}\frac{f(x)}{g(x)}$ exists and equals
$$\lim_{x\to 0}\frac{f'(x)}{g'(x)},\qquad\qquad(*)$$
provided that this last limit $(*)$ exists.

Next, $f'(0)=0$ and $g'(0)=0$, and f' and g' are differentiable on $\mathbb{R}$. Thus f' and g' satisfy the conditions of l'Hôpital's Rule at 0.

It follows, from l'Hôpital's Rule, that the limit $\lim_{x\to 0}\frac{f'(x)}{g'(x)}$ exists and equals

$$\lim_{x \to 0} \frac{f''(x)}{g''(x)}, \qquad\qquad\qquad (**)$$

provided that this last limit $(**)$ exists.

Now

$$f''(x) = -\sin(x^2 + \sin(x^2)) \times (2x + 2x\cos(x^2))^2$$
$$+ \cos(x^2 + \sin(x^2)) \times (2 + 2\cos(x^2) - 4x^2\sin(x^2))$$

and

$$g''(x) = 16\cos 4x.$$

But, by the Combination Rules for continuous functions, the functions f'' and g'' are continuous at 0, so that

$$\lim_{x \to 0} \frac{f''(x)}{g''(x)} = \frac{f''(0)}{g''(0)}$$
$$= \frac{4}{16} = \frac{1}{4}.$$

It follows, from l'Hôpital's Rule, that the limit $(**)$ exists, and that its value is $\frac{1}{4}$. It then follows, from a second application of l'Hôpital's Rule, that the original limit exists, and that its value is $\frac{1}{4}$.

(d) Let

$$f(x) = \sin x - x\cos x \quad \text{and} \quad g(x) = x^3, \quad \text{for} \quad x \in \mathbb{R}.$$

Then f and g are differentiable on $\mathbb{R}$, and

$$f(0) = g(0) = 0,$$

so that f and g satisfy the conditions of l'Hôpital's Rule at 0.

Now

$$f'(x) = x\sin x \quad \text{and} \quad g'(x) = 3x^2,$$

so that

$$\frac{f'(x)}{g'(x)} = \frac{x\sin x}{3x^2}$$
$$= \frac{1}{3} \times \frac{\sin x}{x}.$$

It follows, from l'Hôpital's Rule, that the desired limit $\lim_{x \to 0}\frac{f(x)}{g(x)}$ exists and equals

$$\lim_{x \to 0} \frac{f'(x)}{g'(x)} = \lim_{x \to 0}\left(\frac{1}{3} \times \frac{\sin x}{x}\right)$$
$$= \frac{1}{3} \times \lim_{x \to 0}\left(\frac{\sin x}{x}\right),$$

provided that this last limit exists.

But $\lim_{x \to 0}\left(\frac{\sin x}{x}\right) = 1$, so that the limit $\lim_{x \to 0}\frac{f'(x)}{g'(x)}$ does exist and equals $\frac{1}{3}$. It follows, from l'Hôpital's Rule, that the original limit also exists, and that its value is $\frac{1}{3}$.

Chapter 7

Section 7.1

1. In this case, the function f is continuous, except at $\frac{1}{2}$ and 1.

   On the three subintervals $\left[0, \frac{1}{3}\right]$, $\left[\frac{1}{3}, \frac{3}{4}\right]$ and $\left[\frac{3}{4}, 1\right]$ in P, we have

$$m_1 = f(0) = 0, \quad M_1 = f\left(\frac{1}{3}\right) = \frac{2}{3}, \quad \delta x_1 = \frac{1}{3} - 0 = \frac{1}{3},$$

$$m_2 = f\left(\frac{1}{2}\right) = 0, \quad M_2 = f\left(\frac{3}{4}\right) = \frac{3}{2}, \quad \delta x_2 = \frac{3}{4} - \frac{1}{3} = \frac{5}{12},$$

$$m_3 = f(1) = 1, \quad M_3 = 2\left(= \lim_{x \to 1^-} f(x)\right), \quad \delta x_3 = 1 - \frac{3}{4} = \frac{1}{4}.$$

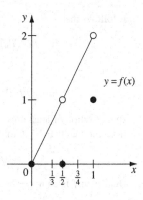

It then follows, from the definitions of $L(f, P)$ and $U(f, P)$, that

$$L(f, P) = \sum_{i=1}^{3} m_i \delta x_i = m_1 \delta x_1 + m_2 \delta x_2 + m_3 \delta x_3$$
$$= 0 \times \frac{1}{3} + 0 \times \frac{5}{12} + 1 \times \frac{1}{4}$$
$$= 0 + 0 + \frac{1}{4} = \frac{1}{4},$$

$$U(f, P) = \sum_{i=1}^{3} M_i \delta x_i = M_1 \delta x_1 + M_2 \delta x_2 + M_3 \delta x_3$$
$$= \frac{2}{3} \times \frac{1}{3} + \frac{3}{2} \times \frac{5}{12} + 2 \times \frac{1}{4}$$
$$= \frac{2}{9} + \frac{5}{8} + \frac{1}{2} = \frac{97}{72}.$$

2. In this case, the function f is increasing and continuous on $[0, 1]$. Thus, on each subinterval in $[0, 1]$, the infimum of f is the value of f at the left end-point of the subinterval and the supremum of f is the value of f at the right end-point of the subinterval.

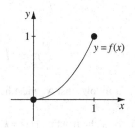

Hence, on the ith subinterval $\left[\frac{i-1}{n}, \frac{i}{n}\right]$ in P_n, for $1 \le i \le n$, we have

$$m_i = f\left(\frac{i-1}{n}\right) = \left(\frac{i-1}{n}\right)^2, \quad M_i = f\left(\frac{i}{n}\right) = \left(\frac{i}{n}\right)^2, \quad \text{and}$$
$$\delta x_i = \frac{i}{n} - \frac{i-1}{n} = \frac{1}{n}.$$

It then follows, from the definitions of $L(f, P_n)$ and $U(f, P_n)$, that

$$L(f, P_n) = \sum_{i=1}^{n} m_i \delta x_i = \sum_{i=1}^{n} \left(\frac{i-1}{n}\right)^2 \times \frac{1}{n}$$
$$= \frac{1}{n^3} \sum_{i=1}^{n} \left(i^2 - 2i + 1\right)$$
$$= \frac{1}{n^3} \left\{ \frac{n(n+1)(2n+1)}{6} - 2\frac{n(n+1)}{2} + n \right\}$$
$$= \frac{1}{n^2} \left\{ \frac{(n+1)(2n+1)}{6} - (n+1) + 1 \right\}$$
$$= \frac{1}{6n^2} \left\{ (2n^2 + 3n + 1) - 6(n+1) + 6 \right\}$$
$$= \frac{1}{6n^2} \left\{ 2n^2 - 3n + 1 \right\} = \frac{1}{3} - \frac{1}{2n} + \frac{1}{6n^2},$$

$$U(f, P_n) = \sum_{i=1}^{n} M_i \delta x_i = \sum_{i=1}^{n} \left(\frac{i}{n}\right)^2 \times \frac{1}{n}$$
$$= \frac{1}{n^3} \sum_{i=1}^{n} i^2$$
$$= \frac{1}{n^3} \times \frac{n(n+1)(2n+1)}{6}$$
$$= \frac{2n^2 + 3n + 1}{6n^2} = \frac{1}{3} + \frac{1}{2n} + \frac{1}{6n^2}.$$

It follows that

$$\lim_{n\to\infty} L(f, P_n) = \lim_{n\to\infty} \left(\frac{1}{3} - \frac{1}{2n} + \frac{1}{6n^2} \right) = \frac{1}{3}$$

and

$$\lim_{n\to\infty} U(f, P_n) = \lim_{n\to\infty} \left(\frac{1}{3} + \frac{1}{2n} + \frac{1}{6n^2} \right) = \frac{1}{3}.$$

3. By definition of least upper bound, we know that $f(x) \leq \sup_{x\in J} f$, for $x \in J$. It follows, from the fact that $I \subseteq J$, that

$$f(x) \leq \sup_{x\in J} f, \quad \text{for } x \in I.$$

Hence $\sup_{x\in J} f$ is an upper bound for f on I, so that $\sup_{x\in I} f \leq \sup_{x\in J} f$.

4. In Problem 2, we showed that for the function $f(x) = x^2$, $x \in [0, 1]$, and the partition $P_n = \left\{ \left[0, \frac{1}{n}\right], \left[\frac{1}{n}, \frac{2}{n}\right], \ldots, \left[\frac{i-1}{n}, \frac{i}{n}\right], \ldots, \left[1 - \frac{1}{n}, 1\right] \right\}$ of $[0, 1]$

$$\lim_{n\to\infty} L(f, P_n) = \frac{1}{3} \quad \text{and} \quad \lim_{n\to\infty} U(f, P_n) = \frac{1}{3}.$$

It follows that

$$\underline{\int_0^1} f \geq \frac{1}{3} \quad \text{and} \quad \overline{\int_0^1} f \leq \frac{1}{3}.$$

But

$$\underline{\int_0^1} f \leq \overline{\int_0^1} f, \quad \text{by part (b) of Theorem 4.}$$

It follows that we must have $\underline{\int_0^1} f = \overline{\int_0^1} f = \frac{1}{3}$, so that f is integrable on $[0, 1]$ and $\int_0^1 f = \frac{1}{3}$.

5. Here the function $f(x) = k$, $x \in [0, 1]$, is the constant function on $[0, 1]$.
   Hence, on the ith subinterval $\left[\frac{i-1}{n}, \frac{i}{n}\right]$ in P_n, for $1 \leq i \leq n$, we have

$$m_i = k, \quad M_i = k, \quad \text{and} \quad \delta x_i = \frac{i}{n} - \frac{i-1}{n} = \frac{1}{n}.$$

It then follows, from the definitions of $L(f, P_n)$ and $U(f, P_n)$, that

$$L(f, P_n) = \sum_{i=1}^n m_i \delta x_i = \sum_{i=1}^n k \times \frac{1}{n}$$

$$= k \sum_{i=1}^n \frac{1}{n} = k,$$

$$U(f, P_n) = \sum_{i=1}^n M_i \delta x_i = \sum_{i=1}^n k \times \frac{1}{n}$$

$$= k \sum_{i=1}^n \frac{1}{n} = k.$$

It follows that

$$\underline{\int_0^1} f \geq k \quad \text{and} \quad \overline{\int_0^1} f \leq k.$$

But

$$\underline{\int_0^1} f \leq \overline{\int_0^1} f, \quad \text{by part (b) of Theorem 4.}$$

It follows that we must have $\underline{\int_0^1} f = \overline{\int_0^1} f = k$, so that f is integrable on $[0, 1]$ and $\int_0^1 f = k$.

6. (a) Let $P_n = \{[x_0, x_1], [x_1, x_2], \ldots, [x_{i-1}, x_i], \ldots, [x_{n-1}, x_n]\}$ be a standard partition of $[0, 1]$.

The function f is constant on the interval $[0, 1)$, so that, for this partition P_n, we have

$$m_i = -2 \quad \text{for all } i,$$

$$M_i = \begin{cases} -2, & 1 \leq i \leq n - 1, \\ 3, & i = n, \end{cases}$$

$$\delta x_i = \frac{1}{n}.$$

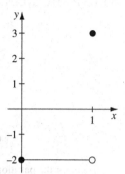

It follows that

$$L(f, P_n) = \sum_{i=1}^{n} m_i \delta x_i = \sum_{i=1}^{n} (-2) \times \frac{1}{n}$$

$$= -2 \sum_{i=1}^{n} \frac{1}{n} = -2,$$

and

$$U(f, P_n) = \sum_{i=1}^{n} M_i \delta x_i = \sum_{i=1}^{n-1} M_i \delta x_i + M_n \delta x_n$$

$$= \sum_{i=1}^{n-1} (-2) \times \frac{1}{n} + 3 \times \frac{1}{n}$$

$$= -2 \sum_{i=1}^{n-1} \frac{1}{n} + \frac{3}{n}$$

$$= -2 \frac{n-1}{n} + \frac{3}{n} = -2 + \frac{5}{n}.$$

Thus, in particular

$$U(f, P_n) - L(f, P_n) = \frac{5}{n}.$$

Now let ε be any given positive number. It follows that, if we choose n such that $\frac{5}{n} < \varepsilon$ (that is, if we choose $n > \frac{5}{\varepsilon}$), then

$$U(f, P_n) - L(f, P_n) \left(= \frac{5}{n} \right) < \varepsilon.$$

It follows, from Riemann's Criterion for integrability, that f is integrable on $[0, 1]$.

Alternatively, let $P = \{[x_0, x_1], [x_1, x_2], \ldots, [x_{i-1}, x_i], \ldots, [x_{n-1}, x_n]\}$ be any partition of $[0, 1]$. Then, since $M_i = m_i$ for $1 \leq i \leq n-1$, we have

$$U(f, P) - L(f, P) = \sum_{i=1}^{n} (M_i - m_i) \delta x_i = (M_n - m_n) \delta x_n$$

$$= (3 - (-2)) \delta x_n = 5 \delta x_n. \quad (*)$$

Now let ε be any given positive number, and choose P to be any partition of $[0, 1]$ such its mesh, $\|P\|$, is less than $\frac{\varepsilon}{5}$; then, in particular, we have $\delta x_n < \frac{\varepsilon}{5}$. It then follows, from the inequality $(*)$, that

$$U(f, P) - L(f, P) = 5 \delta x_n$$

$$< 5 \times \frac{\varepsilon}{5} = \varepsilon.$$

It follows, from Riemann's Criterion for integrability, that f is integrable on $[0, 1]$.

(b) Let $P = \{[x_0, x_1], [x_1, x_2], \ldots, [x_{i-1}, x_i], \ldots, [x_{n-1}, x_n]\}$ be any partition of $[0, 1]$.

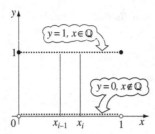

Then, for the partition P, we have

$$m_i = 0 \quad \text{and} \quad M_i = 1,$$

for all i.

It follows that

$$L(f, P) = \sum_{i=1}^{n} m_i \delta x_i = \sum_{i=1}^{n} 0 \times \delta x_i = 0,$$

and

$$U(f, P) = \sum_{i=1}^{n} M_i \delta x_i = \sum_{i=1}^{n} 1 \times \delta x_i$$
$$= \sum_{i=1}^{n} \delta x_i = 1.$$

Thus, for ALL partitions P of $[0, 1]$, we have

$$U(f, P) - L(f, P) = 1.$$

In particular, it follows that, for any positive ε for which $0 < \varepsilon < 1$, there is NO partition P of $[0, 1]$ for which

$$U(f, P) - L(f, P) < \varepsilon.$$

It follows, from Riemann's Criterion for integrability, that f is not integrable on $[0, 1]$.

Section 7.2

1. First, we want to use the Strategy to prove that

$$\inf_{x \in I}\{k + f(x)\} = k + \inf_{x \in I}\{f(x)\}.$$

Since f is bounded on I, $\inf_{x \in I}\{f(x)\}$ exists. In particular

$$f(x) \geq \inf_{x \in I}\{f(x)\}, \quad \text{for all } x \in I.$$

It follows that

$$k + f(x) \geq k + \inf_{x \in I}\{f(x)\}, \quad \text{for all } x \in I,$$

so that $k + \inf_{x \in I}\{f(x)\}$ is a lower bound for $k + f(x)$ on I.

Next, let m' be any number greater than $k + \inf_{x \in I}\{f(x)\}$. It follows that

$$m' - k > \inf_{x \in I}\{f(x)\},$$

so that, in particular, there is some number x in I such that

$$m' - k > f(x).$$

We may reformulate this fact as: there is some number x in I such that

$$m' > k + f(x),$$

so that m' is not a lower bound for $k + f(x)$ on I.

It follows that the greatest lower bound for $k + f(x)$ on I is $k + \inf_{x \in I}\{f(x)\}$; in other words

$$\inf_{x \in I}\{k + f(x)\} = k + \inf_{x \in I}\{f(x)\}.$$

Next, we want to use the Strategy to prove that

$$\sup_{x \in I}\{k + f(x)\} = k + \sup_{x \in I}\{f(x)\}.$$

Since f is bounded on I, $\sup_{x \in I}\{f(x)\}$ exists. In particular

$$f(x) \leq \sup_{x \in I}\{f(x)\}, \quad \text{for all } x \in I.$$

It follows that

$$k + f(x) \leq k + \sup_{x \in I}\{f(x)\}, \quad \text{for all } x \in I,$$

so that $k + \sup_{x \in I}\{f(x)\}$ is an upper bound for $k + f(x)$ on I.

Next, let M' be any number less than $k + \sup_{x \in I}\{f(x)\}$. It follows that

$$M' - k < \sup_{x \in I}\{f(x)\},$$

so that, in particular, there is some number x in I such that

$$M' - k < f(x).$$

We may reformulate this fact as: there is some number x in I such that

$$M' < k + f(x),$$

so that M' is not an upper bound for $k + f(x)$ on I.

It follows that the least upper bound for $k + f(x)$ on I is $k + \sup_{x \in I}\{f(x)\}$; in other words

$$\sup_{x \in I}\{k + f(x)\} = k + \sup_{x \in I}\{f(x)\}.$$

2. The function f is decreasing on the interval $(-2, 0]$, so that

$$\inf_{x \in (-2,0]} f = f(0) = 0 \quad \text{and} \quad \sup_{x \in (-2,0]} f = \lim_{x \to -2^+} f(x) = 4;$$

f is increasing on the interval $[0, 3]$, so that

$$\inf_{x \in [0,3]} f = f(0) = 0 \quad \text{and} \quad \sup_{x \in [0,3]} f = f(3) = 9.$$

Since $I = (-2, 0] \cup [0, 3]$, it follows that

$$\inf_I f = f(0) = 0 \quad \text{and} \quad \sup_I f = \max\left\{f(3), \lim_{x \to -2^+} f(x)\right\} = 9.$$

Since $0 \leq f(x) \leq 9$ and $0 \leq f(y) \leq 9$, so that $-9 \leq -f(y) \leq 0$, we have

$$-9 \leq f(x) - f(y) \leq 9.$$

We will prove that $\inf_{x,y \in I}\{f(x) - f(y)\} = -9$ and $\sup_{x,y \in I}\{f(x) - f(y)\} = 9$.

First, since $f(x) - f(y) \geq -9$, -9 is a lower bound for $f(x) - f(y)$, for $x, y \in I$. To prove that -9 is the greatest lower bound, we now need to prove that:

for each positive number ε, there are X and Y in $I = (-2, 3]$ for which

$$f(X) - f(Y) < -9 + \varepsilon.$$

Now, by the definition of infimum and supremum on I, we know that, since $\frac{1}{2}\varepsilon > 0$, there exist X and Y in $(-2, 3]$ such that

$$f(X) < \tfrac{1}{2}\varepsilon \quad \text{and} \quad f(Y) > 9 - \tfrac{1}{2}\varepsilon.$$

It follows from these two inequalities that

$$f(X) - f(Y) < \left(\tfrac{1}{2}\varepsilon\right) + \left(-9 + \tfrac{1}{2}\varepsilon\right)$$
$$= -9 + \varepsilon.$$

It follows that $\inf_{x,y \in I}\{f(x) - f(y)\} = -9$.

Finally, since $f(x) - f(y) \le 9$, 9 is an upper bound for $f(x) - f(y)$, for $x, y \in I$. To prove that 9 is the least upper bound, we now need to prove that:

for each positive number ε, there are X and Y in $I = (-2, 3]$ for which

$$f(X) - f(Y) > 9 - \varepsilon.$$

Now, by the definition of infimum and supremum on I, we know that, since $\frac{1}{2}\varepsilon > 0$, there exist X and Y in $(-2, 3]$ such that

$$f(X) > 9 - \tfrac{1}{2}\varepsilon \quad \text{and} \quad f(Y) < \tfrac{1}{2}\varepsilon.$$

It follows from these two inequalities that

$$f(X) - f(Y) > \left(9 - \tfrac{1}{2}\varepsilon\right) + \left(-\tfrac{1}{2}\varepsilon\right)$$
$$= 9 - \varepsilon.$$

It follows that $\sup_{x,y \in I}\{f(x) - f(y)\} = 9$.

3. (a) For any x in $I, f(x) \le \sup_I f$, so that, since $\lambda > 0$

$$\lambda f(x) \le \lambda \sup_I f, \quad \text{for all } x \in I.$$

Thus

$$\sup_{x \in I}(\lambda f(x)) \le \lambda \sup_I f.$$

To prove that $\sup_{x \in I}(\lambda f(x)) = \lambda\left(\sup_I f\right)$, we now need to prove that:

for each positive number ε, there is some X in I for which

$$\lambda f(X) > \lambda\left(\sup_I f\right) - \varepsilon. \tag{$*$}$$

Now, since $\varepsilon > 0$ and $\lambda > 0$, we have $\frac{\varepsilon}{\lambda} > 0$. It follows, from the definition of supremum of f on I, that there exists an X in I for which

$$f(X) > \sup_I f - \frac{\varepsilon}{\lambda}.$$

Multiplying both sides by the positive number λ, we obtain the desired result $(*)$.

(b) For any x in $I, f(x) \ge \inf_I f$ so that

$$-f(x) \le -\inf_I f, \quad \text{for all } x \in I.$$

Thus

$$\sup_{x \in I}(-f(x)) \le -\inf_I f.$$

To prove that $\sup_{x \in I}(-f(x)) = -\inf_I f$, we now need to prove that:

for each positive number ε, there is some X in I for which

$$-f(X) > -\inf_I f - \varepsilon. \tag{$*$}$$

For example, for any number X in $(-2, 3]$ with $|X| \le \frac{1}{2}\sqrt{\varepsilon}$, we have

$$f(X) = X^2 \le \frac{1}{4}\varepsilon;$$

we could, for instance, choose $X = \frac{1}{2}\sqrt{\varepsilon}$.

Now, since $\varepsilon > 0$ it follows, from the definition of infimum of f on I, that there exists an X in I for which

$$f(X) < \inf_I f + \varepsilon,$$

so that

$$-f(X) > -\inf_I f - \varepsilon.$$

This is precisely the desired result $(*)$.

Section 7.3

1. (a) $F'(x) = \left(x + \sqrt{x^2 - 4}\right)^{-1} \times \left(1 + \frac{1}{2}(x^2 - 4)^{-\frac{1}{2}}(2x)\right)$

$\qquad = \left(x + \sqrt{x^2 - 4}\right)^{-1} \times (x^2 - 4)^{-\frac{1}{2}} \times \left((x^2 - 4)^{\frac{1}{2}} + x\right)$

$\qquad = (x^2 - 4)^{-\frac{1}{2}}.$

(b) $F'(x) = \frac{1}{2}\sqrt{4 - x^2} + \frac{1}{2}x \times \frac{1}{2}(4 - x^2)^{-\frac{1}{2}}(-2x) + 2\left(1 - \frac{1}{4}x^2\right)^{-\frac{1}{2}}\frac{1}{2}$

$\qquad = (4 - x^2)^{-\frac{1}{2}}\left(\frac{1}{2}(4 - x^2) - \frac{1}{2}x^2 + 2\right)$

$\qquad = (4 - x^2)^{-\frac{1}{2}}(4 - x^2)$

$\qquad = (4 - x^2)^{\frac{1}{2}}.$

2. (a) From the Fundamental Theorem of Calculus and the Table of Standard Primitives, we obtain

$$\int_0^4 (x^2 + 9)^{\frac{1}{2}}dx = \left[\frac{1}{2}x(x^2 + 9)^{\frac{1}{2}} + \frac{9}{2}\log_e\left(x + (x^2 + 9)^{\frac{1}{2}}\right)\right]_0^4$$

$$= 10 + \frac{9}{2}\log_e 9 - \frac{9}{2}\log_e 3$$

$$= 10 + \frac{9}{2}\log_e 3.$$

(b) From the Fundamental Theorem of Calculus and the Table of Standard Primitives, we obtain

$$\int_1^e \log_e x\,dx = [x\log_e x - x]_1^e$$

$$= (e - e) - (0 - 1)$$

$$= 1.$$

3. Using the Table of Standard Primitives and the Combination Rules, we obtain the following primitives:

(a) $F(x) = 4(x\log_e x - x) - \tan^{-1}\left(\frac{x}{2}\right);$

(b) $F(x) = \frac{2}{3}\log_e(\sec 3x) + \frac{1}{5}e^{2x}(2\sin x - \cos x).$

4. (a) Here we use integration by parts. Let

$$f(x) = \log_e x \quad \text{and} \quad g'(x) = x^{\frac{1}{3}},$$

so that

$$f'(x) = \frac{1}{x} \quad \text{and} \quad g(x) = \frac{3}{4}x^{\frac{4}{3}}.$$

It follows that

$$\int x^{\frac{1}{3}} \log_e x \, dx = \frac{3}{4} x^{\frac{4}{3}} \log_e x - \int x^{-1} \frac{3}{4} x^{\frac{4}{3}} dx$$

$$= \frac{3}{4} x^{\frac{4}{3}} \log_e x - \frac{3}{4} \int x^{\frac{1}{3}} dx$$

$$= \frac{3}{4} x^{\frac{4}{3}} \log_e x - \frac{9}{16} x^{\frac{4}{3}}.$$

(b) Here we use integration by parts, twice. On each occasion we differentiate the power function and integrate the trigonometric function.

Hence

$$\int_0^{\frac{\pi}{2}} x^2 \cos x \, dx = \left[x^2 \sin x \right]_0^{\frac{\pi}{2}} - \int_0^{\frac{\pi}{2}} 2x \sin x \, dx$$

$$= \frac{1}{4} \pi^2 - 2 \int_0^{\frac{\pi}{2}} x \sin x \, dx,$$

and

$$\int_0^{\frac{\pi}{2}} x \sin x \, dx = \left[x \times (-\cos x) \right]_0^{\frac{\pi}{2}} - \int_0^{\frac{\pi}{2}} (-\cos x) \, dx$$

$$= 0 + \left[\sin x \right]_0^{\frac{\pi}{2}}$$

$$= 1.$$

It follows that

$$\int_0^{\frac{\pi}{2}} x^2 \cos x \, dx = \frac{1}{4} \pi^2 - 2.$$

5. (a) We follow the strategy just before Example 3, with x in place of t and u in place of x.

Let $u = g(x) = 2 \sin 3x$, $x \in \mathbb{R}$. Then

$$\frac{du}{dx} = 6 \cos 3x, \quad \text{so that } du = 6 \cos 3x \, dx.$$

Hence

$$\int \sin(2 \sin 3x) \cos 3x \, dx = \int \frac{1}{6} \sin u \, du$$

$$= -\frac{1}{6} \cos u$$

$$= -\frac{1}{6} \cos(2 \sin 3x).$$

(b) We follow the strategy just before Example 3, with x in place of t and u in place of x.

Let $u = g(x) = e^x$, $x \in \mathbb{R}$. The function g is one–one on $\mathbb{R}$. Then

$$\frac{du}{dx} = e^x, \quad \text{so that } du = e^x dx;$$

also

when $x = 0$,   then $u = 1$,

when $x = 1$,   then $u = e$.

Hence

$$\int_0^1 \frac{e^x}{(1 + e^x)^2} dx = \int_1^e \frac{du}{(1 + u)^2}$$

$$= \left[\frac{-1}{1 + u} \right]_1^e$$

$$= -\frac{1}{1 + e} + \frac{1}{2}$$

$$= \frac{e - 1}{2(1 + e)}.$$

6. (a) We follow the strategy just before Example 4.

Let $t = g(x) = (x-1)^{\frac{1}{2}}$, $x \in [1, \infty)$. The function g is one–one on $[1, \infty)$.

Then $t^2 = x - 1$, so that $x = t^2 + 1$. It follows that

$$\frac{dx}{dt} = 2t, \quad \text{so that } dx = 2t\,dt.$$

Hence

$$\int \frac{dx}{3(x-1)^{\frac{3}{2}} + x(x-1)^{\frac{1}{2}}} = \int \frac{2t\,dt}{3t^3 + (t^2+1)t}$$

$$= \int \frac{2\,dt}{4t^2 + 1}$$

$$= \tan^{-1}(2t)$$

$$= \tan^{-1}\left(2(x-1)^{\frac{1}{2}}\right).$$

(b) We follow the strategy just before Example 4.

Let $t = g(x) = \sqrt{1 + e^x}$, $x \in [0, \infty)$. The function g is one–one on $[0, \infty)$.

Then $t^2 = 1 + e^x$, so that $e^x = t^2 - 1$ and $x = \log_e(t^2 - 1)$. It follows that:

$$\frac{dx}{dt} = \frac{2t}{t^2 - 1}, \quad \text{so that } dx = \frac{2t}{t^2 - 1}\,dt;$$

also:

when $x = 0$, then $t = \sqrt{2}$,

when $x = \log_e 3$, then $t = 2$.

Hence

$$\int_0^{\log_e 3} e^x \sqrt{1 + e^x}\,dx = \int_{\sqrt{2}}^2 (t^2 - 1) \times t \times \frac{2t}{t^2 - 1}\,dt$$

$$= \int_{\sqrt{2}}^2 2t^2\,dt$$

$$= \left[\frac{2}{3}t^3\right]_{\sqrt{2}}^2$$

$$= \frac{16 - 4\sqrt{2}}{3}.$$

Section 7.4

1. Since

$$x\sin\left(\frac{1}{x^{10}}\right) \le x, \quad \text{for } x \in [1, 3],$$

it follows, from part (a) of Theorem 2, that

$$\int_1^3 x\sin\left(\frac{1}{x^{10}}\right)dx \le \int_1^3 x\,dx$$

$$= \left[\frac{1}{2}x^2\right]_1^3$$

$$= \frac{1}{2}(9 - 1) = 4.$$

2. Since

$$e^{x^2} \le \frac{1}{1 - x^2}$$

$$\le \frac{1}{1 - \frac{1}{4}} = \frac{4}{3}, \quad \text{for } x \in \left[0, \frac{1}{2}\right],$$

it follows, from part (b) of Theorem 2, that

$$\int_0^{\frac{1}{2}} e^{x^2} dx \le \frac{4}{3}\left(\frac{1}{2} - 0\right)$$
$$= \frac{2}{3}.$$

(*Remark*   We could obtain a smaller upper estimate for the integral by applying part (a) of Theorem 2 to the inequality $e^{x^2} \le \frac{1}{1-x^2}$.)

Next, since

$$e^{x^2} \ge 1 + x^2 \ge 1, \quad \text{for } x \in \left[0, \frac{1}{2}\right],$$

it follows, from part (b) of Theorem 2, that

$$\int_0^{\frac{1}{2}} e^{x^2} dx \ge 1 \times \left(\frac{1}{2} - 0\right)$$
$$= \frac{1}{2}.$$

3. (a) Since

$$\left|\sin\left(\frac{1}{x}\right)\right| \le 1, \quad \text{for } x \in [1, 4],$$

and

$$2 + \cos\left(\frac{1}{x}\right) \ge 1, \quad \text{for } x \in [1, 4],$$

it follows that

$$\left|\frac{\sin\left(\frac{1}{x}\right)}{2 + \cos\left(\frac{1}{x}\right)}\right| \le 1, \quad \text{for } x \in [1, 4].$$

It follows, from Theorem 3, that

$$\left|\int_1^4 \frac{\sin\left(\frac{1}{x}\right)}{2 + \cos\left(\frac{1}{x}\right)}\right| \le 1 \times (4 - 1)$$
$$= 3.$$

(b) Since

$$\tan x \ge 0 \quad \text{for } x \in \left[0, \frac{\pi}{4}\right],$$

and

$$3 - \sin(x^2) \ge 2, \quad \text{for } x \in \left[0, \frac{\pi}{4}\right],$$

it follows that

$$\frac{\tan x}{3 - \sin(x^2)} \le \frac{1}{2}\tan x, \quad \text{for } x \in \left[0, \frac{\pi}{4}\right].$$

It follows, from Theorems 2 and 3, that

$$\left|\int_0^{\frac{\pi}{4}} \frac{\tan x}{3 - \sin(x^2)} dx\right| \le \int_0^{\frac{\pi}{4}} \left|\frac{\tan x}{3 - \sin(x^2)}\right| dx$$
$$\le \int_0^{\frac{\pi}{4}} \frac{1}{2}\tan x\, dx$$
$$= \left[\frac{1}{2}\log_e(\sec x)\right]_0^{\frac{\pi}{4}}$$
$$= \frac{1}{2}\left(\log_e\left(\sec\frac{\pi}{4}\right) - \log_e 1\right)$$
$$= \frac{1}{2}\log_e \sqrt{2} = \frac{1}{4}\log_e 2.$$

4. (a)
$$I_0 = \int_0^1 e^x dx$$
$$= [e^x]_0^1 = e - 1.$$

(b) Using integration by parts, we obtain
$$I_n = \int_0^1 e^x x^n dx$$
$$= [e^x x^n]_0^1 - \int_0^1 e^x n x^{n-1} dx$$
$$= e - n I_{n-1}.$$

(c) Using the result of part (b), with $n = 1, 2, 3$ and 4 in turn, we obtain
$$I_1 = e - I_0 = e - (e - 1) = 1,$$
$$I_2 = e - 2I_1 = e - 2,$$
$$I_3 = e - 3I_2 = e - 3(e - 2) = 6 - 2e,$$
$$I_4 = e - 4I_3 = e - 4(6 - 2e) = 9e - 24.$$

5. (a)
$$a_1 = \frac{2}{1} \cdot \frac{2}{3} = \frac{4}{3};$$
$$a_2 = \frac{2}{1} \cdot \frac{2}{3} \cdot \frac{4}{3} \cdot \frac{4}{5} = \frac{64}{45};$$
$$a_3 = \frac{2}{1} \cdot \frac{2}{3} \cdot \frac{4}{3} \cdot \frac{4}{5} \cdot \frac{6}{5} \cdot \frac{6}{7} = \frac{256}{175};$$
$$b_1 = \frac{(1!)^2 2^2}{2!\sqrt{1}} = 2;$$
$$b_2 = \frac{(2!)^2 2^4}{4!\sqrt{2}} = \frac{4}{3}\sqrt{2};$$
$$b_3 = \frac{(3!)^2 2^6}{6!\sqrt{3}} = \frac{16}{15}\sqrt{3}.$$

(b)
$$b_1^2 = 4 \quad \text{and} \quad 3a_1 = 4;$$
$$b_2^2 = \frac{32}{9} \quad \text{and} \quad \frac{5}{2}a_2 = \frac{32}{9};$$
$$b_3^2 = \frac{256}{75} \quad \text{and} \quad \frac{7}{3}a_3 = \frac{256}{75}.$$

(c) Firstly
$$b_n^2 = \frac{(n!)^4 2^{4n}}{((2n)!)^2 n}.$$

We now express a_n in a similar way, tackling the numerator and denominator separately.

The numerator is a product of $2n$ even numbers; so, taking a factor 2 from each term, we obtain
$$2.2.4.4. \ldots .(2n).(2n) = 2^{2n}(1.1.2.2. \ldots .n.n)$$
$$= 2^{2n}(n!)^2.$$

The denominator of a_n cannot be tackled in quite the same way, as all its factors are odd. However, we can relate it to factorials by introducing the missing even factors:

$$1.3.3.5.5. \ldots .(2n-1).(2n+1) = \frac{1.2.2.3.3.4.4. \ldots .(2n-1).(2n).(2n).(2n+1)}{2.2. \quad 4.4. \quad \ldots \quad .(2n).(2n)}$$

$$= \frac{((2n)!)^2(2n+1)}{2^{2n}(n!)^2}.$$

It follows that

$$\begin{aligned}
a_n &= \frac{2^{2n}(n!)^2}{\frac{((2n)!)^2(2n+1)}{2^{2n}(n!)^2}} \\
&= \frac{2^{4n}(n!)^4}{((2n)!)^2(2n+1)} \\
&= b_n^2 \frac{n}{2n+1},
\end{aligned}$$

so that

$$b_n^2 = \frac{2n+1}{n} a_n.$$

6. Let

$$f(x) = \frac{1}{x^p}, \quad \text{for } x \in [1, \infty),$$

where $p > 0$ and $p \neq 1$.

Then f is positive and decreasing, and

$$f(x) \to 0 \quad \text{as } x \to \infty.$$

Also,

$$\begin{aligned}
\int_1^n f &= \int_1^n \frac{dx}{x^p} \\
&= \left[\frac{x^{1-p}}{1-p} \right]_1^n \\
&= \frac{n^{1-p}-1}{1-p}.
\end{aligned} \qquad (*)$$

Now, if $p > 1$, then $1 - p < 0$, so that

$$\int_1^n f = \frac{1}{p-1} - \frac{n^{1-p}}{p-1} < \frac{1}{p-1}.$$

Since the set $\left\{ \int_1^n f : n \in \mathbb{N} \right\}$ is bounded above, it follows, from the Maclaurin Integral Test, that the series converges.

Finally, if $0 < p < 1$, then $1 - p > 0$, so that

$$n^{1-p} \to \infty \quad \text{as } n \to \infty.$$

It follows, from the Maclaurin Integral Test, that the series diverges.

7. Let $t = g(x) = \log_e x$, $x \in (1, \infty)$. The function g is one–one on $(1, \infty)$.

Then $x = e^t$; also

$$\frac{dt}{dx} = \frac{1}{x} \quad \text{and so } dt = \frac{dx}{x}.$$

Hence

$$\begin{aligned}
\int \frac{dx}{x(\log_e x)^2} &= \int \frac{dt}{t^2} \\
&= -\frac{1}{t} = -\frac{1}{\log_e x}.
\end{aligned}$$

Now let

$$f(x) = \frac{1}{x(\log_e x)^2}, \quad x \in [2, \infty).$$

Then f is positive and decreasing on $[2, \infty)$, and

$$f(x) \to 0 \quad \text{as } x \to \infty.$$

Also

$$\int_2^n f = \int_2^n \frac{dx}{x(\log_e x)^2}$$

$$= \left[-\frac{1}{\log_e x} \right]_2^n$$

$$= \frac{1}{\log_e 2} - \frac{1}{\log_e n}$$

$$\leq \frac{1}{\log_e 2}.$$

Since the set $\{\int_2^n f : n \in \mathbb{N}\}$ is bounded above, it follows, from the Maclaurin Integral Test, that the series converges.

8. Let $t = g(x) = \log_e x$, $x \in (1, \infty)$. The function g is one–one on $(1, \infty)$.
   Then $x = e^t$; also

$$\frac{dt}{dx} = \frac{1}{x} \quad \text{and so } dt = \frac{dx}{x}.$$

Hence

$$\int \frac{dx}{x \log_e x} = \int \frac{dt}{t}$$

$$= \log_e t = \log_e(\log_e x).$$

Now let

$$f(x) = \frac{1}{x \log_e x}, \quad x \in [2, \infty).$$

Then f is positive and decreasing on $[2, \infty)$, and

$$f(x) \to 0 \quad \text{as } x \to \infty.$$

Also

$$\int_2^n f = \int_2^n \frac{dx}{x \log_e x}$$

$$= [\log_e(\log_e x)]_2^n$$

$$= \log_e(\log_e n) - \log_e(\log_e 2)$$

$$\to \infty \quad \text{as } n \to \infty.$$

Hence, by the Maclaurin Integral Test, the series diverges.

9. Let $f(x) = \frac{1}{1+x^2}$, $x \in [0, 1]$.
   Then f is positive and decreasing on $[0, 1]$; it follows, from the strategy, that

$$\frac{1}{n} \sum_{i=1}^n f\left(\frac{i}{n}\right) \to \int_0^1 f.$$

Now

$$\frac{1}{n} \sum_{i=1}^n f\left(\frac{i}{n}\right) = \frac{1}{n} \sum_{i=1}^n \frac{1}{1 + \left(\frac{i}{n}\right)^2} = n \sum_{i=1}^n \frac{1}{n^2 + i^2}$$

$$= n\left(\frac{1}{n^2 + 1^2} + \frac{1}{n^2 + 2^2} + \cdots + \frac{1}{n^2 + n^2} \right),$$

and

$$\int_0^1 f = \int_0^1 \frac{1}{1+x^2}\,dx$$

$$= \left[\tan^{-1} x\right]_0^1$$

$$= \tan^{-1} 1 - \tan^{-1} 0 = \frac{1}{4}\pi.$$

It follows that

$$n\left(\frac{1}{n^2+1^2} + \frac{1}{n^2+2^2} + \cdots + \frac{1}{n^2+n^2}\right) \to \frac{1}{4}\pi \quad \text{as } n \to \infty.$$

10. Let $f(x) = \frac{1}{\sqrt{x}}$, $x \in [1, \infty)$.

  Then f is positive and decreasing on $[1, \infty)$, and $f(x) \to 0$ as $x \to \infty$.

  Now

$$\sum_{i=1}^n f(i) = \sum_{i=1}^n \frac{1}{\sqrt{i}}$$

$$= 1 + \frac{1}{\sqrt{2}} + \frac{1}{\sqrt{3}} + \cdots + \frac{1}{\sqrt{n}},$$

and

$$\int_1^n f = \int_1^n x^{-\frac{1}{2}}\,dx$$

$$= \left[2x^{\frac{1}{2}}\right]_1^n$$

$$= 2n^{\frac{1}{2}} - 2 \to \infty \quad \text{as } n \to \infty.$$

It follows, from the strategy, that

$$1 + \frac{1}{\sqrt{2}} + \frac{1}{\sqrt{3}} + \cdots + \frac{1}{\sqrt{n}} \sim 2n^{\frac{1}{2}} - 2 \quad \text{as } n \to \infty,$$

and so, using the remark before the problem, that

$$1 + \frac{1}{\sqrt{2}} + \frac{1}{\sqrt{3}} + \cdots + \frac{1}{\sqrt{n}} \sim 2n^{\frac{1}{2}} \quad \text{as } n \to \infty.$$

Section 7.5

1. $20! \simeq 2.4329 \times 10^{18}$;

   $30! \simeq 2.6525 \times 10^{32}$;

   $40! \simeq 8.1592 \times 10^{47}$;

   $50! \simeq 3.0414 \times 10^{64}$;

   $60! \simeq 8.3210 \times 10^{81}$.

2. First, $f_2(n) \sim f_5(n)$; that is, $\sin\left(\frac{1}{n^2}\right) \sim \frac{1}{n^2}$ as $n \to \infty$.

   We know that

$$\frac{\sin x}{x} \to 1 \quad \text{as } x \to 0;$$

   since the sequence $\left\{\frac{1}{n^2}\right\}$ tends to 0 as $n \to \infty$, it follows that

$$\frac{\sin\left(\frac{1}{n^2}\right)}{\frac{1}{n^2}} \to 1 \quad \text{as } n \to \infty.$$

   In other words, $\sin\left(\frac{1}{n^2}\right) \sim \frac{1}{n^2}$ as $n \to \infty$.

   Second, $f_3(n) \sim f_7(n)$; that is, $1 - \cos\left(\frac{1}{n}\right) \sim \frac{1}{2n^2}$ as $n \to \infty$.

   We can use the formula $\cos x = 1 - 2\sin^2\left(\frac{1}{2}x\right)$ to obtain

$$\frac{1 - \cos x}{x^2} = \frac{2\sin^2\left(\frac{1}{2}x\right)}{x^2}$$

$$= \frac{1}{2} \times \frac{\sin\left(\frac{1}{2}x\right)}{\frac{1}{2}x} \times \frac{\sin\left(\frac{1}{2}x\right)}{\frac{1}{2}x}.$$

Then, since $\frac{1}{2}x \to 0$ as $x \to 0$, we deduce, from the fact that $\frac{\sin x}{x} \to 1$ as $x \to 0$, that

$$\frac{1 - \cos x}{x^2} \to \frac{1}{2} \quad \text{as } x \to 0.$$

Then, since the sequence $\left\{\frac{1}{n}\right\}$ tends to 0 as $n \to \infty$, it follows that

$$\frac{1 - \cos\left(\frac{1}{n}\right)}{\frac{1}{n^2}} \to \frac{1}{2} \quad \text{as } n \to \infty,$$

so that

$$\frac{1 - \cos\left(\frac{1}{n}\right)}{\frac{1}{2n^2}} \to 1 \quad \text{as } n \to \infty;$$

in other words, $1 - \cos\left(\frac{1}{n}\right) \sim \frac{1}{2n^2}$ as $n \to \infty$.

3. Using the Hint, let $f(n) = n^2$ and $g(n) = n$.
   Then

$$(f(n))^{\frac{1}{n}} = \left(n^2\right)^{\frac{1}{n}}$$

$$= \left(n^{\frac{1}{n}}\right)^2$$

$$\to (1)^2 = 1 \quad \text{as } n \to \infty,$$

and

$$(g(n))^{\frac{1}{n}} = (n)^{\frac{1}{n}}$$

$$\to 1 \quad \text{as } n \to \infty;$$

so that $\frac{(f(n))^{\frac{1}{n}}}{(g(n))^{\frac{1}{n}}} \to 1$ as $n \to \infty$. In other words, $(f(n))^{\frac{1}{n}} \sim (g(n))^{\frac{1}{n}}$ as $n \to \infty$.
   However

$$\frac{f(n)}{g(n)} = \frac{n^2}{n}$$

$$= n \not\to 1 \quad \text{as } n \to \infty,$$

so that $f(n) \not\sim g(n)$ as $n \to \infty$.

4. First, using a calculator, we find that the value of the expression $\sqrt{2\pi n}\left(\frac{n}{e}\right)^n$ when $n = 5$ is approximately 118.019.
   Now $5! = 120$, so that the error in the Stirling's Formula approximation is about $120 - 118.019 = 1.981$. The percentage error in this approximation is thus about

$$\frac{1.981}{120} \times 100 \simeq 1.651\%.$$

This approximation is therefore not within 1% of the exact value.

5. Using Stirling's Formula, we obtain

$$\frac{n^n}{n!} \sim \frac{n^n}{\sqrt{2\pi n}\left(\frac{n}{e}\right)^n}$$

$$= \frac{e^n}{\sqrt{2\pi n}};$$

it follows from Example 1 that

$$\left(\frac{n^n}{n!}\right)^{\frac{1}{n}} \sim \frac{e}{\pi^{\frac{1}{2n}}(2n)^{\frac{1}{2n}}} \quad \text{as } n \to \infty.$$

We have seen previously that, for any positive number a

$$a^{\frac{1}{n}} \to 1 \quad \text{and} \quad n^{\frac{1}{n}} \to 1, \quad \text{as } n \to \infty.$$

It follows that

$$\pi^{\frac{1}{2n}} \to 1 \quad \text{and} \quad (2n)^{\frac{1}{2n}} \to 1, \quad \text{as } n \to \infty.$$

It follows that

$$\left(\frac{n^n}{n!}\right)^{\frac{1}{n}} \sim e \quad \text{as } n \to \infty;$$

that is

$$\left(\frac{n^n}{n!}\right)^{\frac{1}{n}} \to e \quad \text{as } n \to \infty.$$

6. (a)

$$\binom{300}{150} \frac{1}{2^{300}} = \frac{300!}{150! \times 150! \times 2^{300}}$$

$$\simeq \frac{\sqrt{600\pi}\left(\frac{300}{e}\right)^{300}}{300\pi\left(\frac{150}{e}\right)^{300} 2^{300}}$$

$$= \frac{1}{30}\sqrt{\frac{6}{\pi}}$$

$$= 0.046 \quad \text{(to two significant figures)}.$$

(b)

$$\frac{300!}{(100!)^3} \times \frac{1}{3^{300}} \simeq \frac{\sqrt{600\pi}\left(\frac{300}{e}\right)^{300}}{(200\pi)^{\frac{3}{2}}\left(\frac{100}{e}\right)^{300} 3^{300}}$$

$$= \frac{\sqrt{3}}{200\pi}$$

$$= 0.0028 \quad \text{(to two significant figures)}.$$

7. Now

$$\binom{4n}{2n} = \frac{(4n)!}{((2n)!)^2} \quad \text{and} \quad \binom{2n}{n} = \frac{(2n)!}{(n!)^2},$$

so that

$$\frac{\binom{4n}{2n}}{\binom{2n}{n}} = \frac{(4n)!(n!)^2}{((2n)!)^3}.$$

It follows, from Stirling's Formula, that

$$\frac{\binom{4n}{2n}}{\binom{2n}{n}} \sim \frac{\sqrt{8\pi n}\left(\frac{4n}{e}\right)^{4n} 2\pi n\left(\frac{n}{e}\right)^{2n}}{(4\pi n)^{\frac{3}{2}}\left(\frac{2n}{e}\right)^{6n}}$$

$$= \frac{4^{4n}}{\sqrt{2} \times 2^{6n}}$$

$$= \frac{2^{2n}}{\sqrt{2}}.$$

Hence $\lambda = 1/\sqrt{2}$.

Chapter 8

Section 8.1

1. The tangent approximation to f at a is

$$f(x) \simeq f(a) + f'(a)(x - a).$$

(a)
$$f(x) = e^x, \quad f(2) = e^2,$$
$$f'(x) = e^x, \quad f'(2) = e^2.$$

Hence the tangent approximation to f at 2 is

$$e^x \simeq e^2 + e^2(x - 2).$$

(b)
$$f(x) = \cos x, \quad f(0) = 1,$$
$$f'(x) = -\sin x, \quad f'(0) = 0.$$

Hence the tangent approximation to f at 0 is

$$\cos x \simeq 1 + 0(x - 0) = 1.$$

2. (a)
$$f(x) = e^x, \quad f(2) = e^2,$$
$$f'(x) = e^x, \quad f'(2) = e^2,$$
$$f''(x) = e^x, \quad f''(2) = e^2,$$
$$f'''(x) = e^x, \quad f'''(2) = e^2.$$

Hence

$$T_1(x) = f(2) + \frac{f'(2)}{1!}(x - 2)$$
$$= e^2 + e^2(x - 2),$$
$$T_2(x) = f(2) + \frac{f'(2)}{1!}(x - 2) + \frac{f''(2)}{2!}(x - 2)^2$$
$$= e^2 + e^2(x - 2) + \frac{1}{2}e^2(x - 2)^2,$$
$$T_3(x) = f(2) + \frac{f'(2)}{1!}(x - 2) + \frac{f''(2)}{2!}(x - 2)^2 + \frac{f'''(2)}{3!}(x - 2)^3$$
$$= e^2 + e^2(x - 2) + \frac{1}{2}e^2(x - 2)^2 + \frac{1}{6}e^2(x - 2)^3.$$

(b)
$$f(x) = \cos x, \quad f(0) = 1,$$
$$f'(x) = -\sin x, \quad f'(0) = 0,$$
$$f''(x) = -\cos x, \quad f''(0) = -1,$$
$$f'''(x) = \sin x, \quad f'''(0) = 0.$$

Hence

$$T_1(x) = f(0) + \frac{f'(0)}{1!}x = 1,$$
$$T_2(x) = f(0) + \frac{f'(0)}{1!}x + \frac{f''(0)}{2!}x^2 = 1 - \frac{1}{2}x^2,$$
$$T_3(x) = f(0) + \frac{f'(0)}{1!}x + \frac{f''(0)}{2!}x^2 + \frac{f'''(0)}{3!}x^3 = 1 - \frac{1}{2}x^2.$$

3. The Taylor polynomial of degree 4 for f at a is

$$T_4(x) = f(a) + \frac{f'(a)}{1!}(x-a) + \frac{f''(a)}{2!}(x-a)^2 + \frac{f'''(a)}{3!}(x-a)^3 + \frac{f^{(4)}(a)}{4!}(x-a)^4.$$

(a)
$$f(x) = 7 - 6x + 5x^2 + x^3, \quad f(1) = 7,$$
$$f'(x) = -6 + 10x + 3x^2, \quad f'(1) = 7,$$
$$f''(x) = 10 + 6x, \quad\quad\quad f''(1) = 16,$$
$$f'''(x) = 6, \quad\quad\quad\quad\quad f'''(1) = 6,$$
$$f^{(4)}(x) = 0, \quad\quad\quad\quad\quad f^{(4)}(1) = 0.$$

Hence

$$T_4(x) = 7 + 7(x-1) + 8(x-1)^2 + (x-1)^3.$$

(b)
$$f(x) = (1-x)^{-1}, \quad\quad f(0) = 1,$$
$$f'(x) = (1-x)^{-2}, \quad\quad f'(0) = 1,$$
$$f''(x) = 2(1-x)^{-3}, \quad\quad f''(0) = 2,$$
$$f'''(x) = 3!(1-x)^{-4}, \quad\quad f'''(0) = 3!,$$
$$f^{(4)}(x) = 4!(1-x)^{-5}, \quad\quad f^{(4)}(0) = 4!.$$

Hence

$$T_4(x) = 1 + x + x^2 + x^3 + x^4.$$

(c)
$$f(x) = \log_e(1+x), \quad\quad f(0) = 0,$$
$$f'(x) = (1+x)^{-1}, \quad\quad f'(0) = 1,$$
$$f''(x) = -(1+x)^{-2}, \quad\quad f''(0) = -1,$$
$$f'''(x) = 2(1+x)^{-3}, \quad\quad f'''(0) = 2,$$
$$f^{(4)}(x) = -3!(1+x)^{-4}, \quad f^{(4)}(0) = -3!.$$

Hence

$$T_4(x) = x - \frac{1}{2}x^2 + \frac{1}{3}x^3 - \frac{1}{4}x^4.$$

(d)
$$f(x) = \sin x, \quad\quad f\left(\frac{\pi}{4}\right) = \frac{1}{\sqrt{2}},$$
$$f'(x) = \cos x, \quad\quad f'\left(\frac{\pi}{4}\right) = \frac{1}{\sqrt{2}},$$
$$f''(x) = -\sin x, \quad\quad f''\left(\frac{\pi}{4}\right) = -\frac{1}{\sqrt{2}},$$
$$f'''(x) = -\cos x, \quad\quad f'''\left(\frac{\pi}{4}\right) = -\frac{1}{\sqrt{2}},$$
$$f^{(4)}(x) - \sin x, \quad\quad f^{(4)}\left(\frac{\pi}{4}\right) = \frac{1}{\sqrt{2}}.$$

Hence

$$T_4(x) = \frac{1}{\sqrt{2}}\left(1 + \left(x - \frac{\pi}{4}\right) - \frac{1}{2}\left(x - \frac{\pi}{4}\right)^2 - \frac{1}{6}\left(x - \frac{\pi}{4}\right)^3 + \frac{1}{24}\left(x - \frac{\pi}{4}\right)^4\right).$$

(e)
$$f(x) = 1 + \frac{1}{2}x - \frac{1}{2}x^2 - \frac{1}{6}x^3 + \frac{1}{4}x^4, \quad f(0) = 1,$$

$$f'(x) = \frac{1}{2} - x - \frac{1}{2}x^2 + x^3, \qquad\qquad f'(0) = \frac{1}{2},$$

$$f''(x) = -1 - x + 3x^2, \qquad\qquad f''(0) = -1,$$

$$f'''(x) = -1 + 3!x, \qquad\qquad f'''(0) = -1,$$

$$f^{(4)}(x) = 3!, \qquad\qquad f^{(4)}(0) = 3!.$$

Hence

$$T_4(x) = 1 + \frac{1}{2}x - \frac{1}{2}x^2 - \frac{1}{6}x^3 + \frac{1}{4}x^4.$$

4.
$$f(x) = \tan x, \qquad\qquad f(0) = 0,$$

$$f'(x) = \sec^2 x, \qquad\qquad f'(0) = 1,$$

$$f''(x) = 2\sec^2 x \tan x, \qquad\qquad f''(0) = 0,$$

$$f'''(x) = 4\sec^2 x \tan^2 x + 2\sec^4 x, \qquad f'''(0) = 2.$$

Hence

$$T_3(x) = x + \frac{1}{3}x^3,$$

so that

$$T_3(0.1) = 0.1 + \frac{1}{3} \times 0.001 = 0.100\overline{3}.$$

Since (by calculator) $\tan 0.1 = 0.10033467\ldots$, the percentage error involved is about

$$\frac{0.10033467 - 0.10033333}{0.10033467} \times 100 \simeq 0.001\%.$$

5. (a)
$$f(x) = (1-x)^{-1}, \qquad f(0) = 1,$$

$$f'(x) = (1-x)^{-2}, \qquad f'(0) = 1,$$

$$f''(x) = 2 \times (1-x)^{-3}, \qquad f''(0) = 2,$$

$$\vdots$$

$$f^{(n)}(x) = (n!) \times (1-x)^{-n-1}, \quad f^{(n)}(0) = n!.$$

Hence

$$T_n(x) = f(0) + f'(0)x + \frac{1}{2!}f''(0)x^2 + \cdots + \frac{1}{n!}f^{(n)}(0)x^n$$

$$= 1 + x + x^2 + \cdots + x^n.$$

(b)
$$f(x) = \log_e(1+x), \qquad\qquad f(0) = 0,$$

$$f'(x) = (1+x)^{-1}, \qquad\qquad f'(0) = 1,$$

$$f''(x) = (-1) \times (1+x)^{-2}, \qquad f''(0) = -1,$$

$$f'''(x) = (-1)(-2) \times (1+x)^{-3}, \qquad f'''(0) = 2,$$

$$\vdots$$

$$f^{(n)}(x) = (-1)^{n+1}(n-1)! \times (1+x)^{-n}, \qquad f^{(n)}(0) = (-1)^{n+1}(n-1)!.$$

Hence

$$T_n(x) = f(0) + f'(0)x + \frac{1}{2!}f''(0)x^2 + \cdots + \frac{1}{n!}f^{(n)}(0)x^n$$

$$= x - \frac{1}{2}x^2 + \frac{1}{3}x^3 - \cdots + \frac{(-1)^{n+1}}{n}x^n.$$

(c)
$$f(x) = e^x, \quad f(0) = 1;$$

and, for each positive integer k

$$f^{(k)}(x) = e^x, \quad f^{(k)}(0) = 1.$$

Hence

$$T_n(x) = f(0) + f'(0)x + \frac{1}{2!}f''(0)x^2 + \cdots + \frac{1}{n!}f^{(n)}(0)x^n$$

$$= 1 + x + \frac{x^2}{2!} + \frac{x^3}{3!} + \cdots + \frac{x^n}{n!}.$$

(d)
$$f(x) = \sin x, \qquad f(0) = 0,$$
$$f'(x) = \cos x, \qquad f'(0) = 1,$$
$$f''(x) = -\sin x, \quad f''(0) = 0,$$
$$f'''(x) = -\cos x, \quad f'''(0) = -1,$$
$$f^{(4)}(x) = \sin x, \qquad f^{(4)}(0) = 0;$$

and, in general, for each positive integer k

$$f^{(k)}(0) = \begin{cases} 0, & \text{if } k \text{ is even,} \\ 1, & \text{if } k \equiv 1 \ (\text{mod } 4), \\ -1, & \text{if } k \equiv 3 \ (\text{mod } 4). \end{cases}$$

Hence

$$T_n(x) = x - \frac{x^3}{3!} + \frac{x^5}{5!} + \cdots + (0 \text{ or } 1 \text{ or } -1)\frac{x^n}{n!},$$

where the coefficient of x^k is $\frac{f^{(k)}(0)}{k!}$, and the value of $f^{(k)}(0)$ is as specified above.

(e)
$$f(x) = \cos x, \qquad f(0) = 1,$$
$$f'(x) = -\sin x, \quad f'(0) = 0,$$
$$f''(x) = -\cos x, \quad f''(0) = -1,$$
$$f'''(x) = \sin x, \qquad f'''(0) = 0,$$
$$f^{(4)}(x) = \cos x, \qquad f^{(4)}(0) = 1;$$

and, in general, for each positive integer k

$$f^{(k)}(0) = \begin{cases} 0, & \text{if } k \text{ is odd,} \\ 1, & \text{if } k \equiv 0 \ (\text{mod } 4), \\ -1, & \text{if } k \equiv 2 \ (\text{mod } 4). \end{cases}$$

Hence

$$T_n(x) = 1 - \frac{x^2}{2!} + \frac{x^4}{4!} - \cdots + (0 \text{ or } 1 \text{ or } -1)\frac{x^n}{n!},$$

where the coefficient of x^k is $\frac{f^{(k)}(0)}{k!}$, and the value of $f^{(k)}(0)$ is as specified above.

Section 8.2

1. Here

$$f(x) = \frac{1}{1-x}, \ f'(x) = \frac{1}{(1-x)^2} \quad \text{and} \quad f''(x) = \frac{2}{(1-x)^3}.$$

Hence

$$R_1(x) = \frac{f''(c)}{2!}x^2 = \frac{x^2}{(1-c)^3}.$$

To calculate the value of c, we use Taylor's Theorem, with $n = 1$

$$f(x) = f(a) + f'(a)(x - a) + R_1(x).$$

This gives

$$f\left(\frac{3}{4}\right) = f(0) + f'(0)\frac{3}{4} + R_1\left(\frac{3}{4}\right);$$

that is

$$4 = 1 + \frac{3}{4} + \frac{\left(\frac{3}{4}\right)^2}{(1 - c)^3};$$

and so

$$\frac{9}{4} = \frac{\left(\frac{3}{4}\right)^2}{(1 - c)^3}.$$

It follows that $(1 - c)^3 = \frac{1}{4}$, so that $1 - c = \left(\frac{1}{4}\right)^{\frac{1}{3}} \simeq 0.630$. Thus $c \simeq 0.370$.

2. When f is a polynomial of degree n or less

$$f^{(n+1)}(c) = 0, \quad \text{so that } R_n(x) = 0.$$

3.
$$f(x) = \cos x, \quad f(0) = 1,$$
$$f'(x) = -\sin x, \quad f'(0) = 0,$$
$$f''(x) = -\cos x, \quad f''(0) = -1,$$
$$f'''(x) = \sin x, \quad f'''(0) = 0,$$
$$f^{(4)}(x) = \cos x.$$

Hence, by Taylor's Theorem with $a = 0, f(x) = \cos x$ and $n = 3$, we have

$$\cos x = 1 - \frac{1}{2}x^2 + R_3(x),$$

where

$$R_3(x) = \frac{f^{(4)}(c)}{4!}x^4.$$

Thus

$$|R_3(x)| \leq \frac{|\cos c|}{24}x^4$$
$$\leq \frac{1}{24}x^4.$$

4.
$$f(x) = \log_e(1 + x), \quad f(0) = 0,$$
$$f'(x) = \frac{1}{1 + x}, \quad f'(0) = 1,$$
$$f''(x) = \frac{-1}{(1 + x)^2}, \quad f''(0) = -1,$$
$$f'''(x) = \frac{2}{(1 + x)^3}.$$

Hence, by Taylor's Theorem with $x = 0.02$

$$\log_e 1.02 = 0 + (1 \times 0.02) - \left(\frac{1}{2} \times 0.02^2\right) + R_2(0.02)$$

$$= 0.0198 + R_2(0.02).$$

Now, for $c \in (0, 0.02)$

$$|f'''(c)| = \left|\frac{2}{(1 + c)^3}\right| \leq 2.$$

Hence, by the Remainder Estimate with $M = 2$, we obtain

$$|R_2(0.02)| \le \frac{2}{3!} \times 0.02^3 = 0.000002\overline{6}.$$

It follows that

$$\log_e 1.02 = 0.0198 \ (\text{to four decimal places}).$$

5. For any positive integer n

$$f^{(n)}(x) = \pm \cos x \quad \text{or} \quad \pm \sin x;$$

hence

$$\left| f^{(n)}(c) \right| \le 1, \quad \text{for all } c \in \mathbb{R},$$

so that $M = 1$.

It follows that the remainder term in the Remainder Estimate satisfies the inequality

$$|R_n(0.2)| \le \frac{1}{(n+1)!} (0.2)^{n+1}.$$

Now

$$\frac{(0.2)^4}{4!} = 0.00006\overline{6} \quad \text{and} \quad \frac{(0.2)^5}{5!} = 0.0000026\ldots;$$

so we should try $n = 4$ in the Remainder Estimate, to be safe.

Here

$$
\begin{aligned}
f(x) &= \cos x, & f(0) &= 1, \\
f'(x) &= -\sin x, & f'(0) &= 0, \\
f''(x) &= -\cos x, & f''(0) &= -1, \\
f'''(x) &= \sin x, & f'''(0) &= 0, \\
f^{(4)}(x) &= \cos x, & f^{(4)}(0) &= 1 \\
f^{(5)}(x) &= -\sin x.
\end{aligned}
$$

Hence, by Taylor's Theorem with $f(x) = \cos x$, $a = 0$, $x = 0.2$ and $n = 4$, we have

$$f(x) = 1 - \frac{1}{2}x^2 + \frac{1}{24}x^4 + R_4(x);$$

in other words

$$
\begin{aligned}
\cos 0.2 &= 1 - \frac{1}{2}(0.2)^2 + \frac{1}{24}(0.2)^4 + R_4(0.2) \\
&= 1 - 0.02 + 0.00006\overline{6} + R_4(0.2) \\
&= 0.9801 \quad (\text{rounded to four decimal places}).
\end{aligned}
$$

6. Using the same function f as in Problem 5, we find that

$$
\begin{aligned}
f(\pi) &= -1, \quad f'(\pi) = 0, \quad f''(\pi) = 1, \\
f'''(\pi) &= 0, \quad f^{(4)}(\pi) = -1.
\end{aligned}
$$

It follows that

$$T_4(x) = -1 + \frac{1}{2}(x - \pi)^2 - \frac{1}{24}(x - \pi)^4.$$

The difference between $T_4(x)$ and $f(x)$ is

$$
\begin{aligned}
R_4(x) &= \frac{f^{(5)}(c)}{5!}(x - \pi)^5 \\
&= \frac{-\sin c}{120}(x - \pi)^5,
\end{aligned}
$$

for some number c between π and x. Hence, for $x \in \left[\frac{3}{4}\pi, \frac{5}{4}\pi\right]$, we have

$$|R_4(x)| \leq \frac{1}{120}\left(\frac{\pi}{4}\right)^5$$
$$= 0.00249\ldots < 3 \times 10^{-3}.$$

Section 8.3

1. (a) Applying the Ratio Test with $a_n = 2^n + 4^n$, we obtain

$$\left|\frac{a_{n+1}}{a_n}\right| = \frac{2^{n+1} + 4^{n+1}}{2^n + 4^n} = \frac{2 + 2^{n+2}}{1 + 2^n}$$
$$= 2\frac{\left(\frac{1}{2}\right)^n + 2}{\left(\frac{1}{2}\right)^n + 1}$$
$$\to 2 \times 2 = 4 \quad \text{as } n \to \infty.$$

Hence the series has radius of convergence $\frac{1}{4}$.

(b) Applying the Ratio Test with $a_n = \frac{(n!)^2}{(2n)!}$, we obtain

$$\left|\frac{a_{n+1}}{a_n}\right| = \frac{(n+1)(n+1)}{(2n+1)(2n+2)} = \frac{\left(1 + \frac{1}{n}\right)\left(1 + \frac{1}{n}\right)}{\left(2 + \frac{1}{n}\right)\left(2 + \frac{2}{n}\right)}$$
$$\to \frac{1}{4} \quad \text{as } n \to \infty.$$

Hence the series has radius of convergence 4.

(c) Applying the Ratio Test with $a_n = n + 2^{-n}$, we obtain

$$\left|\frac{a_{n+1}}{a_n}\right| = \frac{n + 1 + 2^{-n-1}}{n + 2^{-n}} = \frac{1 + \frac{1}{n} + \frac{1}{n2^{n+1}}}{1 + \frac{1}{n2^n}}$$
$$\to 1 \quad \text{as } n \to \infty.$$

Hence the series has radius of convergence 1.

(d) Applying the Ratio Test with $a_n = \frac{1}{(n!)^{\frac{1}{n}}}$, we obtain

$$\left|\frac{a_{n+1}}{a_n}\right| = \frac{(n!)^{\frac{1}{n}}}{((n+1)!)^{\frac{1}{n+1}}} \sim \frac{(\sqrt{2\pi n})^{\frac{1}{n}}\left(\frac{n}{e}\right)}{(\sqrt{2\pi(n+1)})^{\frac{1}{n+1}}\left(\frac{n+1}{e}\right)}$$
$$= \frac{(\sqrt{2\pi})^{\frac{1}{n}}}{(\sqrt{2\pi})^{\frac{1}{n+1}}} \times \frac{(n^{\frac{1}{n}})^{\frac{1}{2}}}{((n+1)^{\frac{1}{n+1}})^{\frac{1}{2}}} \times \frac{n}{n+1}$$
$$\to 1 \quad \text{as } n \to \infty.$$

(Here we have used Stirling's Formula to estimate $n!$ and $(n+1)!$.)
Hence the series has radius of convergence 1.

2. (a) Applying the Ratio Test with $a_n = n$, we obtain

$$\left|\frac{a_{n+1}}{a_n}\right| = \frac{n + 1}{n} \to 1 \quad \text{as } n \to \infty.$$

Hence the series has radius of convergence 1.
   Thus the series converges if $x \in (-1, 1)$, and diverges if $|x| > 1$.
   When $x = \pm 1$, the sequence $\{nx^n\}$ is non-null; hence, by the Non-null Test, the series diverges.
   Hence the interval of convergence of the series is $(-1, 1)$.

(b) Applying the Ratio Test with $a_n = \frac{1}{n3^n}$, we obtain

$$\left|\frac{a_{n+1}}{a_n}\right| = \frac{n3^n}{(n+1)3^{n+1}} = \frac{1}{\left(1+\frac{1}{n}\right)3}$$
$$\to \frac{1}{3} \quad \text{as } n \to \infty.$$

Thus the series converges if $x \in (-3, 3)$, and diverges if $|x| > 3$.

When $x = 3$, the series is $\sum_{n=1}^{\infty} \frac{1}{n}$, which is divergent.

When $x = -3$, the series is $\sum_{n=1}^{\infty} \frac{(-1)^n}{n}$, which is convergent.

Hence the interval of convergence of the series is $[-3, 3)$.

3. Applying the Ratio Test with $a_n = \frac{\alpha(\alpha-1)\cdot\ldots\cdot(\alpha-n+1)}{n!}$, we obtain

$$\left|\frac{a_{n+1}}{a_n}\right| = \left|\frac{\alpha - n}{n+1}\right| = \left|\frac{\frac{\alpha}{n} - 1}{1 + \frac{1}{n}}\right|$$
$$\to 1 \quad \text{as } n \to \infty.$$

Hence the series has radius of convergence 1.

4. From the Hint, we know that $\tan^{-1} x$ is the sum function of the series

$$\sum_{n=0}^{\infty} \frac{(-1)^n}{2n+1}x^{2n+1} = x - \frac{x^3}{3} + \frac{x^5}{5} - \frac{x^7}{7} + \cdots, \quad \text{for } |x| < 1. \qquad (*)$$

From the Hint, we also know that the radius of convergence of this power series is 1.

It follows that, by applying Abel's Limit Theorem to the power series $(*)$, we may obtain the following

$$\lim_{x \to 1^-} \left(\tan^{-1} x\right) = \sum_{n=0}^{\infty} \frac{(-1)^n}{2n+1},$$

provided that this last series is convergent; it is indeed convergent, by the Alternating Test for series.

Since the function $x \mapsto \tan^{-1} x$ is continuous at 1, it follows that

$$\sum_{n=0}^{\infty} \frac{(-1)^n}{2n+1} = \lim_{x \to 1^-} \left(\tan^{-1} x\right)$$
$$= \tan^{-1} 1 = \frac{\pi}{4}.$$

For $x \mapsto \tan^{-1} x$ is continuous on its domain $\mathbb{R}$.

Section 8.4

1. (a) We know that

$$e^x = 1 + x + \frac{x^2}{2!} + \frac{x^3}{3!} + \cdots + \frac{x^n}{n!} + \cdots, \quad \text{for } x \in \mathbb{R},$$

and so

$$e^{-x} = 1 - x + \frac{x^2}{2!} - \frac{x^3}{3!} + \cdots + (-1)^n\frac{x^n}{n!} + \cdots, \quad \text{for } x \in \mathbb{R}.$$

Hence, by the Sum and Multiple Rules for power series

$$\sinh x = \frac{1}{2}\left(e^x - e^{-x}\right)$$
$$= x + \frac{x^3}{3!} + \cdots + \frac{x^{2n+1}}{(2n+1)!} + \cdots, \quad \text{for } x \in \mathbb{R}.$$

This series converges for all x.

(b) We know that

$$\log_e(1-x) = -x - \frac{x^2}{2} - \frac{x^3}{3} - \cdots - \frac{x^n}{n} - \cdots, \quad \text{for } |x| < 1,$$

and

$$\frac{1}{1-x} = 1 + x + x^2 + x^3 + \cdots + x^n + \cdots, \quad \text{for } |x| < 1.$$

Hence, by the Sum and Multiple Rules for power series

$$\log_e(1-x) + \frac{2}{1-x}$$

$$= \left(-x - \frac{x^2}{2} - \frac{x^3}{3} - \cdots - \frac{x^n}{n} - \cdots \right) + \left(2 + 2x + 2x^2 + 2x^3 + \cdots + 2x^n + \cdots \right)$$

$$= 2 + x + \frac{3}{2}x^2 + \frac{5}{3}x^3 + \cdots + \frac{2n-1}{n}x^n + \cdots, \quad \text{for } |x| < 1.$$

The radius of convergence of this series is 1.

2. (a) We know that

$$\sinh x = x + \frac{x^3}{3!} + \cdots + \frac{x^{2n+1}}{(2n+1)!} + \cdots, \quad \text{for } x \in \mathbb{R},$$

and that

$$\sin x = x - \frac{x^3}{3!} + \cdots + \frac{(-1)^n x^{2n+1}}{(2n+1)!} + \cdots, \quad \text{for } x \in \mathbb{R}.$$

Hence, by the Sum and Multiple Rules for power series

$$\sinh x + \sin x$$

$$= \left(x + \frac{x^3}{3!} + \cdots + \frac{x^{2n+1}}{(2n+1)!} + \cdots \right) + \left(x - \frac{x^3}{3!} + \cdots + \frac{(-1)^n x^{2n+1}}{(2n+1)!} + \cdots \right)$$

$$= 2 \left(x + \frac{x^5}{5!} + \cdots + \frac{x^{4n+1}}{(4n+1)!} + \cdots \right), \quad \text{for } x \in \mathbb{R}.$$

This series converges for all x.

(b) We know that

$$\log_e(1+x) = x - \frac{x^2}{2} + \frac{x^3}{3} - \cdots + (-1)^{n+1}\frac{x^n}{n} + \cdots, \quad \text{for } |x| < 1,$$

and so

$$\log_e(1-x) = -x - \frac{x^2}{2} - \frac{x^3}{3} - \cdots - \frac{x^n}{n} - \cdots, \quad \text{for } |x| < 1.$$

Hence, by the Sum and Multiple Rules for power series

$$\log_e\left(\frac{1+x}{1-x}\right)$$

$$= \log_e(1+x) - \log_e(1-x)$$

$$= \left(x - \frac{x^2}{2} + \frac{x^3}{3} - \cdots + (-1)^{n+1}\frac{x^n}{n} + \cdots \right) - \left(-x - \frac{x^2}{2} - \frac{x^3}{3} - \cdots - \frac{x^n}{n} - \cdots \right)$$

$$= 2 \left(x + \frac{x^3}{3} + \cdots + \frac{x^{2n+1}}{2n+1} + \cdots \right), \quad \text{for } |x| < 1.$$

The radius of convergence of this series is 1.

(c) We know that

$$\frac{1}{1-x} = 1 + x + x^2 + x^3 + \cdots + x^n + \cdots, \quad \text{for } |x| < 1.$$

It follows, by replacing x by $-2x^2$, that

$$\frac{1}{1+2x^2} = 1 + \left(-2x^2\right) + \left(-2x^2\right)^2 + \left(-2x^2\right)^3 + \cdots + \left(-2x^2\right)^n + \cdots$$

$$= 1 - 2x^2 + 2^2x^4 - 2^3x^6 + \cdots + (-1)^n 2^n x^{2n} + \cdots,$$

for $|2x^2| < 1$; that is, for $|x| < \frac{1}{\sqrt{2}}$.

The radius of convergence of this series is $\frac{1}{\sqrt{2}}$.

3. (a) We know that

$$\log_e(1+x) = x - \frac{x^2}{2} + \frac{x^3}{3} - \cdots + (-1)^{n+1}\frac{x^n}{n} + \cdots, \quad \text{for } |x| < 1.$$

Hence, by the Product Rule

$$(1+x)\log_e(1+x)$$

$$= \left(x - \frac{x^2}{2} + \frac{x^3}{3} - \cdots + (-1)^{n+1}\frac{x^n}{n} + \cdots\right)$$

$$+ \left(x^2 - \frac{x^3}{2} + \frac{x^4}{3} - \cdots + (-1)^n\frac{x^n}{n-1} + \cdots\right)$$

$$= x + \frac{x^2}{2} - \frac{x^3}{6} - \cdots + (-1)^n\frac{x^n}{n(n-1)} + \cdots, \quad \text{for } |x| < 1.$$

The radius of convergence of this series is 1.

(b) We know that

$$\frac{1}{1-x} = 1 + x + x^2 + x^3 + \cdots + x^n + \cdots, \quad \text{for } |x| < 1,$$

and (from Example 1) that

$$\frac{1+x}{(1-x)^2} = 1 + 3x + 5x^2 + \cdots + (2n+1)x^n + \cdots, \quad \text{for } |x| < 1.$$

Hence, by the Product Rule, we obtain

$$\frac{1+x}{(1-x)^3}$$

$$= \frac{1}{1-x} \times \frac{1+x}{(1-x)^2}$$

$$= \left(1 + x + x^2 + x^3 + \cdots + x^n + \cdots\right)$$

$$\times \left(1 + 3x + 5x^2 + \cdots + (2n+1)x^n + \cdots\right)$$

$$= 1 + (3+1)x + (5+3+1)x^2 + (7+5+3+1)x^3 + \cdots$$

$$+ ((2n+1) + (2n-1) + \cdots + 1)x^n + \cdots$$

$$= 1 + 4x + 9x^2 + 16x^3 + \cdots + (n+1)^2 x^n + \cdots, \quad \text{for } |x| < 1.$$

(We can prove, by Mathematical Induction, that

$$1 + 3 + 5 + \cdots + (2n+1) = \sum_{k=0}^{n}(2k+1) = (n+1)^2, \quad \text{for } n \in \mathbb{N}.)$$

The radius of convergence of this series is 1.

4. (a) We know that

$$(1-x)^{-1} = 1 + x + x^2 + x^3 + \cdots + x^n + \cdots, \quad \text{for } |x| < 1.$$

Hence, by the Differentiation Rule, we obtain

$$(1-x)^{-2} = 1 + 2x + 3x^2 + 4x^3 + \cdots + nx^{n-1} + \cdots, \quad \text{for } |x| < 1.$$

The radius of convergence of this series is 1.

(b) If we differentiate the series in part (a), by the Differentiation Rule for power series, we obtain

$$2(1-x)^{-3} = 2 + 6x + 12x^2 + \cdots + n(n-1)x^{n-2} + \cdots, \quad \text{for } |x| < 1.$$

Hence, by the Multiple Rule, we have

$$(1-x)^{-3} = 1 + 3x + 6x^2 + \cdots + \frac{n(n-1)}{2}x^{n-2} + \cdots, \quad \text{for } |x| < 1.$$

The radius of convergence of this series is 1.

(c) Notice that, for the function $f(x) = \tanh^{-1} x$, $|x| < 1$, we have

$$f'(x) = \frac{1}{1-x^2} = 1 + x^2 + x^4 + \cdots + x^{2n} + \cdots.$$

Hence, by the Integration Rule for power series, the Taylor series for f at 0 is

$$\tanh^{-1} x = c + x + \frac{x^3}{3} + \frac{x^5}{5} + \cdots + \frac{x^{2n+1}}{2n+1} + \cdots, \quad \text{for } |x| < 1;$$

since $f(0) = 0$, it follows that $c = 0$. Hence

$$\tanh^{-1} x = x + \frac{x^3}{3} + \frac{x^5}{5} + \cdots + \frac{x^{2n+1}}{2n+1} + \cdots, \quad \text{for } |x| < 1.$$

The radius of convergence of this series is 1.

5. We know that

$$e^x = 1 + x + \frac{x^2}{2!} + \frac{x^3}{3!} + \cdots + \frac{x^n}{n!} + \cdots, \quad \text{for } x \in \mathbb{R};$$

and we know, from part (a) of Problem 4, that

$$(1-x)^{-2} = 1 + 2x + 3x^2 + 4x^3 + \cdots + (n+1)x^n + \cdots, \quad \text{for } |x| < 1.$$

Hence, by the Product Rule for power series, we have

$$e^x(1-x)^{-2} = \left(1 + x + \frac{x^2}{2} + \frac{x^3}{6} + \cdots\right) \times \left(1 + 2x + 3x^2 + 4x^3 + \cdots\right)$$

$$= 1 + (2+1)x + \left(3 + 2 + \frac{1}{2}\right)x^2 + \cdots$$

$$= 1 + 3x + \frac{11}{2}x^2 + \cdots, \quad \text{for } |x| < 1.$$

The radius of convergence of this series is 1.

6. We are given that

$$f(x) = x + \frac{x^3}{1.3} + \frac{x^5}{1.3.5} + \cdots + \frac{x^{2n+1}}{1.3\ldots.(2n+1)} + \cdots, \quad \text{for } x \in \mathbb{R}.$$

(a) By the Differentiation Rule, we can differentiate the power series term-by-term; this gives

$$f'(x) = 1 + \frac{x^2}{1} + \frac{x^4}{1.3} + \cdots + \frac{x^{2n}}{1.3\ldots.(2n-1)} + \cdots, \quad \text{for } x \in \mathbb{R}.$$

(b) It follows, from the definition of f, that

$$xf(x) = x^2 + \frac{x^4}{1.3} + \frac{x^6}{1.3.5} + \cdots + \frac{x^{2n+2}}{1.3\ldots.(2n+1)} + \cdots, \quad \text{for } x \in \mathbb{R}.$$

Hence, by the Combination Rules for power series,

$$f'(x) - xf(x) = 1,$$

since all the other terms cancel out.

7. Since

$$e^x = 1 + x + \frac{x^2}{2!} + \frac{x^3}{3!} + \cdots + \frac{x^n}{n!} + \cdots, \quad \text{for } x \in \mathbb{R},$$

we may deduce that

$$e^{-x^2} = 1 - x^2 + \frac{x^4}{2!} - \frac{x^6}{3!} + \cdots + (-1)^n \frac{x^{2n}}{n!} + \cdots, \quad \text{for } x \in \mathbb{R}.$$

It follows, from the Integration Rule for power series, that

$$\int_0^1 e^{-x^2} dx = \left[x - \frac{x^3}{3} + \frac{x^5}{5 \times 2!} - \frac{x^7}{7 \times 3!} + \cdots + (-1)^n \frac{x^{2n+1}}{(2n+1) \times n!} + \cdots \right]_0^1$$

$$= 1 - \frac{1}{3} + \frac{1}{10} - \frac{1}{42} + \cdots + (-1)^n \frac{1}{(2n+1) \times n!} + \cdots.$$

8. By the General Binomial Theorem

$$(1 + 6x)^{\frac{1}{4}} = \sum_{n=0}^{\infty} \binom{\frac{1}{4}}{n} x^n, \quad \text{for } |6x| < 1,$$

where

$$\binom{\frac{1}{4}}{n} = \frac{\left(\frac{1}{4}\right)\left(-\frac{3}{4}\right)\left(-\frac{7}{4}\right) \cdots \left(\frac{1}{4} - n + 1\right)}{n!}.$$

It follows that

$$(1 + 6x)^{\frac{1}{4}} = 1 + \frac{\left(\frac{1}{4}\right)}{1}(6x) + \frac{\left(\frac{1}{4}\right)\left(-\frac{3}{4}\right)}{2}(6x)^2 + \frac{\left(\frac{1}{4}\right)\left(-\frac{3}{4}\right)\left(-\frac{7}{4}\right)}{6}(6x)^3 + \cdots$$

$$= 1 + \frac{3}{2}x - \frac{27}{8}x^2 + \frac{189}{16}x^3 - \cdots, \quad \text{for } |x| < \frac{1}{6}.$$

The radius of convergence of this series is $\frac{1}{6}$.

9. (a) By the General Binomial Theorem

$$(1 - x)^{-\frac{1}{2}} = \sum_{n=0}^{\infty} \binom{-\frac{1}{2}}{n}(-x)^n, \quad \text{for } |x| < 1,$$

where

$$\binom{-\frac{1}{2}}{n} = \frac{\left(-\frac{1}{2}\right)\left(-\frac{3}{2}\right)\left(-\frac{5}{2}\right) \cdots \left(-\frac{1}{2} - n + 1\right)}{n!}.$$

It follows that

$$(1 - x)^{-\frac{1}{2}} = 1 + \frac{1}{2}x + \frac{3}{8}x^2 + \cdots + (-1)^n \binom{-\frac{1}{2}}{n} x^n + \cdots, \text{ for } |x| < 1.$$

The radius of convergence of this series is 1.

(b) We know that

$$\frac{d}{dx} \sin^{-1} x = \frac{1}{\sqrt{1 - x^2}}, \quad \text{for } x \in (-1, 1).$$

Then, by the result of part (a), with x replaced by x^2, we obtain

$$\frac{1}{\sqrt{1 - x^2}} = 1 + \frac{1}{2}x^2 + \frac{3}{8}x^4 + \cdots + (-1)^n \binom{-\frac{1}{2}}{n} x^{2n} + \cdots, \quad \text{for } |x| < 1.$$

Hence, by the Integration Rule for power series

$$\sin^{-1} x = c + x + \frac{1}{6}x^3 + \frac{3}{40}x^5 + \cdots$$

$$+ (-1)^n \binom{-\frac{1}{2}}{n} \frac{x^{2n+1}}{2n + 1} + \cdots, \text{ for } |x| < 1.$$

Putting $x = 0$ into this equation, we see that $c = 0$. It follows that

$$\sin^{-1} x = x + \frac{1}{6}x^3 + \frac{3}{40}x^5 + \cdots$$
$$+ (-1)^n \binom{-\frac{1}{2}}{n} \frac{x^{2n+1}}{2n+1} + \cdots, \quad \text{for } |x| < 1.$$

The radius of convergence of this series is 1.

Section 8.5

1. We use the Addition Formula

$$\tan^{-1} x + \tan^{-1} y = \tan^{-1}\left(\frac{x+y}{1-xy}\right), \qquad (\ast)$$

which holds provided that $\tan^{-1} x + \tan^{-1} y$ lies in the interval $\left(-\frac{\pi}{2}, \frac{\pi}{2}\right)$.

First, we deduce, from the Addition Formula $(\ast)$, that

$$\tan^{-1}\left(\frac{1}{3}\right) + \tan^{-1}\left(\frac{1}{4}\right) = \tan^{-1}\left(\frac{\frac{1}{3}+\frac{1}{4}}{1-\frac{1}{3}\times\frac{1}{4}}\right)$$
$$= \tan^{-1}\left(\frac{4+3}{12-1}\right)$$
$$= \tan^{-1}\left(\frac{7}{11}\right).$$

This equation holds, since

$$\tan^{-1}\left(\frac{1}{3}\right) + \tan^{-1}\left(\frac{1}{4}\right) \simeq 0.3218 + 0.2450 = 0.5668,$$

and $0.5668 \in \left(-\frac{\pi}{2}, \frac{\pi}{2}\right) \simeq (-1.5708, 1.5708)$.

Next, we deduce, from the Addition Formula $(\ast)$, that

$$\tan^{-1}\left(\frac{1}{3}\right) + \tan^{-1}\left(\frac{1}{4}\right) + \tan^{-1}\left(\frac{2}{9}\right) = \tan^{-1}\left(\frac{7}{11}\right) + \tan^{-1}\left(\frac{2}{9}\right)$$
$$= \tan^{-1}\left(\frac{\frac{7}{11}+\frac{2}{9}}{1-\frac{7}{11}\times\frac{2}{9}}\right)$$
$$= \tan^{-1}\left(\frac{63+22}{99-14}\right)$$
$$= \tan^{-1}(1) = \frac{\pi}{4}.$$

This equation holds, since

$$\tan^{-1}\left(\frac{7}{11}\right) + \tan^{-1}\left(\frac{2}{9}\right) \simeq 0.5667 + 0.2187 = 0.7854,$$

and $0.7854 \in \left(-\frac{\pi}{2}, \frac{\pi}{2}\right) \simeq (-1.5708, 1.5708)$.

2. First

$$I_0 = \int_{-1}^{1} \cos\left(\frac{1}{2}\pi x\right) dx$$
$$= \left[\frac{2}{\pi}\sin\left(\frac{1}{2}\pi x\right)\right]_{-1}^{1} = \frac{4}{\pi}.$$

It follows that $\pi I_0 = 4$.

Next, using integration by parts twice, we obtain

$$
\begin{aligned}
I_1 &= \int_{-1}^{1} (1 - x^2) \cos\left(\frac{1}{2}\pi x\right) dx \\
&= \left[(1 - x^2)\frac{2}{\pi}\sin\left(\frac{1}{2}\pi x\right) \right]_{-1}^{1} - \int_{-1}^{1} (-2x)\frac{2}{\pi}\sin\left(\frac{1}{2}\pi x\right) dx \\
&= \frac{4}{\pi} \int_{-1}^{1} x\sin\left(\frac{1}{2}\pi x\right) dx \\
&= \frac{4}{\pi}\left[x\left(-\frac{2}{\pi}\right)\cos\left(\frac{1}{2}\pi x\right) \right]_{-1}^{1} - \frac{4}{\pi}\int_{-1}^{1} \left(-\frac{2}{\pi}\right)\cos\left(\frac{1}{2}\pi x\right) dx \\
&= \frac{8}{\pi^2}\left[\frac{2}{\pi}\sin\left(\frac{1}{2}\pi x\right) \right]_{-1}^{1} \\
&= \frac{32}{\pi^3}.
\end{aligned}
$$

It follows that $\pi^3 I_1 = 32$.

Index

Printed in the United States
By Bookmasters